# Advances in Intelligent Systems and Computing

Volume 289

AF180592

*Series editor*

Janusz Kacprzyk, Polish Academy of Sciences, Warsaw, Poland
e-mail: kacprzyk@ibspan.waw.pl

For further volumes:
http://www.springer.com/series/11156

*About this Series*

The series "Advances in Intelligent Systems and Computing" contains publications on theory, applications, and design methods of Intelligent Systems and Intelligent Computing. Virtually all disciplines such as engineering, natural sciences, computer and information science, ICT, economics, business, e-commerce, environment, healthcare, life science are covered. The list of topics spans all the areas of modern intelligent systems and computing.

The publications within "Advances in Intelligent Systems and Computing" are primarily textbooks and proceedings of important conferences, symposia and congresses. They cover significant recent developments in the field, both of a foundational and applicable character. An important characteristic feature of the series is the short publication time and world-wide distribution. This permits a rapid and broad dissemination of research results.

Ivan Zelinka · Ponnuthurai Nagaratnam Suganthan
Guanrong Chen · Václav Snášel
Ajith Abraham · Otto E. Rössler
Editors

# Nostradamus 2014: Prediction, Modeling and Analysis of Complex Systems

Springer

*Editors*
Ivan Zelinka
Department of Computer Science
Faculty of Electrical Eng. & Comp. Sci.
VŠB-TUO
Ostrava-Poruba
Czech Republic

Václav Snášel
Department of Computer Science
Faculty of Electrical Eng. & Comp. Sci.
VŠB-TUO
Ostrava-Poruba
Czech Republic

Ponnuthurai Nagaratnam Suganthan
Nanyang Technological University
Singapore
Singapore

Ajith Abraham
Scientific Network for Innovation and Research Excellence
Machine Intelligence Research Labs (MIR Labs)
Auburn Washington
USA

Guanrong Chen
Department of Electronic Engineering
City University of Hong Kong
Hong Kong
China

Otto E. Rössler
Institute for Physical and Theoretical Chemistry
University of Tübingen
Tübingen Baden-Württemberg
Germany

ISSN 2194-5357 ISSN 2194-5365 (electronic)
ISBN 978-3-319-07400-9 ISBN 978-3-319-07401-6 (eBook)
DOI 10.1007/978-3-319-07401-6
Springer Cham Heidelberg New York Dordrecht London

Library of Congress Control Number: 2014940405



Printed on acid-free paper

Springer is part of Springer Science+Business Media (www.springer.com)

# Preface

This proceeding book of the Nostradamus conference (http://nostradamus-conference.org) contains accepted papers presented at this event in June 2014. Nostradamus conference was held in the third largest city in the Czech Republic - Ostrava (http://www.ostrava.cz/en).

Conference topics are focused on classical as well as modern methods for modelling, prediction, identification and simulation of complex systems with applications in science, engineering and economy. Topics are (but not limited to): *prediction by classical and novel methods, predictive control, deterministic chaos and its control, complex systems, modeling and prediction of its dynamics, interdisciplinary fusion of chaos, randomness and evolution* and much more.

The prediction of behavior of complex systems, analysis and modeling of its structure is a vitally important problem in engineering, economy and generally in science today. Examples of such systems can be seen in the world around us (including our bodies) and of course in almost every scientific discipline including such "exotic" domains as the earth's atmosphere, turbulent fluids, economics (exchange rate and stock markets), population growth, physics (control of plasma), information flow in social networks and its dynamics, chemistry and complex networks. To understand such complex dynamics, which often exhibit strange behavior, and to use it in research or industrial applications, it is paramount to create its models. For this purpose there is rich spectra of methods, from classical such as ARMA models or Box Jenkins method to modern ones like evolutionary computation, neural networks, fuzzy logic, geometry, deterministic chaos amongst others.

The main aim of the conference is to create periodical possibility for students, academics and researchers to exchange their ideas and novel methods. This conference will establish a forum for the presentation and discussion of recent trends in the area of applications of various modern as well as classical methods for researchers, students and academics.

The accepted selection of papers was extremely rigorously reviewed in order to maintain the high quality of the conference that is supported by grant no. CZ.1.07/2.3.00/20.0072 funded by Operational Programme Education for Competitiveness, co-financed by ESF. Regular as well as student's papers (1/4 of all papers) has been submitted to the conference, and in accordance with ESF support as well as with conference topics guidelines, has been accepted after a positive review. Based on accepted papers structure and topics, the proceeding book consists of sections as: *Chaos, Evolution and Complexity* discusses topics from the field of evolutionary algorithms, deterministic chaos, its complex dynamics and mutual intersections of all three topics (chaos powered swarm intelligence algorithms, complexity in bio-inspired algorithms dynamics, chaos control, etc.). Section *Nature-Inspired Algorithms and Nonlinear Systems* contain participations about the use of bio-inspired algorithms on various complex and/or nonlinear problems (forecasting, EEG signal modeling, battleship game strategy, symbolic regression powered by chaos, etc.). Section *Nonlinear and Predictive Control* and *Nonlinear Dynamics and Complex Systems* contains papers about controlling modeling and analysis of complex and nonlinear systems and the last section, *Various Topics*, contains a few borderline papers (fractal geometry application on process modeling and theory of basic law of physics) that appear to be interesting and keeping with the theme of the conference topics.

For this year, as a follow-up of the conference, we anticipate further publication of selected papers in a special issue of the prestigious journal Swarm and Evolutionary Computation, Computer Science (Hidawi), special book in Emergence Complexity and Computation series and more.

We would like to thank the members of the Program Committees and reviewers for their hard work. We believe that Nostradamus conference represents a high standard conference in the domain of prediction and modeling of complex systems. Nostradamus 2014 enjoyed outstanding keynote lectures by distinguished guest speakers: **René Lozi** (France), **Ponnuthurai Nagaratnam Suganthan** (Singapore) and **Lars Nolle** (Germany).

Particular thanks goes as well to the Workshop main Sponsors, IT4Innovations, VŠB-Technical University of Ostrava, MIR labs (USA), Centre for Chaos and Complex Networks (Hong Kong), Journal of Unconventional Computing (UK). Special thanks belong to Ministry of Education of the Czech Republic. This conference was supported by the Development of human resources in research and development of latest soft computing methods and their application in practice project, reg. no. CZ.1.07/2.3.00/20.0 072 funded by Operational Programme Education for Competitiveness, co-financed by ESF and state budget of the Czech Republic.

We would like to thank all the contributing authors, as well as the members of the Program Committees and the Local Organizing Committee for their hard and highly valuable work. Their work has definitely contributed to the success of the Nostradamus conference.

The Editors

Ivan Zelinka
Ponnuthurai Nagaratnam Suganthan
Guanrong Chen
Václav Snášel
Ajith Abraham
Otto E. Rössler

This conference was supported by the Development of human resources in research and development of latest soft computing methods and their application in practice project, reg. no. CZ.1.07/2.3.00/20.0072 funded by Operational Programme Education for Competitiveness, co-financed by ESF and state budget of the Czech Republic.

esf european social fund in the czech republic
EUROPEAN UNION
MINISTRY OF EDUCATION, YOUTH AND SPORTS
OP Education for Competitiveness

INVESTMENTS IN EDUCATION DEVELOPMENT

# Organization

## International Conference Committee

Edward Ott, USA
Ivan Zelinka, Czech Republic
Guanrong Chen, Hong Kong
Otto E. Rössler, Germany
Sergej Čelikovsky, Czech Republic
Mohammed Chadli, France
Ajith Abraham, MIR Labs, USA
Vaclav Snasel, Czech Republic
Emilio Corchado, Spain
Andy Adamatzky, UK
Jiří Pospíchal, Slovakia
Jouni Lampinen, Finland
Juan Carlos Burguillo-Rial, Spain
Pandian Vasant, Malysia
Petr Dostál, Czech Republic
Davendra Donald, Fiji, Czech Republic
Bernabé Dorronsoro, Luxembourg
Oplatková Zuzana, Czech Republic
Linqiang Pan, China
Šenkeřík Roman, Czech Republic
Fečkan Michal, Slovakia
Jašek Roman, Czech Republic
Joanna Kolodziej, Poland
Radek Matoušek, Czech Republic
Hendrik Richter, Germany
Zdeněk Beran, Czech Republic
Ana Peleteiro, Spain
Vadim Strijov, Russia
Oldřich Zmeškal, Czech Republic
Masoud Mohammadian, Australia
Miguel A. F. Sanjuan, Spain
Gennady Leonov, Russia
Nikolay Kuznetsov, Russia
René Lozi, France
Petr Saloun, Czech Republic

## Local Conference Committee

Jan Martinovič
Lenka Skanderová
Jan Platoš
Eliška Odchodková
Martin Milata
Pavel Krömer
Michal Krumnikl
Miloš Kudělka
Pavel Moravec
Jiří Dvorský
Tilkova Ludmila
Kvapulinska Petra

# Contents

## Proposed Content of Nostradamus 2014

### Chaos, Evolution and Complexity

**Chaos Level Measurement in Logistic Map Used as the Chaotic Numbers Generator in Differential Evolution** . . . . . . . . . . . . . . . . . . . . . . . . . . . . . . . . 1
*Lenka Skanderova, Ivan Zelinka, Tran Trong Dao, Duy Vo Hoang*

**Estimations of Initial Errors Growth in Weather Prediction by Low-dimensional Atmospheric Model** . . . . . . . . . . . . . . . . . . . . . . . . . . . . 11
*Hynek Bednář, Aleš Raidl, Jiří Mikšovský*

**Prediction Based Context Data Dissemination and Storage Model for Cooperative Vehicular Networks** . . . . . . . . . . . . . . . . . . . . . . . . . . . . . . . . . . . . . . 21
*Mindaugas Kurmis, Dale Dzemydiene, Arunas Andziulis, Miroslav Voznak, Sergej Jakovlev, Zydrunas Lukosius, Gediminas Gricius*

**SOM Based Multi-agent Hydro Meteorological Data Collection System** . . . . 31
*Gediminas Gricius, Darius Drungilas, Arunas Andziulis, Dale Dzemydiene, Miroslav Voznak*

**Estimation of Fractal Dimension and Statistical Tools for Surface Evaluation** . . . . . . . . . . . . . . . . . . . . . . . . . . . . . . . . . . . . . . . . . . . . . . . 43
*Vlastimil Hotař, Petr Salač, Ondřej Matúšek*

**Statistics of Fractal Systems** . . . . . . . . . . . . . . . . . . . . . . . . . . . . . . . . . . . . . 55
*Oldrich Zmeskal, Stanislav Nespurek, Michal Vesely, Petr Dzik*

**Modelling Queues in Transportation Networks Using P Systems** . . . . . . . . . . 65
*Zbyněk Janoška, Jiří Dvorský*

**Tuning the Lozi Map in Chaos Driven PSO Inspired by the Multi-chaotic Approach** . . . . . . . . . . . . . . . . . . . . . . . . . . . . . . . . . . . . . . . . . . . . . . . . . 79
*Michal Pluhacek, Roman Senkerik, Ivan Zelinka, Donald Davendra*

**Chaos Driven PSO with Ensemble of Priority Factors** .................... 89
*Michal Pluhacek, Roman Senkerik, Ivan Zelinka, Donald Davendra*

**Multi-chaotic Differential Evolution: Determining the Switching Time** ..... 99
*Roman Senkerik, Michal Pluhacek, Ivan Zelinka, Donald Davendra, Zuzana Kominkova Oplatkova*

## Nature-Inspired Algorithms and Nonlinear Systems

**Comparison of Pseudorandom Numbers Generators and Chaotic Numbers Generators Used in Differential Evolution** .................... 111
*Lenka Skanderova, Adam Řehoř*

**Analytic Programming Powered by Chaotic Dynamics** .................. 123
*Ivan Zelinka, Lenka Skanderova, Petr Šaloun, Roman Senkerik, Tran Trong Dao, Duy Vo Hoang*

**A New Approach to Modeling of Bio-inspired Information Diffusion with Ant Colony Optimization in Complex Networks** .................................................... 131
*Reisa Rahmatu Dewi, Tae-Hyong Kim*

**Scheduling the Flowshop with Zero Intermediate Storage Using Chaotic Discrete Artificial Bee Algorithm** .................................... 141
*Magdalena Metlická, Donald Davendra*

**Better and Faster Spectra Analysis Using Analytical Programming on CUDA** ........................................................... 153
*Peter Drábik, Petr Šaloun, Ivan Zelinka, Marie Vraná*

**Complex Network Analysis of Discrete Self-organising Migrating Algorithm** ...................................................... 161
*Donald Davendra, Ivan Zelinka, Roman Senkerik, Michal Pluhacek*

## Nonlinear and Predictive Control

**GUNT RT 010 Experimental Unit Modelling and Predictive Control Application** .................................................... 175
*Daniel Honc, František Dušek, Rahul Sharma*

**Primary Method of Quadratic Programming in Multivariable Predictive Control with Constraints** ..................................... 185
*Tomáš Barot, Marek Kubalčík*

**Control of Concentration inside CSTR Using Nonlinear Adaptive Controller** ..................................................... 195
*Jiri Vojtesek, Petr Dostál*

**LMI Approach of Constrained Fuzzy Model Predictive Control of DC-DC Boost Converter** ........ 205
*S. Bououden, M. Chadli, Ivan Zelinka*

**A Note about Robust Stabilization of Chaotic Hénon System Using Grammatical Evolution** ........ 219
*Radomil Matousek, Ladislav Dobrovsky, Petr Minar, Katerina Mouralova*

## Nonlinear Dynamics and Complex Systems

**Possibility of Dynamical Process Linearization Using Piecewise-Linear Neural Network** ........ 229
*Petr Dolezel, Martin Mariška*

**Bilinear Time Series Model as an Alternative Way of Speaker Modeling** .... 241
*Oskar Kochana, Patrycja Ksiazek, Michal Olszak, Ewa Bielinska*

**On the Growth of Large Independent Sets in Scale-Free Networks** ........ 251
*David Chalupa, Jiří Pospíchal*

**Time Series Prediction Based on Averaging Values via Neural Networks** .... 261
*Eva Volna, Martin Kotyrba*

**ECG Prediction Based on Classification via Neural Networks** ........ 271
*Eva Volna, Martin Kotyrba*

**Measurable Error Compensation with GPC in a Heat-Exchanger with a Traffic Delay** ........ 281
*Stanislav Talaš, Vladimír Bobál*

**Global and Local Environment State Information as Neural Network Input by Solving the Battleship Game** ........ 291
*Ladislav Clementis*

**Multi Agent Environment for Modelling and Testing of Cooperative Behaviour of Agents** ........ 301
*Martin Mariška, Petr Dolezel*

**Traffic Simulation Study: Traffic Intersection in the Village of Dobrá** ...... 307
*Jan Vrobel, Michal Janošek*

**Mathematical Models of Multivariable Systems** ........ 317
*Vladimír Jehlička*

**Risk Modeling in Process Industries by Stochastic Petri Nets** ........ 327
*Radim Briš, Ondřej Grunt*

**Impact of Hurst Exponent on Indicator Based Trading Strategies** ........ 337
*Tomáš Vantuch*

**Entropy and Market Prediction with Technical Indicators** . . . . . . . . . . . . . . . 347
*Marian Bielik*

**Dynamics and Efficiency of an Impact Damper** . . . . . . . . . . . . . . . . . . . . . . . . 355
*Marek Lampart, Jaroslav Zapoměl*

**Chaos in Nitrogen Dioxide Concentration Time Series and Its Prediction** . . . 365
*Radko Kříž*

## Various Topics

**Fuzzy Clustering and Loan Risk Prediction** . . . . . . . . . . . . . . . . . . . . . . . . . 377
*Petr Dostál, Stanislav Škapa*

**Sensor Fusion: An Application to Localization and Obstacle Avoidance in Robotics Using Multiple IR Sensors** . . . . . . . . . . . . . . . . . . . . . . . . . . . . . . . . 385
*Rahul Sharma, Honc Daniel, František Dušek*

**Estimation of Angles Yaw, Pitch and Roll in a Special Transformation Problem** . . . . . . . . . . . . . . . . . . . . . . . . . . . . . . . . . . . . . . . . . . . . . . . . . . . 393
*Martin Svoboda, Jaroslav Marek, Jana Heckenbergerová*

**Multi-Channel Multi-Objective Routing Metric for Vehicular Ad-Hoc Networks** . . . . . . . . . . . . . . . . . . . . . . . . . . . . . . . . . . . . . . . . . . . . . . . . . . 401
*Peppino Fazio, Cesare Sottile, Mauro Tropea, Floriano De Rango, Miroslav Voznak*

**Artificial intelligence in ISES Measureserver® for Remote Experiment Control** . . . . . . . . . . . . . . . . . . . . . . . . . . . . . . . . . . . . . . . . . . . . . . . . . . . 411
*Michal Gerža, František Schauer, Ivan Zelinka*

**Artificial Intelligence Elements in Data Mining from Remote Experiments** . . . . . . . . . . . . . . . . . . . . . . . . . . . . . . . . . . . . . . . . . . . . . . . . . 421
*Lukas Pálka, Franz Schauer, Ivan Zelinka*

**Fractal Models of Atoms and Molecules** . . . . . . . . . . . . . . . . . . . . . . . . . . . . 429
*Pavel Ošmera senior, Pavel Ošmera junior*

**Classification Methods Accuracy for Speech Emotion Recognition System** . . . . . . . . . . . . . . . . . . . . . . . . . . . . . . . . . . . . . . . . . . . . . . . . . . . . 439
*Pavol Partila, Jaromir Tovarek, Miroslav Voznak, Jakub Safarik*

**Complex Analysis of EEG Signal for Biometrical Classification Purposes** . . . 449
*Jaromir Svejda, Roman Zak, Roman Senkerik, Roman Jasek*

**Author Index** . . . . . . . . . . . . . . . . . . . . . . . . . . . . . . . . . . . . . . . . . . . . . . . . 461

# Chaos Level Measurement in Logistic Map Used as the Chaotic Numbers Generator in Differential Evolution

Lenka Skanderova[1], Ivan Zelinka[1,2], Tran Trong Dao[2], and Duy Vo Hoang[2]

[1] Department of Computer Science, VSB - Technical university of Ostrava, 17. listopadu 15/2172, 708 33 Ostrava - Poruba, Czech Republic
{lenka.skanderova,ivan.zelinka}@vsb.cz
[2] MERLIN, Ton Duc Thang University, 19 Nguyen Huu Tho Str., Dist. 7, Ho Chi Minh City, Vietnam
{trantrongdao,vohoangduy}@tdt.edu.vn

**Abstract.** In present time some researchers use chaotic numbers generators in evolutionary algorithms like differential evolution, SOMA or particle swarm optimization. These chaotic numbers generators are based on chaotic discrete systems which replace pseudorandom numbers generators like Mersenne Twister, Xorshift etc. In this paper we will investigate the influence of chaos level in logistic map which is used as chaotic numbers generator to the convergence's speed of differential evolution to the global minimum of testing functions.

## 1 Introduction

Differential evolution (DE) uses pseudorandom numbers generators in many steps of the algorithm. At first pseudorandom numbers generator (PSNG) is used when the first population is created – parameters of individual are generated randomly in lower and upper bounds. Then DE needs PSNG in random choosing of three different parents, then PSNG is needed in crossing, etc. In this paper PSNG is replaced by chaotic discrete system – Logistic map. We know that the level of chaos is given by Lyapunov exponents. If Lyapunov exponent is greater than zero, system's behavior can be described as chaotic. This paper will deal with influence of chaos level to convergence's speed of DE. In section 2 we explain motivation of this paper, in section 3 experiments design is described. In section 4 we can see results of experiments and section 5 summarize findings.

### 1.1 Differential Evolution

In DE we will see all principles of evolutionary algorithms – natural selection, crossing and mutation. The principle of DE can be described like this:

- First population is generated randomly. Number of individuals is given by the parameter $NP$. The values of parameters can be only from interval [lower

I. Zelinka et al. (eds.), *Nostradamus 2014: Prediction, Modeling and Analysis of Complex Systems*, Advances in Intelligent Systems and Computing 289,
DOI: 10.1007/978-3-319-07401-6_1, 

bound, upper bound]. Individuals are evaluated – fitness value is computed according to the cost function. Fitness value says how the individual is good for population.
– Difference between the first chosen parent $x_{r1,j}$ and parent $x_{r2,j}$ is multiplied by mutation constant $F$ and the third parent $X_{r3,j}$ is added to the result. The noise vector is created.
– New individual creation: Random number $r$ from interval [0,1] is generated. If $r < CR$, where $CR$ is crossing probability, parameter from the noise vector is taken to the new individual. If $r > CR$ parameter from the actual individual is taken. Fitness value of a new individual is computed. If the fitness value is better than fitness value of actual individual, the actual individual will be replaced by new individual in next generation. If not, actual individual will be taken to the next generation [1].

In this paper DE called DE/rand/1/bin is used.

In present time DE appears in many areas of research. In [3] authors deal with multi objective optimization by an adaptive DE. In [4] DE plays an essential role in identification time-delayed fractional order chaos. In [5] authors present a novel Particle Swarm Optimization (PSO) based on a non-homogenous Markov chain and DE and in [6] authors analyze the behavior of DE algorithm applied to the objective function, which are transformed by means of local searches. Authors of [7] use distributed DE in detecting moving objects from a video sequence. In [8] authors describe repairing the crossover rate in adaptive DE.

DE in connection with chaos and chaotic systems is mentioned for example in [9] – [13].

### 1.2 Chaos Level

Chaos level is given by Lyapunov exponent's value. Lyapunov exponent is computed for an orbit. We know that the Lyapunov exponent can be undefined for some orbits. In [2] authors says: "*In particular, an orbit containing a point $x_i$ with $f'(x) = 0$ causes the Lyapunov exponent to be undefined.*"

**Definition 1.** *Let $f$ be a smooth map of the real line $\Re$. The Lyapunov number $L(x_1)$ of the orbit $x_1, x_2, x_3 \ldots$ is defined as*

$$L(x_1) = \lim_{x \to \infty} |f'(x_1)| \ldots |f'(x_n)|)^{\frac{1}{n}} \tag{1}$$

*if this limit exists. The Lyapunov exponent $h(x_1)$ is defined as*

$$h(x_1) = \lim_{x \to \infty} (\frac{1}{n})[\ln |f'(x_1)| + \ldots + \ln |f'(x_n)|] \tag{2}$$

*if this limit exists. Notice that $h$ exists if and only if $L$ exists and is nonzero, and $\ln(L) = h$ [2].*

**Definition 2.** *Let $f$ be a map of the real line $\Re$, and let $x_1, x_2, \ldots$ be a bounded orbit of $f$. The orbit is chaotic if:*

- $\{x_1, x_2, \ldots\}$ *is not asymptotically periodic*
- *the Lyapunov exponent is greater than zero [2].*

### 1.3 Chaos Level in Logistic Map

Logistic map is defined by Eg.3.

$$x_{n+1} = ax(1 - x_n) \tag{3}$$

In our research we will change value of parameter $a$ and for each value of $a$ the Lyapunov exponent will be computed. Then we will observe influence of the computed Lyapunov exponent value to the DE convergence's speed to the global minimum.

Logistic map appears for example in [14], where period 3 and chaos for unimodal maps are studied. In [15] Logistic map is mentioned in connection with chaos optimization algorithms based on chaotic maps with different probability distribution. In [16] authors describe logistic neural networks and their chaotic pattern recognition properties and in [17] discrete fractional Logistic map and its chaos is investigated.

## 2 Motivation

As it was mentioned above, the main goal of this research was to investigate differential evolution convergence's speed reliance on Lyapunov exponent's values. Chaos is defined by a Lyapunov exponent greater than zero [2]. In this paper we observe differential evolution convergence's speed when Lyapunov exponent acquires different values.

## 3 Experiment Design

Precise setting of DE parameters is mentioned in the Table 1, where $NP$ means number of individuals in population, $D$ dimension (number of parameters of the individual), *Generations* means number of generation cycles, $F$ mutation constant and $CR$ crossing probability. In our research Schwefel's (see Eq.4), Griewangk's (see Eq.5), Rastrigin's (see Eq.6), Egg Holder's (see Eq.7) and Rana's (see Eq.8) functions have been used as cost functions. Schwefel's global minimum is $f(x) = -415.9829D$ where $D$ denotes dimension, for Rastrigin's and Griewangk's the global minimum is $f(x) = 0$. For Egg Holder's and Rana's functions there is not common formula for easy calculation of global minimum value. For experiments HP Pavilion dv7-6050 with processor Intel Core i7 with frequency 2 GHz, 4 GB RAM and graphic card AMD Radeon HD 6770M and Microsoft Visual Studio 2010 have been used. The experiments have been processed by Mathematica 8.

$$\sum_{i=1}^{D} -x_i \sin\left(\sqrt{|x_i|}\right) \tag{4}$$

$$1 + \sum_{i=1}^{D} \frac{x_i^2}{4000} - \prod_{i=1}^{D} \cos\left(\frac{x_i}{\sqrt{i}}\right) \tag{5}$$

$$10D + \sum_{i=1}^{D} x_i^2 - 10cos\left(2\pi x_i\right) \tag{6}$$

$$\begin{aligned} &\textstyle\sum_{i=1}^{D-1} \Big(-x_i \sin(\sqrt{|x_i - x_{i+1} - 47|}) - \\ &-(x_{i+1} + 47) \sin(\sqrt{|x_{i+1} + 47 + \frac{x_i}{2}|})\Big) \end{aligned} \tag{7}$$

$$\begin{aligned} \sum_{i=1}^{n-1} \Big[ &(x_{i+1} + 1) \cos\left(\sqrt{|x_{i+1} - x_i + 1|}\right) + \\ &+ \sin\left(\sqrt{|x_{i+1} + x_i + 1|}\right) + \\ &+ x_i \cos\left(\sqrt{|x_{i+1} + x_i + 1|}\right) \\ &\sin\left(\sqrt{|x_{i+1} - x_i + 1|}\right)\Big] \end{aligned} \tag{8}$$

**Table 1.** DE setting

| Parameter | Value |
|---|---|
| $NP$ | 50 |
| $D$ | 20 |
| $Generations$ | 1800 |
| $F$ | 0.9 |
| $CR$ | 0.4 |

At first parameter $a$ of Logistic map had been set to the beginning value $a = 3.5$ and then it was increased by 0.01. For each value of parameter $a$ one hundred experiments have been generated. For each cost function 5100 experiments have been generated. Initial value of $x$ has been set to $x = 0.02$. This value has been chosen randomly.

# 4 Results

In Table 2 we can see resultant values of parameter $a$, Lyapunov exponent, average fitness and median fitness for all cost functions. These results are mentioned in connection with the biggest convergence's speed of DE. In Tables 3 and 4 we can find cost functions median fitness values intervals for $a \in [3.50, 3.60]$ and $a \in [3.61, 4.00]$. Dependence of Lyapunov exponents on parameter $a$ is shown in Fig.1. Minimum, maximum and median fitness values reached during experiments are shown in Fig.2 for Schwefel's function, in Fig.3 for Griewangk's function, in Fig.4 for Rastrigin's function, in Fig.5 Egg Holder's function and in Fig.6 for Rana's function.

**Table 2.** The fastest convergence of DE for Schwefel's, Griewangk's, Rastrigin's, Egg Holder's and Rana's functions

| | Schwefel | Griewangk | Rastrigin | Egg Holder | Rana |
|---|---|---|---|---|---|
| **Parameter** $a$ | 3.94 | 3.77 | 3.76 | 3.98 | 3.6 |
| **Lyapunov exp.** | 0.540 | 0.398 | 0.383 | 0.596 | 0.178 |
| **Average fitness** | -6554.91 | 0.843 | 68.53 | -6477.80 | -4177.43 |
| **Median fitness** | -6541.40 | 0.848 | 61.25 | -8920.75 | -4198.46 |

**Table 3.** Cost functions (Schwefel's, Griewangk's and Rastrigin's) median fitness values interval

| Interval of $a$ | Schwefel | Griewangk | Rastrigin |
|---|---|---|---|
| $a \in [3.50, 3.60]$ | [-6399.96, -6307.58] | [0.849,0.862] | [68.58, 70.66] |
| $a \in [3.61, 4.00]$ | [-6541.40,-6186.59] | [0.843, 0.864] | [61.25, 71.37] |

**Table 4.** Cost functions (Egg Holder's, Rana's) median fitness values interval

| Interval of $a$ | Egg Holder | Rana |
|---|---|---|
| $a \in [3.50, 3.60]$ | [-6507.70, -6412.98] | [-4211.54, -4142.86] |
| $a \in [3.61, 4.00]$ | [-6587.57, -6390.79] | [-4223.99,-4105.34] |

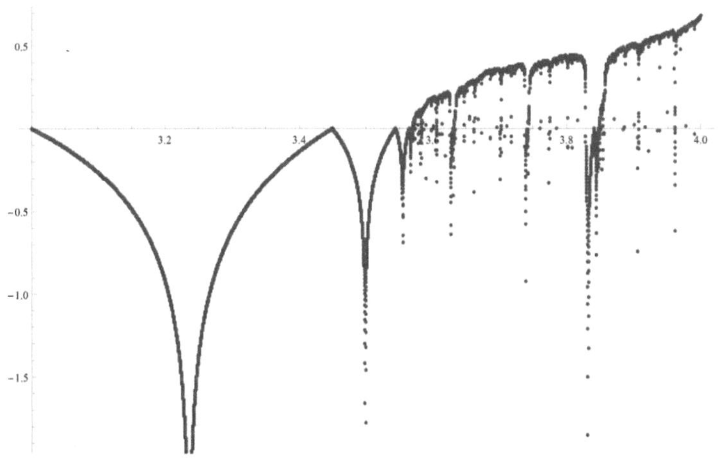


**Fig. 1.** Lyapunov exponents for Logistic map. X-axis is values of parameter $a$ of Logistic map, y-axes is Lyapunov exponents.

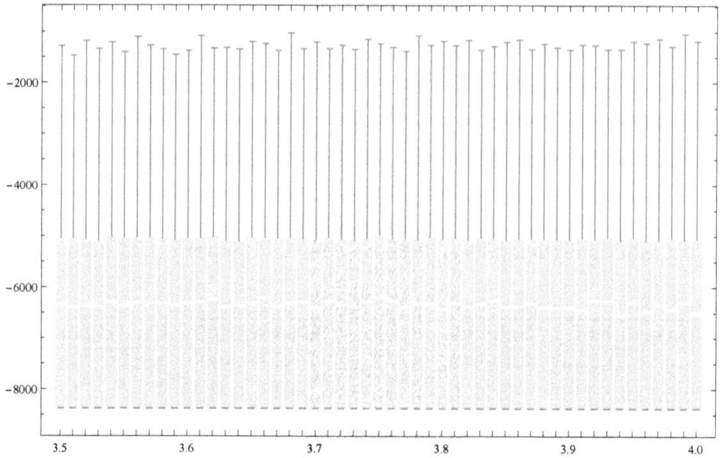


**Fig. 2.** Experiments results for Schwefel's function

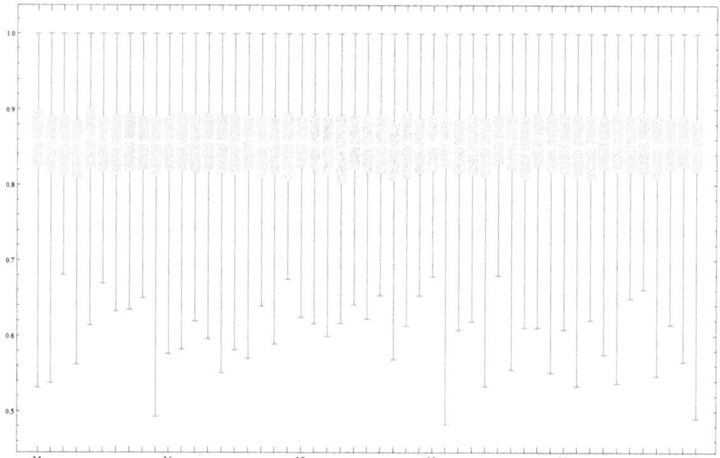

**Fig. 3.** Experiments results for Griewangk's function

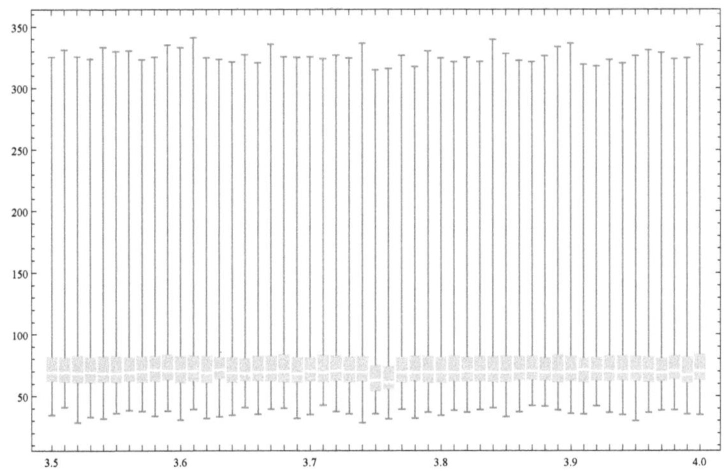


**Fig. 4.** Experiments results for Rastrigin's function

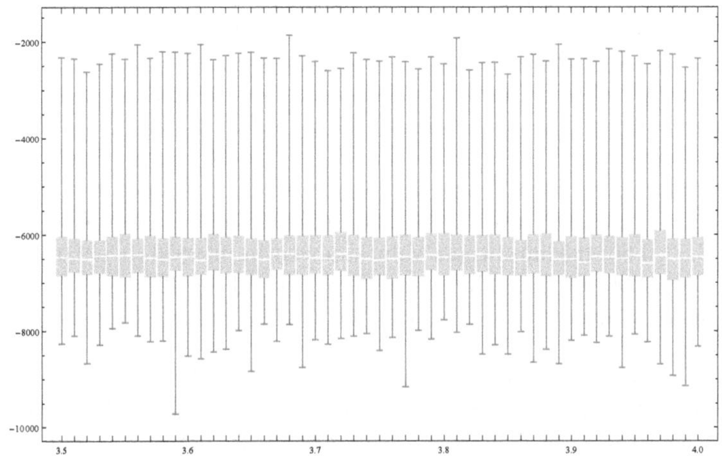


**Fig. 5.** Experiments results for Egg Holder's function

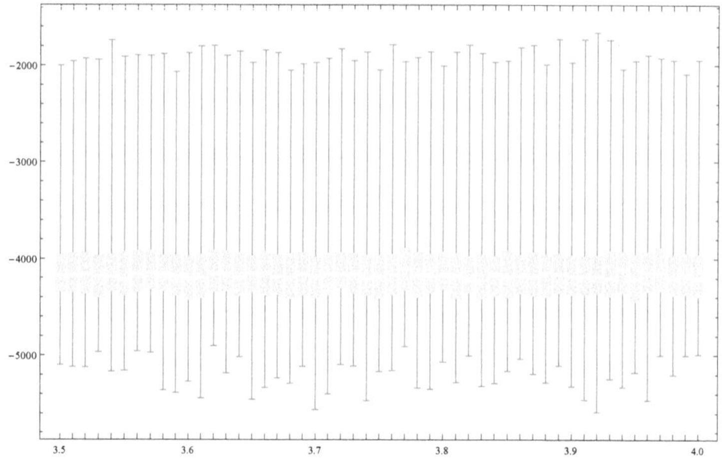


**Fig. 6.** Experiments results for Rana's function

# 5 Conclusion

From results mentioned in section 4 we can make some conclusions:

- **Schwefel's Function:** When we look at the Fig.2 we can see that DE has reached global minimum in all cases. When parameter $a$ had been set to $a = 3.94$ DE convergence's speed was the biggest. In this case Lyapunov exponent gained the value 0.540 and average fitness value gained value -6554.91, median fitness value was then -6541.40. When $a \in [3.50, 3.60]$ median fitness values moved in interval [-6399.96, -6307.58]. When $a \in [3.61, 4.00]$ median fitness values moved in interval [-6541.40,-6186.59], see Table 3.
- **Griewangk's Function:** In Fig.3 there are results for Griewangk's function. It is clear that DE has not reached the global minimum in any case. It is probably caused by DE's setting. The biggest convergence's speed was observed when $a = 3.77$ and Lyapunov exponent gained the value 0.398, average fitness value was 0.843 and median fitness value was 0.848. When $a \in [3.50, 3.60]$ median fitness values moved in interval [0.849,0.862]. When $a \in [3.61, 4.00]$ fitness values moved in interval [0.843, 0.864], see Table 3. The smallest fitness value was reached when $a = 3.81$, fitness value gained the value 0.483.
- **Rastrigin's Function:** In Fig.4 we can see that DE has not reached global minimum. It is probably caused by DE's setting. The smallest fitness value was reached when $a = 3.74$, its value was 28.44. When $a = 3.76$ Lyapunov exponent gained value 0.383 and DE's convergence's speed was the biggest, the average fitness value was 68.53 and median fitness value 61.25. When $a \in [3.50, 3.60]$ median fitness values moved in interval [68.58, 70.66]. On the other hand when $a \in [3.61, 4.00]$ median fitness values moved in interval [61.25, 71.37], see Table 3.

- **Egg Holder's Function:** The results for Egg Holder's function are showed in Fig.5. We know that for Egg Holder's function there is not described global minimum in the literature. The smallest fitness value was reached when $a = 3.59$. Lyapunov exponent for this value of parameter $a$ has the value 0.138. When $a = 3.98$ DE's convergence's speed was the biggest, average fitness value gained the value -6477.80 and median fitness value -8920.75. Lyapunov exponent for $a = 3.98$ is 0.596. When $a \in [3.50, 3.60]$ median fitness values moved in interval [-6507.70, -6412.98]. When $a \in [3.61, 4.00]$ median fitness values moved in interval [-6587.57, -6390.79], see Table 4.
- **Rana's Function:** As well as Egg Holder's function for Rana's function there is not described global minimum in the literature. The results for Rana's function are in Fig.6. The smallest fitness value -5588.24 was reached when $a = 3.92$. In this case Lyapunov exponent gained the value 0.517 for $a = 3.92$. When $a = 3.6$, Lyapunov exponent gained the value 0.178 and the convergence's speed of DE was the biggest. Average fitness value was -4177.43 and median fitness value -4198.46. When $a \in [3.50, 3.60]$ median fitness values moved in interval [-4211.54, -4142.86]. When $a \in [3.61, 4.00]$ median fitness values moved in interval [-4223.99,-4105.34], see Table 4.
- When we look at the results mentioned above, we can make conclusion that DE convergence's speed was the biggest when Lyapunov exponent had gained values greater than zero for all testing functions. For Schwefel's function its value was 0.540 ($a = 3.94$), for Griewangk's 0.398 ($a = 3.77$), for Rastrigin's 0.383 ($a = 3.76$), for Egg Holder's 0.138 ($a = 3.98$) and for Rana's 0.517 ($a = 3.6$). For all functions DE convergence's speed was the biggest when $a \geq 3.6$.

In the future, we would like to extend our research by adding other cost functions as Michalwicz's, Rosenbrock's, Patological's etc. We would like to try these experiments to other evolutionary algorithms like PSO, Self-organizing migrating algorithm etc.

**Acknowledgement.** The following grants are acknowledged for the financial support provided for this research: Grant Agency of the Czech Republic - GACR P103/13/08195S, is partially supported by Grant of SGS No. SP2014/159, VSB - Technical University of Ostrava, Czech Republic, by the Development of human resources in research and development of latest soft computing methods and their application in practice project, reg. no. CZ.1.07/2.3.00/20.0072 funded by Operational Programme Education for Competitiveness. Special thanks also belong to the research group MERLIN of Ton Duc Thang University, Ho Chi Minh City, Vietnam.

## References

1. Zelinka, I., Celikovsky, S., Richter, H., Chen, G. (eds.): Evolutionary Algorithms and Chaotic Systems. SCI, vol. 267. Springer, Heidelberg (2010)

2. Alligood, K., Sauer, T.D., Yorke, J.A.: CHaos - an introduction to dynamical systems. Textobooks in Mathematical Sciences, vol. 1197. Springer - Verlag New York, Inc. (1996) ISBN 0- 987-94677-2
3. Venske, S.M., et al.: ADEMO/D: Multiobjective optimization by an adaptive differential evolution algorithm. Neurocomputing 127, 65–77 (2014)
4. Gao, F., et al.: Identification time-delayed fractional order chaos with functional extrema model via differential evolution. Expert Systems with Applications 41, 1601–1608 (2014)
5. Zeng, N.Y., et al.: A novel switching local evolutionary PSO for quantitative analysis of lateral flow immunoassay. Expert Systems with Applications 41, 1708–1715 (2014)
6. Locatelli, M., et al.: Differential evolution methods based on local searches. Computers & Operations Research 43, 169–180 (2014)
7. Ghosh, A., et al.: Moving object detection using Markov Random Field and Distributed Differential Evolution. Applied Soft Computing 15 (February 2014)
8. Gong, W.Y., et al.: Repairing the crossover rate in adaptive differential evolution. Applied Soft Computing 15, 149–168 (2014)
9. Oplatkova, Z.K., et al.: Analytic programming in the task of evolutionary synthesis of a controller for high order oscillations stabilization of discrete chaotic systems. Computers & Mathematics with Applications 66, 177–189 (2013)
10. Senkerik, R., et al.: Investigation on the Differential Evolution Driven by Selected Six Chaotic Systems in the Task of Reactor Geometry Optimization. In: 2013 IEEE Congress on Evolutionary Computation (CEC), pp. 3087–3094 (2013)
11. Senkerik, R., Davendra, D., Zelinka, I., Pluhacek, M., Kominkova Oplatkova, Z.: Chaos Driven Differential Evolution with Lozi Map in the Task of Chemical Reactor Optimization. In: Rutkowski, L., Korytkowski, M., Scherer, R., Tadeusiewicz, R., Zadeh, L.A., Zurada, J.M. (eds.) ICAISC 2013, Part II. LNCS, vol. 7895, pp. 56–66. Springer, Heidelberg (2013)
12. Senkerik, R.: On the Evolutionary Optimization of Chaos Control - A Brief Survey. In: Zelinka, I., Snasel, V., Rössler, O.E., Abraham, A., Corchado, E.S. (eds.) Nostradamus: Mod. Meth. of Prediction, Modeling. AISC, vol. 192, pp. 35–48. Springer, Heidelberg (2013)
13. Senkerik, R., Oplatkova, Z., Zelinka, I.: Evolutionary Synthesis of Control Rules by Means of Analytic Programming for the Purpose of High Order Oscillations Stabilization of Evolutionary Synthesized Chaotic System. In: Zelinka, I., Snasel, V., Rössler, O.E., Abraham, A., Corchado, E.S. (eds.) Nostradamus: Mod. Meth. of Prediction, Modeling. AISC, vol. 192, pp. 191–201. Springer, Heidelberg (2013)
14. Cheng, K.J., et al.: Period 3 and Chaos for Unimodal Maps. Discrete and Continuous Dynamical Systems 34, 1933–1949 (2014)
15. Yang, D.X., et al.: Chaos optimization algorithms based on chaotic maps with different probability distribution and search speed for global optimization. Communications in Nonlinear Science and Numerical Simulation 19, 1229–1246 (2014)
16. Ke, Q., Oommen, B.J.: Logistic Neural Networks: Their chaotic and pattern recognition properties. Neurocomputing 125, 184–194 (2014)
17. Wu, G.C., Baleanu, D.: Discrete fractional logistic map and its chaos. Nonlinear Dynamics 75, 283–287 (2014)

# Estimations of Initial Errors Growth in Weather Prediction by Low-dimensional Atmospheric Model

Hynek Bednář, Aleš Raidl, and Jiří Mikšovský

Department of Meteorology and Environment Protection, Faculty of Mathematics and Physics, Charles University in Prague
{Hynek.Bednar,Ales.Raidl,Jiri.Miksovsky}@mff.cuni.cz

**Abstract.** Initial errors in weather prediction grow in time. As errors become larger, their growth slows down and then stops at an asymptotic value. Time of reaching this value represents the limit of predictability. Other time limits that measure the error growth are doubling time $\tau_d$, and times when the forecast error reaches *95%*, *71%*, *50%*, and *25%* of the limit of predictability. This paper studies asymptotic value and time limits in a low-dimensional atmospheric model for five initial errors, using ensemble prediction method as well as error approximation by quadratic and logarithmic hypothesis. We show that quadratic hypothesis approximates the model data better for almost all initial errors and time lengths. We also demonstrate that both hypotheses can be further improved to achieve even better match of the asymptotic value and time limits with the model.

**Keywords:** Chaos, Atmosphere, Prediction, Error growth.

## 1 Introduction

Forecast errors in numerical weather prediction models (NWPM) grow in time because of the inaccuracy of the initial state and the model imperfections. Due to the nonlinear terms in the governing equations the forecast error will saturate after some time. Time of saturation or *the limit of predictability of deterministic forecast* in NWPM is defined by [1] as time when the prediction state diverges as much from the verifying state as a randomly chosen, but possible state (dynamically and statistically). Forecasters use other time limits (TL) to measure this growth. Forecast-error doubling time $\tau_d$ is time when initial error doubles its size. $\tau_{95\%}$, $\tau_{71\%}$, $\tau_{50\%}$ and $\tau_{25\%}$ are the times when the forecast error reaches *95%*, *71%*, *50%*, and *25%* of the limit of predictability. The time limit $\tau_{71\%}$ is the time when the forecast error exceeds $1/\sqrt{2}$ of the saturation or asymptotic value (AV) and by [2] this limit corresponds to the level of climatic variability. Lorenz [3] calculated forecast error growth of NWPM by comparing the integrations of model started from slightly different initial states. Present-day calculations use the approach developed by Lorenz [4], where we can obtain two types of error growth. The first is called *lower bound* and is calculated as the root mean-square error (RMSE) between forecast data of increasing lead times and

I. Zelinka et al. (eds.), *Nostradamus 2014: Prediction, Modeling and Analysis of Complex Systems*, Advances in Intelligent Systems and Computing 289,
DOI: 10.1007/978-3-319-07401-6_2, © Springer International Publishing Switzerland 2014

analysis data valid at the same time. The second is called *upper bound* and is calculated as the root mean-square (RMS) difference between pairs of forecasts, valid at the same time but with times differing by some fixed time interval. This second method compares only model equations and therefore it represents growth without model error. The innovation to upper bound, that is also used, is calculated as the RMS difference between forecast and control forecast with higher resolution of the model (*perfect model framework*).

*Quadratic hypothesis* (QH) was the first attempt that was made by Lorenz [3] to quantify and qualify the error growth. QH is based on the assumption that, if the principal nonlinear terms in the atmospheric equations are quadratic, then the nonlinear terms in the equations governing the field of errors are also quadratic. Dalcher and Kalnay [5] added a model error to Lorenz's QH. A version that is used by recent researchers is Simons's modification [6] of [5]. The Lorenz's QH is therefore suitable for upper bound of error growth and the Simons's modification for lower bound. Trevisan *et al.* [7] came out with idea that logarithmic term is more valid than quadratic and linear term in the equations governing the field of errors, but this *logarithmic hypothesis* (LH) has never been used in NWPM computations.

*Ensemble prediction systems* (EPS) are used in order to estimate forecast uncertainties. They consist of a given number of deterministic forecasts where each individual forecast starts from slightly different initial states. EPS also includes a stochastic scheme designed to simulate the random model errors due to parameterized physical processes. Recent studies of predictability and forecast error growth (for example [8-11]) are mostly done by models of European Centre for Medium Range Weather Forecasts (ECMWF) and the Global Ensemble Forecast System (GEFS) from the National Centers for Environmental Prediction (NCEP). They include deterministic and ensemble forecast with *1* to *70* members of ensemble (Operational model of ECMWF uses *50* members plus control forecast. More detailed study [10] uses *5* members plus control forecast). The initial conditions of ensemble members are defined by linear combination of the fastest singular vectors. Horizontal resolution with spectral truncation varies from T95 to T1279 and number of vertical level varies from *19* to *91* (analyses use higher resolution than forecasts). The output data are interpolated to *1°* latitude x *1°* longitude or *2.5°* latitude x *2.5°* longitude resolution separately for the Northern Hemisphere (*20°*, *90°*) and Southern Hemisphere (*-90°*, *-20°*). Forecast is usually run for *90* days at winter (DJF) or summer (JJA) season with *0* (analysis) to *10*, *15* (ECMWF) or *16* days (NCEP) of *forecast length* (FL) at *6* or *12* hours intervals. The most often used variable for analyzing the forecast error is geopotential height at *500* hPa (Z500). Others are geopotential height at *1000* hPa (Z1000) and the *850* hPa temperatures (T850). To describe the forecast error growth over the calculated forecast length, the Simons's modification [6] of Lorenz's QH [3] is used.

The questions that have arisen from studies of predictability and forecast error growth and that represent the key issues addressed in this work are: Is the LH [7] better approximation of initial error growth than QH [3]? If so, how much difference it creates in time limits that measure the forecast error growth? How precisely does the approximations describe forecast error growth over the FL (*10*, *15* or *16* days)? How the approximations obtained from model values with various number of ensemble members differ from each other? Lorenz's low-dimensional atmospheric model (model II) [12] will be used. For a more comprehensive introduction to the problem of weather predictability we refer reader to the book by Palmer and Hagedorn [13]. After this introduction, section *2* describes the model and experimental design, *3* describes ensemble prediction method, *4* introduces quadratic and logarithmic hypotheses, *5* sets experimental designs. Section *6* presents the results and their discussion and section *7* summarizes the conclusions.

## 2 Model

Because of the limitations of NWPM that enable us to answer the key questions and because we want to derivate the impact of initial error (perfect model framework), we use modification [13] of low-dimensional atmospheric model (L96). L96 [14] is a nonlinear model, with $N$ variables $X_1,\ldots, X_N$ connected by governing equations

$$dX_n/dt = -X_{n-2}X_{n-1} + X_{n+1}X_{n-1} - X_n + F. \tag{1}$$

$X_{n-2}$, $X_{n-1}$, $X_n$, $X_{n+1}$ are *unspecified (i.e., unrelated to actual physical variables) scalar meteorological quantities*, $F$ is a constant representing external forcing and $t$ is time. The index is cyclic so that $X_{n-N} = X_{n+N} = X_n$ and variables can be viewed as existing around a circle. Nonlinear terms of (1) simulate advection. Linear terms represent mechanical and thermal dissipation. The model quantitatively, to a certain extent, describes weather systems, but, unlike the well-known Lorenz's model of atmospheric convection [15], it cannot be derived from any atmospheric dynamic equations. The motivation was to formulate the simplest possible set of dissipative chaotically behaving differential equations that share some properties with the "real" atmosphere. NWPM interpolate the output data mostly to *1°* latitude x *1°* longitude. In L96 it means *N=360*. Such a high resolution would create large number of waves with similar maxima ("highs") and minima ("lows"), however, to share some properties with the "real" atmosphere, we would rather have *5* to *7* main highs and lows that correspond to planetary waves (Rossby waves) and a number of smaller waves that correspond to synoptic waves. Therefore we introduce spatial continuity modification (L05II) [12] of L96. Equation (1) is rewritten to the form:

$$dX_n/dt = [X,X]_{L,n} - X_n + F, \tag{2}$$

where

$$[X,X]_{L,n} = \sum_{j=-J}^{J}{}' \sum_{i=-J}^{J}{}' \left(-X_{n-2L-i}X_{n-L-j} + X_{n-L+j-i}X_{n+L+j}\right)/L^2.$$

If $L$ is even, $\sum'$ denotes a modified summation, in which the first and last terms are to be divided by *2*. If $L$ is odd, $\sum'$ denotes an ordinary summation. Generally $L$ is much smaller than $N$ and $J = L/2$ if $K$ is even and $J = (L-1)/2$ if $L$ is odd. For our computation we choose $N = 360$, so each sector covers $1°$ degrees of longitude. To keep a desirable number of main highs a lows, Lorenz suggested to keep ration $N/L = 30$ and therefore $L = 12$. Parameter $F = 15$ is selected as a compromise between too long doubling time (smaller $F$) and unanticipated shorter waves (larger $F$). We first choose arbitrary values of the variables, and, using a fourth order Runge-Kutta method with a time step $\Delta t = 0.05$ or *6 hours*, we integrate forward for *14400 steps*, or *10 years*. We then use the final values, which should be free of transient effect. For this setting and by the method of numerical calculation presented in [16] the global largest Lyapunov exponent is $\lambda_{max} = 0.32$. By the definition [3]: „A bounded dynamical system with a positive Lyapunov exponent is chaotic". Because the value of the largest Lyapunov exponent is positive and the system under study is bounded, it is chaotic; Strictly speaking, we also need to exclude the asymptotically periodic behavior, but such a task is impossible to fulfill for the numerical simulation. The choice of parameters $F$ and *time unit* = *5 days* is made to obtain the similar value of the largest Lyapunov exponent as state of the art NWPM.

## 3 Ensemble Prediction Method

The ensemble prediction method (EPM) employed is similar to [14] and is used to calculate average initial error growth. We make an initial "run" by choosing error $e_{n0}$ and letting $X^{/}_{n0} = X_{n0} + e_{n0}$ be the "observed" initial value of $N$ variables. We then integrate forward from the true and the observed initial state, for between *25* to *37.5* days ($K=100$ to $K=150$ steps). This time length covers initial error growth till the limit of predictability. We obtain $N$ sequences $X_{n0},\ldots,X_{nK}$ and $X^{/}_{n0},\ldots,X^{/}_{nK}$ , after which we let $e_{nk} = X^{/}_{nk} - X_{nk}$ for all values of $k$ and $n$. In NWPM, forecast error growth is obtained from an average of values from *90* days and from various number of ensemble member. To simulate that, we make a total of $M_1 = 100$, $M_2 = 250$ and $M_3 = 500$ runs in the above described manner. In each run, new values of $X_{n0}$ are set as the old values of $X_{nK}$. Finally, we let $e^2(\tau) = 1/N\left(e^2_{1k} + \ldots + e^2_{Nk}\right)$ be the average of the $N$ values, where $\tau = k\Delta t$ is the predictable range and $\log E^2(\tau) = 1/M\left(\log e^2(\tau)_1 + \ldots + \log e^2(\tau)_M\right)$ is the average of $M$ values. Logarithmic average is chosen because of its suitability for comparison with growth governed by the largest Lyapunov exponent. For further information see [17-19].

## 4 Quadratic and Logarithmic Hypothesis

According to Lorenz [14], there is an eventual cessation of the exponential growth due to processes represented by nonlinear terms in the weather governing equations. Most important are the quadratic terms, which represent the advection of the temperature and velocity fields. Under the assumption, that the principal nonlinear terms in the atmospheric equations are quadratic, nonlinear terms in equations governing the field of errors are also quadratic. To describe this, Lorenz [14] defined QH

$$dE(t)/dt = aE(t) - bE(t)^2, \tag{3}$$

where $E(t)$ is a distance at time $t$ between two originally nearby trajectories and $a$, $b$ are constants. As an alternative, Trevisan *et al.* [7] introduced LH

$$dE(t)/dt = -cE(t)\ln(gE(t)), \tag{4}$$

where $E(t)$ is a distance at time $t$ between two originally nearby trajectories and $c$, $g$ are constants. The explanation follows, if we let $Q(t) = \ln\left(\overline{E(t)}\right)$, $\bar{E}$ is the normalized $E$, then $dQ(t)/dt = a\left(1 - e^{Q(t)}\right)$ represents the QH. In [7] it is assumed that linear fit $dQ(t)/dt = -aQ(t)$ is superior to the QH. In [7, 20, 21], it is shown on low-dimensional models that, if the initial error is sufficiently small and therefore the early stages of error growth are exponential, QH is superior. If the initial error is not small enough, it is better to use LH. Generally, whether an error is small enough to guarantee the exponential growth depends on specific meteorological conditions and/or model under study.

## 5 Experimental Design

We want to achieve the conditions as similar to NWPM as possible. The size of initial error for NWPM (perfect model framework) is by [9] approximately between *2%* and *0.01%* of AV of the forecast error for control forecast and between *10%* and *3%* of AV for ensemble members. Different values of AV fraction are a result of varying resolution and because it is calculated for different variables (Z500, Z1000 and T850). In our case the AV is $E_{asym} = 8.4$. This is calculated by four independent methods with same results. The first method is numerical computation of ensemble prediction approach. Second and third methods are calculated as:

$$E_{asym} = \overline{\left(f_1 - X_{avr}\right)^2} + \overline{\left(f_2 - X_{avr}\right)^2} = 2\overline{\left(f_1 - X_{avr}\right)^2}, \tag{5}$$

where $f_1$ is "forecast" from $X_{n0}$, $f_2$ from $X'_{n0}$ and $X_{avr}$ is average value of $X_n$. The bars above the (5) members mean the average value. The explanation for (5) can be found in [6, 10]. The fourth method is based on assumption [2] that variability of $X_n$ is *71%* of $E_{asym}$.

Recalculation of initial errors to L05II model leads to the sizes between *0.001* and *0.84*. For initial error sizes $e_{n0}$ we therefore choose randomly from five normal distributions $ND(\mu;\ \sigma)$. $ND_1 = (0;\ 0.1)$, $ND_2 = (0;\ 0.2)$, $ND_3 = (0;\ 0.4)$, $ND_4 = (0;\ 0.6)$, $ND_5 = (0;\ 1)$, where $\mu$ is mean and $\sigma$ is standard deviations. These choices of $e_{n0}$ are made, because [20, 21] shows that change over QH and LH takes place between

$e_{n0} = 0.1$ and $e_{n0} = 1$ for L96. NWPM routinely define initial conditions of ensemble members by the fastest singular vectors. We do not adopt it, because by [10, 22] it affects early stages of forecast error growth and in our case, we want to have model data as close as possible to the tested hypotheses. From these initial conditions the average initial error growth $E$ is calculated from ensemble prediction method by the fourth order Runge-Kutta integration schema with a time step $\Delta t = 0.05$ or *6 hours* for $M_1 = M_2 = 100$, $M_3 = M_4 = 250$ and $M_5 = 500$. Because we want to study agreement of (3) and (4) with model data, we make differences of model data $y_k = \left(E(\tau+\Delta t) - E(\tau)\right)/\Delta t$ at points $x_k = \left(E(\tau) + E(\tau+\Delta t)\right)/2$, $k = 1,..,K$ and $K = 56$ ($\tau_1 = 14$ *days*), $K = 76$ ($\tau_2 = 19$ *days*), $K =$ *limit of predictability* ($\tau_3$), and we calculate parameters $a$, $b$, $c$, and $g$. The choice of the first two values of $K$ is made, because we want to keep ratio $\tau_{95\%}/\mathit{forecast\ lenght}$ the same for NWPM and L05II. The reason for the third value is obvious.

The solutions of (3) and (4) are

$$E(t) = \frac{a}{\left((a/e_0) - b\right)\exp(-at) + b}, \tag{6}$$

$$E(t) = \left(ge_0\right)^{\exp(-ct)}/g. \tag{7}$$

We have five types of normal distributions for reaching sizes of initial error $ND_{1,..,5}$, five settings for EPM $M_{1,..,5}$ ,three FL $\tau_{1,..3}$ and three ways of getting data of initial error growth: EPM, (6) and (7). To answer the key questions we compute TL for all combinations. We take $M_3$, $\tau_3$, EPM as the most reliable dataset in our experiment for all $e_0$ and we calculate differences with other combinations at the same TL.

# 6 Result and Discussion

Table 1 shows with darker grey lines the resulting values ($\bar{M}$ , $\tau_3$, EPM) for all TL and for all $e_0$ represented by $ND_{1,..,5}$. $\bar{M}$ is average value of $M_{1,..,5}$ and we use it, because the difference between $M_{1,..,5}$ is of the order of $0.1$ and $M_3$ and $M_4$ do not show closer values to $M_5$ than $M_1$ and $M_2$. The difference of average values $\bar{M}$ of (7) from ($\bar{M}$ , $\tau_3$, EPM) (lighter grey lines in Table 2) is higher or equal than difference of ((6), $\bar{M}$ ) (not shadowed lines in Table 2) for $ND_{1,..,4}$. Only for some cases in $ND_5$ is the difference of (6) higher than for (7). Average values $\overline{ND}$ over $ND_{1,..,5}$ (bottom right part of Table 1) and Fig. 1 show that there is almost constant difference between the model data and data received from (7) for $\tau_d$, $\tau_{71\%}$, $\tau_{50\%}$ and $\tau_{25\%}$. This is caused by negative growth rate for the first day, but turning into increase thereafter. At around two days, the difference reaches the same value as it had initially. This behavior causes the above mentioned differences. NWPM also show this type of behavior [23]. After the subtraction of two day from time limits $\tau_d$, $\tau_{71\%}$, $\tau_{50\%}$ , $\tau_{25\%}$ and some cases for $\tau_{95\%}$ we reach close values (error of order $0.1$) of approximation (6) to the model data for all prediction times limit $\tau_{1,..,3}$. Exception from this is difference $\tau_1$ - $\tau_3$ for $\tau_{95\%}$, where the difference is of the order of days (between *1* and *3* days). One may argue that because of subtraction of *2* days we should recalculate the approximations. We did that and the results are close to ones with subtraction. It is also good to mention that $\tau_1$ is always higher than $\tau_{25\%}$ and lower than $\tau_{95\%}$, and $\tau_2$ is always higher than $\tau_{71\%}$.

**Table 1.** Average values over $M_{1,..,5}$ of time limits $\tau_d$, $\tau_{95\%}$, $\tau_{71\%}$, $\tau_{50\%}$, $\tau_{25\%}$ in days for model values (EPM), for all normal distributions $ND_{1,..,5}$ and prediction time length $\tau_3$ (darker grey) and difference between this model values and data received from (6) and (7) with parameters $a$, $b$, $c$, $g$ calculated from approximations of (3) and (4) (lighter grey) for $\tau_1$, $\tau_2$, $\tau_3$.

| | $ND_1$ | | | | | $ND_2$ | | | | |
|---|---|---|---|---|---|---|---|---|---|---|
| (days) | $\tau_d$ | $\tau_{25\%}$ | $\tau_{50\%}$ | $\tau_{71\%}$ | $\tau_{95\%}$ | $\tau_d$ | $\tau_{25\%}$ | $\tau_{50\%}$ | $\tau_{71\%}$ | $\tau_{95\%}$ |
| *(EPM, $\tau_3$)* | 4.8 | 12.9 | 16.3 | 19.1 | 26 | 4.8 | 10.6 | 14 | 16.9 | 24.3 |
| *((6),$\tau_3$) - (EPM, $\tau_3$)* | -2.5 | -2.6 | -2.8 | -2.5 | -1.7 | -2.6 | -2.5 | -2.4 | -2.3 | -2.7 |
| *((7),$\tau_3$) - (EPM, $\tau_3$)* | -3.8 | -6.5 | -6.3 | -5.6 | -4 | -3.8 | -5.35 | -5.1 | -4.4 | -4.2 |
| *((6),$\tau_2$) - (EPM, $\tau_3$)* | -2.5 | -2.8 | -2.7 | -2.6 | -2.8 | -2.6 | -2.6 | -2.6 | -2.6 | -3.2 |
| *((7),$\tau_2$) - (EPM, $\tau_3$)* | -3.4 | -4.7 | -4.4 | -4.6 | -8.1 | -3.6 | -4.35 | -4 | -3.7 | -6.8 |
| *((6),$\tau_1$) - (EPM, $\tau_3$)* | -2,6 | -3.1 | -2.9 | -2.6 | -3.2 | -2.7 | -2.7 | -2.7 | -2.6 | -5.3 |
| *((7),$\tau_1$) - (EPM, $\tau_3$)* | -3.2 | -3.8 | -3.6 | -4.2 | -9 | -3 | -3.1 | -3.1 | -3.7 | -9.2 |
| | $ND_3$ | | | | | $ND_4$ | | | | |
| (days) | $\tau_d$ | $\tau_{25\%}$ | $\tau_{50\%}$ | $\tau_{71\%}$ | $\tau_{95\%}$ | $\tau_d$ | $\tau_{25\%}$ | $\tau_{50\%}$ | $\tau_{71\%}$ | $\tau_{95\%}$ |
| *(EPM, $\tau_3$)* | 4.8 | 8.3 | 11.6 | 14.4 | 21.6 | 4.8 | 6.9 | 10.2 | 12.7 | 19.7 |
| *((6),$\tau_3$) - (EPM, $\tau_3$)* | -2.6 | -2.3 | -2.4 | -2.2 | -2.7 | -2.3 | -2.4 | -2.2 | -2 | -2.6 |
| *((7),$\tau_3$) - (EPM, $\tau_3$)* | -3.4 | -4 | -3.9 | -3.2 | -2.6 | -3.4 | -3.5 | -3.3 | -2.2 | -1.6 |
| *((6),$\tau_2$) - (EPM, $\tau_3$)* | -2.4 | -2.4 | -2.1 | -2.1 | -3 | -2.4 | -2.4 | -2.3 | -2.1 | -2.7 |
| *((7),$\tau_2$) - (EPM, $\tau_3$)* | -3.3 | -3.6 | -3.1 | -2.7 | -5 | -3.1 | -3.3 | -2.8 | -2.1 | -3.5 |
| *((6),$\tau_1$) - (EPM, $\tau_3$)* | -2.6 | -2.4 | -2.4 | -2.3 | -4.5 | -2.4 | -2.2 | -2.1 | -1.8 | -4.1 |
| *((7),$\tau_1$) - (EPM, $\tau_3$)* | -2.9 | -2.9 | -2.5 | -2.5 | -5.8 | -2.5 | -2.5 | -2.1 | -1.8 | -5.7 |
| | $ND_5$ | | | | | $\overline{ND}$ | | | | |
| (days) | $\tau_d$ | $\tau_{25\%}$ | $\tau_{50\%}$ | $\tau_{71\%}$ | $\tau_{95\%}$ | $\tau_d$ | $\tau_{25\%}$ | $\tau_{50\%}$ | $\tau_{71\%}$ | $\tau_{95\%}$ |
| *(EPM, $\tau_3$)* | 4.8 | 5.1 | 8.1 | 11 | 18.3 | | | | | |
| *((6),$\tau_3$) - (EPM, $\tau_3$)* | -2.2 | -2.2 | -2.1 | -2.1 | -2.9 | -2.4 ±0.2 | -2.4 ±0.1 | -2.4 ±0.2 | -2.2 ±0.1 | -2.5 ±0.3 |
| *((7),$\tau_3$) - (EPM, $\tau_3$)* | -2.8 | -2.9 | -2.3 | -1.9 | -1.6 | -3.4 ±0.3 | -4.5 ±1.2 | -4.2 ±1.2 | -3.5 ±1.2 | -2.8 ±1 |
| *((6),$\tau_2$) - (EPM, $\tau_3$)* | -2.2 | -2.5 | -2.1 | -2.1 | -2.7 | -2.4 ±0.1 | -2.5 ±0.1 | -2.4 ±0.2 | -2.3 ±0.2 | -2.9 ±0.2 |
| *((7),$\tau_2$) - (EPM, $\tau_3$)* | -2.7 | -2.7 | -2.1 | -1.8 | -2.8 | -3.2 ±0.3 | -3.7 ±0.6 | -3.3 ±0.7 | -3 ±0.9 | -5.2 ±1.8 |
| *((6),$\tau_1$) - (EPM, $\tau_3$)* | -2.2 | -2.2 | -1.9 | -2.1 | -4.3 | -2.5 ±0.2 | -2.5 ±0.3 | -2.4 ±0.3 | -2.3 ±0.3 | -4.3 ±0.5 |
| *((7),$\tau_1$) - (EPM, $\tau_3$)* | -2.4 | -2.5 | -1.9 | -2 | -4.8 | -2.8 ±0.3 | -3 ±0.4 | -2.6 ±0.6 | -2.8 ±0.9 | -6.9 ±1.8 |

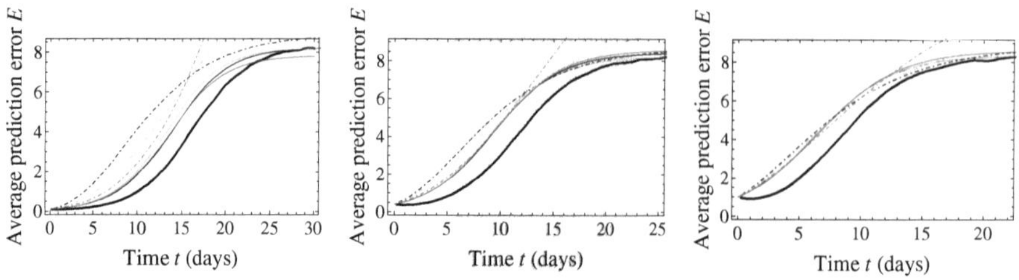


**Fig. 1.** Time variations of the average prediction error $E$ for $ND_1$ (left picture), $ND_3$ (central picture) and $ND_5$ (right picture). The thick line represents data from EPS, the thin lines from (6) and the dotdashed lines from (7). The light grey lines represent the data extrapolate from time length $\tau_1$, the grey lines represent the data extrapolate from time length $\tau_2$ and black lines represent the data in time length $\tau_3$.

**Table 2.** Average values over $M_{1,..,5}$ of $E_{asym,\,\tau 1,..,3}$ calculated from approximations (3) and (4) (grey lines) for all initial conditions $ND_{1,..,5}$ and for all prediction time lengths $\tau_{1,..,3}$ (grey columns) and difference $\varepsilon_{1,..,3}$ of $E_{asym,\,\tau 1,..,3}$ - $E_{asym}$.

| | $E_{asym,\,\tau 3}$ | $\varepsilon_3$ | $E_{asym,\,\tau 2}$ | $\varepsilon_2$ | $E_{asym,\,\tau 1}$ | $\varepsilon_1$ |
|---|---|---|---|---|---|---|
| $(ND_1,(3))$ | 8.3 ± 0.1 | -0.1 ± 0.1 | 8.1 ± 0.2 | -0.3 ± 0.2 | 8.2 ± 0.2 | -0.2 ± 0.2 |
| $(ND_1,(4))$ | 8.8 ± 0.1 | 0.4 ± 0.1 | 15.2 ± 0.5 | 6.8 ± 1 | 45 ± 11 | 37 ± 11 |
| $(ND_2,(3))$ | 8.3 ± 0 | -0.1 ± 0 | 8.2 ± 0.1 | -0.2 ± 0.1 | 8.7 ± 0.8 | 0.3 ± 0.9 |
| $(ND_2,(4))$ | 8.8 ± 0 | 0.4 ± 0.1 | 11.4 ± 0.5 | 3 ± 1 | 60 ± 21 | 52 ± 21 |
| $(ND_3,(3))$ | 8.3 ± 0 | -0.1 ± 0 | 8.5 ± 0.2 | 0.1 ± 0.2 | 8.3 ± 0.2 | -0.1 ± 0.3 |
| $(ND_3,(4))$ | 8.8 ± 0 | 0.4 ± 0.1 | 10.4 ± 0.3 | 2 ± 1 | 19 ± 1 | 10 ± 1 |
| $(ND_4,(3))$ | 8.4 ± 0 | 0 ± 0 | 8.4 ± 0 | 0 ± 0 | 8,9 ± 0 | 0.5 ± 0 |
| $(ND_4,(4))$ | 8.8 ± 0 | 0.4 ± 0.1 | 9.8 ± 0.1 | 1.4 ± 0.4 | 17 ± 1 | 8 ± 1 |
| $(ND_5,(3))$ | 8.4 ± 0 | 0 ± 0 | 8.3 ± 0.1 | -0.1 ± 0.1 | 8.6 ± 0 | 0.2 ± 0.1 |
| $(ND_5,(4))$ | 8.7 ± 0.1 | 0.3 ± 0.1 | 9.2 ± 0.1 | 0.8 ± 0.1 | 11.7 ± 0.5 | 3 ± 1 |

LH (4) does not give good fit to the model data. In Table 1, Table 2 and Fig. 2 we can see it and we can also see results closer to model values as initial error increases. That is in good agreement with [7, 20, 21], but in this case there is no initial condition where (4) would approximate the model data more closely than (3). Main reason for this we can find in the definition of (4) that starts in $(x_1;y_1) = (0;0)$ (Fig. 2), but model data starts at the point that is close to the size of initial error on the $x$-axis (Fig. 2). For example in the right picture of Fig. 2 we can see that $(x_1;y_1) = (0.9;0.02)$. Therefore it would be more appropriate to introduce modification of (4)

$$dE(t)/dt = -vE(t)\ln(wE(t)) + z, \tag{8}$$

where $E(t)$ is again a distance at time $t$ between two originally nearby trajectories and $v$, $w$, $z$ are constants.

Table 2 focuses on average values over $M_{1,..,5}$ of $E_{asym,\ \tau 1,..,3}$ . This value is for example used to find out if the variability of the model is equal to the variability of the atmosphere [10]. The differences $\varepsilon_{1,..,3}$ from model values $E_{asym}$ indicate really poor approximation of (4) and error of order of 0.1 for eq. (3). For the latter equation, $\varepsilon_3$ lies between *-0.1* and *0* (relatively against $E_{asym}$ it means between *-1.2 %* and *0 %*), $\varepsilon_2$ between *-0.3* and *0.1* (between *-3.6 %* and *1.2 %*) and $\varepsilon_1$ between *-0.2* and *0.5*. *(-2.4 %* and *6%*).

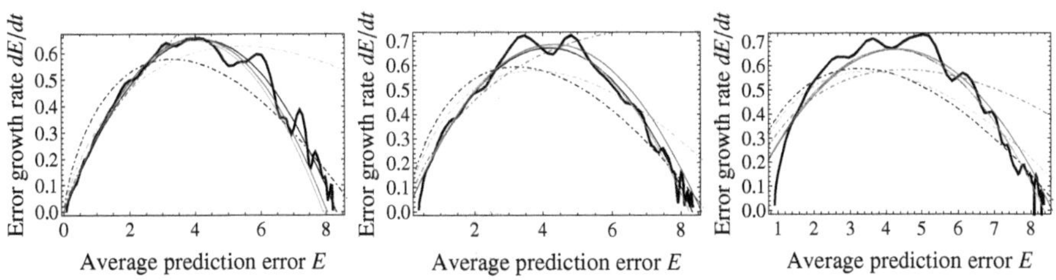


**Fig. 2.** The error growth rate $dE/dt$ versus $E$ for $ND_1$ (the left picture), $ND_3$ (the central picture) and $ND_5$ (the right picture). The thick line represents data from EPS, the thin lines from (3) and the dotdashed lines from (4). The light grey lines represent the data extrapolate from time length $\tau_1$, the grey lines represent the data extrapolate from time length $\tau_2$ and black lines represent the data in time length $\tau_3$.

## 7 Conclusion

This paper studies errors of estimations of time limits and asymptotic value of initial errors growth in low-dimensional atmospheric model L05II introduced by Lorenz [13] with the parameters, that are as close to NWPM as possible. Five types of initial conditions are represented by five normal distributions. Five settings of EPM showed the differences of order 0.1 and therefore the average value was chosen as model data. Quadratic hypothesis approximates these data more closely than logarithmic hypothesis for almost all initial errors, forecast lengths and time limits, even though the difference between logarithmic hypothesis and model data decrease with the increase of the initial error. It is shown, how the validity of both hypotheses can be improved and that the difference of improved quadratic hypothesis with model data is of order of *0.1* days, with the exception of shorter forecast lengths and the biggest initial error, where the difference can reach two days. The differences from model data of asymptotic value of initial error growth can vary between *-3.6 %* and *6%* of this value. As the forecast lengths decrease the differences increase.

**Acknowledgements.** The authors were supported by project no. SVV-2013-267308.

## References

1. Lorenz, E.N.: The Predictability of Flow which Possesses Many Scale of Motion. Tellus 21, 289–307 (1969)
2. Savijärvi, H.: Error Growth in a Large Numerical Forecast System. Mon. Weather Rev. 123, 212–221 (1995)
3. Lorenz, E.N.: Atmospheric predictability as revealed by naturally occurring analogs. J. Atmos. Sci. 26, 636–646 (1969)
4. Lorenz, E.N.: Atmospheric predictability experiments with a large numerical model. Tellus 34, 505–513 (1982)
5. Dalcher, A., Kalney, E.: Error Growth and Predictability in Operational ECMWF Analyses. Tellus 39A, 474–491 (1987)
6. Simmons, A.J., Mureau, R., Petroliagis, T.: Error Growth and Estimates of Predictability from the ECMWF Forecasting System. Q. J. R. Meteorol. Soc. 121, 1739–1771 (1995)
7. Trevisan, A., Malguzzi, P., Fantini, M.: A Note on Lorenz's Law for the Growth of Large and Small Errors in the Atmosphere. J. Atmos. Sci. 49, 713–719 (1992)
8. Froude, L.S.R., Bengtsson, L., Hodges, K.I.: Atmospheric Predictability Revisited. Tellus A 65, 1–13 (2013)
9. Buzzia, R.: Horizontal Resolution Impact on Short- and Long-range Forecast Error. Q. J. R. Meteorol. Soc. 136, 1020–1035 (2010)
10. Bengtosson, L.K., Magnusson, L., Källén, E.: Independent Estimations of the Asymptotic Variability in an ensemble Forecast System. Mon. Weather Rev. 136, 4105–4112 (2008)
11. Ma, J., Zhu, Y., Wobus, R., Wang, P.: An Effective Configuration of Ensemble Size and Horizontal Resolution for the NCEP GEFS. Advances in Atm. Sci. 29, 782–794 (2012)
12. Lorenz, E.N.: Designing Chaotic Models. J. Atmos. Sci. 62, 1574–1587 (2005)
13. Palmer, T., Hagedorm, R.: Predictability of Weather and Climate. Cambridge University Press, Cambridge (2006)
14. Lorenz, E.N.: Predictability: A Problem Partly Solved. In: Proc. Seminar on Predictability, vol. 1, pp. 1–18. CMWF, Reading (1996); Reprinted in Palmer T., Hagegorn R.: Predictability of Weather and Climate, 718 p. Cambridge University Press (2006)
15. Lorenz, E.N.: Deterministic Nonperiodic Flow. J. Atmos. Sci. 20, 130–141 (1963)
16. Sprott, J.C.: Chaos and Time-Series Analysis. Oxford University Press, New York (2003)
17. Trevisan, A.: Impact of transient error growth on global average predictability measures. J. Atmos. Sci. 50, 1016–1028 (1993)
18. Ruiqing, D., Jianping, L.: Comparisons of Two Ensemble Mean Methods in Measuring the Average Error Growth and the Predictability. Acta Meteorologica Sinica 25, 395–404 (2011)
19. Benzi, R., Carnevale, F.C.: A Possible Measure of Local Predictability. J. Atmos. Sci. 46, 3595–3598 (1989)
20. Bednar, H., Raidl, A., Miksovsky, J.: Initial Errors Growth in Chaotic Low-dimensional Weather Prediction Model. In: Zelinka, I., Chen, G., Rössler, O.E., Snasel, V., Abraham, A. (eds.) Nostradamus 2013: Prediction, Model. & Analysis. AISC, vol. 210, pp. 333–342. Springer, Heidelberg (2013)
21. Bednář, H., Raidl, A., Mikšovský, J.: Initial Error Growth and Predictability of Chaotic Low-dimensional Atmospheric Model. International Journal of Automation and Computing (in press, 2014)
22. Trevisan, A., Pancotti, F., Molteni, F.: Ensemble Prediction in a Model with Flow Regimes. Quart. J. Roy. Meteor. Soc. 127, 343–358 (2001)
23. Bengtsson, L., Hodges, K.: I. A Note on Atmospheric Predictability. Tellus 58A, 154–157 (2006)

# Prediction Based Context Data Dissemination and Storage Model for Cooperative Vehicular Networks

Mindaugas Kurmis[1,2], Dale Dzemydiene[1], Arunas Andziulis[2], Miroslav Voznak[3], Sergej Jakovlev[2], Zydrunas Lukosius[2], and Gediminas Gricius[2]

[1] Vilnius University, Institute of Mathematics and Informatics, Akademijos str. 4, LT-08663 Vilnius, Lithuania
mindaugas.kurmis@mii.vu.lt

[2] Klaipeda University, Department of Informatics engineering, Bijunu str. 17, LT-91225, Klaipeda, Lithuania
arunas.iik.ku@gmail.com

[3] VSB - Technical University of Ostrava, Department of Telecommunications, 17th November 15, 708 33 Ostrava-Poruba, Czech Republic
miroslav.voznak@vsb.cz

**Abstract.** The vehicle as the context information source generates a huge amount of different information including from physical vehicle and environment sensors. The implementation of an efficient and scalable model for information dissemination in VANETs confronts with major problems. In this dynamic environment, an increasing number of context dissemination messages are increasing channels utilization which affects the network performance. This article discusses analyses and assesses the key proposals how to deal with the context data dissemination and how to decrease the amounts of transferred and stored data in vehicular cooperation environment. This is one of the most important topics of the pervasive computing.

**Keywords:** context data dissemination, storage, vehicular communication net-works, VANET.

## 1 Introduction

As the human mobility increasing day to day, the vehicle becomes a very important component of human life. The ITSs (Intelligent Transport Systems) en-compass a broad range of advanced information and communication technologies, which are applied in transport infrastructures and vehicular networks. They are expected to offer fundamental breakthroughs in enhancing road safety, reducing congestion, improving driving comfort and protecting environment, to name a few [1]. Future hybrid Vehicular Ad-Hoc Networks (VANET) will utilize both long range communications such as cellular networks, as well as short range communication technologies such as Wireless Fidelity (Wi-Fi) and Dedicated Short Range Communications (DSRC) [2]. These types of communications allow vehicles to share different kinds of information, for example, safety information for the purpose of accident prevention, post-accident

I. Zelinka et al. (eds.), *Nostradamus 2014: Prediction, Modeling and Analysis of Complex Systems*, Advances in Intelligent Systems and Computing 289,
DOI: 10.1007/978-3-319-07401-6_3, © Springer International Publishing Switzerland 2014

investigation or traffic jams. Other type of information can be disseminated such as traveller related information which is considered as non-safety information. The intention behind distributing and sharing this information is to provide a safety message to warn drivers about expected hazards in order to decrease the number of accidents and save people's lives, or to provide passengers with pleasant journeys [3]. Differently from other pervasive computing devices vehicles have specific requirements and does not have strict energy constraints so it can be equipped with powerful computational resources, wireless transmitters and various sensors [4]. The vehicle must not distract drivers attention during driving it must provide user with services autonomous and without user intervention. To provide the necessary services at the right time in the right place and in the right way it is necessary to adapt services and their support to user needs [5]. One of the ways to increase the user-vehicle interface autonomy and efficiency is to understand the context in which user and the vehicle are at the moment, also to know what context was in the past and to predict the context of the future.

To know the context it can be utilized various sensors and information sources of the vehicle, user and the environment. The vehicle as the context information source generates a huge amount of different information including from physical vehicle and environment sensors: GPS, speed, acceleration, temperature, radar, video, etc. and virtual sensors as road information, warnings, interaction with other vehicles, calls, etc.

The communication between vehicles and the Road side unit (RSU) and the infrastructure form three types of domains [3]. **In-vehicle domain** consists of an on board unit (OBU) and one or multiple Application units (AU)). **Ad-hoc domain** is composed of vehicles equipped with OBUs. Vehicles communicate with other vehicles forming a MANET, which allows communication between vehicles in a fully distributed manner with decentralized coordination. In **Infrastructural domain** the RSU can connect to the infrastructural networks or to the Internet, allowing the OBU to access the infrastructure network (Fig. 1).

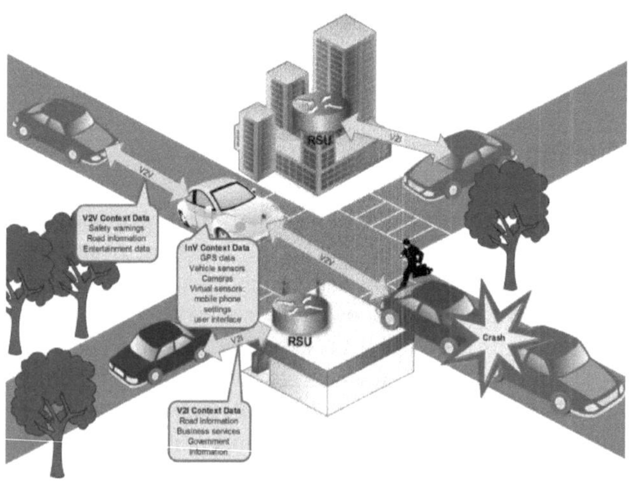


**Fig. 1.** Wireless communication types and context data producers in VANETs

One of the main problems in VANETs is how disseminate to other cooperating vehicles and access huge amounts of different types of context information from such a complex system in real-time or even before the occurred event. The existing context data dissemination and storage systems should be extended to support modelling and prediction of its dynamics in distributed algorithms for storage of real-time acquisition data with system scalability in mind. One of the solutions is presented in our research – the prediction based context data dissemination and storage model for cooperative vehicular networks.

## 2 Vehicular Communication Networks (VANETs) Applications

The VANET applications can be categorized in two categories: safety applications and comfort/entertainment applications:

**The safety applications** enhance the protection of passengers by sending and receiving information pertinent to vehicle safety. Generally, these alerts, such as cooperative collision warning, lane change warning, emergency video streaming, and incident management, are directly sent to the drivers or are received by the automatic active safety system [6].

**The comfort/entertainment** category of applications is referred to as non-safety applications, and aim to improve drivers and passengers comfort levels (make the journey more pleasant) and enhance traffic efficiency. They can provide drivers or passengers with weather and traffic information and detail the location of the nearest restaurant, petrol station or hotel and their prices. Passengers can play online games, access the internet and send or receive instant messages while the vehicle is connected to the infrastructure network [3]. The investments by the vehicle manufacturers show that these applications are becoming extremely popular.

## 3 Related Works

During the last years there was a huge interest in context data dissemination and storage in cooperating vehicular networks research. Liu and Lee investigate timely and adaptive data dissemination in the dynamically changing traffic environment and present the analytical theoretical model of the effects of the dynamic traffic factors. [1]. Ali et al. examine query starvation and bandwidth utilization problem in multi-item queries in wireless broadcasting systems. [7]. Delot et al. present a system for data sharing in vehicular networks Vehicular Event Sharing with a mobile Peer-to-peer Architecture (VESPA). In this system, a technique based on the concept of Encounter Probability is proposed for vehicles to share information using vehicle-to-vehicle communications. [8]. Barberis and Malnati present the design and evaluation of a collaborative system for content diffusion and retrieval among traveling vehicles. This system relies on multicast epidemic dissemination of messages and exploits vehicles mobility and their local storage capabilities [9]. Despite the fact that there is increasing number of research in this are there is still lack of solutions for the efficient way how to predict which data is needed to store in local, which data should be forwarded to other vehicles and which should be stored in the hybrid vehicle cloud environment databases.

## 4 Methodologies and the Model

Fig. 2 illustrates a high-level view of our context dissemination and aggregation architecture and its logical flows of the contexts. In our approach the environment consists of the vehicle communicating with other vehicles directly using safety messages. The cooperating and communicating vehicles forms a hybrid VANET cloud system. Through the system it is exchanged entertainment related context data. The actual vehicle stores the acquired data in its local databases – safety DB and entertainment DB. Using the data from the local DB and from hybrid VANET cloud the vehicle reasoning engine reasons about the current, past and future situations and selects needed services from services cloud system. The services is adopted to user need and supported to the user.

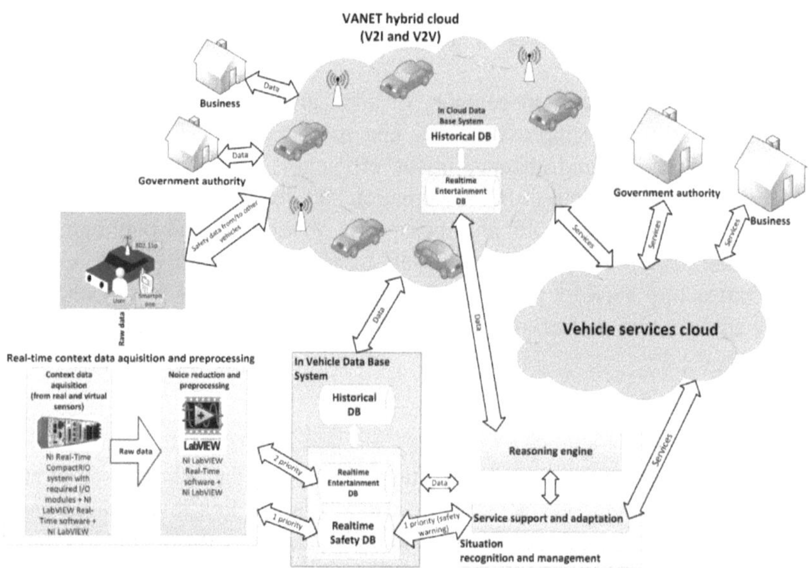


**Fig. 2.** High-level view of our context dissemination and aggregation architecture

Due to not strict requirements of energy consumption in the proposed system it can be used more different sensors (physical and virtual) which generates huge amounts of context data. The data have to be acquired in the real time and stored in the databases in efficient way thus using the methods of artificial intelligence it should be able to predict self-system and other systems performing in the transportation system dynamics and to model safety situations in real time and non-safety in near real-time. An example of the various sensors (physical and virtual) for the potential usage for the context acquisition is shown in the Table 1.

**Table 1.** An excerpt of example of the sensors used for the context acquisition and storage

| | Sensor | Sample rate (S/s) | No of Channels | No of Devices | Bandwidth (KB/s) | Information source | Data exchange | Historical data? | Safety (Sf)or entertainment data (E)? |
|---|---|---|---|---|---|---|---|---|---|
| 1 | GPS coordinates | 100 | 1 | 1 | 0.1 | Vehicle | inV | + | Sf, E |
| 2 | Speed | 10 | 1 | 1 | 0.01 | Vehicle | inV | + | Sf, E |
| 3 | Reminders | 1 | 1 | 1 | 0.001 | Smartphone | V2M | - | E |
| 4 | User preferences | 1 | 1 | 1 | 0.001 | Smartphone | V2M | + | E |
| 5 | Road information | 10 | 3 | 1 | 0.03 | Other vehicles, government, environment | V2I, V2V, V2M | + | Sf |
| 6 | Warnings | 1000 event | 3 | 1 | 3 | Other vehicles, government, environment | V2I, V2V, V2M | - | Sf |
| 7 | Interaction with other vehicles | 1000 event | 12 | 1 | 12 | Environment | V2I, V2V, V2M | + | Sf, E |

To achieve the higher efficiency of the data being saved in the local DB we propose to save the utility of the local context in a matrix ($M_L$) for *l* data messages (*m*) from n sensors (*s*) (1).

$$M_L = \begin{pmatrix} d_{11} & d_{12} & \dots & d_{1n} \\ d_{21} & d_{22} & \dots & d_{2n} \\ \dots & \dots & \dots & \dots \\ d_{l1} & d_{l2} & \dots & d_{\ln} \end{pmatrix} \tag{1}$$

The predicted utility of the contextual data messages can be weighted in a function which assigns a value to each data message of the data sensor. The value is calculated by the equation (2):

$$\begin{aligned} d_{L_{ij}} &= (Ty_1 + H_1 + Ex_1)m_1 cr_1 Pr_1, (Ty_2 + H_2 + Ex_2)m_2 cr_2 Pr_1, \dots, \\ & (Ty_n + H_n + Ex_n)m_j cr_j Pr_j \end{aligned} \tag{2}$$

Where *Ty* is the type of context data in the interval [1, 2, 3] (1 – entertainment related, 2 – entertainment and safety related, 3 – safety related). *H* is the parameter in the interval [0, 1] showing if the data should be used for historical saving (1) or not (0). Ex is the parameter in the interval [1-4] showing the data exchange domain (1 – V2M, 2 – InV, 3 – V2I, 4 – V2V) and *cr* is the coordinates of the data generation location. The priority of the message (*Pr*) is calculated by the $\mathrm{Pr}_j = 1 + \frac{\mathrm{I_j}}{\mathrm{A_j}}$ normalized with values falling in a predetermined interval [1, 2, 3], where 3 means that the message priority is critical and it must be sent immediately and stored in the corresponding DB, 2 means that the message have medium priority, and 1 means that the message is not important and can be rejected.

$I_j$ is the importance of the message in a predetermined interval [0, 1] where 0 is the safety related message and 1 is the infotainment related message. $\mathrm{A_j}$ is the message age function normalized with values falling in a predetermined interval [1, 2, 3] which is calculated by the (3), where $\mathrm{T_M}$ is the subtraction from the current time and the message creation time.

$$A = \begin{cases} 1, \text{if } \mathrm{T_M} > 5s \\ 2, \text{if } 1 < \mathrm{T_M} < 5s \\ 3 \text{ if } \mathrm{T_M} < 1s \end{cases} \tag{3}$$

To reduce the bandwidth and achieve a better efficiency of the data being provided for the exchange with other vehicles we propose to store the utility of the context in a matrix ($M_O$) for *l* data messages (*m*) for the *n* of vehicles (*v*) (4).

$$M_O = \begin{pmatrix} d_{11} & d_{12} & \dots & d_{1n} \\ d_{21} & d_{22} & \dots & d_{2n} \\ \dots & \dots & \dots & \dots \\ d_{l1} & d_{l2} & \dots & d_{\mathrm{ln}} \end{pmatrix} \tag{4}$$

The predicted utility of the contextual data messages can be weighted in a function which assign a value to each data message ($m_l$) intended to send to the vehicle ($v_n$). The value is calculated by the equation (5):

$$\begin{aligned} d_{O_{ij}} &= (Ty_1 + Exc_1 + Z_1)m_1 cr_1 Pr_1 n_1, (Ty_2 + Exc_2 + Z_2)m_2 cr_2 Pr_2 n_2, \dots, \\ &(Ty_n + Exc_n + Z_2)m_j cr_j Pr_j n_j \end{aligned} \tag{5}$$

Where *Exc* is the parameter in the interval [1-4] of the special set of non-confidential data and showing the data exchange domain (1 – V2M, 2 – InV, 3 – V2I, 4 – V2V), n is showing number of cooperating vehicles in the cluster, *Z* is the parameter of prediction of the communication channel availability and calculated by the following:

$$Z_t = \frac{1 + \left( \frac{C_t + D_t}{2} \right)}{Tr} \tag{6}$$

Where C is the collision parameter calculated by the: $C=1-\left(\frac{1}{1+c_{t-1}}\right)$, where D is the drop parameter calculated by the $D=1-\left(\frac{1}{1+d_{t-1}}\right)$ and *Tr* is the throughput parameter and calculated by the $Tr=1+\left(\frac{tr_{t-1}}{100}\right)$.

The utility of the context data for the exchange with the hybrid VANET cloud is stored in the matrix $M_C$ for *l* data messages (*m*) for the *r* of the receiving entities (7).

$$M_C=\begin{pmatrix} d_{11} & d_{12} & \dots & d_{1n} \\ d_{21} & d_{22} & \dots & d_{2n} \\ \dots & \dots & \dots & \dots \\ d_{l1} & d_{l2} & \dots & d_{\mathrm{ln}} \end{pmatrix} \tag{7}$$

The predicted utility of the contextual data messages for the exchange with the hybrid VANET cloud can be weighted in a function which assign a value to each data message $m_l$ intended to send to the receiving entity $r_n$. The value is calculated by the equation (8).

$$\begin{aligned} d_{C_{ij}} &= (Tye_1+Hx_1+Exc_1+Z_1)m_1cr_1Pr_1,(Ty_2+Hx_2+Exc_2+Z_1)m_2cr_2Pr_2,\dots, \\ &(Ty_n+Hx_n+Exc_n+Z_1)m_jcr_jPr_j \end{aligned} \tag{8}$$

Where *Tye* is the reduced *Ty* parameter in the interval [1, 2] (1 – entertainment related data, 2 – entertainment and safety related data), *Hx* is the special set of non-confidential data parameter in the interval [0, 1] showing if the data should be used for historical saving (1) or not (0).

## 5 Results and Discussion

In this section we briefly present our simulation and modelling results. First we introduce the simulation scenarios and then the evaluation of our solution.

The evaluation of the proposed model is carried out by means of simulations and numerical methods. For the experiments it was used the data from the simula-tion environment NCTUns [10], It was chosen as using the existent Linux TCP/UDP/IP protocol stack providing high-accuracy results; it can be used with any actual Unix application on a simulated node without additional modifications. In experimental scenario a (Fig. 3) the network model is created where the data from the vehicles is sending to the VANET hybrid cloud DB server. The modelled network consists of the DB server, 802.11p RSU and 1 to 10 vehicles equipped with the 802.11 OBUs. In the experimental scenario b, the data is transferred in both ways – from vehicles to DB and from DB to vehicles. The simulations have carried out for 60 s. For the link layer bit rate it was used 27 Mb/s, the packet size – 1000 B.

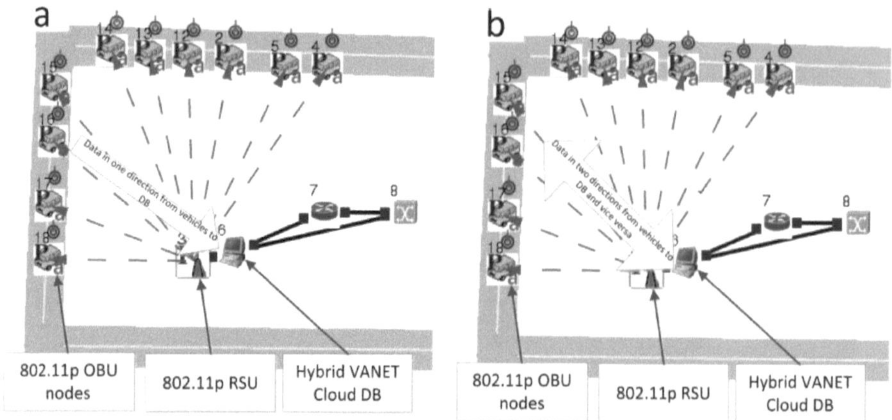


**Fig. 3.** Simulation scenarios of the context data exchange using the 802.11p

## 5.1 Evaluation of the Solution

The Fig. 4 shows the savings of the data throughput with a different number of vehicles in the network. The results show that the developed method saves a large number of bandwidth and there is a huge potential by calibrating and adjusting the prediction functions parameters.

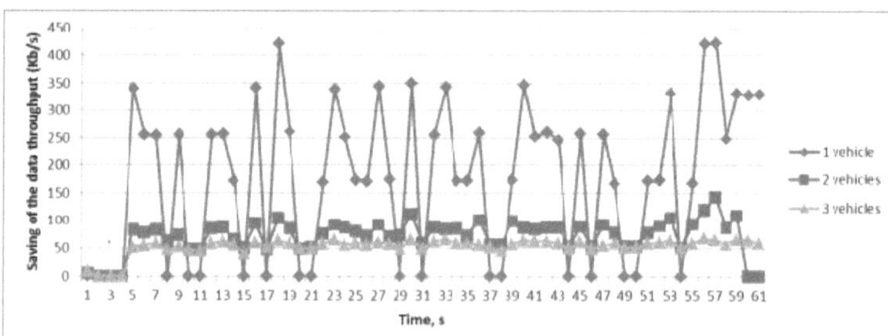


**Fig. 4.** Savings of the data throughput with different number of vehicles

Fig. 5 shows the average results in terms of accumulated prediction utility val-ue, Exc, normalized Z and Ty parameters change over time. Using the b simulation model it was performed the experiments and evaluated how the prediction of the communication channel availability parameter Z changes over time. As the pa-rameter is inversely proportional it can be seen that as the number of vehicles in-creases the parameter is also increasing. It means that the collision and dropped packets is increasing thus leading to poorer channel availability.

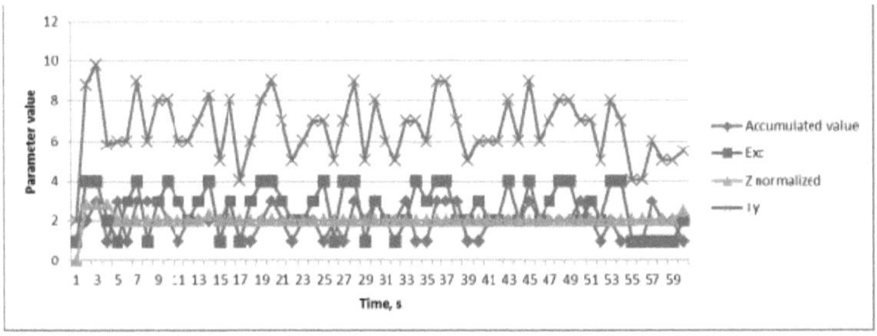


**Fig. 5.** Accumulated, Exc, normalized Z and Ty parameters change over time

Fig. 6 clearly presents the best result in terms of used throughput efficiency at the cost of having simulation results with our model implemented and without it. Also the influence of different number of vehicles can also be seen. The results show that using the two way context data sending with 1 to 10 vehicles using our model we get about 23% savings in required bandwidth. Using one way sending we get 22% savings with 1 vehicle, 47% savings using 5 vehicles and 69% savings with 10 vehicles.

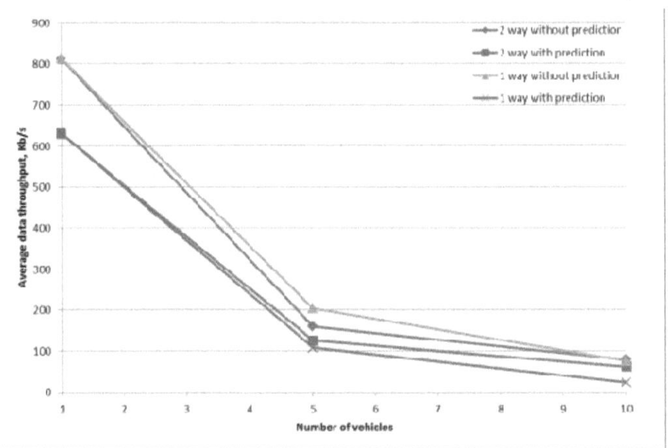


**Fig. 6.** Average data throughput from the vehicles (1, 5, 10) with and without the developed prediction model using the 802.11p

# 6 Conclusions

We have analyzed a complex problem of context data dissemination in vehicular communication networks. Three models were proposed for the context data dissemination bandwidth reduction and cooperating usage of the channel availability. The solutions were evaluated by the simulation models and numerical methods. Results suggest that an environment with our implemented solutions and a network with the larger number of vehicles can perform much better than without it and it is showing 22% (with 1 vehicle) to 69% (with 10 vehicles) savings in required bandwidth.

Thus it can be used in real life applications where large number of data is needed to be stored locally, exchanged with other vehicles and vehicular hybrid cloud.

**Acknowledgments.** This work was supported by the Latvia-Lithuania Cross Border Cooperation Programme within the project "JRTC Extension in Area of Development of Distributed Real-Time Signal Processing and Control Systems", code LLIV-215. Also it was partially supported by the Development of human resources in research and development of latest soft computing methods and their application in practice project, reg. no. CZ.1.07/2.3.00/20.0072 funded by Operational Programme Education for Competitiveness, co-financed by ESF and state budget of the Czech Republic.

## References

1. Savran, A., Tasaltin, R., Becerikli, Y.: Intelligent adaptive nonlinear flight control for a high performance aircraft with neural networks. ISA Transactions 45(2), 225–247 (2006) CrossRef
2. Ali, G., Chan, E., Li, W.: Supporting real-time multiple data items query in multi-RSU vehicular ad hoc networks (VANETs). Journal of Systems and Software 86(8), 2127–2142 (2006)
3. Al-Sultan, S., et al.: A comprehensive survey on vehicular Ad Hoc network. Journal of Network and Computer Applications 37, 380–392 (2014)
4. Barberis, C., Malnati, G.: Design and evaluation of a collaborative system for content diffusion and retrieval in vehicular networks. IEEE Transactions on Consumer Electronics 57(1), 105–112 (2011)
5. Cheng, H.T., Shan, H., Zhuang, W.: Infotainment and road safety service support in vehicular networking: From a communication perspective. Mechanical Systems and Signal Processing 25(6), 2020–2038 (2011)
6. Delot, T., Cenerario, N., Ilarri, S.: Vehicular event sharing with a mobile peer-to-peer architecture. Transportation Research Part C: Emerging Technologies 18(4), 584–598 (2010)
7. Kurmis, M., et al.: Development of the Real Time Situation Identification Model for Adaptive Service Support in Vehicular Communication Networks Domain. Advances in Electrical and Electronic Engineering 11(5), 342–348 (2013)
8. Liu, K., Lee, V.C.S.: Adaptive data dissemination for time-constrained messages in dynamic vehicular networks. Transportation Research Part C: Emerging Technologies 21(1), 214–229 (2012)
9. Sharef, B.T., Alsaqour, R.A., Ismail, M.: Vehicular communication ad hoc routing protocols: A survey. Journal of Network and Computer Applications 40, 363–396 (2014)
10. Talebifard, P., Leung, V.C.M.: Towards a content-centric approach to crowd-sensing in vehicular clouds. Journal of Systems Architecture 59(10), 976–984 (2013)

# SOM Based Multi-agent Hydro Meteorological Data Collection System

Gediminas Gricius[1], Darius Drungilas[2], Arunas Andziulis[2], Dale Dzemydiene[1], and Miroslav Voznak[3]

[1] Institute of Mathematics and Informatics, Vilnius University, Akademijos St. 4, LT-08663 Vilnius, Lithuania
gediminas@ik.ku.lt, daledz@mruni.eu
[2] Department of Informatics Engineering, Faculty of Marine Engineering, Klaipeda University, Bijunu St. 17-206, LT- 91225 Klaipeda, Lithuania
dorition@gmail.com, arunas.iik.ku@gmail.com
[3] Department of Telecommunications, Faculty of Electrical Engineering and Computer Science, VSB-Technical University of Ostrava, 17. Listopadu 15, 708 00 Ostrava, Czech Republic
miroslav.voznak@vsb.cz

**Abstract.** The paper presents the possibilities of development the hydro meteoro-logical data collection system (HMDCS) involving advanced technologies such as multi agent based interaction and data collection between several monitoring system's nodes (i.e. buoys) based on self-organizing maps (SOM). The require-ments for such system development are rather complex and are attached to allow-ing the real-time monitoring, control, and prediction of the negative consequences of contamination of surface water recourses and making their evaluation by effec-tiveness in monitoring of Baltic Sea surface water. The experiment is based on the design an inexpensive, but reliable Baltic Sea autonomous monitoring network (buoys), which would be able, continuously monitor and collect temperature, waviness, and other required data. Moreover, it makes ability to monitor all the data from the costal-based station with limited transition speed by setting different tasks for agent based buoy system according to the SOM.

**Keywords:** wireless networking system, hydro meteorological sensors, multi agent systems, embedded systems.

## 1 Introduction

There is a variety of tools to monitor and evaluate the Baltic Sea hydro meteorological data, but most of received information has low spatial coverage and low level of detail in time [2]. Sea wave height, water temperature, underwater noise data is used for many practical applications usually obtained from three sources: buoy measurements, model calculations, and ship observations. Compared to other data acquisition methods, the buoy measurement is the most reliable and readily data source available continuously for years [13]. Basically the network of buoys involves mapping of

I. Zelinka et al. (eds.), *Nostradamus 2014: Prediction, Modeling and Analysis of Complex Systems*, Advances in Intelligent Systems and Computing 289,
DOI: 10.1007/978-3-319-07401-6_4, © Springer International Publishing Switzerland 2014

temperature, wave height and underwater noise at a buoy location using the date retrieved at other buoys locations [10]. However, many hydro meteorological data measurements from sea buoys can be lost due to malfunctioning, maintenance, connection problems or dubious data recorded by the buoy. In order to ensure greater reliability of data collection it is necessary to develop the distributed information system, predicting complex situations and supporting the decision-making processes. Information provided from such system is important for decision makers and are needed to ensure the provision of information for decision-making institutions [3], [4], [7]. An important feature of such a buoys network is the ability to monitor, collect and evaluate wide spatial coverage and real time hydro meteorological data of the Baltic Sea [4]. Hydro meteorological information system is faced with great data flows, but the data often is excess, depending on the observed region of the water. Therefore, the current traditional methods are no longer sufficient to ensure the rapid collection of data and valuable information extraction

The purpose of this study is to show possibilities of development the hydro meteorological data collection system (HMDCS) involving advanced technologies such as multi agent based interaction and data collection between several monitoring system's nodes (i.e. buoys) based on self-organizing maps (SOM). The experiment is based on the design an inexpensive, but reliable Baltic Sea autonomous monitoring network (buoys), which would be able, continuously monitor and collect temperature, waviness, and other required data. Moreover, it makes ability to monitor all the data from the costal-based station with limited transition speed by setting different tasks for agent based buoy system according to the SOM.

## 2 Sea Hydro Meteorological Data Monitoring

Nowadays, there are numerous and varied designs for autonomous systems used for meteorological and oceanographic monitoring with different integration degrees. The buoy network system used in Canary Islands is one of them [1]. It has a control center that manages the transmission communications, and provides data in a useful form to diverse socioeconomic important sectors which make an exhaustive use of the littoral in the, and need data from the buoys to well manage the coastal environment. These buoys monitor water temperature, salinity, dissolved oxygen, hydrocarbons, and other characteristics, which allow to measure equipped other sensors such as fluorometer, turbidimeter and also each buoy is able to communicate via GSM modem. Following a programmed sampling rate (every hour), the ECU send to the central receiver unit a SMS message, which includes a sensor data set, GPS position and battery level. However, deeper analysis of the data has showed that such a sampling rate is not sufficient which means that data transmit protocol has to be re-viewed.

In order to provide greater hydro-meteorological data monitoring reliability and faster data retrieval there are proposed variety of sensory systems networks [8], [9], and [12]. There are proposed communication technologies that enable communication between sensor nodes [12], the systems for communication between maritime platforms like vessels, commercial ships or buoys [9], real-time monitoring of the underwater environment where an acoustic mesh network is located between the underwater senor networks and central monitoring system [8]. The proposed models can solve various problems, but require more flexible solutions for complex data transfer problems. This problem can be solved by developing active autonomous sensor multi-agent based system, which according to the situation is able to combine the data processing methods.

# 3 Hydro Meteorological Data Sensory System

## 3.1 Temperature Data Collection

During the investigation stage of the HMDCS development, several types of temperature sensors were compared. The comparison possibilities are made analyzing their parameters according to the technical specification presented in datasheets.

After comparative analysis of temperature sensors, we have selected digital sensor DS18B20. This digital temperature sensor can measurement temperatures within range from -55°C to +125°C by 12-bit precision, with accuracy – 0.50°C [11]. However, after additional calculations, it is possible to reduce the temperature measurement error of up to 0.10°C. The most attractive is the fact, that these sensors have already been calibrated at the factory and their accuracy error is ±0.5°C in the range from -10°C to +85°C and ±2°C error over the operating range (55°C to +125°C). Sensor supply voltage is in the range of +3 to +5.5 V. In standby mode, current consumption is close to zero (less than 1 μA), while the temperature conversion will be used during the current is about 1 mA. The measurement process lasts no more than 0.7 sec. The DS18B20 communicates over a 1-Wire® bus that by definition requires only one data line (and ground) for communication with a central microprocessor. In addition, the DS18B20 can derive power directly from the data line ("parasite power"), eliminating the need for an external power supply. Each DS18B20 has a unique 64-bit serial code, which allows multiple DS18B20s to function on the same 1-Wire bus. Thus, it is simple to use one microprocessor to control many DS18B20s distributed over a large area. This part has already become the corner stone of many data logging and temperature control projects.

## 3.2 Waviness Measurements

At present, the sea and the oceans waviness measurements uses variety of methods, depending on the geographic region, measuring accuracy, and common tasks [5]. The main and most commonly used are:

- Ultrasound based sensors:
  - Pros: suitable for measuring waves with a height of over 5 meters
  - Cons: significant measurement errors
- Rheostat-type structures:
  - Pros: allows you to get a fairly accurate data
  - Cons: because of its design features cannot be long-lasting

- Satellite image analysis:
  - Cons: due to big error can used only for ocean waviness measurement,
- GPS system :
  - Cons: not suitable for measuring waves with a height of 0.5-2.0 meters range,
- Accelerometer and gyroscope design:
  - Pros: small measurement errors, easy implementation.

For our experiment couple of accelerometer and gyroscope was used. Based on the experience of other scientists [2], accelerometer data were processed by removing the component of gravity, according to the formulas (1) – (13).

$$\begin{bmatrix} X_E \\ Y_E \\ Z_E \end{bmatrix} = \begin{bmatrix} a_1 & b_1 & c_1 \\ a_2 & b_2 & c_2 \\ a_3 & b_3 & c_3 \end{bmatrix} \begin{bmatrix} X_S \\ Y_S \\ Z_S \end{bmatrix} \tag{1}$$

Here $X_S$, $Y_S$, $Z_S$ represent the accelerations measured in the sensor frame, $X_E$, $Y_E$, $Z_E$ are the accelerations rotated into the earth coordinate frame; and the direction cosines for the above transformation are in terms of the Euler attitude angles.

The coefficients a, b, and c are calculated by following formulas:

$$a_1 = cos\theta cos\psi \tag{2}$$

$$b_1 = sin\varphi sin\theta cos\psi - cos\varphi sin\psi \tag{3}$$

$$c_1 = sin\varphi sin\theta cos\psi + cos\varphi sin\psi \tag{4}$$

$$a_2 = cos\theta sin\psi \tag{5}$$

$$b_2 = sin\varphi sin\theta cos\psi + cos\varphi sin\psi \tag{6}$$

$$c_2 = sin\varphi sin\theta cos\psi - cos\varphi sin\psi \tag{7}$$

$$a_3 = -sin\theta \tag{8}$$

$$b_3 = sin\varphi cos\theta \tag{9}$$

$$c_3 = cos\varphi cos\theta \tag{10}$$

Here θ, ψ and φ are data from gyroscope. After the accelerations have been rotated into the earth frame, the earth-referenced accelerations of the buoy are given by

$$A_x = -gX_E \tag{11}$$

$$\mathrm{A_y} = -gY_E \tag{12}$$

$$A_Z = g(1 - Z_e) \tag{13}$$

Where Ax, Ay, and Az are no gravitational accelerations along the earth oriented x, y, and z axes.

### 3.3 Data Transmissions

Comparison of most popular data transmission protocols such as: Bluetooth, UWB, ZigBee, Wi-Fi and other where done [4]. We have decided that the best transmission protocol, for such type task (low cost, low power, mesh network support) ZigBee was preferable. So, this mash type network protocol has been used for developing of the HMDCS buoys network. ZigBee is an open standard for short range wireless networks based on the Physical Layer and the Media Access Control from IEEE 802.15.4, focusing on minimizing the overall power consumption and at the same time maximizing network reliability [14].

ZigBee protocol offers three kinds of devices to form PAN (personal area network):

1. End-devices, which periodically collect data and transmit them.
2. Routers. They collect data from end-devices and forward them to the destination (like another router or to the final coordinator).
3. Coordinator. One of the routers in a PAN is usually configured as coordinator. Its main function are parameterization and management of the PAN and the collection of the networks data.

In our case, we have used so-called "Full function Devices" which collects data and works as router and Coordinator that manages PAN network and sends collected data via GSM to costal station (Fig. 1). Following ZigBee network configuration were used for data transmitting to the costal station.

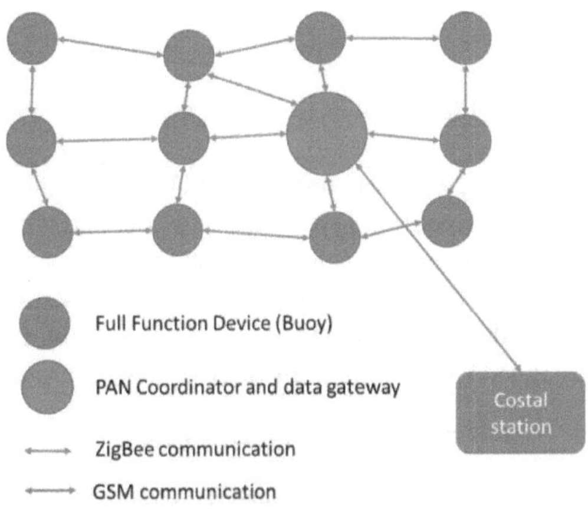


**Fig. 1.** Mash network

## 4 Agent Action Distribution Using SOM

The proposed multi-agent sensory system is based on the goal of the task distribution for agents according to the action similarities. This can be implemented applying selforganising map neural networks (SOM). SOM defines a two-dimensional nonlinear manifold as a regular array of discrete points. In this way, the application of an un-supervised learning allows multidimensional vector represent in two-dimensional output space. SOM output layer neurons retain topological structure according to internal data structure. The typical SOM neural network architecture is shown in Fig. 2. The input nodes represent parameter vector, which according to the similarity is projected in the two-dimensional output space - competitive layer. The input layer represents the parameters of agents target selection, and the competitive layer represents the autonomous agents based sensory system.

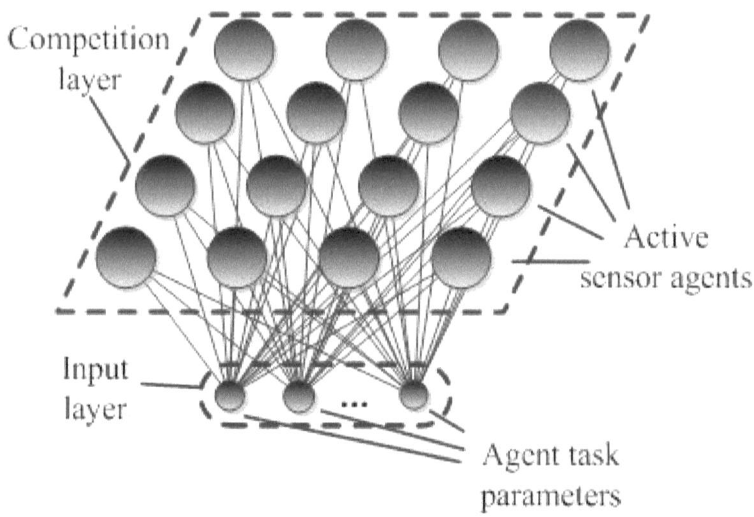


**Fig. 2.** SOM for autonomous agents based sensory system

In order to get the topological structure of the SOM the training process should be applied. Each unit in the competition layer array is associated with a parametric reference vector weight of dimension n. Each input vector is compared with the reference vector weight wj of each unit. The best match, with the smallest Euclidean distance is defined as response, and the input is mapped onto this location. Initially, all reference vector weights are assigned to small random values and they are updated as [6]:

$$\Delta w_j = \alpha_n(t)h_j(g,t)(x_i - w_j(t)) \tag{14}$$

where α(t) is the learning rate at time t and hj(g, t) is the neighborhood function from winner unit neuron g to neuron j at time t. In general, neighborhood function decreases monotonically as a function of the distance from neuron g to neuron n. This decreasing property is a necessary condition for convergence [6].

SOM competition layer nodes correspond to individual agents as active sensory nodes, which are able to process data at a different level (filtering, sampling, transfer and other). From these characteristics depends the capacity of wireless network, data capture excess in central database, and so on. Assuming that each agent as an active buoy sensor node performs different actions, the central unit can distribute tasks for the agents in accordance with their capabilities and required information. In this case, we use three parameters, which determine the actions performed by agents - the significance of the measurement data, hydro-meteorological characteristics of interest and the number of sampling rate. Under these settings, agents distribute the action according to the common goal. For example if we need the raw data – the task will be forwarded to agents that have a high data transfer bandwidth but do not have the filtering capabilities.

## 5 Multi-agent System Model for Hydro Meteorological Sensory System

For proper buoys operation multi-agent type system was designed. Agent software was developed using multi agent framework and works internally in the buoy. The Fig. 3 shows one buoy agent example. Buoy agent has main goal: measure data and different tasks are given by posting newMeasurmentGoal message from coordinator (SOM network). Buoy agent can read new data using capability Measure (Fig. 4). After sensors has read the data the messages onReadWTemper (for water temperature), oReadOTemper (for weather temperature) and onReadWaveHg (for wave height) occurs.

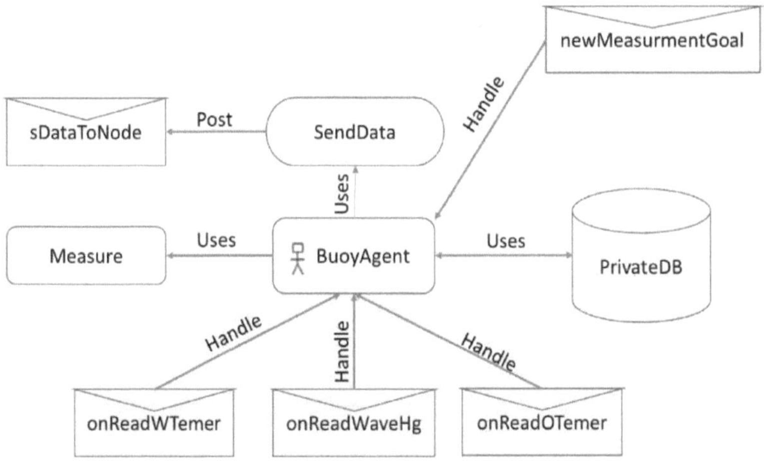


**Fig. 3.** Buoy Agent schematic

Buoy agent stores data in local DB and if it is necessary, it is able to post it to the other agents via ZigBee network using the plan SendData.

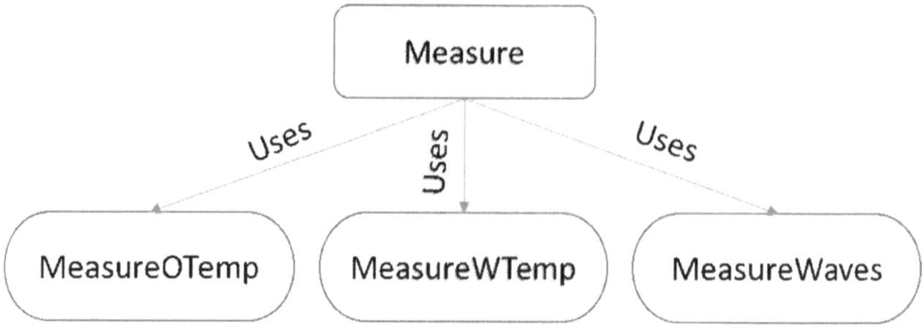


**Fig. 4.** Measure data capability

# 6 Results and Discussion

For sea waves height five different measurement methods were analyzed: using ultrasonic sensor, rheostat type sensor, accelerometer and gyroscope sensors, satellite photos and GPS data [5]. For data transmission from buoys to main station where analyzed different transmission methods and protocols, but most focused on mash type wireless networks and agent based communication methods [11],[15].

For testing purposes experimental buoy sensory system was developed. The core component of the prototype is Arduino Mega platform with ATmega2560 microcontroller which operating at 16 MHz clock frequency, (Fig. 5 - 1). The experimental buoy system is powered by solar power supply, which also recharges Ni-Mh batteries, which allow buoy sensory system operate at the night (Fig. 5 - 3, 9, 10). Buoy status are shown on LCD display (Fig. 5 – 2). XBee Pro modules (Fig. 5- 4) implement communication via ZigBee protocol, which have 10mW transmission power and according to the specifications, expected distance is about 1-1.5 km in the outdoor. Temperature measurements (underwater and weather) are implemented using DS18B20 sensors array connected in to the 1-wire network (Fig. 5 - 5). Data loging to MMC (Fig. 5 – 7). The wave height is measured using MPU6050 (Fig. 5 – 6) accelerometer/gyroscope and calculated by provided method.

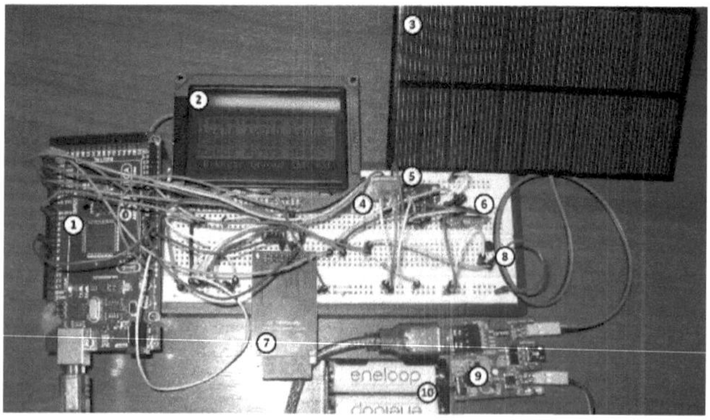


**Fig. 5.** Buoy electronic system prototype

The constructed prototype was placed in hermetic housing and tested in offshore Baltic Sea. The construction design and electronics solutions looks much promising: one buoy electronics cost only about 100 EUR, experimentally tested point to point network with 10mW Xbee modules in open sea have transmit distance for at least 900 m (it is enough for building buoy mash type network).

According to the Baltic Sea Monitoring Data Base [16] it was established hydro meteorological data collection mash type network, which allows the performance evaluation of each sensor node. This evaluation allows costal central station to distribute the agent performance according to the amount of required data. The Fig. 6 shows the distribution of sensory nodes priorities using SOM neural network.

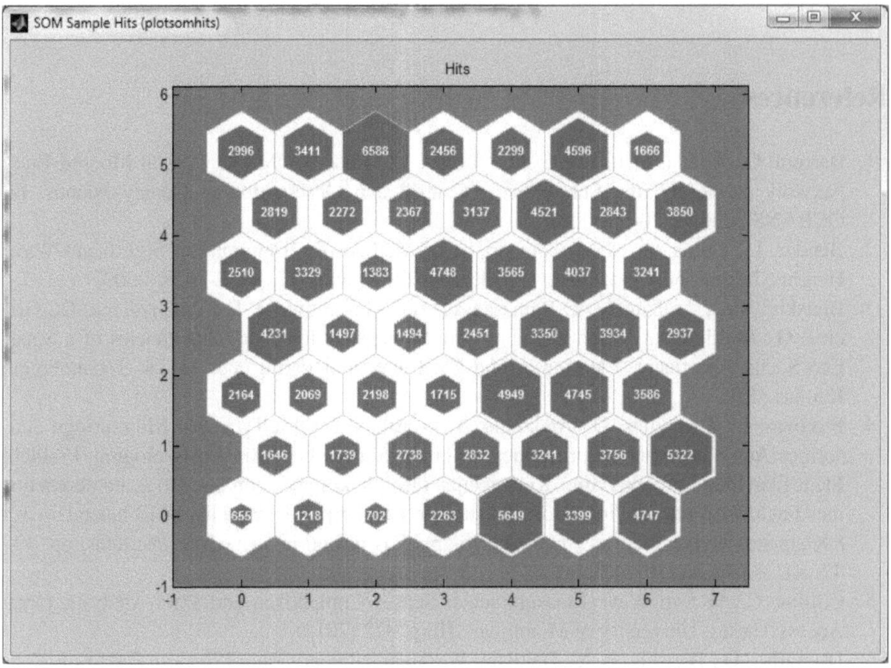


**Fig. 6.** The distributed sensory nodes priorities using SOM neural network

Each sensor node priority defines the importance of the measurements and the amount of data transmitted – i.e. sensor node with higher priority requires agent transmit larger amounts of data, which should allow more accurate assessment of the interested sea region.

## 7 Conclusions

This paper presents the possibilities of development the hydro meteorological data collection system (HMDCS) involving advanced technologies such as multi agent based interaction and data collection between several monitoring system's nodes (i.e. buoys) based on self-organizing maps (SOM). The construction design and electronics solutions of HMDCS looks much promising because of inexpensive, but reliable Baltic Sea autonomous monitoring network (buoys), which would be able, continuously monitor and collect temperature, waviness, and other required data. The multi-agent type system was designed enabling to monitor the data from the costal-based station with limited transition speed by setting different tasks for agent based buoy system according to the SOM.

**Acknowledgments.** This work was supported by the Latvia-Lithuania cross border cooperation programme within the project "JRTC Extension in Area of Development of Distributed Real-Time Signal Processing and Control Systems", project code LLIV-215.

## References

1. Barrera, C., Rueda, M.J., Elgue, J.C., Llinas, O.: Red ACOMAR: Coastal Moored Buoy Network for Real-Time Surveillance, Control and Observation in Canary Islands. In: OCEANS 2006, pp. 1–5 (2006)
2. Bender, L., Guinasso, N.: A Comparison of Methods for Determining Significant Wave Heights. Journal of Atmospheric and Oceanic Technology 27, 1012–1028 (2009)
3. Bielskis, A.A., Andziulis, A., Ramasauskas, O., Guseinoviene, E., Dzemydiene, D., Gricius, G.: Multi-Agent Based E-Social Care Support System For Inhabitancies of a Smart Eco-Social Apartment. Electronics and Electrical Engineering, pp. 11–14. Technologija, Kaunas (2011)
4. Bykovas, D., Drungilas, D., Andziulis, A., Venskus, J.: Jūrų Tyrimų ir Monitoringo Sensorinės Automatizuotos Informacinės Sistemos, Skirtos Išankstiniam Ekologinių Problem Identifikavimui, Projektavimo Koncepcija. Jūros ir krantų tyrimai - 2013: konferencijos medžiaga. 7-oji nacionalinė jūros mokslų ir technologijų konferencija, 2013 balandžio 3-5 / Klaipėdos Universiteto Baltijos pajūrio aplinkos tyrimų ir planavimo institutas, pp. 39–42. KU, Klaipėda (2013)
5. Collins, C.: In Situ Wave Measurements: Sensor Comparison and Data Analysis. Open Access Theses, University of Miami, vol. III, p. 372 (2012)
6. Drungilas, D., Bielskis, A.A., Denisov, V.: An Intelligent Control System Based on Non-Invasive Man Machine Interaction. In: Innovations in Computing Sciences and Software Engineering, pp. 63–68 (2010)
7. Dzemydienė, D.: Sprendimų Paramos Sistemos Galimybės Vertinti Vandens Taršos Procesus. Jūros ir krantų tyrimai - 2013 : konferencijos medžiaga. 7-oji nacionalinė jūros mokslų ir technologijų konferencija, 2013 balandžio 3-5 / Klaipėdos Universiteto Baltijos pa-jūrio aplinkos tyrimų ir planavimo institutas, pp. 69–72. KU, Klaipėda (2013)
8. Laarhuis, J.H.: MaritimeManet: Mobile ad-hoc networking at sea. In: 2010 International Water-Side Security Conference, pp. 1–6 (2010)
9. Li, T.L.T.: Multi-sink opportunistic routing protocol for underwater mesh network. In: 2008 International Conference on Communications, Circuits and Systems, pp. 405–409 (2008)
10. Londhe, S.N.: Development of Wave Buoy Network Using Soft Computing Techniques. In: OCEANS 2008 - MTS/IEEE Kobe Techno-Ocean, pp. 1–8 (2008)
11. Maxim Integrated: DS18B20, Datasheat (2008)

12. Mirza, M.A., Shakir, M.Z., Slim-Alouini, M.A.: GPS-free Passive Acoustic Localization Scheme for Underwater Wireless Sensor Networks. In: 2011 IEEE Eighth International Conference on Mobile Ad-Hoc and Sensor Systems, pp. 879–884 (2011)
13. Panchang, V., Zhao, L., Demirbilek, Z.: Estimation Of Extreme Wave Heights Using GEOSAT Measurements. Ocean Engineering, 205–225 (1999)
14. Sieber, A., Cocco, M., Markert, J., Wagner, M.F., Bedini, R., Dario, P.: ZigBee based buoy network platform for environmental monitoring and preservation: Temperature profiling for better understanding of Mucilage massive blooming. In: 2008 International Workshop on Intelligent Solutions in Embedded Systems, pp. 1–14 (2008)
15. Texas Instruments: LM35 Precision Centigrade Temperature Sensors. Datasheat (2013)
16. International Council for the Exploration of the Sea, Baltic Sea monitoring data, `http://ocean.ices.dk/Helcom/Helcom.aspx`

# Estimation of Fractal Dimension and Statistical Tools for Surface Evaluation

Vlastimil Hotař[1], Petr Salač[2], and Ondřej Matúšek[1]

[1] Technical University of Liberec,
Department of Glass Producing Machines and Robotics,
Studentská 1402/2, 461 17 Liberec 1
[2] Technical University of Liberec,
Department of Mathematics and Didactics of Mathematics,
Studentská 1402/2, 461 17 Liberec 1
{vlastimil.hotar,petr.salac,ondrej.matusek}@tul.cz

**Abstract.** Fractal geometry is a useful tool for describing the complexity of structured data. However, fractal geometry does not substitute for other tools like statistics and should be used with other parameters for general analysis. Hence, the sensitivity to changes in complex data is studied here. For this purpose, samples with structured surfaces produced with different technologies and properties were measured and evaluated with many types of parameters. The parameters were compared and a selection of the results is presented in this article.

## 1 Introduction

Due to the continuously increasing pressure from competitors to improve the quality of products, there is a demand for objective measurement and control methods for materials, processes and production processes. However, it is almost impossible by conventional methods to describe many structures (e.g. defects, surfaces, cracks, time series from dynamic processes) because they are complex and irregular. One approach is the application of fractal geometry which is successfully used in science. Even though its applications in industry are quite rare and experimental, fractal geometry in conjunction with statistics, can be used as a useful and powerful tool for an explicit, objective and automatic description of production process data (laboratory, off-line and potentially on-line). Here, we carry out research into the mentioned tools on applications in industry [1 - 5]. For reliable usage of the methodology and analysis used, their properties and limitations have first to be defined. For this purpose we analysed 14 surfaces produced by 5 different processes and in different conditions, tab. 1. Fig. 1 shows 28 samples (with 14 surfaces). The samples were measured by 3 methods: by a surface roughness tester, by image-capturing with an electron microscope, and by image-capturing of metallographic samples using an optical microscope, fig. 2. The measurements were analysed using the developed methodology with 30 parameters. This article focuses on data from a surface roughness tester. Data analyses from others measurement will be published later.

I. Zelinka et al. (eds.), *Nostradamus 2014: Prediction, Modeling and Analysis of Complex Systems*, Advances in Intelligent Systems and Computing 289,
DOI: 10.1007/978-3-319-07401-6_5, © Springer International Publishing Switzerland 2014

The aim of the research is; to compare standard and non-standard parameters, to find the optimal parameters for a complete analysis, and to specify the sensitivity to directionality of samples for these types of surfaces.

**Table 1.** List of analysed samples with their production properties

| Sample | Technology of surfaces production |
|---|---|
| 1 | Polished surface to maximum gloss |
| 2 | Ballotini (glass beads) blasting, grain size F120 (mean diameter 0.109 mm) |
| 3 | Corundum blasting, grain size F36 (mean diameter 0.525 mm) |
| 4 | Corundum blasting, grain size F12 (mean diameter 1.765 mm) |
| 5 | Electro-erosion machining 29A |
| 6 | Electro-erosion machining 42A |
| 7 | Electro-erosion machining 54A |
| 8 | Sandpaper, K400 |
| 9 | Emery cloth, 120 |
| 10 | Emery cloth, 80 |
| 11 | Vertical milling machine, milling cutter 20 mm, 120 rpm, feed 30 mm/min |
| 12 | Grinding wheel, 98A 60 J 9 V C40 |
| 13 | Grinding wheel, 96A 36P 5V |
| 14 | Vertical milling machine, milling cutter 20 mm, 120 rpm, feed 240 mm/min |

**Fig. 1.** Analysed samples with machined surfaces

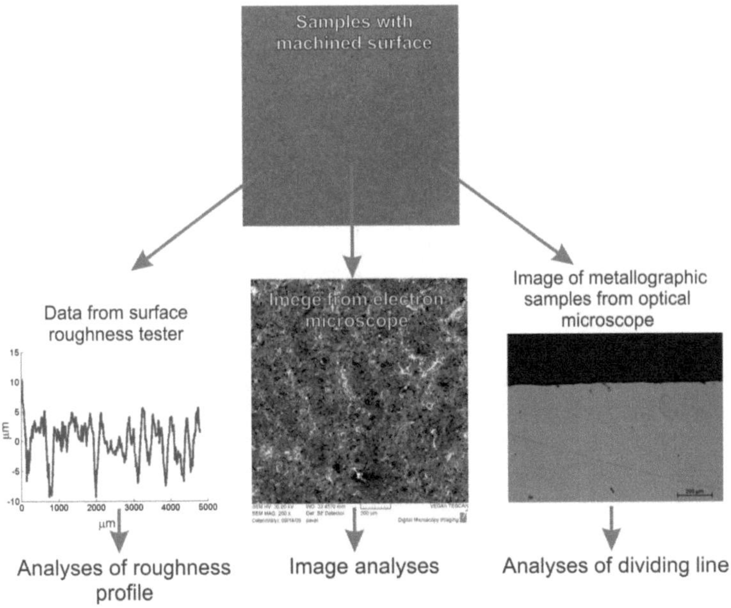


**Fig. 2.** Measurement of samples, obtained data and analyses

## 2 Methodology and Tools Used

The unfiltered reading (raw data) from a surface roughness tester is called a profile (curve). The profile can be evaluated by various methods. The parameters obtained can be divided into three groups:

ι) *parameters of amplitude,* useful for depth characterization (*Std* - Standard Deviation, *Ra* - Average Roughness, *Rt* - Maximum Roughness, *Rz* – Mean Roughness Depth, etc.).

ιι) *parameters of frequency,* used to describe surface profile spacing parameters and for corrugation frequency characterization (e.g. *Sm* - Mean Spacing),

ιιι) *parameters of complexity and deformation,* estimation of fractal dimension by Compass Dimension (*Dc*) [6,7,8], by EEE method [9] or Relative Length ($L_R$) and Proportional Length ($L_P$) of the profile.

Average Roughness, Maximum Roughness, Mean Roughness Depth and Mean Spacing are surface profile parameters defined by standard ISO 4287-1997 [10].

Average Roughness (*Ra*) is also known as the Arithmetical Mean Roughness. The Average Roughness is the area between the roughness profile and its mean line, or the integral of the absolute value of the roughness profile height over the evaluation length:

$$R_a = \frac{1}{l}\int_0^l |z(x)|dx \qquad (1)$$

(where $l$ is the evaluation length, $z$ is the deviation from the center line $m$, fig. 3). When evaluated from digital data, the integral is normally approximated by a trapezoidal rule:

$$R_a = \frac{1}{n}\sum_{i=1}^{n} |z_i| \qquad (2)$$

(where $n$ is the number of measurements). Graphically, the average roughness is the area (yellow in fig. 3) between the roughness profile and its centre line $m$ divided by the evaluation length. In this field of research, a filtered profile is not being used. For this reason the Average Roughness is called *Pa.*

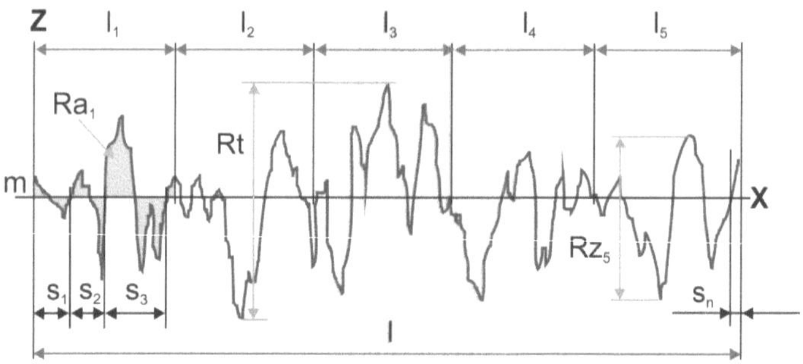


**Fig. 3.** Parameters *Ra, Rt, Rz, Sm;* with the centre line *m*

Maximum Roughness (*Rt*) also Maximum Height, or Total Roughness, is the vertical distance from the deepest trough to the highest peak, fig. 3. For the unfiltered profile, Maximum Roughness is denoted *Pt.*

Mean Roughness Depth ($Rz_5$) is the arithmetic mean of the single distance from the deepest trough to the highest peak from 5 sampling lengths ($l_1$-$l_5$), fig. 3. For the unfiltered profile, Mean Roughness Depth is denoted $Pz_5$.

*Sm* is the Mean spacing between peaks, now with a peak defined relative to the mean line. A peak must cross above the mean line and then cross back below it. If the width of each peak is denoted as *Si*, then the mean spacing is the average width of a peak over the evaluation length, fig 3:

$$Sm = \frac{1}{n}\sum_{i=1}^{n} S_i \,. \qquad (3)$$

The estimated Compass Dimension expresses the degree of complexity of the profile by means of a single number [6]. A compass method [6,7,8] is based on measuring the profile (curve) using different ruler sizes (Fig. 4 A) according to the equation:

$$L_i(r_i) = N_i(r_i) \cdot r_i \qquad (4)$$

*Where* $L_i$ is the length in i-step of the measurement, $r_i$ is the ruler size and $N_i$ is the number of steps needed for the measurement. If the profile is fractal, and hence the estimated fractal dimension is larger than the topological dimension, then the length measured increases as the ruler size is reduced. The logarithmic dependence between $\log_2 N(r_i)$ and $\log_2 r_i$ is called the Richardson-Mandelbrot plot (Fig. 4 B). The Compass Dimension is then determined from the slope *s* of the regression line:

$$D_C = 1 - s = 1 - \frac{\Delta \log_2 L(r)}{\Delta \log_2 r} \tag{5}$$

For better comparison of the results, the dimension is multiplied by 1000 ($D_{C\ 1000}$). The fractal dimension can also be estimated by a different method [7,8].

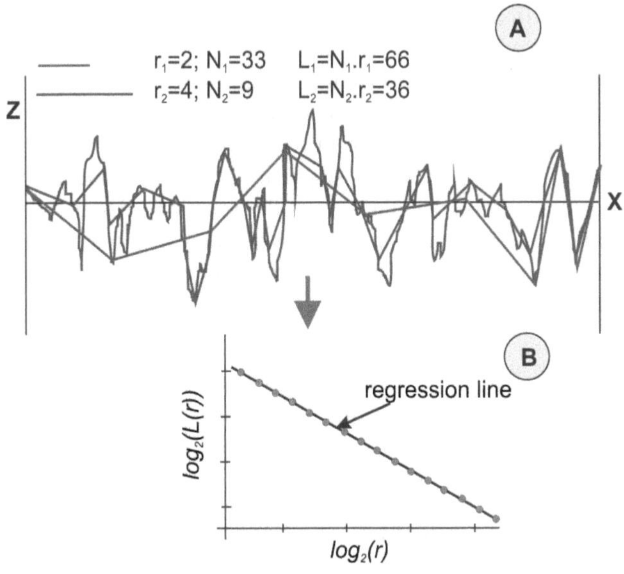


**Fig. 4.** Estimation of the fractal dimension by the compass method

The rate of profile deformation can be evaluated from its Relative Length $L_R$. This fast and reliable method measures the ratio of the profile length $l_{PIXEL}$ (red curve in fig. 3) using the smallest ruler (1 pixel) $r_{PIXEL}$ and the length of the projection $l$ (fig. 3)

$$L_R = \frac{l_{PIXEL}}{l}. \tag{6}$$

Another similar approach is to compute the Proportional Length of the profile $L_P$. The Proportional Length is the ratio of the profile length measured with a defined ruler $l_r$ (e.g. green line in fig. 4 A) and the length measured with the maximum ruler $l_{r\,max}$ (the length between the first and the last point of the profile)

$$L_P = \frac{l_r}{l_{r\max}} \tag{7}$$

The EEE method (**E**valuation of length changes with **E**limination of insignificant **E**xtremes) [9] stems from an estimation of the fractal dimension, so it measures changes of lengths in sequential steps. The method does not use a fixed "ruler" for its measurement in every step, but the line is defined by local extremes (maxima and minima). The method is based on length evaluation of a profile (curve or signal).

The profile is defined by measured values and they are isolated points $x_1, x_2, \ldots, x_n$ in the range $z(x_1), z(x_2), \ldots, z(x_n)$. The points represent local extremes (maxima and minima). On the profile, unnecessary extremes are classified with a defined ruler and a new simplified function is defined by the remaining points. A relative length $L_{R\,1}$ of the new function is measured and the result is saved.

The procedure for the elimination of insignificant extremes is applied to the simplified function (profile). The function obtained is also measured and the process is re-iterated. The last function is formed from the global maximum and minimum of all functions, at which point the analysis is stopped. The steps $i$ of the analysis are plotted against the computed relative lengths $L_{R\ i}$ of the functions. The relation between the relative lengths $L_{R\ i}$ and the steps of elimination $i$ are evaluated by a suitable regression function that can be: a regression line, a quadratic function or a hyperbolic function. In the case of using a regression line, the dimension can be computed from the equation:

$$D_{EEE} = 1 + |s| \tag{8}$$

For better comparison of the results the dimension is multiplied by 1000 ($D_{EEE\ 1000}$). More information can be found in [9].

## 3 Measurement of Samples

The surface roughness tester Mitutoyo SV 2000 was used for taking measurements. All samples (2 samples with the same surface) were measured in 9 positions, each position in 3 directions, $x$, $y$, and transversely. The length of measurement is 4800 μm, and sampling interval is 0.5 μm. All data obtained is in the form of unfiltered profiles. A software tool for a data evaluation was developed in Matlab.

## 4 Results

The samples analyzed have clearly different structural characters. In fig. 1, the samples are ordered from smooth to the most structured surface (from the left to the right). The two upper lines represent the blasted and electro-eroded surfaces (random surfaces) and the two bottom lines represent the classically machined surfaces. Fig. 5 to fig. 9 show the results of the analyses for the surfaces from the measurement of the profiles in one direction. A correlation between the chosen parameters is clearly visible (*Pa*, $D_{EEE\ 1000}$, and $L_R$). $D_{C\ 1000}$ parameter correlates lower and *Sm* parameter does not correlate. To evaluate the parameters objectively, the Pearson's correlation coefficients were computed, see tab. 2 (the parameters are normally distributed). The aim is to specify the appropriate parameters for fast and reliable analysis for industrial data evaluation [1] (for example, production control or quality monitoring). Only the chosen parameters should be used for a complete (global) analysis of the data in order to reduce processing time. Some parameters linearly correlate with others (they provide similar information about the data), tab. 2. If the situation is simplified and a linear correlation is assumed, we can specify suitable parameters for evaluation of these types of data as: Average Roughness, *Pa* (*parameter of amplitude*), Mean Spacing, *Sm* (*parameter of frequency*), Compass Dimension, $D_{C\ 1000}$ (*parameter of complexity and deformation*). These 3 parameters provide diverse information about the data.

A decisive number (a testing number, a quality number) is required in several applications. Only one number is evaluated from these 3 parameters by weight coefficients and can be compared with a specified quality scale. The weight coefficients have to be specified using an appropriate methodology [1, 2].

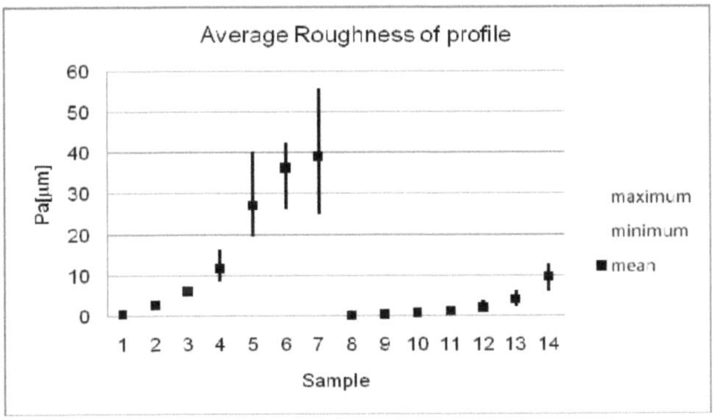


**Fig. 5.** Results of *Pa* parameter

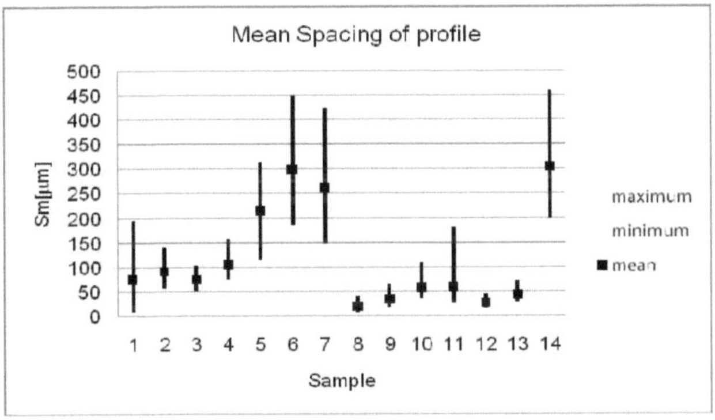


**Fig. 6.** Results of *Sm* parameter

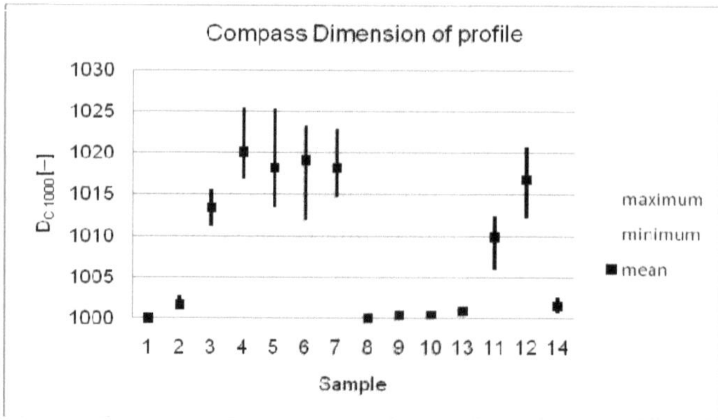


**Fig. 7.** Results of fractal dimension estimation ($D_{C\ 1000}$)

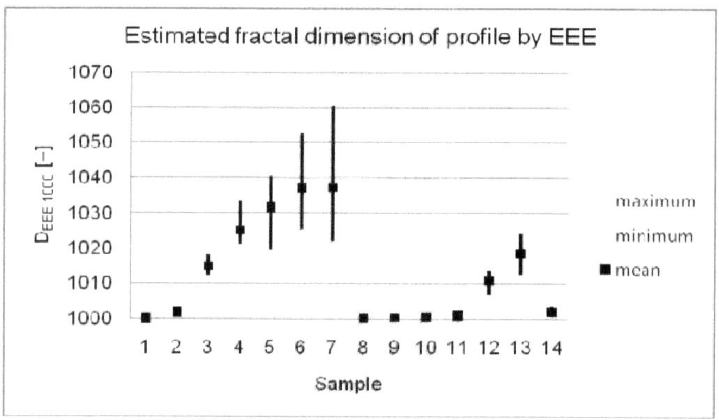


**Fig. 8.** Results of fractal dimension estimation ($D_{EEE\ 1000}$)

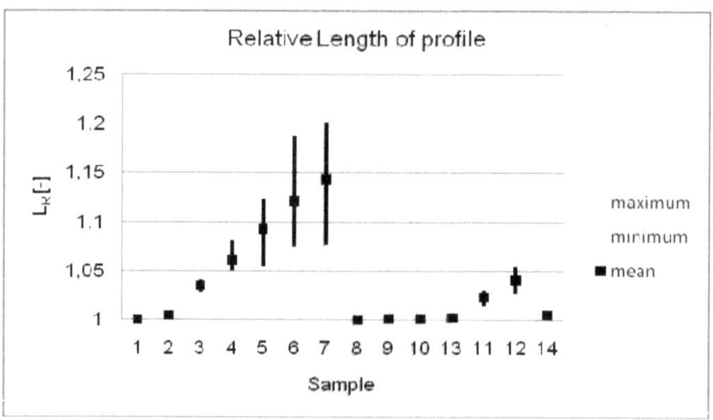


**Fig. 9.** Results of relative length measurement ($L_R$)

**Table 2.** Correlation coefficients of parameters

| | $L_P$ Proportional Length, [-] | $L_R$ Relative Length, [-] | $D_{EEE\ 1000}$ EEE Dimension, [-] | $D_{C\ 1000}$ Compass Dimension, [-] | $Sm$ Mean Spacing, [μm] | $Pz_5$ Mean Roughness Depth, [μm] | $Pt$ Maximum Roughness, [μm] | $Pa$ Average Roughness, [μm] | $Std$ Standard Deviation, [μm] |
|---|---|---|---|---|---|---|---|---|---|
| Standard Deviation, $Std$ [μm] | 0,92 | 0,92 | 0,90 | 0,74 | 0,37 | 0,98 | 0,99 | 1,00 | 1,00 |
| Average Roughness, $Pa$ [μm] | 0,92 | 0,92 | 0,89 | 0,74 | 0,37 | 0,98 | 0,98 | 1,00 | |
| Maximum Roughness, $Pt$ [μm] | 0,93 | 0,93 | 0,92 | 0,79 | 0,34 | 0,99 | 1,00 | | |
| Mean Roughness Depth, $Pz_5$ [μm] | 0,95 | 0,95 | 0,94 | 0,81 | 0,33 | 1,00 | | | |
| Mean Spacing, $Sm$ [μm] | 0,25 | 0,25 | 0,22 | 0,15 | 1,00 | | | | |
| Compass Dimension, $D_{C\ 1000}$ [-] | 0,85 | 0,85 | 0,94 | 1,00 | | | | | |
| EEE Dimension, $D_{EEE\ 1000}$ [-] | 0,96 | 0,96 | 1,00 | | | | | | |
| Relative Length, $L_R$ [-] | 1,00 | 1,00 | | | | | | | |
| Proportional Length, $L_P$ [-] | 1,00 | | | | | | | | |

Measurements were taken at 9 different measurement points in the *x, y* and transverse direction for each of the 28 samples examined. This was done for all (9) methods presented.

The mean values of the data obtained from individual samples of *x-axis* directions ($\mu_1$), Y-axis directions ($\mu_2$) and the transverse directions ($\mu_3$) were compared for each sample. Conformity of the mean values was tested by one-way analysis of variance (ANOVA) [11] at significance level $\alpha = 0.05$ using Matlab software. Thus

$$H_0 : \mu_1 = \mu_2 = \mu_3 \qquad (9)$$
$$H_1 : non\ H_0$$

The test results are shown in tab. 3, where 0 means a rejection $H_0$ and a benefit $H_1$. 1 does not constitute rejection $H_0$.Samples 1 to 7 were prepared by technologies that produce random structures. Samples 8 to 14 were produced by a standard machining method that generates directionally visible structures (fig. 1). Samples 8, 9, 10, 12, and 13 have linearly oriented structures. Samples 11 and 14 have rotationally oriented structures, because of the milling technology. Parameters $D_{C\ 1000}$, $D_{EEE\ 1000}$, $L_R$, and $L_P$

show good results in recognition of the directionality. The only exception was for samples 11. These samples have a smooth rotationally-oriented structure that is identified as a random structure.

**Table 3.** One-way analysis of variance (ANOVA), Null hypothesis: $H_0$

| | Standard Deviation, *Std* [µm] | Average Roughness, *Pa* [µm] | Maximum Roughness, *Pt* [µm] | Mean Roughness Depth, Pz5 [µm] | Mean Spacing, *Sm* [µm] | Compass Dimension, $D_{C\ 1000}$ [-] | EEE Dimension, $D_{EEE\ 1000}$ [-] | Relative Length, $L_R$ [-] | Proportional Length, $L_P$ [-] |
|---|---|---|---|---|---|---|---|---|---|
| 1 A | 1 | 1 | 1 | 1 | 1 | 1 | 1 | 1 | 1 |
| 1 B | 1 | 1 | 1 | 1 | 1 | 1 | 1 | 1 | 1 |
| 2 A | 1 | 1 | 1 | 1 | 1 | 1 | 1 | 1 | 1 |
| 2 B | 1 | 1 | 1 | 1 | 1 | 1 | 1 | 1 | 1 |
| 3 A | 1 | 1 | 1 | 1 | 1 | 1 | 1 | 1 | 1 |
| 3 B | 1 | 0 | 1 | 1 | 1 | 1 | 1 | 1 | 1 |
| 4 A | 1 | 1 | 1 | 1 | 1 | 1 | 1 | 1 | 1 |
| 4 B | 1 | 1 | 1 | 1 | 1 | 1 | 1 | 1 | 1 |
| 5 A | 1 | 1 | 1 | 1 | 0 | 1 | 1 | 1 | 1 |
| 5 B | 1 | 0 | 1 | 1 | 1 | 1 | 1 | 1 | 1 |
| 6 A | 0 | 1 | 1 | 1 | 1 | 1 | 1 | 1 | 1 |
| 6 B | 1 | 1 | 1 | 1 | 1 | 1 | 1 | 1 | 1 |
| 7 A | 1 | 1 | 1 | 1 | 1 | 1 | 1 | 1 | 1 |
| 7 B | 1 | 1 | 1 | 1 | 1 | 1 | 1 | 1 | 1 |
| 8 A | 1 | 1 | 1 | 1 | 0 | 0 | 0 | 0 | 0 |
| 8 B | 1 | 0 | 1 | 1 | 1 | 0 | 0 | 0 | 0 |
| 9 A | 1 | 1 | 1 | 1 | 0 | 0 | 0 | 0 | 0 |
| 9 B | 0 | 1 | 0 | 1 | 0 | 0 | 0 | 0 | 0 |
| 10 A | 1 | 0 | 1 | 0 | 0 | 0 | 0 | 0 | 0 |
| 10 B | 1 | 0 | 1 | 0 | 0 | 0 | 0 | 0 | 0 |
| 11 A | 1 | 1 | 1 | 1 | 0 | 1 | 1 | 1 | 1 |
| 11 B | 1 | 1 | 1 | 1 | 1 | 1 | 1 | 1 | 1 |
| 12 A | 0 | 0 | 0 | 0 | 0 | 0 | 0 | 0 | 0 |
| 12 B | 1 | 1 | 1 | 1 | 0 | 0 | 0 | 0 | 0 |
| 13 A | 1 | 1 | 1 | 1 | 0 | 0 | 0 | 0 | 0 |
| 13 B | 1 | 1 | 1 | 1 | 0 | 0 | 0 | 0 | 0 |
| 14 A | 1 | 1 | 1 | 1 | 1 | 0 | 0 | 0 | 0 |
| 14 B | 1 | 1 | 1 | 1 | 1 | 0 | 0 | 0 | 0 |

## 5 Conclusions

The samples were measured by a surface roughness tester and analysed. Several parameters can be specified to evaluate these data types: Average Roughness, *Pa* (*parameter of amplitude*), Mean Spacing, *Sm* (*parameter of frequency*), Compass Dimension, $D_{C\ 1000}$ (*parameter of complexity and deformation*). These 3 parameters provide diverse information about the data and can be used for a complete data analysis.

Parameters Compass Dimension ($D_{C\ 1000}$), EEE Dimension ($D_{EEE\ 1000}$), Relative Length ($L_R$) and Proportional Length ($L_P$) can be used for linear structure recognition of the presented data. Based on these results, it can be inferred that the tools represented here are suitable for surface description.

**Acknowledgments.** The results of this project LO1201 were obtained through the financial support of the Ministry of Education, Youth and Sports in the framework of the targeted support of the "National Programme for Sustainability I" ,the OPR&DI project Centre for Nanomaterials, Advanced Technologies and Innovation CZ.1.05/2.1.00/01.0005 and by the grant of Students grant contest of the Technical University of Liberec, number SGS 21006/115, which use special-purpose support for the university research and is financed by the Ministry of Education, Czech Republic.

## References

[1] Hotař, V.: Fractal geometry for industrial data evaluation. Computers & Mathematics with Applications (2013)
[2] Hotař, V., Novotný, F., Reinischová, H.: Objective evaluation of corrugation test. Glass Technology: Eur. J. Glass Sci. Technol. Part A 52(6), 197–202 (2011)
[3] Hotař, V., Novotný, F.: Surface Profile Evaluation by Fractal Dimension and Statistic Tools. In: Proceedings: 11th International Conference on Fracture, p. 588. CCI Centro Congressi Internazionale s.r.l., Turin (2005)
[4] Hotar, V., Novotny, F.: Evaluation of Surface Defects by Fractal Geometry and Statistical Analysis. Glastech. Ber. Glass Sci. Technol. 77C, 230–237 (2004)
[5] Hotař, A., Kratochvíl, P., Hotař, V.: The Corrosion Resistance of Fe3Al Based Iron Aluminides in Molten Glasses. Kov. Materiály – Met. Mater., 247–252 (2009)
[6] Mandelbrot, B.B.: The fractal geometry of nature. W. H. Freeman and Co., New York (1982)
[7] Peitgen, H.O., Juergens, H., Saupe, D.: Chaos and Fractals: New Frontiers of Science. Springer, Heidelberg (1992)
[8] Evertsz, C.J.G., Peitgen, H.O., And Voss, R.F.: Fractal Geometry and Analysis. World Scientific Publishing Co. Pte. Ltd., Singapore (1996)
[9] Hotař, V.: EEE Method: Improved Approach of Compass Dimension Calculation. In: Zelinka, I., Chen, G., Rössler, O.E., Snasel, V., Abraham, A. (eds.) Nostradamus 2013: Prediction, Model. & Analysis. AISC, vol. 210, pp. 343–351. Springer, Heidelberg (2013)
[10] ISO 4287:1997, Geometrical Product Specifications (GPS) - Surface texture: Profile method - Terms, definitions and surface texture parameters. Geneva: International Organization for Standardization (1997)
[11] Statistics I: Introduction to ANOVA, Regression, and Logistic Regression. Sas Inst. (2005)

# Statistics of Fractal Systems

Oldrich Zmeskal, Stanislav Nespurek, Michal Vesely, and Petr Dzik

Faculty of Chemistry, Brno University of Technology, Czech Republic

**Abstract.** Distribution functions are used for the description of energy distribution of elementary particles, atoms, and molecules in dynamic systems. These distribution functions depend on the energy of the system and on its properties. The paper focuses on the generalization of the relationships commonly used to study the statistical properties of particles in 3D space so that they become generally applicable onto an $E$-dimensional space. These relationships can then be applied e.g. for studying the properties of the particles in 2D and in 1D space.

Two approaches are discussed to describe the classic (Maxwell Boltzmann) and quantum (Fermi-Dirac, Einstein-Bose) distribution functions. The first approach is based on standard theory of probability, the second one on the fractal theory. We have shown that both approaches lead to the same results for defined boundary conditions. But the validity of the second one, i.e. the fractal approach, is much more general.

**Keywords:** fractal physics, classic and quantum statistics, classical theory of statistics, fractal theory of statistics.

## 1 Introduction

Distribution functions can be conveniently used for the description of energy distribution of elements (elementary particles, atoms, molecules) in dynamic systems. These distribution functions depend on the energy (or temperature) of the system and on the the properties of the elements (distinguishability, spin). The paper focuses on the generalization of the relationships commonly used to study the statistical properties of particles in 3D (volume) so that they become generally applicable onto an $E$-dimensional space. These relationships can then be applied, e.g. for studying the properties of the particles in 2D (such as surface effects), and in 1D space (e.g. quantum wires).

Two approaches are used to describe the classic (Maxwell Boltzmann) and quantum (Fermi-Dirac, Einstein-Bose) distribution functions [1]. The first approach is based on the standard theory of probability, the second one on the fractal theory. It is shown that both approaches lead to the same results for defined boundary conditions. But the validity of the second one, i.e. the fractal approach, is much more general.

According to it, the distribution function can be defined generally in $E$-dimensional space (motion in a straight line, the planar surface, in space ...). In this paper, the relationship between the fractal dimension $D$ and the particle distribution functions (e.g. photon, electron, and molecule) is described. We also present how to derive all three

I. Zelinka et al. (eds.), *Nostradamus 2014: Prediction, Modeling and Analysis of Complex Systems*, Advances in Intelligent Systems and Computing 289,
DOI: 10.1007/978-3-319-07401-6_6, © Springer International Publishing Switzerland 2014

(classical and quantum) distribution functions under defined conditions from the general fractal relationship.

## 2 Energy Distribution Functions - Classical Approach

The energy distribution function f($E$) expresses the probability that a particle occupies certain energy state $E$. There are three distinctly different distribution functions in nature. The first (Maxwell Boltzmann - MB) is used to describe the distribution of classic particles (e.g. molecules), the second and third ones are used to describe the distribution of quantum particles, fermions (Fermi-Dirac – FD) and bosons (Einstein-Bose - EB).

Let us start with the ***Maxwell Boltzmann distribution function*** (defined for identical but distinguishable particles)

$$f_{\mathrm{MB}}(E) = \frac{1}{\exp[(E - E_0)/k_{\mathrm{B}}T]}. \tag{1}$$

The dependence of the molecule concentration $n$ of an ideal gas on the energy $E$ is an example of this classical approach

$$\frac{\mathrm{d}n}{\mathrm{d}E} = g(E) f_{\mathrm{MB}}(E) = \frac{A}{k_{\mathrm{B}}T} \cdot \frac{[(E - E_0)/k_{\mathrm{B}}T]^{1/2}}{\exp[(E - E_0)/k_{\mathrm{B}}T]} \tag{2}$$

where $g(E)$ is an energy distribution of possible states, $E - E_0 = mv^2/2$ is the kinetics energy, and $E_{\mathrm{sk}} = mv_{sk}^2/2 = k_{\mathrm{B}}T/2$ is average kinetic energy of molecule, $v_{\mathrm{sk}}$ is root-mean-square speed, $m$ is a molecule mass, and $T$ is a thermodynamic temperature. The Maxwell–Boltzmann speed distribution function is then [2]

$$\frac{\mathrm{d}n}{\mathrm{d}v} = A\frac{v^2}{v_{\mathrm{sk}}^3} \exp\left(-\frac{v^2}{v_{\mathrm{sk}}^2}\right) \tag{3}$$

By integrating of Eq. (2) we derive

$$p = nk_{\mathrm{B}}T \,, \tag{4}$$

which is a relation defining the gas pressure (the ideal gas equation of unit volume of the gas respectively).

Next, let us consider the ***Fermi Dirac distribution function*** (defined for identical indistinguishable particles with half integer spin - fermions)

$$f_{\mathrm{FD}}(E) = \frac{1}{\exp[(E - E_0)/k_{\mathrm{B}}T] + 1}. \tag{5}$$

The dependence of the electron concentration $n$ in a metals or semiconductors is an example of such quantum distribution function

$$\frac{\mathrm{d}n}{\mathrm{d}E} = g(E) f_{\mathrm{FD}}(E) = \frac{C}{k_{\mathrm{B}}T} \cdot \frac{\left[(E - E_0)/k_{\mathrm{B}}T\right]^{1/2}}{\exp\left[(E - E_0)/k_{\mathrm{B}}T\right] + 1}, \tag{6}$$

where $E - E_0 = h\nu = hc/\lambda$ is an energy of electrons with frequency $\nu$ and wavelength $\lambda$.

The distribution of density of energy $w_{\mathrm{E}}$ of energy is then

$$w_{\mathrm{E}} = \frac{\mathrm{d}w_{\mathrm{s}}}{\mathrm{d}E} = C\left(\frac{k_{\mathrm{B}}T}{hc}\right)^{1/2} \frac{(h\nu/k_{\mathrm{B}}T)^{1/2}}{\exp(h\nu/k_{\mathrm{B}}T) - 1}, \tag{7}$$

where $w_{\mathrm{s}} = nk_{\mathrm{B}}T$ is density of energy.

Further, let us mention the ***Einstein-Bose distribution function*** (defined for identical indistinguishable particles with integer spin - bosons)

$$f_{\mathrm{EB}}(E) = \frac{1}{\exp\left[(E - E_0)/k_{\mathrm{B}}T\right] - 1}. \tag{8}$$

The dependence of the photon concentration $n$ of thermal radiation on the energy $E$ represents an example of such quantum distribution function:

$$\frac{\mathrm{d}n}{\mathrm{d}E} = g(E) f_{\mathrm{EB}}(E) = \frac{B}{k_{\mathrm{B}}T} \cdot \frac{\left[(E - E_0)/k_{\mathrm{B}}T\right]^{3}}{\exp\left[(E - E_0)/k_{\mathrm{B}}T\right] - 1}, \tag{9}$$

$E - E_0 = h\nu = hc/\lambda$ is an energy of photons with frequency $\nu$ and wavelength $\lambda$.

The distribution of density of energy $w_{\mathrm{E}}$ of energy is then Planck's radiation law

$$w_{\mathrm{E}} = \frac{\mathrm{d}w_{\mathrm{s}}}{\mathrm{d}E} = B\left(\frac{k_{\mathrm{B}}T}{hc}\right)^{3} \frac{(h\nu/k_{\mathrm{B}}T)^{3}}{\exp(h\nu/k_{\mathrm{B}}T) - 1}, \tag{10}$$

where $w_{\mathrm{s}} = nk_{\mathrm{B}}T$ is density of energy. More often are used the next formulations of Planck's radiation law

$$w_{\nu} = \frac{\mathrm{d}w_{s}}{\mathrm{d}\nu} = \frac{\lambda^2}{c}\frac{\mathrm{d}w_{s}}{\mathrm{d}\lambda} = \frac{2\pi h\nu^3}{c^3\left[\exp(h\nu/kT) - 1\right]} = \frac{2\pi h}{\lambda^3\left[\exp(hc/\lambda kT) - 1\right]}. \tag{11}$$

Considering $\mathrm{d}w_{\mathrm{E}}/\mathrm{d}E = \mathrm{d}^2 w_s/\mathrm{d}E^2 = 0$ one can derive relation between the maximal wavelength of the function $\lambda_m = c/\nu_m$ and temperature $T_m$ (so-called Wien's displacement law). The value $\lambda_{\nu m} T_m \approx 5.0941\times 10^{-3}\,\mathrm{m}$.

Other example is an Einstein model of specific heat

$$C_{\mathrm{V}} = \frac{\mathrm{d}U}{\mathrm{d}T} = 3k_{\mathrm{B}}N\left(\frac{E-E_0}{k_{\mathrm{B}}T}\right)^2 f_{\mathrm{EB}}(E)\left[1-f_{\mathrm{EB}}(E)\right], \quad (12)$$

where $f_{\mathrm{EB}}(E)$ is Einstein Bose distribution function (5).

# 3 Fractal Approach

## 3.1 Fractal Physical Field

The following text allows unifying the description of the heat transfer properties assuming that structures transferring the heat are of fractal nature [3].

In the previous papers [4] and [5], the density of quantity $\rho(r)$ (e.g. mass density, density of electric charge, or density of heat radiation) was defined as

$$\rho(r) = c_{\mathrm{N}}Kr^{D-E}, \quad (13)$$

where $c_{\mathrm{N}}$ is an elementary quantity (e.g. mass unit, elementary charge or Boltzmann constant).

Field intensity and potential of radial physical field can be calculated from terms

$$F_{\mathrm{r}} = \frac{c_{\mathrm{N}}c_{\mathrm{V}}K}{D}r^{D-E+1}, \quad V_{\mathrm{r}} = -\frac{c_{\mathrm{N}}c_{\mathrm{V}}K}{D(D-E+2)}r^{D-E+2} \quad (14)$$

on the elementary cell dimension $r$. The last equations are very important for describing fractal physical fields not only with radial configuration but also in general cases. For a 3D space ($E = 3$), these equations then describe the spherical field for a point source of physical quantity ($D = 0$), the cylindrical field for line source ($D = 1$), the planar field for surface source ($D = 2$) and finally the volume field for homogeneously distributed sources ($D = 3$).

The flow of energy can be derived from the density of quantity (13) and from the intensity of physical field (14)

$$i(r) = \rho(r)F_r = c_{\mathrm{V}}c_{\mathrm{N}}^2\frac{K^2r^{2D-2E+1}}{D}, \quad (15)$$

The density of energy can be derived by similar

$$w(r) = \rho(r)V_r = c_{\mathrm{V}}c_{\mathrm{N}}^2\frac{K^2r^{2(D-E+1)}}{D(E-D-2)}, \quad (16)$$

where $w(r)/i(r) = r/(D-E+2)$.

A special case takes place when the effective density of energy is for equipotential field

$$w_0 = -\frac{F_r^2}{c_V} = -c_V c_N^2 \frac{K^2 r^{2(D-E+1)}}{D^2}, \tag{17}$$

Then the frequency and the fractal dimension is connected by the following equation

$$\alpha_w = \frac{w}{w_0} = \frac{\nu}{\nu_0} = \frac{\lambda_0}{\lambda} = \frac{D}{D-E+2}, \tag{18}$$

where $\alpha_w$ is so called inverse coupling constant of energy density, $\nu_0$ is the frequency for effective density of field, and $\lambda_0$ is corresponding wave length. The fractal dimension can be from this equation derived as

$$D = \frac{\alpha_w (E-2)}{\alpha_w - 1}, \quad D - E + 2 = \frac{E-2}{\alpha_w - 1} \tag{19}$$

### 3.2 Application to Thermal Field

In this case, the physical quantity $c_N$ used in previous equations stands for elementary electric charge ($c_N = k_B$) and the material constant $c_V$ is proportional to the electric permittivity ($c_V = \hbar c / k_B^2$), see [6] and [7]. Application examples of mathematical equations (13), (14), (15) and (16) to thermal field are given in Table 1.

**Table 1.** Physical quantities used for thermal field

| heat capacity ($J.K^{-1}m^{-3}$) | heat flow ($K.m^{-1}$) | temperature (K) |
|---|---|---|
| $C(r) = k_B n(r) = k_B K r^{D-E}$ | $q_r(r) = \frac{\hbar c}{k_B} \frac{K r^{D-E+1}}{D}$ | $T_r(r) = -\frac{\hbar c}{k_B} \cdot \frac{K r^{D-E+2}}{D(D-E+2)}$ |
| flow of energy ($J.m^{-4}$) | density of energy ($J.m^{-3}$) | |
| $i(r) = C(r) q_r(r) = \hbar c \frac{K^2 r^{2(D-E)+1}}{D}$ | $w(r) = C(r) T_r(r) = -\hbar c \frac{K^2 r^{2(D-E+1)}}{D(D-E+2)}$ | |

It is defined by heat capacity, heat flow, and temperature. By combination of these quantities it is possible define other two: density of energy and flow of energy. These fractal quantities (with fractal dimension $D$) are generally defined in $E$-dimensional space, the unit of fractal measure $[K] = \mathrm{m}^{E-D-3}$ was chosen so that the physical constants valid.

The dependence of density and flow of energy on the temperature can be evaluated by simple combination of equations from Table 1

$$w(T_{\mathrm{r}}) = K\,k_{\mathrm{B}}T_{\mathrm{r}}\left[\frac{k_{\mathrm{B}}T_{\mathrm{r}}}{K\hbar c}D(E-D-2)\right]^{\frac{D-E}{D-E+2}}, \tag{20}$$

$$i(T_{\mathrm{r}}) = \frac{K^2\hbar c}{D}\left[\frac{k_{\mathrm{B}}T_{\mathrm{r}}}{K\hbar c}D(E-D-2)\right]^{\frac{2(D-E)+1}{D-E+2}}. \tag{21}$$

# 4 Discussion of Statistics in $E$-dimensional Space

## 4.1 The Heat Radiation in 3D

Let's first consider the properties of heat transfer in the three-dimensional space ( $E = 3$ ) just for the fractal dimensions of the heat source. This kind of heat transport occurs when the fractal dimensions lie in the interval $D \in \langle 0,1 \rangle$. In this case, the heat density (20) can be written using a generalised Planck radiation law

$$w(T_{\mathrm{r}}) = K\,k_{\mathrm{B}}T_{\mathrm{r}}\left[\frac{k_{\mathrm{B}}T_{\mathrm{r}}}{K\hbar c}D(1-D)\right]^{\frac{D-3}{D-1}}. \tag{22}$$

In this case, the inverse coupling constant (18) will be defined as $\alpha_{\mathrm{w}} = D/(1-D) > 0$ (fractal structure is very sparse) and the distribution function will be ***Einstein – Bose***: $\mathrm{f}(\alpha_{\mathrm{EB}}) = 1/(\alpha_{\mathrm{EB}} - 1)$ where coefficient $\alpha_{\mathrm{EB}}$ represents the Boltzmann factor $\alpha_{\mathrm{w}} = \alpha_{\mathrm{EB}} = \exp(\Delta E/k_{\mathrm{B}}T)$

$$w(T_{\mathrm{r}}) = K\,k_{\mathrm{B}}T_{\mathrm{r}}\left[\frac{k_{\mathrm{B}}T_{\mathrm{r}}}{K\hbar c}\frac{\alpha_{\mathrm{EB}}}{(\alpha_{\mathrm{EB}}-1)^2}\right]^{3-2\alpha_{\mathrm{EB}}}. \tag{23}$$

## 4.2 Heat Conduction and Heat Flow in 3D

This kind of heat transport occurs when the fractal dimensions lie in the interval $D \in \langle 1,3 \rangle$. In this case the $D(1-D)$ in Eq. (22) is negative and relation for energy density, which in this case represents the gas pressure, i.e. $w(T_{\mathrm{r}}) = p(T_{\mathrm{r}})$ ), has to be rewritten into

$$p(T_{\mathrm{r}}) = K\,k_{\mathrm{B}}T_{\mathrm{r}}\left(-1\right)^{\frac{D-3}{D-1}}\left[\frac{k_{\mathrm{B}}T_r\,D(D-1)}{K\hbar c}\right]^{\frac{D-3}{D-1}}, \tag{24}$$

where using the Moivre's theorem

$$(-1)^{\frac{D-3}{D-1}} = \exp\left(\mathrm{j}\pi \frac{D-3}{D-1}\right) = \cos\left(\frac{D-3}{D-1}\pi\right) + \mathrm{j}\cdot\sin\left(\frac{D-3}{D-1}\pi\right). \tag{25}$$

When the fractal dimension is $D = 3$ the Eq. (23) can be expressed as

$$p_{\mathrm{id}} = \frac{RT_{\mathrm{r}}}{V_{\mathrm{M}}} = K k_{\mathrm{B}} T_{\mathrm{r}}, \tag{26}$$

where $V_{\mathrm{M}} = N_{\mathrm{A}}/K$ is the molar gas volume, $R = k_{\mathrm{B}} N_{\mathrm{A}}$ is the molar gas constant, $N_{\mathrm{A}}$ is the Avogadro constant, which represents the ideal gas equation.

To describe the behaviour of real gases it is more practical to use the so-called compressibility factor. With the help of (24) and (26) it can be defined by the equation

$$Z = \frac{p}{p_{\mathrm{id}}} = \frac{pV_{\mathrm{M}}}{RT_{\mathrm{r}}} = (-1)^{\frac{D-3}{D-1}} \left[\frac{k_{\mathrm{B}} T_{\mathrm{r}}\, D(D-1)}{K\hbar c}\right]^{\frac{D-3}{D-1}}. \tag{27}$$

In this case, the inverse coupling constant will be $\alpha_{\mathrm{w}} = D/(1-D) < 0$ (fractal structure is very thick) and the distribution function will be ***Fermi-Dirac***: $\mathrm{f}(\alpha_{\mathrm{FD}}) = 1/(\alpha_{\mathrm{FD}} + 1)$ where coefficient $\alpha_{\mathrm{FD}}$ represents the Boltzmann factor $\alpha_{\mathrm{w}} = -\alpha_{\mathrm{FD}} = \exp(\Delta E/k_{\mathrm{B}} T)$. Then the Eq. (24) can be rewritten to the form

$$p(T_{\mathrm{r}}) = K k_{\mathrm{B}} T_{\mathrm{r}} \left[\frac{k_{\mathrm{B}} T_r}{K\hbar c} \frac{\alpha_{\mathrm{FD}}}{(\alpha_{\mathrm{FD}} + 1)^2}\right]^{3+2\alpha_{\mathrm{FD}}}. \tag{28}$$

## 4.3 Heat Conduction and Heat Flow in 2D

This kind of heat transport occurs when the fractal dimensions lie in the interval $D \in \langle 0, 2\rangle$. In this case the $-D^2$ in Eq. (22) is negative and relation for energy density, which in this case represents the e.g. gas pressure, i.e. $w(T_{\mathrm{r}}) = p(T_{\mathrm{r}})$ ), has to be rewritten into

$$p(T_{\mathrm{r}}) = K k_{\mathrm{B}} T_{\mathrm{r}} (-1)^{\frac{D-2}{D}} \left[\frac{k_{\mathrm{B}} T_{\mathrm{r}}}{K\hbar c} D^2\right]^{\frac{D-2}{D}}, \tag{29}$$

where using the Moivre's theorem

$$(-1)^{\frac{D-2}{D}} = \exp\left(\mathrm{j}\pi \frac{D-2}{D}\right) = \cos\left(\frac{D-2}{D}\pi\right) + \mathrm{j}\cdot\sin\left(\frac{D-2}{D}\pi\right). \tag{30}$$

In this case, the inverse coupling constant will be $\alpha_{\mathrm{w}} = 1$. The distribution function will be constant; it means that all states on the surface have the same probability.

### 4.4 Heat Conduction and Heat Flow in 1D

This kind of heat transport occurs when the fractal dimensions lie in the interval $D \in \langle 0,1 \rangle$. In this case the $D(D+1)$ in Eq. (22) is negative and relation for energy density, which in this case represents the gas pressure, i.e. $w(T_{\mathrm{r}}) = p(T_{\mathrm{r}})$), has to be rewritten into

$$p(T_{\mathrm{r}}) = K k_{\mathrm{B}} T_{\mathrm{r}} \, (-1)^{\frac{D-1}{D+1}} \left[ \frac{k_{\mathrm{B}} T_r \, D(D+1)}{K \hbar c} \right]^{\frac{D-1}{D+1}}, \tag{31}$$

where using the Moivre's theorem

$$(-1)^{\frac{D-1}{D+1}} = \exp\left( \mathrm{j}\pi \frac{D-1}{D+1} \right) = \cos\left( \frac{D-1}{D+1} \pi \right) + \mathrm{j} \cdot \sin\left( \frac{D-1}{D+1} \pi \right). \tag{32}$$

In this case, the inverse coupling constant (18) will be defined as $\alpha_{\mathrm{w}} = D/(1+D) > 0$ and the distribution function will be ***Einstein – Bose***: $\mathrm{f}(\alpha_{\mathrm{EB}}) = 1/(\alpha_{\mathrm{EB}} - 1)$ where coefficient $\alpha_{\mathrm{EB}}$ represents the Boltzmann factor $\alpha_{\mathrm{w}} = \alpha_{\mathrm{EB}} = \exp(\Delta E / k_{\mathrm{B}} T)$

$$p(T_{\mathrm{r}}) = K k_{\mathrm{B}} T_{\mathrm{r}} \left[ \frac{k_{\mathrm{B}} T_{\mathrm{r}}}{K \hbar c} \frac{\alpha_{\mathrm{EB}}}{(\alpha_{\mathrm{EB}} - 1)^2} \right]^{2\alpha_{\mathrm{EB}} - 1}. \tag{33}$$

## 5 Conclusion

*Fermion (electron) distribution function* is occurring in 3D space for $D \in \langle 1,3 \rangle$, in 2D space for $D \in \langle 0,2 \rangle$ and in 1D space $D \in \langle 0,1 \rangle$. Then the coupling constant is defined in 3D as $\alpha_{\mathrm{w}} = D/(1-D)$ - it is less than zero, in 2D as $\alpha_{\mathrm{w}} = 1$ and in 1D space as $\alpha_{\mathrm{w}} = D/(1+D)$ - it is higher than zero.

*Boson (photon) distribution function* is occurred in 3D space for $D \in \langle 0,1 \rangle$, in 2D and 1D spaces not coupling constant defined. In 3D is this value equal to $\alpha_{\mathrm{w}} = D/(1-D)$ - it is higher than zero.

The similar results can be obtained by analyzing of electrical field with focus to space charge limited currents [8] and entropy of fractal systems [9].

**Acknowledgements.** This work was supported by the project from the Technology Agency of the CR (Grant TA03010548), and from "Centre for Materials Research at FCH BUT" No. CZ.1.05/2.1.00/01.0012 supported by ERDF.

## References

[1] Beiser, A.: Perspectives of Modern Physics. McGraw-Hill, New York (1969)
[2] Maxwell Boltzmann distribution, in Wikipedia, `http://en.wikipedia.org/wiki/Maxwell-Boltzmann_distribution`
[3] Mandelbrot, B.B.: Fractal Geometry of Nature. W. H. Freeman and Co., New York (1983)
[4] Zmeskal, O., Nezadal, M., Buchnicek, M.: Fractal–Cantorian Geometry, Hausdorff Dimension and the Fundamental Laws of Physics. Chaos, Solitons & Fractals 17, 113–119 (2003)
[5] Zmeskal, O., Nezadal, M., Buchnicek, M.: Field and potential of fractal–Cantorian structures and El Naschie's $\varepsilon(\infty)$ theory. Chaos, Solitons & Fractals 19, 1013–1022 (2004)
[6] Zmeskal, O., Buchnicek, M., Vala, M.: Thermal Properties of bodies in fractal and cantorian physics. Chaos, Solitons & Fractals 25, 941–954 (2005)
[7] Zmeskal, O., Vala, M., Weiter, M., Stefkova, P.: Fractal-cantorian geometry of space-time. Chaos, Solitons & Fractals 42, 1878–1892 (2009)
[8] Zmeskal, O., Nespurek, S., Weiter, M.: Space-charge-limited currents: An E-infinity Cantorian approach. Chaos, Solitons & Fractals 34, 143–156 (2007)
[9] Zmeskal, O., Dzik, P., Vesely, M.: Entropy of fractal systems, Computers and mathematics with applications (2013), doi:10.1016/j.camwa.2013.01.017

# Modelling Queues in Transportation Networks Using P Systems

Zbyněk Janoška[1] and Jiří Dvorský[2]

[1] Department of Geoinformatics, Faculty of Science, Palacký University Olomouc, Třída Svobody 26, 771 46, Olomouc, Czech Republic
zbynek.janoska@google.com
[2] Department of Computer Science, VŠB – Technical university of Ostrava, 17. listopadu 15, 708 33 Ostrava – Poruba, Czech republic
jiri.dvorsky@vsb.cz

**Abstract.** This paper proposes variant of P system for passenger flow modelling in transportation networks. Mobile membranes are used as vehicles, which enable transportation of passengers within the network. Performance of the system is shown on four examples, which examine the queueing mechanisms and queue propagation in transportation networks. Both artificial and real transportation systems are used in simulations and results are discussed.

**Keywords:** P systems, mobile membranes, transportation, queue, Prague Metro.

## 1 Introduction

Transportation has been a prominent topic in membrane computing for several years. P systems with symport/antiport rules were first introduced by [12] and gained immediate attention in the community. The basic idea of symport/antiport rules is to allow transportation of pairs of objects through the membrane, either in the same or in different direction. Great number of research articles were devoted to computational power of such systems (e.g. [1],[2],[5]). P systems with carriers offer different approach to transportation in P systems – objects are communicated through membranes using vehicles [11]. This model however did not gain as much attention as symport/antiport rules. Yet another approach to transportation was presented in [3] as P systems with transportation and diffusion channels. They allow transportation of objects between membranes only if specific channels are present. Unlike in symport/antiport systems, a single object can be communicated through membrane and unlike in P systems with carriers, transported objects are not bound with the vehicle. Recently, an attempt to generalise communication strategies in P systems was presented by [14].

In this paper we focus on modelling transport of passengers in transportation network using mobile membranes as vehicles. Our primary aim is to develop P system based model for passenger flow simulation in public transportation

I. Zelinka et al. (eds.), *Nostradamus 2014: Prediction, Modeling and Analysis of Complex Systems*, Advances in Intelligent Systems and Computing 289,
DOI: 10.1007/978-3-319-07401-6_7, © Springer International Publishing Switzerland 2014

networks and we carry on with research initiated in our previous works [6,9]. In presented model, the communication of objects through membrane structure is achieved using mobile membranes, not symport/antiport or carriers. This formalization enables more expressive representation of modelled phenomena and straightforward design of the model. Proposed model works with two types of membranes. First type of membranes are stations, which are ordered in a graph. Second type of membranes are vehicles, which are communicated between the stations and carry passengers (multisets of objects).

This paper specifically focuses on mechanism of queues in transportation systems and several simulations, which explore the propagation of the queue in network, are presented.

The paper is structured as follows: in section 2, a formal description of model is given, in section 3, queueing mechanisms are examined on four examples. Two examples are using hypothetical network and two are using Prague Metro system, which was subject of simulation in previous work [9]. Section 4 discusses some problems and limitations of proposed model and suggests future direction of research. Section 5 contains short summary.

## 2 Model Description

There exist several levels of traffic modelling, ranging from macro-models, describing traffic flow only in terms of populations of object, to micro-models, where every individual object in the system is examined in detail [7]. In mezo-models, some parts of system are described in detail, while description of other parts is simplistic. Such a model is suitable for focusing on certain part of traffic flow phenomena, while retaining robust and computationally efficient. In passenger flow modelling, meso-scale models enable detailed description of passenger movement and behavior, while omitting unnecessary detailed description of vehicle movement. This class of models is therefore recommended as most suitable for passenger flow modelling [13].

With regards to this recommendation, proposed model is designed to capture detailed behavior of passengers, while flow of vehicles is brief and simplistic.

Real world system consists of several components, which must be represented in terms of P systems. Stations are considered as membranes. Network of stations is represented as a graph, similar to neural P systems [8]. Vehicles are represented by membranes, but unlike classical membranes in P systems, they are mobile - their position in the system changes as the system evolves. This evolution is handled by a set of rules [10]. Finally, the passengers are represented as objects. Their behavior follows set of rules, which change according to the position of passengers (inside vehicle or at station). Two classes of membranes – stations and vehicles – are recognized, with each of them serving different purpose. Stations are ordered in a static graph structure, which is not evolved during computation. Passengers enter and leave the system through stations. Vehicles are mobile membranes (which can be created or dissolved at some of the stations) and serve as the only mean of transportation for passengers. Vehicles can be in two states – stopped or moving, which is denoted using membrane polarization. Passengers can enter or leave vehicles, while they are inside any station and stopped. Movement between stations is not instantaneous and takes some time, therefore all rules have time constant assigned to them and idea of pending rules [4] is employed.

We believe this representation is very expressive – it is easy to see stations in transportation systems as membranes, which are entered and left by passengers – objects. Vehicles serve as membranes too – their function, same as function of living membranes – is to protect its content and serve as a mean of selective transportation (not all objects can enter the membrane and membrane can be entered only on specific occasions). This expressiveness (compared to description using i.e. set of differential equations, commonly used in transportation modelling) together with massive parallelism of model – are two main advantages of P systems for transportation modelling.

Formally, P systems for passenger flow simulation in public transportation networks are following construct:

$$\Pi = (O, l, syn, R), \tag{1}$$

where:

- $O = \{passenger_{\{a,\ldots,b\}}, empty\} | a, b \in \{l - \{vehicle_1, \ldots, vehicle_n\}\}\}$ is a set of objects, where *passenger* represents travelling passenger and *empty* represents empty seat inside a vehicle. To each *passenger*, a sequence $\{a, \ldots, b\}$ is assigned. This sequence is an ordered sequence of all stations, which passenger visits on his route from station $a$ (current station) to station $b$ (end station of an individual route).
- $l = \{1, \ldots, k, vehicle_1, \ldots, vehicle_n\}$ is a set of membranes with $k$ stations and $n$ vehicles.
- $syn \subseteq \{(i, j, t) | i, j \in \{l - \{vehicle_1, \ldots, vehicle_n\}\}, i \neq j, t \in \mathbb{N}\}$ is a set of synapses, representing topology of a network. Each synapse consists of two labels of stations $i, j$ and time $t$ necessary to transport vehicle from station $i$ to station $j$.
- $R$ is a set of rules, which describe the behavior of both types membranes and objects inside them. In next sections, following notation will be used: vehicle going to station $k$ will be denoted by $[_k \ ]_k$, hence $k$ is label of next, not current station. Each vehicle can be in two states - stopped or moving. Membrane polarization is used to distinguish between the two, therefore moving vehicle to station $k$ is denoted as $[_k^+ \ ]_k^+$. Station with label $m$ will be denoted as $(_m \ )_m$. Default time necessary for application of every rule is one time unit. Following set of rules is used to describe the evolution of the system.
  1. $passenger_{\{a,b,\ldots,x\}} \ [_a \ empty \ ]_a^- \rightarrow [_a \ passenger_{\{b,\ldots,x\}} \ ]_a^-$ is rule describing passenger entering a vehicle. Passenger, whose next stop is $a$ enters a vehicle going to station $a$, if there is an empty seat (*empty*) and the vehicle is stopped (negative polarization). Once inside the vehicle, passengers next station changes to $b$.

2. $[_a\ passenger_{\{a,b,\ldots,x\}}\ ]_a^- \rightarrow [_a\ passenger_{\{b,\ldots,x\}}\ ]_a^-$ is rule describing passenger staying inside a vehicle. Passenger, whose next stop is $a$ and who is already in a vehicle going to $a$, stays inside and his next stop changes to $b$.
3. $[_a\ passenger_{\{\}}\ ]_a^- \rightarrow [_a\ empty\ ]_a^-$ describes situation, where passenger inside a vehicle has no next station, hence is in his final destination and leaves the system. An *empty* object is created inside a vehicle.
4. $[_a\ passenger_{\{b,c,\ldots,x\}}\ ]_a^- \rightarrow [_a\ empty\ ]_a^-\ passenger_{\{b,c,\ldots,x\}}$ describes passenger leaving the vehicle at transfer station. If passenger, whose next stop is $b$, is inside vehicle going to $a$, he leaves the vehicle and *empty* seat appears inside a vehicle. The passenger stays at the current station.
5. $(_i\ [_j\ \ ]_j^+\ )_i \xrightarrow{t} (_j\ [_k\ \ ]_k^-\ )_j$ is rule describing movement of vehicles inside a network. Moving vehicle at station $i$, whose next station is $j$, is moved to station $j$, its next station is changed to $k$ and the vehicle stops. Application of the rule takes $t$ time units.
6. $(_i\ [_j\ \ ]_j^-\ )_i \rightarrow (_i\ [_j\ \ ]_j^+\ )_i$ changes vehicle from stopped to moving state.
7. $(_i\ [\ empty^n\ ]^-\ )_i \rightarrow (_i\ \ )_i$ is rule describing situation, where vehicle reaches its final destination (i.e. does not have next stop). Such a vehicle is removed from the system (membrane is dissolved). *empty* objects inside the membrane are removed from the system.
8. $(_i\ \ )_i \rightarrow (_i\ [_a\ empty^n\ ]_a^-\ )_i$ is rule describing generation of vehicles at start stations – vehicle going to station $a$ is created at station $i$. Inside a vehicle, there is $n$ empty seats. The vehicle is stopped, therefore passengers can enter the vehicle immediately.
9. $(_i\ \ )_i \rightarrow (_i\ passenger_{\{a,b,\ldots,x\}}\ )_i$ describes arrival of passenger to the station $i$. For each passenger, who arrives at the station, a sequence of stations to visit $a, b, \ldots, x$ is generated.

Schedule of vehicles is explicitly given for each of the start station. Global clock for the whole system is assumed, and schedule for station is an array of integers, denoting the time units from beginnining. Each time a global clock reaches time in the schedule, new vehicle is created in the station. Distances between stations are part of the membrane system definition, not part of the schedule. One minute is used as a time unit in following simulations. The system can not regulate the schedule of vehicles or numbers of generated passengers during the computation – both schedule and membrane system are static.

Rules 1 to 4 and rule 6 are applied in the same computational step, with rule 4 having higher priority than rules 1 and 4. Priority relation exists between rules 1, 4 and 6: **Rule 4 > Rules 1, 6**. Prioritizing rule 4 allows all passengers to first leave the vehicle and create empty seats, and than applying rules 1 (passengers entering the vehicle) and 6 (changing the state of vehicle from stopped to moving). Rules 2 (passengers staying inside vehicle) and 3 (passengers getting out of vehicle at end stations) have no influence on applicability of other rules and are applied in the same computational step.

Number $n$ of empty seats inside a newly created vehicle is set to 500 in both testing network and Prague Metro network.

# 3 Results

Four simulations were conducted in order to test the performance of the model. First two simulations use simple test network (see Fig. 1) and two usc real network of Prague Metro system (see Fig. 4). Aim of the simulations is to create queues and to observe their propagation through the network. In all four scenarios, following methodology is employed: the topology of stations is given and static. The generation of vehicles is governed by a schedule, which is *a priori* known, as well as the numbers of passengers, generated at stations at given time. The queues are created by inserting increased number of passengers to one of the stations. The propagation of queues in network is observed with emphasis on the behavior at transfer stations. From practical point of view, it is important to know, whether a queue can "transfer" to different line, or whether it stays on the line, where it was created.

The R project for statistical computing was used to perform simulations. Several simulation were performed, however figures below show only one randomly chosen simulation, since averaging leads to smoothed pattern, which does not correspond with the behavior of every single simulation run.

## 3.1 Random End Station in Test Network

Test network consists of 17 stations, which are distributed along two lines – lines $A$ and $B$ both have 7 stations with station in the middle being transfer station. Distance between all consequent stations is 2 time units. Vehicles are moving along lines $A$ and $B$ in both directions, which creates total of four lines: *A-original*, *A-reverse*, *B-original*, *B-reverse*. The sequence of visited stations is *A1 – A2 – A3 – A4 – Transfer station – A5 – A6 – A7 – A8* for line *A-original*, reverse order of stations for line *A-reverse* and similarly for lines *B-original* and *B-reverse*. Capacity of vehicles is 500 seats and new vehicle is created at end stations every four time units. At every time unit, $m$ passengers are generated at every station, where $m$ is randomly taken from Poisson distribution with mean 20. End stations for each generated passenger are randomly chosen from all other stations in the system.

First simulation explores scenario, where at time step 500, 10 000 passengers appears at station *A1*. End stations are generated randomly for all 10 000 passengers. Capacity of the vehicles is not sufficient to take in all the passengers and as a result, the queue emerges. Results for selected stations are in Fig. 2.

Prior to insertion of passengers, the behavior of the system is cyclic and stable. There are no queues emerging at any station. After insertion of 10 000 passengers at station *A1*, a massive queue is instantly created. This queue is dissolved during approximately 100 steps and the dissolution is linear. At station *A2*, the queue is created gradualy, and its dissolution is very quick (16 steps). Time between the creation and dissolution of the queue is slightly greater than at station *A1*.

At *transfer station*, the character of queue changes. Instead of increased numbers of passengers at station, a greater variability in their numbers is observable. Dissolution of the queue takes almost 150 time units. After the *transfer station*,

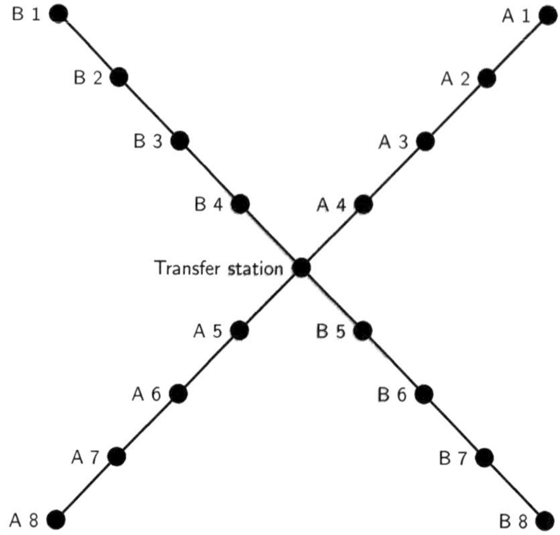


**Fig. 1.** Testing network

the queue dissolves at all stations in all directions. The behavior is stable and cyclic at all stations (see station *A5* in Fig. 2).

### 3.2 One End Station in Test Network

Second simulation uses the same network as first simulation and the only difference is, that all 10 000 passengers generated at station *A1* have end station *A8* (at first simulation, the stations were generated randomly).

Results at stations *A1* and *A2* are identical as in first simulation (see Fig. 3), but character of the queue at *transfer station* is different. Queue with the shape of turned over letter V emerges. Length of the queue is 170 time units and its dissolution is linear, once it reaches its peak. At station A5, the queue is still present, however the numbers of waiting passengers are substantially lower (up to 150 passengers, compared to 3 000 passengers at transfer station). Queue is still observable at station *A6*, however at station *A7*, it is dissolved. At lines *B* (stations *B4* and *B5*), the queue was not observed.

In both simulations with test network, a propagation of the queue was observed. In case of randomly chosen end stations, the queue propagated from station A1 to transfer station, where it dissolved. In case of all 10 000 passengers travelling from station *A1* to station *A8*, the queue was still present after transfer station, however the numbers of waiting passengers were lower. In neither scenario, transfer of the queue to another line was observed. This may be due to randomly generated end stations for passengers from other stations. Even if fully occupied vehicle enters stations *B5* or *B4*, the numbers of passengers travelling from these stations to outer part of the network are on average lower than numbers of passengers traveling from other parts of the network to

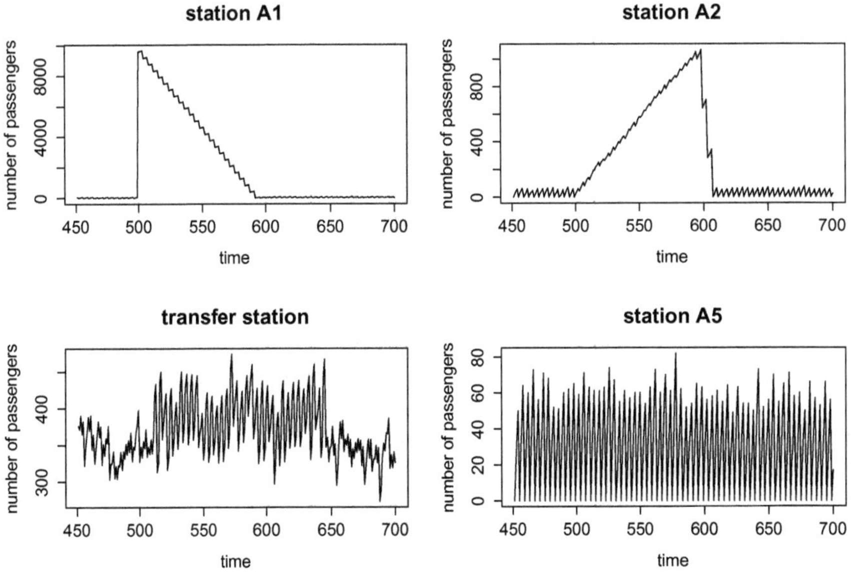


**Fig. 2.** Queues at selected stations for simulation 1

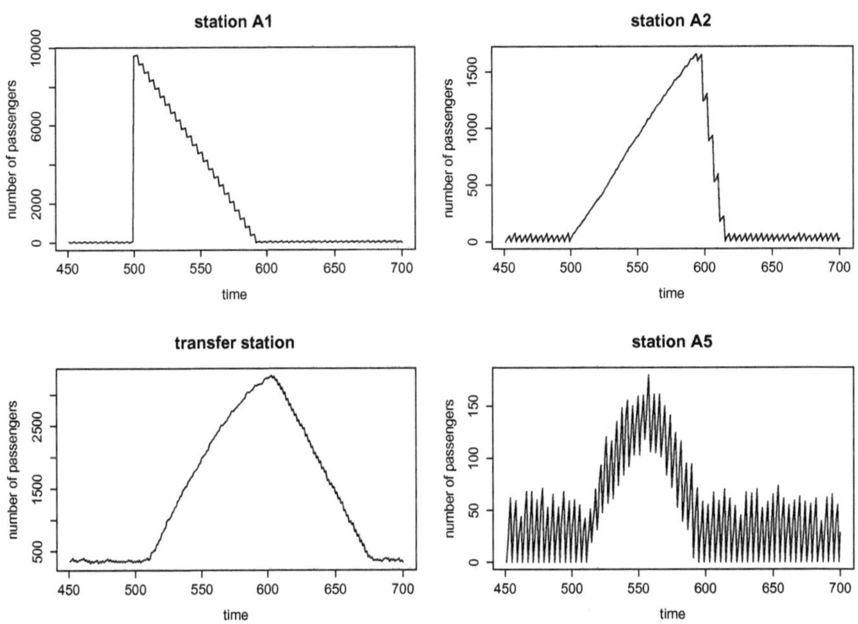


**Fig. 3.** Queues at selected stations for simulation 2

these stations (in these specific scenarios the number of passengers travelling to these stations is on average 4.3 times bigger than the number of passengers travelling from these stations to outer part of the system).

### 3.3 Simulations in Prague Metro system

Prague Metro system consists of three lines, labeled A (Dejvická – Depo Hostivař), B (Zličín – Černý Most) and C (Letňany – Háje), which intersect at three stations (A–B - Můstek, A–C - Muzeum, B–C - Florenc). Prague Metro consists of 53 stations in total and approximately 1.5 millions of people are transported every day. Trains are in service from approximately 4:40 a.m. to 24 a.m. (midnight) and their frequency changes in time. In five-year periods, survey of passenger occupancy is performed with last survey taking place at 2008. Prague Public Transit Company provided detailed data about transit intensity, which were used in this case study. Current (April 2013) schedule of trains was used as basis for generation of trains at start stations. Estimation of passenger flow during the day and preliminary results were published in [9].

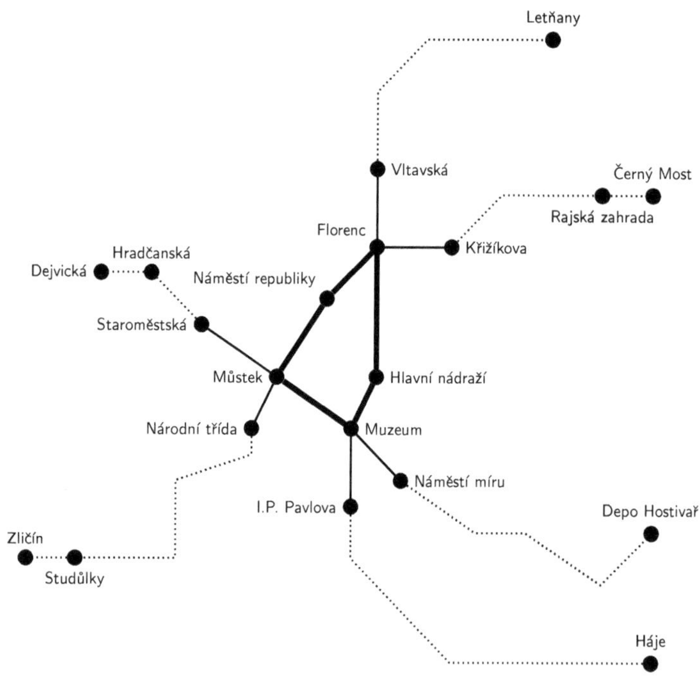


**Fig. 4.** Prague metro – schematic map

Both simulations explored situation, when increased number of passengers travel from one part of a system to another. In first case, we inserted a huge number of passengers to station Zličín and set their end station as Černý Most (passengers were transported through line B). In second simulation we explored

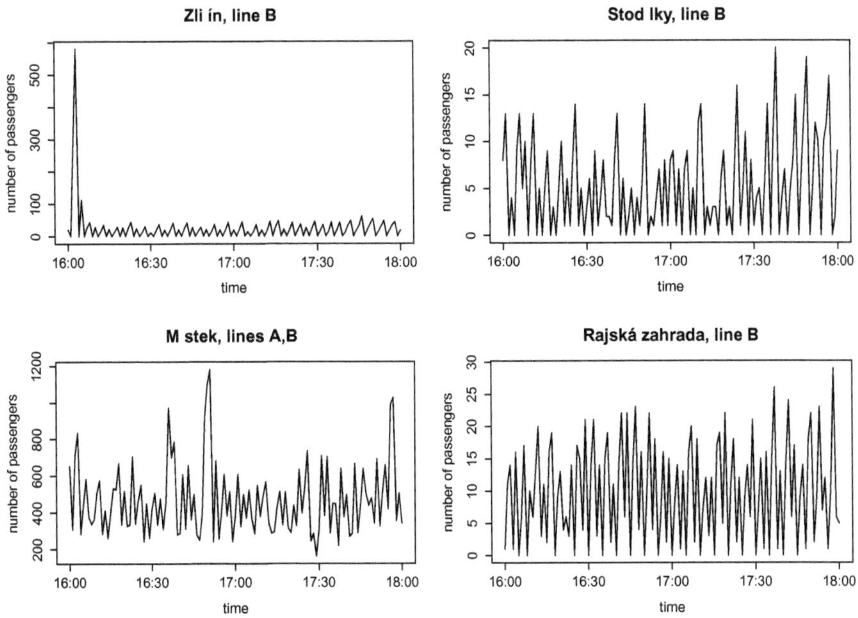


**Fig. 5.** Insertion of 1 000 passengers to Zličín station

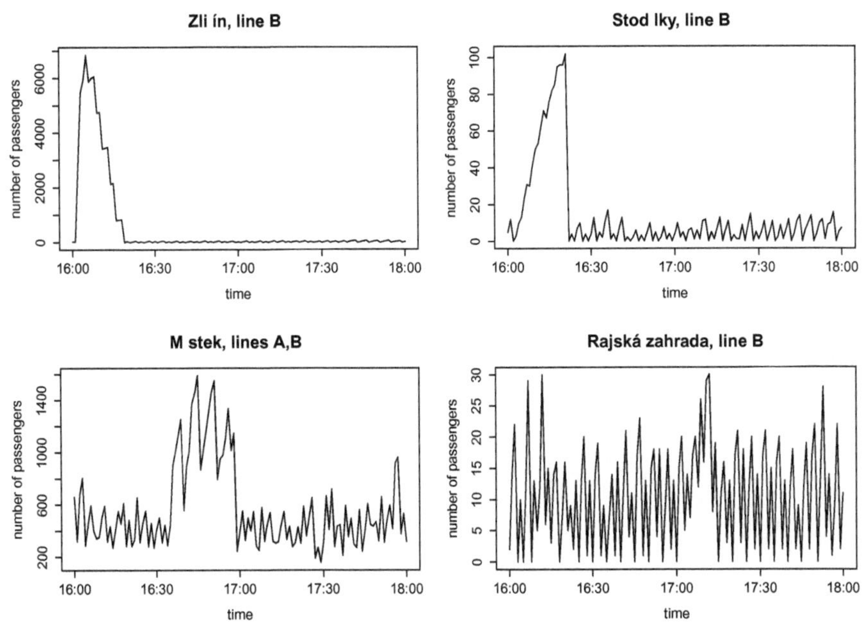


**Fig. 6.** Insertion of 10 000 passengers to Zličín station

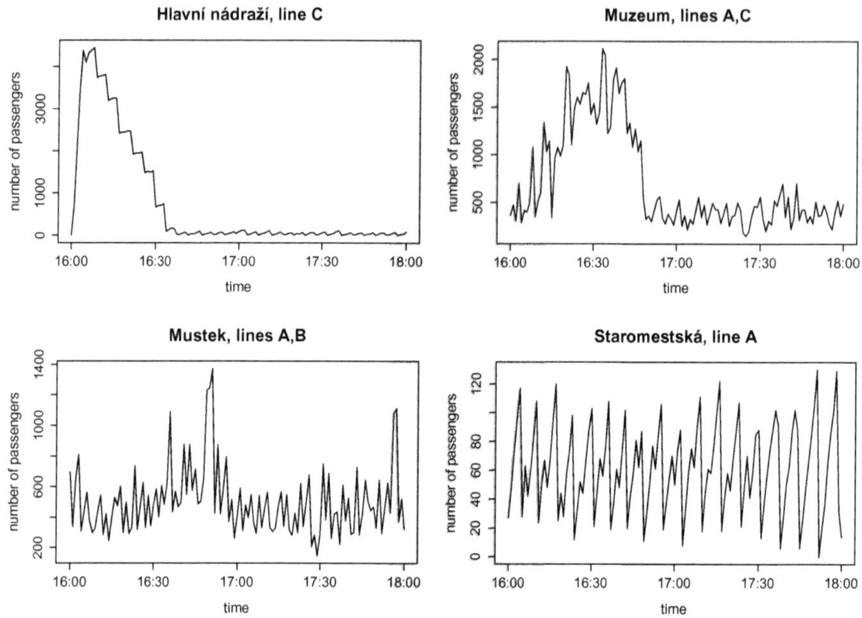


**Fig. 7.** 5 000 fans going to the game

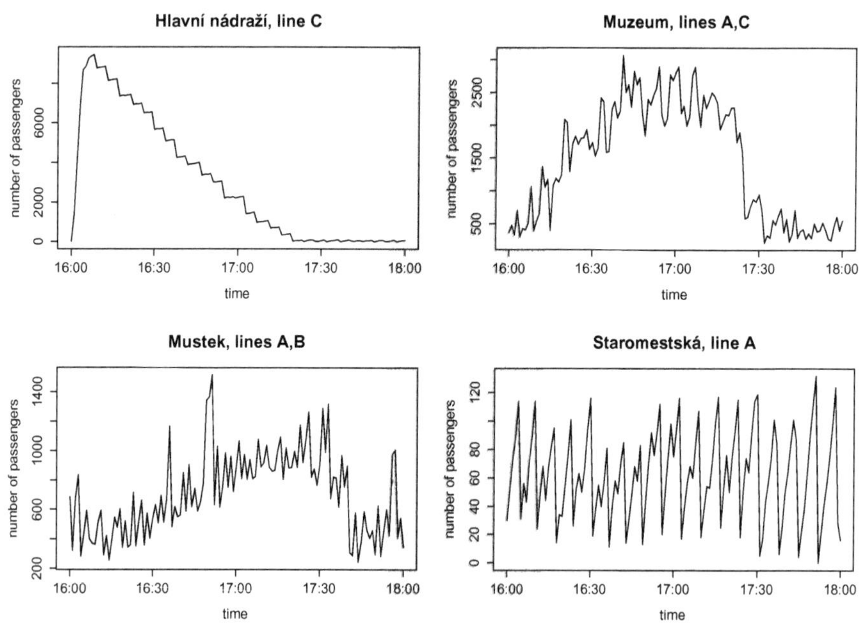


**Fig. 8.** 10 000 fans going to the game

somehow more realistic situation of football fans traveling from Prague main train station (Hlavní Nádraží on map) to Hradčanská station near the football stadium.

In first simulation, we examined queueing behavior in cases when 1 000 and 10 000 passengers are inserted to Zličín station (see Figs. 5 and 6). Passengers were introduced to the station at 16:00 and their end station was set to Černý Most, therefore they were about to travel through whole line B. In case when only 1 000 passengers were introduced to the system, the queue was dissolved during 5 minutes and no propagation of the queue was observed, not even on station immediately after the first station (Fig. 5 top right).

In case of 10 000 passengers inserted into the system, a massive queue propagated through the whole system. On second station, only up to 100 passengers were waiting in the queue, but at Můstek station, which is transfer station between lines A and B, there were almost 1 500 passengers in the queue and its dissolution took 25 minutes (at the beginning, the queue was only 20 minutes long). Last station, where the queue was observable, was Rajská zahrada station, which is 23th and penultimate station on line B. In correspondence with results on test network, transfer of the queue to other lines of the system was not observed. Maximal number of passengers waiting in the queue was 1500 at Můstek station, and the queue was observable yet one hour after it was created. Another simulation where 5 000 passengers were introduced into the system was performed. The queue had also up to 1 500 passengers at Můstek station and was was observable up to 18th station on the B line (results are not show here for the sake of brevity).

The second scenario explored the situation, when increased number of football fans travels from Hlavní Nádraží (main train station) to Hradčanská station near football stadium. 1 000, 5 000 and 10 000 passengers were introduced into the system during 10 minutes long interval. This scenario is somehow different from others, because passengers change trains at Muzeum station (main train station is on line C, Hradčanská station is on line A).

Figures 7 and 8 show results of simulation for 5 000 and 10 000 fans. In case of 1 000 fans, a small queue is observed on the second station, but is dissolves during 5 minutes and does not appear in any other station.

In case of 5 000 fans, the queue propagates to Muzeum station (transfer station) and then on new line. Increase of the number of passengers can be seen at first three station (Hlavní Nádraží, Muzeum and Můstek). At station Staroměstská, there is no sign of queue, as well as no queue is observable at any other station on either of the lines. The same applies for situation, when 10 000 fans are introduced to the system., but on a greater scale. The shape of the queue changes as it propagates. At first station, it grows rapidly and dissolves slowly, at later stations the growth slows down and the dissolution is quicker. Again, the queue is restricted to the route from main train station to the football stadium and can not be seen at other lines.

## 4 Discussion

Proposed model is designed to simulate transportation of passengers in transportation networks, however can be seen in a more general way as a model of transportation of objects inside a network. Proposed methodology differs from ideas already known from P systems. P systems with symport/antiport rules do not incorporate the idea of vehicles, which carry the objects between membranes, P systems with carriers consider these vehicles to be objects, while more realistic representation (with respect to transportation modelling) is to consider them as membranes, which is approach presented in this paper. Primary aim of this article is to simulate creation and propagation of queues inside networks and computational power of proposed model is not examined. Main disadvantage from computational point of view is need to store route for each individual passenger. This is to make the model more realistic – passengers in public transportation systems have *a priori* known origin and destination stations as well as route which they need to take. If the model were to be used to examine flows in network in general (like in the first two simulations in test network), different approach could be used to set end stations of passengers (i.e. application of rules describing passenger behavior – leaving the vehicle, staying in the vehicle or changing vehicles – could be probabilistic). For the application in passenger flow simulation, proposed model of assigning route to each passenger at the beginning of his route seems to be the most expressive.

The model considers only two types of behavior of passengers – getting on and off the train. Other types of behavior like waiting, changing the destination or route during the trip etc. are not considered. More types of objects, which are transported, can be included, but for current application the design of model with one type of transported object is sufficient. Graph structure of stations is static, but dynamic feature can be added through alternation of costs (i.e. time intervals) between stations.

For real world applications, the crucial problem is the estimation of traffic flow between the stations and its variation in time. In case of Prague Metro, the flow between the stations is known, however the variation in time is not and was estimated from train schedule (more on this topic in [9]). The results of simulations in Prague Metro should therefore be taken with caution.

Unfortunately, it is not possible to compare the results of the simulation to the real flow in Prague Metro, since only total number of passengers travelling from station A to station B is known and not the distribution of the flow in time. Simulations in both networks had the aim to examine queueing mechanisms in networks and specifically the queue propagation in network. Queues were created by adding great number of objects at one of the stations, but different approach could be taken. One possibility would be to slowly increase the numbers of generated passengers over time, another to create queues at transport stations, where flows from several directions meet. Results of simulations show two important findings: the shape and length of the queue changes as it propagates; and the queues do not transfer to other lines. Especially the second finding has a practical value, because increased traffic on certain line does not require increased vehicle frequency on other lines. These findings are however restricted to the two networks, which were examined and should not be generalized.

## 5 Summary

This paper proposes model for simulation of passenger flow in public transportation networks. Focus is on application of the model for real world phenomena and not on computational power. The model uses different formalization of transport than currently used variants of P systems and combines mobile membranes with structuring membranes in a graph.

Four simulations are performed, two in test network with random flow of passengers and two in Prague Metro system using empirical data of traffic flow. Simulations explore creation and propagation of queues inside a network. Observed behavior in tested networks shows, that queues do not transfer to other parts of network on transfer stations, however this behavior should not be generalized.

Proposed model is robust and simple and could be a valuable tool for exploration of traffic flow in networks, not only public transportation systems, but in any network, where objects are transported using vehicles with given schedules.

**Acknowledgement.** This work was supported by the project CZ.1.07/2.2.00/28.0078 "InDOG" - Innovation of PhD Geoinformatics and Cartography study with support of modern technological trends which is co-financed from European Social Fund and State financial resources of the Czech Republic.

This work was also supported by the European Regional Development Fund in the IT4Innovations Centre of Excellence project (CZ.1.05/1.1.00/02.0070) and by the Development of human resources in research and development of latest soft computing methods and their application in practice project, reg. no. CZ.1.07/2.3.00/20.0072 funded by Operational Programme Education for Competitiveness, co-financed by ESF and state budget of the Czech Republic.

## References

1. Alhazov, A., Freund, R., Rogozhin, Y.: Computational Power of Symport/Antiport: History, Advances, and Open Problems. In: Freund, R., Păun, G., Rozenberg, G., Salomaa, A. (eds.) WMC 2005. LNCS, vol. 3850, pp. 1–30. Springer, Heidelberg (2006)
2. Alhazov, A., Margenstern, M., Rogozhin, V., Rogozhin, Y., Verlan, S.: Communicative P Systems with Minimal Cooperation. In: Mauri, G., Păun, G., Jesús Pérez-Jímenez, M., Rozenberg, G., Salomaa, A. (eds.) WMC 2004. LNCS, vol. 3365, pp. 161–177. Springer, Heidelberg (2005)
3. Barbuti, R., Maggiolo-Schettini, A., Milazzo, P., Tini, S.: P systems with Transport and Diffusion Channels. Fundamenta Informaticae XX, 1–15 (2009)
4. Barbuti, R., Maggiolo-Schettini, A., Milazzo, P., Tesei, L.: Timed P Automata. Electronic Notes in Theoretical Computer Science 227, 21–36 (2009)

5. Cavaliere, M.: Evolution–Communication P Systems. In: Păun, G., Rozenberg, G., Salomaa, A., Zandron, C. (eds.) WMC 2002. LNCS, vol. 2597, pp. 134–145. Springer, Heidelberg (2003)
6. Dvorský, J., Janoška, Z., Vojáček, L.: P Systems for Traffic Flow Simulation. In: Cortesi, A., Chaki, N., Saeed, K., Wierzchoń, S. (eds.) CISIM 2012. LNCS, vol. 7564, pp. 405–415. Springer, Heidelberg (2012)
7. Hoogendoorn, S.P., Bovy, P.H.L.: State-of-the-art of Vehicular Traffic Flow Modelling. Delft University of Technology, Delft (2001)
8. Ionescu, M., Păun, G., Yokomori, T.: Spiking Neural P Systems. Fundamenta Informaticae 71(2,3), 279–308 (2006)
9. Janoška, Z., Dvorský, J.: P system based model of passenger flow in public transportation systems: a case study of Prague Metro. Presented at 13th Annual International Workshop on Databases, Texts, Specifications, and Objects, Písek, Czech Republic, April 17-19 (2013)
10. Krishna, S.N., Păun, G.: P Systems with Mobile Membranes, pp. 279–308. Kluwer Academic Publishers, Hingham (2005)
11. Martín-Vide, C., Păun, G., Rozenberg, G.: Membrane systems with carriers. Theoretical Computer Science 270(1,2), 779–796 (2002)
12. Păun, A., Păun, G.: The power of communication: P systems with symport/antiport. Journal of Computer and System Sciences 20(3), 295–305 (2002)
13. Peeta, S., Ziliaskopoulos, A.: Foundations of Dynamic Traffic Assignment: The Past, the Present and the Future. Networks and Spatial Economics 1(1/4), 233–266 (2001)
14. Verlan, S., Bernardini, F., Gheorghe, M., Margenstern, M.: Generalized communicating P systems. Theoretical Computer Science 404(1,2), 170–184 (2008)

# Tuning the Lozi Map in Chaos Driven PSO Inspired by the Multi-chaotic Approach

Michal Pluhacek[1], Roman Senkerik[1],
Ivan Zelinka[2], and Donald Davendra[2]

[1] Faculty of Applied Informatics
Tomas Bata University in Zlin
T.G. Masaryka 5555, 760 01 Zlin, Czech Republic
{pluhacek,senkerik}@fai.utb.cz
[2] Faculty of Electrical Engineering and Computer Science
VŠB-Technical University of Ostrava
17. listopadu 15, 708 33 Ostrava-Poruba, Czech Republic
{ivan.zelinka,donald.davendra}@vsb.cz

**Abstract.** In this paper a previous successful research on chaos enhanced particle swarm optimization algorithm (PSO) is expanded. The possibility of adaptive change of control parameters of chaotic systems that is used as a pseudo-random number generator for the velocity calculation in PSO algorithm is investigated. To evaluate the performance of newly designed algorithm the CEC´13 benchmark set was used.

**Keywords:** Particle swarm optimization, chaos, Lozi map, PSO, Evolutionary algorithm.

## 1 Introduction

The Particle Swarm Optimization algorithm (PSO) [1 - 4] is one of the most widely used Evolutionary Computation Techniques (ECT's). In past years it was proposed that the implementation of chaotic sequences as chaotic pseudo-random number generators (CPRNG's) could significantly improve the performance of ECT's such as PSO on many optimization tasks [5 -10]. In this research it is examined the effect of different parameter setting of Lozi chaotic map on the CPRNG and subsequently on the performance of chaos enhanced PSO algorithm. Furthermore the tuned CPRNG is implemented into the previously designed adaptive multi-chaotic approach [9] as the second CPRNG. The performance of the proposed algorithm is tested at the CEC´13 benchmark set.

## 2 Particle Swarm Optimization Algorithm

A brief description of PSO algorithm follows in this section. The PSO algorithm takes inspiration from the natural swarm behavior of birds and fish. It was firstly introduced

I. Zelinka et al. (eds.), *Nostradamus 2014: Prediction, Modeling and Analysis of Complex Systems*, Advances in Intelligent Systems and Computing 289,
DOI: 10.1007/978-3-319-07401-6_8, © Springer International Publishing Switzerland 2014

by Eberhart and Kennedy in 1995 [1]. Each particle in the population represents a candidate solution for the optimization problem that is defined by the cost function (CF). In each iteration of the algorithm, a new location (combination of CF parameters) for the particle is calculated based on its previous location and velocity vector (velocity vector contains particle velocity for each dimension of the problem).

According to the method of selection of the swarm or subswarm for best solution information spreading, the PSO algorithms are noted as global PSO (GPSO) or local PSO (LPSO). Within this research the PSO algorithm with global topology (GPSO) [6] was utilized. The chaotic PRNG is used in the main GPSO formula (1), which determines a new "velocity", thus directly affects the position of each particle in the next iteration.

$$v_{ij}^{t+1} = w \cdot v_{ij}^{t} + c_1 \cdot Rand \cdot (pBest_{ij} - x_{ij}^{t}) + c_2 \cdot Rand \cdot (gBest_j - x_{ij}^{t}) \tag{1}$$

Where:
$v_i^{t+1}$ - New velocity of the *ith* particle in iteration $t$+1.
$w$ – Inertia weight value.
$v_i^{t}$ - Current velocity of the *ith* particle in iteration $t$.
$c_1$, $c_2$ - Priority factors
$pBest_i$ – Local (personal) best solution found by the *ith* particle.
*gBest* - Best solution found in a population.
$x_{ij}^{t}$ - Current position of the *ith* particle (component $j$ of dimension $D$) in iteration $t$.
*Rand* – Pseudo random number, interval (0, 1). CPRNG is applied only here.

The maximum velocity was limited to 0.2 times the range as it is usual. The new position of each particle is then given by (2), where $x_i^{t+1}$ is the new particle position:

$$x_i^{t+1} = x_i^{t} + v_i^{t+1} \tag{2}$$

Finally the linear decreasing inertia weight [3, 4] is used in the typically referred GPSO design that was used in this study. The dynamic inertia weight is meant to slow the particles over time thus to improve the local search capability in the later phase of the optimization. The inertia weight has two control parameters wstart and wend. A new w for each iteration is given by (3), where t stands for current iteration number and n stands for the total number of iterations. The values used in this study were wstart = 0.9 and wend = 0.4.

$$w = w_{start} - \frac{((w_{start} - w_{end}) \cdot t)}{n} \tag{3}$$

## 3 Lozi Chaotic Map

The Lozi map is a simple discrete two-dimensional chaotic map. The map equations are given in (4). The typical parameter values are: $a$ = 1.7 and $b$ = 0.5 with the respect to [12]. For these values, the system exhibits typical chaotic behavior and with this parameter setting it is used in the most research papers and other literature sources. This setting was also used in the previous research [8 - 10]. The $x$, $y$ plot of Lozi map with the aforementioned typical setting is depicted in Fig. 1 (left), whereas the Fig. 1 (right) shows the distribution of CPRNG based on the Lozi map. Random sequence sample of the CPRNG is presented in Fig. 2.

$$\begin{aligned} X_{n+1} &= 1 - a\left|X_n\right| + bY_n \\ Y_{n+1} &= X_n \end{aligned} \tag{4}$$

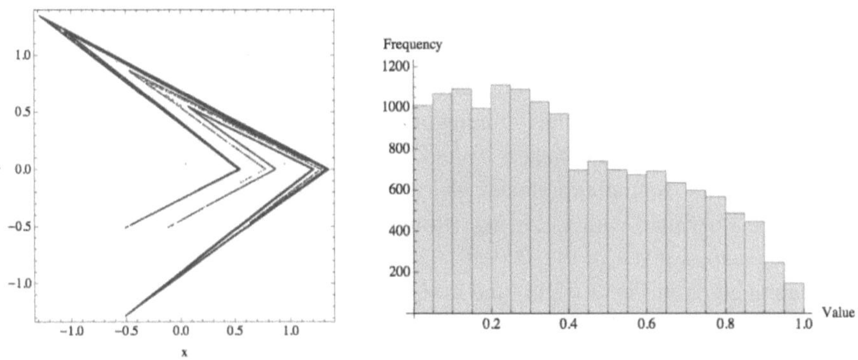


**Fig. 1.** (left) $x,y$ plot of the Lozi map; (right) CPRNG based on Lozi map – distribution histogram transferred into the range ⟨0, 1⟩ (15000 samples)

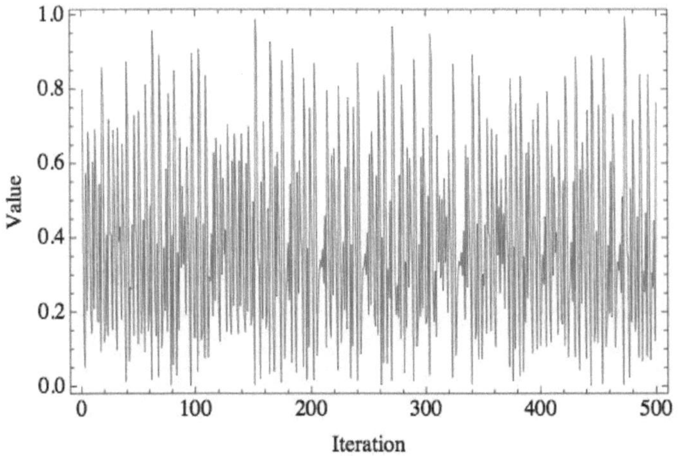


**Fig. 2.** Random sequence sample – Lozi map based CPRNG

## 4 Tuning Experiment

In the first experiment, the impact of different control parameters on the performance of chaos enhanced PSO was examined. The goal was to improve the performance on multi-modal problems. For this reason the Rotated Schwefel's Function (*f(15)*) from the CEC`13 Benchmark set [12] was used with dimension setting 10. Number of particles was set to 40 and number of iterations was 2500. The Lozi map parameters were changed with the step of 0.05 in following way: parameter $a$: from 1.3 to 1.7; parameter $b$: from 0.1 to 0.6. For the each combination 100 repeated runs of the chaos-embedded PSO algorithm were performed.

For some parameter setting the Lozi map does no exhibit expected chaotic behavior or the dynamics is reduced only to two-value sequence (bifurcation). However with respect to [13] all possible combinations in given range were tested.

Summary of the mean results of tuning experiment is given in Table 1. The best result and corresponding parameters values are highlighted by bold. Based on the results of the first tuning experiment, the updated (tuned) CPRNG was constructed utilizing the Lozi map with following setting: $a$ = 1.5 and $b$ = 0.45. The $x$,$y$ plot of tuned Lozi map is given in Fig. 3 (left). The distribution of CPRNG based on tuned Lozi map is given in Fig. 3 (right) and sample sequence of tuned CPRNG is depicted in Fig. 4.

**Table 1.** Tuning experiment - mean results for *f(15)*;100 runs

| a/b | 0.1 | 0.15 | 0.2 | 0.25 | 0.3 | 0.35 | 0.4 | **0.45** | 0.5 | 0.55 | 0.6 |
|---|---|---|---|---|---|---|---|---|---|---|---|
| 1.3 | 911 | 830 | 704 | 1344 | 1638 | 1569 | 1383 | 1389 | 1467 | 1356 | 1402 |
| 1.35 | 851 | 820 | 716 | 665 | 1057 | 1497 | 1383 | 1437 | 1481 | 1391 | 1285 |
| 1.4 | 753 | 703 | 689 | 676 | 655 | 838 | 1397 | 1416 | 1424 | 1229 | 1325 |
| 1.45 | 658 | 715 | 724 | 713 | 547 | 617 | 593 | 1214 | 1373 | 1276 | 1270 |
| **1.5** | 703 | 760 | 757 | 731 | 671 | 749 | 701 | **534** | 1203 | 1171 | 1217 |
| 1.55 | 725 | 732 | 699 | 752 | 702 | 713 | 750 | 762 | 625 | 1118 | 1204 |
| 1.6 | 679 | 826 | 737 | 701 | 643 | 807 | 657 | 643 | 692 | 745 | 1066 |
| 1.65 | 761 | 680 | 756 | 681 | 760 | 798 | 715 | 706 | 812 | 708 | 685 |
| 1.7 | 615 | 771 | 752 | 715 | 696 | 696 | 774 | 724 | 747 | 701 | 690 |

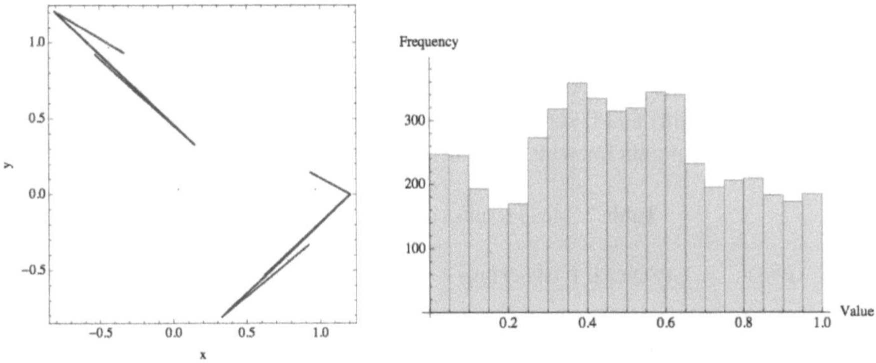


**Fig. 3.** (left) $x$,$y$ plot of the updated Lozi map; (right) CPRNG based on updated Lozi map – distribution histogram transferred into the range ⟨0, 1⟩ (15000 samples)

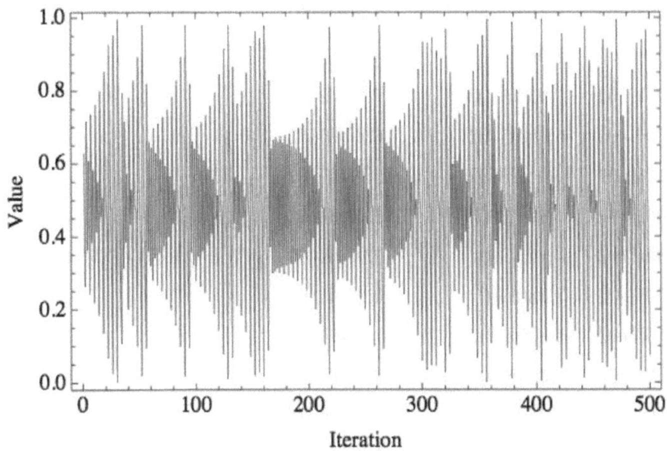


**Fig. 4.** Random sequence sample – Lozi map based CPRNG - tuned

# 5 Performance Experiment 1

Within this first performance experiment, three different PSO variants based on the original global PSO with linear decreasing inertia weight [1-4] were used:

- GPSO with canonical pseudo random number generator – noted GPSO
- GPSO with CPRNG based on Lozi map (a = 1.7, b =0.5) – noted GPSO Lozi 1
- GPSO with CPRNG based on Lozi map (a = 1.5, b =0.45) – noted GPSO Lozi 2

As aforementioned the CPRNG is used only for the velocity calculation (See (1)). The CEC`13 benchmark set [20] was evaluated for the dimension setting 10 with population size 40 and number of iterations 2500. Results are summarized and compared in

Table 2. The bold numbers represents the best mean results. Furthermore the performance of pairs of algorithms is compared, where 1 stands for "win" of the "algorithm 1" (the first from the pair - left); number 2 stand for "win" of algorithm 2 (the second from the pair - right) and 0 stands for draw. The final score is also given in Tables 1 and 2 as a sum of points for wins (1 point) and draws (0.5 point).

**Table 2.** Mean results comparison – Experiment 1

| *f(x)* | GPSO | GPSO Lozi 1 | GPSO Lozi 2 | A1 vs. A2 | A1 vs. A3 | A2 vs. A3 |
|---|---|---|---|---|---|---|
| *f(1)* | **-1.400E+03** | **-1.400E+03** | **-1.400E+03** | 0 | 0 | 0 |
| *f(2)* | 1.599E+05 | 1.750E+05 | **7.729E+04** | 1 | 2 | 2 |
| *f(3)* | **1.739E+06** | 7.130E+06 | 4.203E+06 | 1 | 1 | 2 |
| *f(4)* | -7.450E+02 | **-8.150E+02** | -2.130E+02 | 2 | 1 | 1 |
| *f(5)* | **-1.000E+03** | **-1.000E+03** | **-1.000E+03** | 0 | 0 | 0 |
| *f(6)* | -8.890E+02 | -8.900E+02 | **-8.920E+02** | 2 | 2 | 2 |
| *f(7)* | **-7.970E+02** | -7.950E+02 | -7.960E+02 | 1 | 1 | 2 |
| *f(8)* | **-6.800E+02** | **-6.800E+02** | **-6.800E+02** | 0 | 0 | 0 |
| *f(9)* | **-5.970E+02** | -5.960E+02 | **-5.970E+02** | 1 | 0 | 2 |
| *f(10)* | **-5.000E+02** | -4.990E+02 | **-5.000E+02** | 1 | 0 | 2 |
| *f(11)* | **-3.980E+02** | -3.970E+02 | **-3.980E+02** | 1 | 0 | 2 |
| *f(12)* | -2.870E+02 | -2.880E+02 | **-2.910E+02** | 2 | 2 | 2 |
| *f(13)* | -1.810E+02 | -1.820E+02 | **-1.880E+02** | 2 | 2 | 2 |
| *f(14)* | 1.420E+02 | **1.200E+02** | 1.600E+02 | 2 | 1 | 1 |
| *f(15)* | 5.960E+02 | 7.280E+02 | **5.790E+02** | 1 | 2 | 2 |
| *f(16)* | **2.010E+02** | **2.010E+02** | **2.010E+02** | 0 | 0 | 0 |
| *f(17)* | 3.140E+02 | **3.130E+02** | 3.150E+02 | 2 | 1 | 1 |
| *f(18)* | 4.340E+02 | **4.220E+02** | 4.260E+02 | 2 | 2 | 1 |
| *f(19)* | **5.010E+02** | **5.010E+02** | **5.010E+02** | 0 | 0 | 0 |
| *f(20)* | **6.020E+02** | 6.030E+02 | **6.020E+02** | 1 | 0 | 2 |
| *f(21)* | **1.088E+03** | 1.093E+03 | 1.100E+03 | 1 | 1 | 1 |
| *f(22)* | **1.000E+03** | 1.077E+03 | 1.049E+03 | 1 | 1 | 2 |
| *f(23)* | 1.656E+03 | 1.660E+03 | **1.649E+03** | 1 | 2 | 2 |
| *f(24)* | **1.205E+03** | 1.210E+03 | 1.207E+03 | 1 | 1 | 2 |
| *f(25)* | 1.307E+03 | 1.309E+03 | **1.305E+03** | 1 | 2 | 2 |
| *f(26)* | 1.355E+03 | 1.353E+03 | **1.333E+03** | 2 | 2 | 2 |
| *f(27)* | **1.632E+03** | 1.637E+03 | 1.637E+03 | 1 | 1 | 0 |
| *f(28)* | 1.747E+03 | **1.745E+03** | 1.762E+03 | 2 | 1 | 1 |
| Best: | 11 | 10 | 13 | Score: 16.5 : 11.5 | 14.5 : 13.5 | 9.0 : 19.0 |

According to results presented in Table 2, the PSO with tuned Lozi map based CPRNG (GPSO Lozi 2) managed to obtain the highest number of the best mean results (13) and in direct comparison outperformed the GPSO Lozi 1 version with the final score 19:9. Based on these results a second experiment was designed and is presented in the following section.

# 6 Performance Experiment 2

In this experiment the (optimized) tuned Lozi map is used as a part of multi-chaotic PSO approach [9], where two different CPRNGs are used and alternated. Initially the first CPRNG (*gen1*) is used until the stagnation of *gBest* value occurs for longer than 1/10 of total number of iterations. At this point the first CPRNG is replaced (switched over) with the second CPRNG (*gen2*). For full details about this technique, please see [9]. During this second experiment the tuned Lozi map based CPRNGs was integrated into the multi-chaotic approach in following way:

The CPRNG based on Lozi map with typical setting was used as gen1 and the tuned Lozi map based CPRNG served as gen2. This design is compared with the original design [9] where the Lozi map with typical setting served as *gen1* and CPRNG based on Burgers chaotic map [9,11] was employed as *gen2*. Therefore three PSO variants are compared in this experiment:

- GPSO with single CPRNG (single-chaotic) based on tuned (optimized) Lozi map ($a = 1.5$, $b = 0.45$) (see experiment 1); noted GPSO Lozi 2
- Multi-chaotic GPSO; $gen_1$ = Lozi ($a = 1.7$, $b = 0.5$) $gen_2$= Lozi ($a = 1.5$, $b = 0.45$); noted GPSO Multi 1
- Multi-chaotic GPSO; $gen_1$ = Lozi (a = 1.7, b = 0.5) $gen_2$= Burgers; Original design from [9]; noted GPSO Multi 2

All algorithms were set identically to the performance experiment 1 as in the previous section. The mean results are summarized and presented in Table 3.

According to results given in Table 3, it seems that using two CPRNGs based on similar map with different setting in the multi-chaotic scheme may prove effective for some types of optimization problems. Nevertheless the single-chaotic approach with the optimized CPRNG parameters structure has demonstrated overall the best performance.

**Table 3.** Mean results comparison

| *f(x)* | GPSO Lozi2 | GPSO Multi 1 | GPSO Multi 2 | | A1 vs. A2 | A1 vs. A3 | A2 vs. A3 |
|---|---|---|---|---|---|---|---|
| *f(1)* | **-1.400E+03** | **-1.400E+03** | **-1.400E+03** | | 0 | 0 | 0 |
| *f(2)* | **7.729E+04** | 1.079E+05 | 1.367E+05 | | 1 | 1 | 1 |
| *f(3)* | 4.200E+06 | **1.894E+06** | 2.160E+06 | | 2 | 2 | 1 |
| *f(4)* | -2.130E+02 | **-8.210E+02** | -7.820E+02 | | 2 | 2 | 1 |
| *f(5)* | **-1.000E+03** | **-1.000E+03** | **-1.000E+03** | | 0 | 0 | 0 |
| *f(6)* | -8.920E+02 | -8.910E+02 | **-8.930E+02** | | 1 | 2 | 2 |
| *f(7)* | **-7.960E+02** | -7.940E+02 | -7.950E+02 | | 1 | 1 | 2 |
| *f(8)* | **-6.800E+02** | **-6.800E+02** | **-6.800E+02** | | 0 | 0 | 0 |
| *f(9)* | **-5.970E+02** | -5.960E+02 | **-5.970E+02** | | 1 | 0 | 2 |
| *f(10)* | **-5.000E+02** | **-5.000E+02** | -4.990E+02 | | 0 | 1 | 1 |
| *f(11)* | -3.980E+02 | -3.970E+02 | **-3.990E+02** | | 1 | 2 | 2 |
| *f(12)* | **-2.910E+02** | -2.840E+02 | -2.860E+02 | | 1 | 1 | 2 |
| *f(13)* | **-1.880E+02** | -1.840E+02 | -1.790E+02 | | 1 | 1 | 1 |
| *f(14)* | 1.600E+02 | **1.200E+02** | 1.370E+02 | | 2 | 2 | 1 |
| *f(15)* | **5.790E+02** | 6.930E+02 | 7.620E+02 | | 1 | 1 | 1 |
| *f(16)* | **2.010E+02** | **2.010E+02** | **2.010E+02** | | 0 | 0 | 0 |
| *f(17)* | 3.150E+02 | 3.140E+02 | **3.130E+02** | | 2 | 2 | 2 |
| *f(18)* | 4.260E+02 | **4.240E+02** | 4.270E+02 | | 2 | 1 | 1 |
| *f(19)* | **5.010E+02** | **5.010E+02** | **5.010E+02** | | 0 | 0 | 0 |
| *f(20)* | **6.020E+02** | 6.030E+02 | **6.020E+02** | | 1 | 0 | 2 |
| *f(21)* | 1.100E+03 | 1.100E+03 | **1.096E+03** | | 0 | 2 | 2 |
| *f(22)* | 1.049E+03 | 1.039E+03 | **1.036E+03** | | 2 | 2 | 2 |
| *f(23)* | **1.649E+03** | 1.773E+03 | 1.685E+03 | | 1 | 1 | 2 |
| *f(24)* | 1.207E+03 | **1.205E+03** | **1.205E+03** | | 2 | 2 | 0 |
| *f(25)* | **1.305E+03** | 1.310E+03 | 1.307E+03 | | 1 | 1 | 2 |
| *f(26)* | **1.333E+03** | 1.374E+03 | 1.350E+03 | | 1 | 1 | 2 |
| *f(27)* | **1.637E+03** | 1.660E+03 | 1.652E+03 | | 1 | 1 | 2 |
| *f(28)* | **1.762E+03** | **1.762E+03** | **1.762E+03** | | 0 | 0 | 0 |
| Best: | 15 | 10 | 11 | Score: | 17.0 : 11.0 | 15.0 : 13.0 | 11.5 : 16.5 |

## 7 Conclusion

In this initial research study the influence of Lozi chaotic map control parameter setting on the performance of chaotic PSO algorithm was deeply investigated. In the first experimental part the tuning experiment was evaluated. It seems that the setting of control parameter value may significantly alter the performance of PSO algorithm with CPRNG based on the Lozi map. According to the results of tuning experiment the performance experiment 1 was designed and the performance of PSO with tuned CPRNG was compared with canonical version and CPRNG with typical setting from literature. The performance of PSO algorithm was improved when the tuned CPRNG was used.

Afterwards second performance experiment was designed to investigate the possibility of using two "different" CPRNGs based on same chaotic map with different setting. According to presented data it seems that the adaptive change of Lozi map setting according to the performance of PSO is an interesting an promising way for future research. The future studies will focus on detailed clarification of change in behavior of particles according to the CPRNG setting and output sequences. Results presented in this study are very encouraging for future research in this area.

**Acknowledgments.** This work was supported by Grant Agency of the Czech Republic - GACR P103/13/08195S, Grant of SGS No. SP2014/159 and SP2014/170, VŠB - Technical University of Ostrava, Czech Republic, by the Development of human resources in research and development of latest soft computing methods and their application in practice project, reg. no. CZ.1.07/2.3.00/20.0072 funded by Operational Programme Education for Competitiveness, co-financed by ESF and state budget of the Czech Republic, by European Regional Development Fund under the project CEBIA-Tech No. CZ.1.05/2.1.00/03.0089 and by Internal Grant Agency of Tomas Bata University under the project No. IGA/FAI/2014/010.

## References

1. Kennedy, J., Eberhart, R.: Particle swarm optimization. In: IEEE International Conference on Neural Networks, pp. 1942–1948 (1995)
2. Kennedy, J., Eberhart, R.C., Shi, Y.: Swarm Intelligence. Morgan Kaufmann Publishers (2001)
3. Nickabadi, A., Ebadzadeh, M.M., Safabakhsh, R.: A novel particle swarm optimization algorithm with adaptive inertia weight. Applied Soft Computing 11(4), 3658–3670 (2011)
4. Yuhui, S., Eberhart, R.: A modified particle swarm optimizer. In: IEEE World Congress on Computational Intelligence, May 4-9, pp. 69–73 (1998)
5. Caponetto, R., Fortuna, L., Fazzino, S., Xibilia, M.G.: Chaotic sequences to improve the performance of evolutionary algorithms. IEEE Transactions on Evolutionary Computation 7(3), 289–304 (2003)
6. Araujo, E., Coelho, L.: Particle swarm approaches using Lozi map chaotic sequences to fuzzy modelling of an experimental thermal-vacuum system. Applied Soft Computing 8(4), 1354–1364 (2008)
7. Alatas, B., Akin, E., Ozer, B.A.: Chaos embedded particle swarm optimization algorithms. Chaos, Solitons & Fractals 40(4), 1715–1734 (2009) ISSN 0960-0779

8. Pluhacek, M., Senkerik, R., Davendra, D., Kominkova Oplatkova, Z., Zelinka, I.: On the behavior and performance of chaos driven PSO algorithm with inertia weight. Computers & Mathematics with Applications 66, 122–134 (2013)
9. Pluhacek, M., Senkerik, R., Zelinka, I.: Particle swarm optimization algorithm driven by multichaotic number generator. Soft Comput. 18(4), 631–639 (2014), doi:10.1007/s00500-014-1222-z
10. Pluhacek, M., Senkerik, R., Davendra, D., Zelinka, I., Designing, P.I.D., Controller For, D.C.: Designing PID Controller For DC Motor System By Means of Enhanced PSO Algorithm with Discrete Chaotic Lozi Map. In: Proceedings of the 26th European Conference on Modelling and Simulation, ECMS 2012, pp. 405–409 (2012) ISBN 978-0-9564944-4-3
11. Sprott, J.C.: Chaos and Time-Series Analysis. Oxford University Press (2003)
12. Liang, J.J., Qu, B.-Y., Suganthan, P.N., Hernández-Díaz Alfredo, G.: Problem Definitions and Evaluation Criteria for the CEC 2013 Special Session and Competition on Real-Parameter Optimization. Technical Report 201212, Computational Intelligence Laboratory, Zhengzhou University, Zhengzhou China and Technical Report, Nanyang Technological University, Singapore (January 2013)
13. Zelinka, I., Senkerik, R., Pluhacek, M.: Do evolutionary algorithms indeed require randomness? In: 2013 IEEE Congress on Evolutionary Computation (CEC), June 20-23, pp. 2283–2289 (2013), doi:10.1109/CEC.2 013.6557841

# Chaos Driven PSO with Ensemble of Priority Factors

Michal Pluhacek[1], Roman Senkerik[1],
Ivan Zelinka[2], and Donald Davendra[2]

[1] Faculty of Applied Informatics
Tomas Bata University in Zlin
T.G. Masaryka 5555, 760 01 Zlin, Czech Republic
{pluhacek,senkerik}@fai.utb.cz
[2] Faculty of Electrical Engineering and Computer Science
VŠB-Technical University of Ostrava
17. listopadu 15, 708 33 Ostrava-Poruba, Czech Republic
{ivan.zelinka,donald.davendra}@vsb.cz

**Abstract.** In this paper a new approach for PSO algorithm driven by chaotic pseudorandom number generator is investigated. The ensemble learning method that has been successfully implemented in many evolutionary computational techniques is applied here for the selection of priority factors in the velocity calculations formula. The goal is to improve the performance of chaos driven PSO. The promising results are compared with previously published results of SPSO-2011 on the CEC´ 13 benchmark set.

**Keywords:** Particle swarm optimization, chaos, PSO, Evolutionary algorithm, optimization, Ensemble learning.

## 1 Introduction

The Particle Swarm Optimization algorithm (PSO) [1, 2] is one of the most widely used Evolutionary Computation Techniques (ECT's). The PSO is also one of the best known representatives of the "Swarm intelligence" [2]. The main area of application is in global optimization. In recent years, the demand for fast and effective optimizers is increasing and as a reaction for this, the PSO is intensively studied and regularly modified [3 - 9]. In this paper, two modern trends in ECTs modification are connected into a new PSO modification. The first one represents the Ensemble learning method that has been successfully used in Differential Evolution [10, 11]. The second modification is the implementation of chaotic sequences as the chaotic pseudo-random number generators (CPRNG's) for ECT's [12 -19]. In this paper, the Ensemble of priority factors is implemented into PSO algorithm driven by CPRNG's based on two chaotic systems: The Burgers map and Dissipative standard map. The performance of the proposed algorithm is tested on the CEC´13 benchmark set [20] and compared with the previously published results of state-of-art representative SPSO-2011 [21].

I. Zelinka et al. (eds.), *Nostradamus 2014: Prediction, Modeling and Analysis of Complex Systems*, Advances in Intelligent Systems and Computing 289,
DOI: 10.1007/978-3-319-07401-6_9, © Springer International Publishing Switzerland 2014

## 2 Particle Swarm Optimization

A brief description of PSO algorithm follows in this section. The PSO algorithm takes inspiration from the natural swarm behavior of birds and fish. It was firstly introduced by Eberhart and Kennedy in 1995 [1]. Each particle in the population represents a candidate solution for the optimization problem that is defined by the cost function (CF). In each iteration of the algorithm, a new location (combination of CF parameters) for the particle is calculated based on its previous location and velocity vector (velocity vector contains particle velocity for each dimension of the problem).

According to the method of selection of the swarm or subswarm for best solution information spreading, the PSO algorithms are noted as global PSO (GPSO) or local PSO (LPSO). Within this research the PSO algorithm with global topology (GPSO) [6] was utilized. The chaotic PRNG is used in the main GPSO formula (1), which determines a new "velocity", thus directly affects the position of each particle in the next iteration.

$$v_{ij}^{t+1} = w \cdot v_{ij}^{t} + c_1 \cdot Rand \cdot (pBest_{ij} - x_{ij}^{t}) + c_2 \cdot Rand \cdot (gBest_j - x_{ij}^{t}) \quad (1)$$

Where:
$v_i^{t+1}$ - New velocity of the *ith* particle in iteration $t$+1.
$w$ – Inertia weight value.
$v_i^t$ - Current velocity of the *ith* particle in iteration $t$.
$c_1$, $c_2$ - Priority factors
$pBest_i$ – Local (personal) best solution found by the *ith* particle.
*gBest* - Best solution found in a population.
$x_{ij}^t$ - Current position of the *ith* particle (component $j$ of dimension $D$) in iteration $t$.
*Rand* – Pseudo random number, interval (0, 1). CPRNG is applied only here.

The maximum velocity was limited to 0.2 times the range as it is usual. The new position of each particle is then given by (2), where $x_i^{t+1}$ is the new particle position:

$$x_i^{t+1} = x_i^t + v_i^{t+1} \quad (2)$$

Finally the linear decreasing inertia weight [5, 7] is used in the typically referred GPSO design that was used in this study. The dynamic inertia weight is meant to slow the particles over time thus to improve the local search capability in the later phase of the optimization. The inertia weight has two control parameters $w_{start}$ and $w_{end}$. A new $w$ for each iteration is given by (3), where t stands for current iteration number and n stands for the total number of iterations. The values used in this study were $w_{start}$ = 0.9 and $w_{end}$ = 0.4.

$$w = w_{start} - \frac{((w_{start} - w_{end}) \cdot t)}{n} \quad (3)$$

# 3 Ensemble Learning Implementation

In this study the ensemble learning [10, 11] technique was used for the selection of priority factor values for each particle. Thus the priority factor values can be different for each particle in this design.

As a first step within this new initial designed ensemble implementation, the vector of possible priority values is defined (4).

$$\mathbf{c} = \{0.2, 0.5, 1, 1, 2, 2, 2, 2\} \tag{4}$$

Subsequently two expanded vectors $\mathbf{c}_1$ (5) and $\mathbf{c}_2$ (6) are created from **c**. These vectors contain priority factor values for each particle that are initially randomly selected from **c**. Where $rc_{1\text{-}1}$ is the priority factor 1 for the first particle and $rc_{1\text{-}n}$ is the priority factor 1 for the last particle etc. During the run of the PSO algorithm the performance of each particle is evaluated repeatedly. If the best solution found by the particle (*pBest*) has not changed for five iterations, a new pair of priority factors for that particle has been generated otherwise the priority factors have retained.

$$\mathbf{c}_1 = \{0.2, 0.5, 1, 1, 2, 2, 2, 2, rc_{1\text{-}1}, \ldots\ldots\ldots\ldots, rc_{1\text{-}n}\} \tag{5}$$

$$\mathbf{c}_2 = \{0.2, 0.5, 1, 1, 2, 2, 2, 2, rc_{2\text{-}1}, \ldots\ldots\ldots\ldots, rc_{2\text{-}n}\} \tag{6}$$

where:
$rc_1$ – randomly selected number from **c**
$rc_2$ – randomly selected number from **c**
$n$ – number of particles

Definition of the vectors **c**, $\mathbf{c}_1$ and $\mathbf{c}_2$ is completed only once during the initial population creation. The new priority factors during the search (learning) process are randomly selected from $\mathbf{c}_1$ and $\mathbf{c}_2$ vectors.

# 4 Chaotic Maps

In this section two discrete dissipative chaotic systems (maps) are described. These two chaotic maps were used as CPRNG's for the velocity calculation in PSO (See (1)). The choice was based on previous research [15 - 18]

## 4.1 Burgers Chaotic Map

The Burgers map (See Fig. 1) is a discretization of a pair of coupled differential equations The map equations are given in (7) with control parameters $a$ = 0.75 and $b$ = 1.75 as suggested in [19].

$$\begin{aligned} X_{n+1} &= aX_n - Y_n^2 \\ Y_{n+1} &= bY_n + X_nY_n \end{aligned} \tag{7}$$

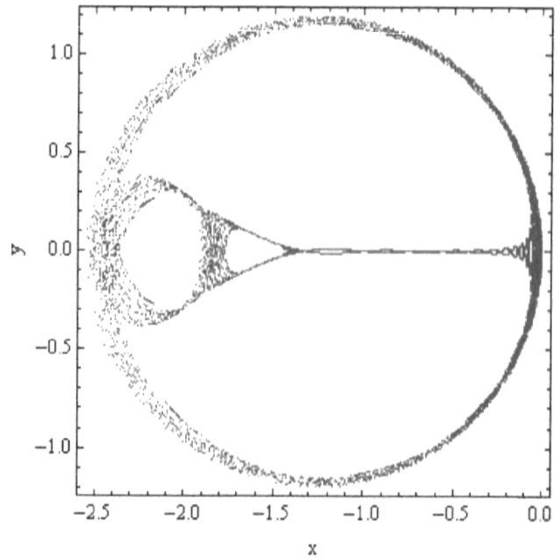


**Fig. 1.** x,y plot of Burgers map

### 4.2 Dissipative Standard Map

The Dissipative standard map is a two-dimensional chaotic map [19]. The parameters used in this work are $b$ = 0.6 and $k$ = 8.8 based on previous experiments [15-18] and suggestions in literature [19]. The $x,y$ plot of Dissipative standard map is given in Fig. 2.The map equations are given in (8).

$$\begin{aligned} X_{n+1} &= X_n + Y_{n+1} (\mathrm{mod} 2\pi) \\ Y_{n+1} &= bY_n + k \sin X_n (\mathrm{mod} 2\pi) \end{aligned} \tag{8}$$

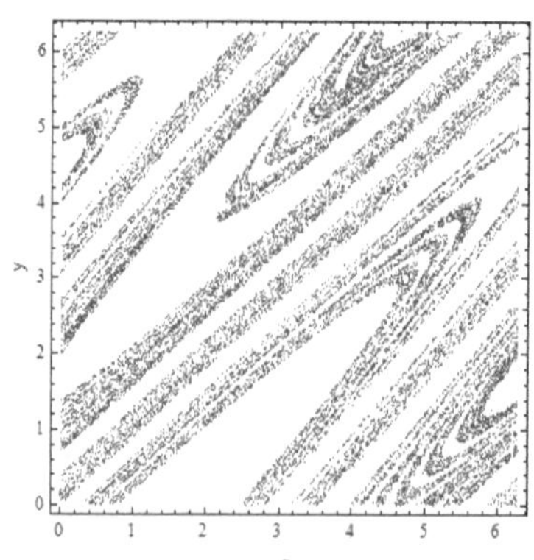


**Fig. 2.** x,y plot of Dissipative standard map

## 5 Experiment Setup

In this study, three different PSO variants with Ensemble of priority factors were comprised:

- GPSO with canonical pseudo random number generator – noted GPSO-E
- GPSO with CPRNG based on Burgers map – noted GPSO-E Burger
- GPSO with CPRNG based on Dissipative standard map – noted GPSO-E Disi

The CEC`13 benchmark set [20] was evaluated for the dimension setting 10 with the population size 40 and number of iterations 2500. For comparison reasons with SPSO-2011 both mean and median *gBest* values were recorded [21].

## 6 Results

In this section the results are summarized. Table 1 contains the median results for all PSO variants with Ensemble of priority factors. These median results are presented alongside the total number of the best results obtained. Furthermore the performance of pairs of algorithms is compared, where 1 stands for "win" of the "algorithm 1" (the first from the pair - left); number 2 stand for "win" of algorithm 2 (the second from the pair - right) and 0 stands for draw. The final score is also given in Table 1 as a sum of points for wins (1 point) and draws (0.5 point).

In Table 2, the comparison of median results for the GPSO-E Burger and SPSO-2011 is given. Furthermore an illustrative example of *gBest* mean history is depicted in Fig. 3. The bold numbers within all tables represents the best results.

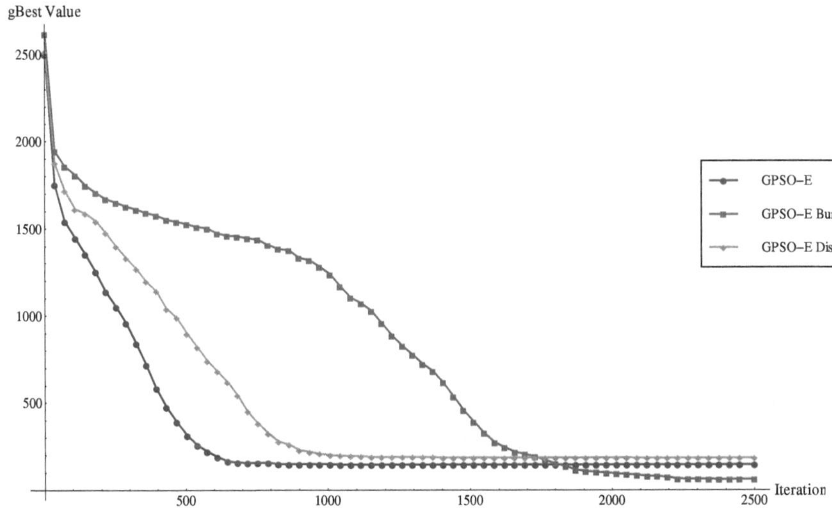


**Fig. 3.** Mean *gBest* history for *f(14)* – PSO variants with Ensemble of priority factors

**Table 1.** Median results comparison GPSO-E, GPSO-E Burger and GPSO-E Disi

| *f(x)* | GPSO-E | GPSO-E Burger | GPSO-E Disi | | A1 vs. A2 | A1 vs. A3 | A2 vs. A3 |
|---|---|---|---|---|---|---|---|
| *f(1)* | **-1.400E+03** | **-1.400E+03** | **-1.400E+03** | | 0 | 0 | 0 |
| *f(2)* | **6.828E+04** | 2.706E+05 | 8.939E+04 | | 1 | 1 | 2 |
| *f(3)* | **1.385E+05** | 8.310E+05 | 1.460E+05 | | 1 | 1 | 2 |
| *f(4)* | -7.950E+02 | -7.940E+02 | **-9.510E+02** | | 1 | 2 | 2 |
| *f(5)* | **-1.000E+03** | **-1.000E+03** | **-1.000E+03** | | 0 | 0 | 0 |
| *f(6)* | -8.890E+02 | **-8.900E+02** | **-8.900E+02** | | 2 | 2 | 0 |
| *f(7)* | -7.960E+02 | **-7.970E+02** | **-7.970E+02** | | 2 | 2 | 0 |
| *f(8)* | **-6.800E+02** | **-6.800E+02** | **-6.800E+02** | | 0 | 0 | 0 |
| *f(9)* | -5.960E+02 | **-5.970E+02** | **-5.970E+02** | | 2 | 2 | 0 |
| *f(10)* | **-4.990E+02** | **-4.990E+02** | **-4.990E+02** | | 0 | 0 | 0 |
| *f(11)* | -3.960E+02 | **-3.970E+02** | -3.950E+02 | | 2 | 1 | 1 |
| *f(12)* | -2.870E+02 | -2.870E+02 | **-2.880E+02** | | 0 | 2 | 2 |
| *f(13)* | **-1.820E+02** | **-1.820E+02** | **-1.820E+02** | | 0 | 0 | 0 |
| *f(14)* | 1.640E+02 | **4.700E+01** | 1.770E+02 | | 2 | 1 | 1 |
| *f(15)* | 7.350E+02 | 8.750E+02 | **6.770E+02** | | 1 | 2 | 2 |
| *f(16)* | **2.010E+02** | **2.010E+02** | **2.010E+02** | | 0 | 0 | 0 |
| *f(17)* | 3.150E+02 | **3.140E+02** | 3.150E+02 | | 2 | 0 | 1 |
| *f(18)* | **4.200E+02** | 4.290E+02 | 4.220E+02 | | 1 | 1 | 2 |
| *f(19)* | **5.010E+02** | **5.010E+02** | **5.010E+02** | | 0 | 0 | 0 |
| *f(20)* | 6.030E+02 | **6.020E+02** | **6.020E+02** | | 2 | 2 | 0 |
| *f(21)* | **1.100E+03** | **1.100E+03** | **1.100E+03** | | 0 | 0 | 0 |
| *f(22)* | 1.171E+03 | **9.600E+02** | 1.078E+03 | | 2 | 2 | 1 |
| *f(23)* | 1.698E+03 | **1.601E+03** | 1.693E+03 | | 2 | 2 | 1 |
| *f(24)* | 1.209E+03 | 1.208E+03 | **1.206E+03** | | 2 | 2 | 2 |
| *f(25)* | 1.308E+03 | **1.305E+03** | 1.306E+03 | | 2 | 2 | 1 |
| *f(26)* | **1.317E+03** | **1.317E+03** | 1.400E+03 | | 0 | 1 | 1 |
| *f(27)* | **1.628E+03** | 1.640E+03 | 1.634E+03 | | 1 | 1 | 2 |
| *f(28)* | **1.700E+03** | 1.702E+03 | **1.700E+03** | | 1 | 0 | 2 |
| Best: | 12 | 14 | 12 | Score: | 12.0 : 16.0 | 12.0 : 16.0 | 13.0 : 15.0 |

From Table 1, it follows that the highest number of best results was obtained by the GPSO-E Burger version. Consequently the performance of GPSO-E Burger was compared to the SPSO-2011 [21] (see Table 2).

**Table 2.** Median results comparison SPSO-2011 [21] and GPSO-E Burger

| *f(x)* | SPSO-2011 | GPSO-E Burger | Difference |
|---|---|---|---|
| *f(1)* | **-1.40E+03** | **-1.40E+03** | 0.00E+00 |
| *f(2)* | 6.79E+05 | **2.71E+05** | 4.08E+05 |
| *f(3)* | 4.37E+08 | **8.31E+05** | 4.36E+08 |
| *f(4)* | 4.99E+04 | **-7.94E+02** | 5.07E+04 |
| *f(5)* | **-1.00E+03** | **-1.00E+03** | 0.00E+00 |
| *f(6)* | -8.57E+02 | **-8.90E+02** | 3.35E+01 |
| *f(7)* | -7.14E+02 | **-7.97E+02** | 8.34E+01 |
| *f(8)* | -6.79E+02 | **-6.80E+02** | 1.10E+00 |
| *f(9)* | -5.46E+02 | **-5.97E+02** | 5.10E+01 |
| *f(10)* | **-5.00E+02** | -4.99E+02 | 6.00E-01 |
| *f(11)* | -1.70E+02 | **-3.97E+02** | 2.27E+02 |
| *f(12)* | -6.52E+01 | **-2.87E+02** | 2.22E+02 |
| *f(13)* | 2.28E+02 | **-1.82E+02** | 4.10E+02 |
| *f(14)* | 7.16E+03 | **4.70E+01** | 7.11E+03 |
| *f(15)* | 8.02E+03 | **8.75E+02** | 7.15E+03 |
| *f(16)* | 2.02E+02 | **2.01E+02** | 1.00E+00 |
| *f(17)* | 6.11E+02 | **3.14E+02** | 2.97E+02 |
| *f(18)* | 6.91E+02 | **4.29E+02** | 2.62E+02 |
| *f(19)* | 5.37E+02 | **5.01E+02** | 3.62E+01 |
| *f(20)* | 6.23E+02 | **6.02E+02** | 2.07E+01 |
| *f(21)* | 1.54E+03 | **1.10E+03** | 4.36E+02 |
| *f(22)* | 9.72E+03 | **9.60E+02** | 8.76E+03 |
| *f(23)* | 1.13E+04 | **1.60E+03** | 9.66E+03 |
| *f(24)* | 1.34E+03 | **1.21E+03** | 1.36E+02 |
| *f(25)* | 1.50E+03 | **1.31E+03** | 1.97E+02 |
| *f(26)* | 1.63E+03 | **1.32E+03** | 3.11E+02 |
| *f(27)* | 2.98E+03 | **1.64E+03** | 1.34E+03 |
| *f(28)* | 1.80E+03 | **1.70E+03** | 9.80E+01 |

# 7 Conclusion

In this initial research study altogether three different version of PSO with global topology and ensemble of priority factors were tested and compared on the CEC`13 benchmark suite. From Table 1, it follows that the highest number of the best results was obtained by means of version with CPRNG based on Burgers map. However it seems that for many benchmark test problems the performance of all three versions is similar or very comparable. The overall best performing version GPSO-E Burger is compared in Table 2 with the state-of-art representative SPSO-2011. It is clear that for the majority of benchmark functions, the performance of GPSO-E Burger is superior to the performance of SPSO-2011. The median values are used given that there are no mean values presented in the original SPSO-2011 results paper [21]. It seems that through utilization of the ensemble of priority factors it is possible to achieve good performance of chaos driven PSO algorithm on the given benchmark set. Future research will focus on different approaches for ensemble implementation, further on testing of higher dimensional settings and finally on the comparisons with different promising state-of-art representatives of PSO algorithm.

**Acknowledgments.** This work was supported by Grant Agency of the Czech Republic - GACR P103/13/08195S, Grants of SGS No. SP2014/159 and SP2014/170, VŠB - Technical University of Ostrava, Czech Republic, by the Development of human resources in research and development of latest soft computing methods and their application in practice project, reg. no. CZ.1.07/2.3.00/20.0072 funded by Operational Programme Education for Competitiveness, co-financed by ESF and state budget of the Czech Republic, by European Regional Development Fund under the project CEBIA-Tech No. CZ.1.05/2.1.00/03.0089 and by Internal Grant Agency of Tomas Bata University under the project No. IGA/FAI/2014/010.

## References

1. Kennedy, J., Eberhart, R.: Particle swarm optimization. In: IEEE International Conference on Neural Networks, pp. 1942–1948 (1995)
2. Kennedy, J., Eberhart, R.C., Shi, Y.: Swarm Intelligence. Morgan Kaufmann Publishers (2001)
3. Liang, J., Suganthan, P.N.: Dynamic multi-swarm particle swarm optimizer. In: Swarm Intelligence Symposium, SIS 2005, pp. 124–129 (2005)
4. Liang, J.J., Qin, A.K., Suganthan, P.N., Baskar, S.: Comprehensive learning particle swarm optimizer for global optimization of multimodal functions. IEEE Transactions on Evolutionary Computation 10(3), 281–295 (2006)
5. Nickabadi, A., Ebadzadeh, M.M., Safabakhsh, R.: A novel particle swarm optimization algorithm with adaptive inertia weight. Applied Soft Computing 11(4), 3658–3670 (2011)
6. Zhi-Hui, Z., Jun, Z., Yun, L., Yu-hui, S.: Orthogonal Learning Particle Swarm Optimization. IEEE Transactions on Evolutionary Computatio 15(6), 832–847 (2011)
7. Yuhui, S., Eberhart, R.: A modified particle swarm optimizer. In: IEEE World Congress on Computational Intelligence, May 4-9, pp. 69–73 (1998)
8. Kennedy, J., Mendes, R.: Population structure and particle swarm performance. In: Proceedings of the 2002 Congress on Evolutionary Computation, CEC 2002, pp. 1671–1676 (2002)
9. van den Bergh, F., Engelbrecht, A.P.: A study of particle swarm optimization particle trajectories. Information Sciences 176(8), 937–971 (2006)

10. Mallipeddi, R., Suganthan, P.N.: Differential Evolution Algorithm with Ensemble of Parameters and Mutation and Crossover Strategies. In: Panigrahi, B.K., Das, S., Suganthan, P.N., Dash, S.S. (eds.) SEMCCO 2010. LNCS, vol. 6466, pp. 71–78. Springer, Heidelberg (2010)
11. Mallipeddi, R., Suganthan, P.N.: Differential Evolution Algorithm with Ensemble of populations for Global Numerical Optimization. OPSEARCH 46(2), 184–213 (2009)
12. Caponetto, R., Fortuna, L., Fazzino, S., Xibilia, M.G.: Chaotic sequences to improve the performance of evolutionary algorithms. IEEE Transactions on Evolutionary Computation 7(3), 289–304 (2003)
13. Araujo, E., Coelho, L.: Particle swarm approaches using Lozi map chaotic sequences to fuzzy modelling of an experimental thermal-vacuum system. Applied Soft Computing 8(4), 1354–1364 (2008)
14. Alatas, B., Akin, E., Ozer, B.A.: Chaos embedded particle swarm optimization algorithms. Chaos, Solitons & Fractals 40(4), 1715–1734 (2009)
15. Pluhacek, M., Senkerik, R., Davendra, D., Kominkova Oplatkova, Z., Zelinka, I.: On the behavior and performance of chaos driven PSO algorithm with inertia weight. Computers & Mathematics with Applications 66, 122–134 (2013)
16. Pluhacek, M., Senkerik, R., Zelinka, I.: Particle swarm optimization algorithm driven by multichaotic number generator. Soft Comput. 18(4), 631–639 (2014), doi:10.1007/s00500-014-1222-z
17. Pluhacek, M., Senkerik, R., Zelinka, I., Davendra, D.: On the Performance of Enhanced PSO Algorithm with Dissipative Chaotic Map in the Task of High Dimensional Optimization Problems. In: Zelinka, I., Chen, G., Rössler, O.E., Snasel, V., Abraham, A. (eds.) Nostradamus 2013: Prediction, Model. & Analysis. AISC, vol. 210, pp. 89–99. Springer, Heidelberg (2013)
18. Pluhacek, M., Senkerik, R., Davendra, D., Zelinka, I.: Designing PID controller for DC motor by means of enhanced PSO algorithm with dissipative chaotic map. In: Snasel, V., Abraham, A., Corchado, E.S. (eds.) SOCO Models in Industrial & Environmental Appl. AISC, vol. 188, pp. 475–483. Springer, Heidelberg (2013)
19. Sprott, J.C.: Chaos and Time-Series Analysis. Oxford University Press (2003)
20. Liang, J.J., Qu, B.-Y., Suganthan, P.N., Hernández-Díaz Alfredo, G.: Problem Definitions and Evaluation Criteria for the CEC 2013 Special Session and Competition on Real-Parameter Optimization. Technical Report 201212, Computational Intelligence Laboratory, Zhengzhou University, Zhengzhou China and Technical Report, Nanyang Technological University, Singapore (January 2013)
21. Zambrano-Bigiarini, M., Clerc, M., Rojas, R.: Standard Particle Swarm Optimisation 2011 at CEC-2013: A baseline for future PSO improvements. In: 2013 IEEE Congress on Evolutionary Computation (CEC), June 20-23, pp. 2337–2344 (2013)

# Multi-chaotic Differential Evolution: Determining the Switching Time

Roman Senkerik[1], Michal Pluhacek[1], Ivan Zelinka[2], Donald Davendra[2], and Zuzana Kominkova Oplatkova[1]

[1] Tomas Bata University in Zlin , Faculty of Applied Informatics, Nam T.G. Masaryka 5555, 760 01 Zlin, Czech Republic
{senkerik,pluhacek,oplatkova}@fai.utb.cz
[2] Technical University of Ostrava, Faculty of Electrical Engineering and Computer Science, 17. listopadu 15,708 33 Ostrava-Poruba, Czech Republic
{ivan.zelinka,donald.davendra}@fai.utb.cz

**Abstract.** This research deals with the hybridization of the two softcomputing fields, which are chaos theory and evolutionary computation. This paper aims on the deeper investigations on the multi-chaos-driven evolutionary algorithm Differential Evolution (DE) concept. This research is aimed at the embedding and alternating of set of two discrete dissipative chaotic systems in the form of chaos pseudo random number generators for the DE. In this paper the novel initial concept of DE/rand/1/bin strategy driven alternately by two chaotic maps (systems) is deeply investigated in terms of determining the optimal switching moment of two different chaotic systems. From the previous research, it follows that very promising results were obtained through the utilization of different chaotic maps, which have unique properties with connection to DE. The idea is then to connect these two different influences to the performance of DE into the one multi-chaotic concept. Repeated simulations were performed on the selected shifted benchmark function in higher dimensions. Finally, the obtained results are compared with canonical DE.

**Keywords:** Differential Evolution, Deterministic chaos, Dissipative systems.

## 1 Introduction

This research deals with the hybridization of the two softcomputing fields, which are chaos theory and evolutionary computation techniques (ECT's). These days the evolutionary algorithms (EA's) are known as powerful tool for almost any difficult and complex optimization problem. Differential Evolution (DE) [1] is one of the most potent heuristics available.

A number of DE variants have been recently developed with the emphasis on adaptivity/selfadaptivity [2], ensemble approach [3] or utilization for discrete domain problems. Together with this persistent development in such mainstream research topics, the basic concept of chaos driven DE have been introduced.

I. Zelinka et al. (eds.), *Nostradamus 2014: Prediction, Modeling and Analysis of Complex Systems*, Advances in Intelligent Systems and Computing 289,
DOI: 10.1007/978-3-319-07401-6_10, © Springer International Publishing Switzerland 2014

Recent research in chaos driven heuristics has been fueled with the predisposition that unlike stochastic approaches, a chaotic approach is able to bypass local optima stagnation. This one clause is of deep importance to evolutionary algorithms. A chaotic approach generally uses the chaotic map in the place of a pseudo random number generator [4]. This causes the heuristic to map unique regions, since the chaotic map iterates to new regions. The task is then to select a very good chaotic map as the pseudo random number generator.

Several papers have been recently focused on the connection of DE and chaotic dynamics either in the form of hybridizing of DE with chaotic searching algorithm [5] or in the form of chaotic mutation factor and dynamically changing weighting and crossover factor in self-adaptive chaos differential evolution (SACDE) [6].

Nevertheless the focus of our research is the direct embedding of chaotic systems in the form of chaos pseudo random number generator (CPRNG) into the DE (ChaosDE) [7] and the deeper investigation on the influence of natural chaotic dynamics implementation into evolutionary algorithm to its overall performance.

Also the PSO (Particle Swarm Optimization) algorithm with elements of chaos was introduced as CPSO [8]. The concept of ChaosDE proved itself to be a powerful heuristic also in combinatorial problems domain [9]. At the same time the chaos embedded PSO with inertia weigh strategy was closely investigated [10], followed by the introduction of a PSO strategy driven alternately by two chaotic systems [11]. Recently the chaos driven firefly algorithm has been introduced [12].

The organization of this paper is following: Firstly, the motivation for this research is proposed. The next sections are focused on the description of the concept of chaos driven DE and the used test function. Results and conclusion follow afterwards.

## 2 Motivation

This research is an extension and continuation of the previous successful initial experiment with chaos driven DE (ChaosDE) [13], [14] with test functions in higher dimensions. In this paper the concept of DE/rand/1/bin strategy is deeply investigated in terms of determining the optimal switching moment of two different driving chaotic systems. To be more precise this research is focused on the influence of the time, when two different CPRNGs are alternated to the performance of DE.

From the previous research it follows, that very promising results were obtained through the utilization of Delayed Logistic, Lozi, Burgers and Tinkerbell maps. The last two mentioned chaotic maps have unique properties with connection to DE: strong progress towards global extreme, but weak overall statistical results, like average cost function (CF) value and std. dev., and tendency to premature stagnation. While through the utilization of the Lozi and Delayed Logistic map the continuously stable and very satisfactory performance of ChaosDE was achieved. The idea is then to connect these two different influences to the performance of DE into the one multi-chaotic concept.

## 3 Differential Evolution

DE is a population-based optimization method that works on real-number-coded individuals [1]. DE is quite robust, fast, and effective, with global optimization ability. It does not require the objective function to be differentiable, and it works well even with noisy and time-dependent objective functions. Due to a limited space and the aims of this paper, the detailed description of well known DE algorithm basic principles is insignificant and hence omitted. Please refer to [1], [16] for the detailed description of the used DERand1Bin strategy (both for ChaosDE and Canonical DE) as well as for the complete description of all other strategies.

## 4 The Concept of ChaosDE

The general idea of ChaosDE and CPRNG is to replace the default pseudorandom number generator (PRNG) with the discrete chaotic map. As the discrete chaotic map is a set of equations with a static start position, we created a random start position of the map, in order to have different start position for different experiments (runs of EA's). This random position is initialized with the default PRNG, as a one-off randomizer. Once the start position of the chaotic map has been obtained, the map generates the next sequence using its current position.

### 4.1 Chaotic Maps

This section contains the description of discrete dissipative chaotic maps used as the chaotic pseudo random generators for DE. In this research, direct output iterations of the chaotic maps were used for the generation of real numbers in the process of crossover based on the user defined CR value and for the generation of the integer values used for selection of individuals. Following chaotic maps were used: Burgers (1), and Lozi map (2).

The Burgers mapping is a discretization of a pair of coupled differential equations which were used by Burgers [13] to illustrate the relevance of the concept of bifurcation to the study of hydrodynamics flows. The map equations are given in (4) with control parameters $a = 0.75$ and $b = 1.75$ as suggested in [14].

$$\begin{aligned} X_{n+1} &= aX_n - Y_n^2 \\ Y_{n+1} &= bY_n + X_n Y_n \end{aligned} \tag{1}$$

The Lozi map is a discrete two-dimensional chaotic map. The map equations are given in (5). The parameters used in this work are: $a = 1.7$ and $b = 0.5$ as suggested in [14]. For these values, the system exhibits typical chaotic behavior and with this parameter setting it is used in the most research papers and other literature sources.

$$
\begin{aligned}
X_{n+1} &= 1 - a|X_n| + bY_n \\
Y_{n+1} &= X_n
\end{aligned}
\tag{2}
$$

### 4.2 Graphical Examples – Lozi Map and Burgers map

Illustrative histograms of the distribution of real numbers transferred into the range <0 - 1> generated by means of studied chaotic maps are in Figures 1 and 3, whereas Figures 2 and 4 show the example of dynamical sequencing during the generating of pseudo number numbers by means of both studied CPRNGs.

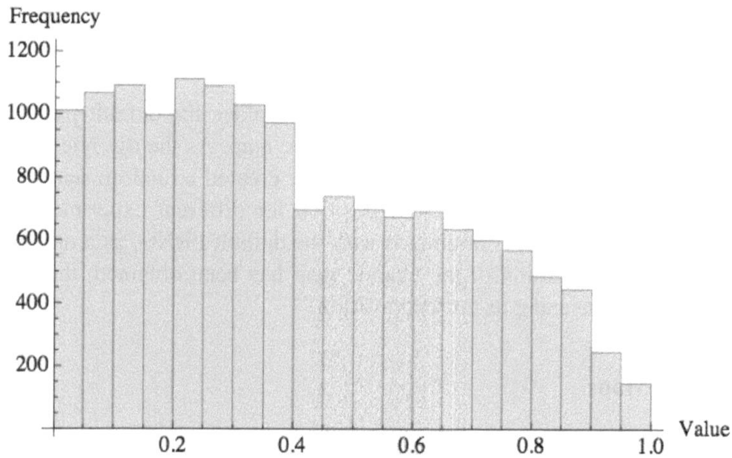


**Fig. 1.** Histogram of the distribution of real numbers transferred into the range <0 - 1> generated by means of the chaotic Lozi map – 5000 samples

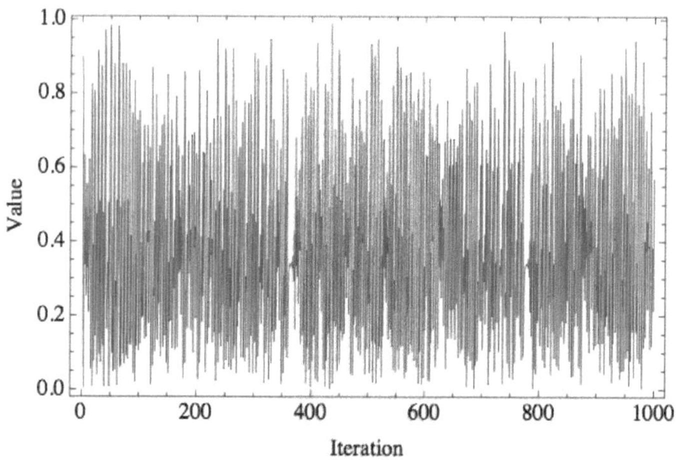


**Fig. 2.** Example of the chaotic dynamics: range <0 - 1> generated by means of the Lozi map

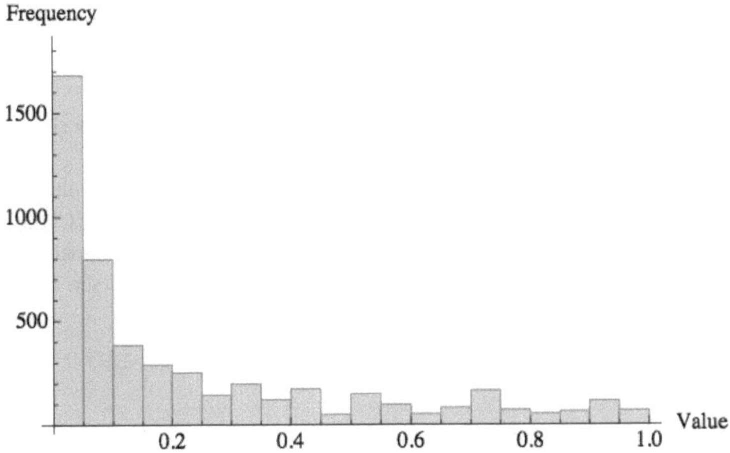


**Fig. 3.** Histogram of the distribution of real numbers transferred into the range <0 - 1> generated by means of the chaotic Burgers map – 5000 samples

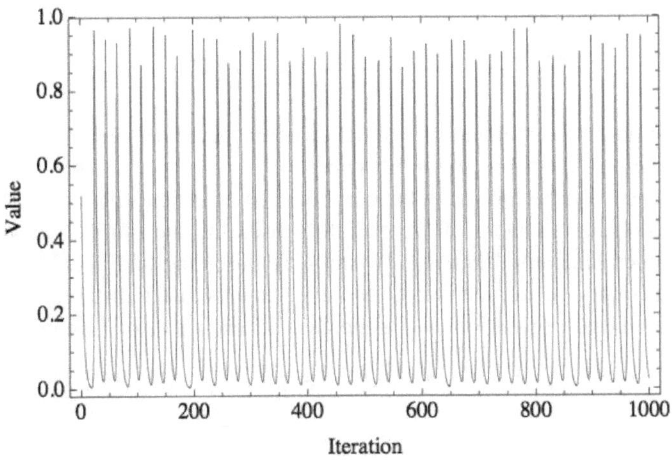


**Fig. 4.** Example of the chaotic dynamics: range <0 - 1> generated by means of the Burgers map

# 5 Experiment Design

For the purpose of detailed DE performance investigation within this initial research, only one shifted Ackley's original function in the form (3) was used.

$$f(x) = -20\exp\left(-0.02\sqrt{\frac{1}{D}\sum_{i=1}^{D}(x_i - s_i)^2}\right) - \exp\left(\frac{1}{D}\sum_{i=1}^{D}\cos 2\pi(x_i - s_i)\right) + \\ + 20 + \exp(1) \tag{3}$$

Function minimum: Position for $E_n$: $(x_1, x_2 \ldots x_n) = s$; Value for $E_n$: $y = 0$
Function interval: <-30, 30>.

Where $s_i$ is a random number from the 90% range of function interval; $s$ vector is randomly generated before each run of the optimization process.

The novelty of this research represents the simulation of the DE performance dependency on switching of the driving chaotic systems.

In this paper, the canonical DE strategy DERand1Bin and the Multi-Chaos DERand1Bin strategy driven alternately by two different chaotic maps (ChaosDE concept) were used. Parameter settings for both canonical DE and ChaosDE were obtained analytically based on numerous experiments and simulations (see Table 1).

**Table 1.** Parameter set up for canonical DE and ChaosDE

| DE Parameter | Value |
|---|---|
| Popsize | 75 |
| F | 0.8 |
| Cr | 0.8 |
| Dimensions | 30 |
| Generations | 1500 |

Investigation on the moment of manual switching over between two driving chaotic maps represents the main aim of this paper.

Experiments were performed in the combined environments of *Wolfram Mathematica* and *C language*, canonical DE therefore used the built-in *C language* pseudo random number generator *Mersenne Twister C* representing traditional pseudorandom number generators in comparisons. All experiments used different initialization, i.e. different initial population was generated in each run of Canonical or Chaos driven DE.

## 6 Experiment Results

This initial research utilizes the maximum number of generations fixed at 3000 generations. This allowed the possibility to analyze the progress of DE within a limited number of generations and cost function evaluations.

The statistical results of the experiments are shown in Table 2, which represent the simple statistics for cost function (CF) values, e.g. average, median, maximum values, standard deviations and minimum values representing the best individual solution for all 50 repeated runs of canonical DE and several versions of ChaosDE and Multi-ChaosDE.

Table 3 compares the progress of several versions of ChaosDE, Multi-ChaosDE and Canonical DE. This table contains the average CF values for the generation No. 750, 1500, 2250 and 3000 from all 50 runs. The bold values within the both Tables 2 and 3 depict the best obtained results. Following versions of Multi-ChaosDE were studied:

- *Burgers-Lozi-Switch-500*: Start with Burgers map CPRNG, switch to the Lozi map CPRNG after 500 generations.
- *Burgers-Lozi-Switch-1500*: Start with Burgers map CPRNG, switch to the Lozi map CPRNG after 1500 generations.
- *Lozi-Burgers-Switch-500*: Start with Lozi map CPRNG, switch to the Burgers map CPRNG after 500 generations.
- *Lozi-Burgers-Switch-1500*: Start with Lozi map CPRNG, switch to the Burgers map CPRNG after 1500 generations.

The graphical comparison of the time evolution of average CF values for all 50 runs of ChaosDE/Multi-ChaosDE and canonical DERand1Bin strategy is depicted in Fig. 5. Finally the Figures 6 – 10 confirm the robustness of Multi-ChaosDE in finding the best solutions for all 50 runs.

Obtained numerical results given in Tables 2 and 3 and graphical comparisons in Figures 5 - 10 support the claim that all Multi-Chaos versions have given better overall results in comparison with the canonical DE version. From the presented data it follows that Multi-Chaos DE versions driven from the start by Lozi map and followed by the Burgers Map have given the best overall results.

**Table 2.** Simple results statistics for the shifted Ackley's original function – 30D

| DE Version | Avg CF | Median CF | Max CF | Min CF | StdDev |
|---|---|---|---|---|---|
| Canonical DE | 3.809154 | 3.81851 | 4.644817 | 3.073961 | 0.400671 |
| Burger-Lozi-Switch-500 | 0.000822 | 0.00073 | 0.002665 | 0.000145 | 0.000532 |
| Burger-Lozi-Switch-1500 | 0.018648 | 1.73E-05 | 0.931305 | 3.91E-06 | 0.131703 |
| Lozi-Burger-Switch-500 | 0.11797 | **4.51E-07** | 1.501747 | **8.05E-08** | 0.362244 |
| Lozi-Burger-Switch-1500 | **1.57E-05** | 1.1E-05 | **5.54E-05** | 2.23E-06 | **1.3E-05** |

**Table 3.** Comparison of progress towards the min. for the shifted Ackley's original function

| DE Version | Generation No. 750 | Generation No. 1500 | Generation No. 2250 | Generation No. 3000 |
|---|---|---|---|---|
| Canonical DE | 12.98826 | 8.475582 | 5.534645 | 3.809154 |
| Burger-Lozi-Switch-500 | 2.836886 | 0.202722 | 0.010288 | 0.000822 |
| Burger-Lozi-Switch-1500 | **1.504561** | **0.022077** | 0.01887 | 0.018648 |
| Lozi-Burger-Switch-500 | 4.366281 | 0.137792 | 0.118069 | 0.11797 |
| Lozi-Burger-Switch-1500 | 7.747079 | 1.597677 | **0.003364** | **1.57E-05** |

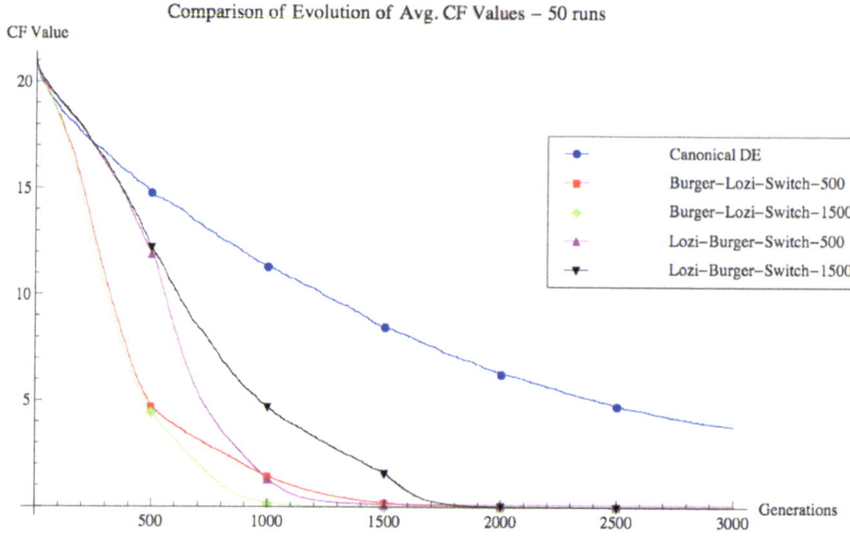


**Fig. 5.** Comparison of the time evolution of avg. CF values for the all 50 runs of Canonical DE, and all four versions of Multi-ChaosDE; shifted Ackley's original function, $D = 30$

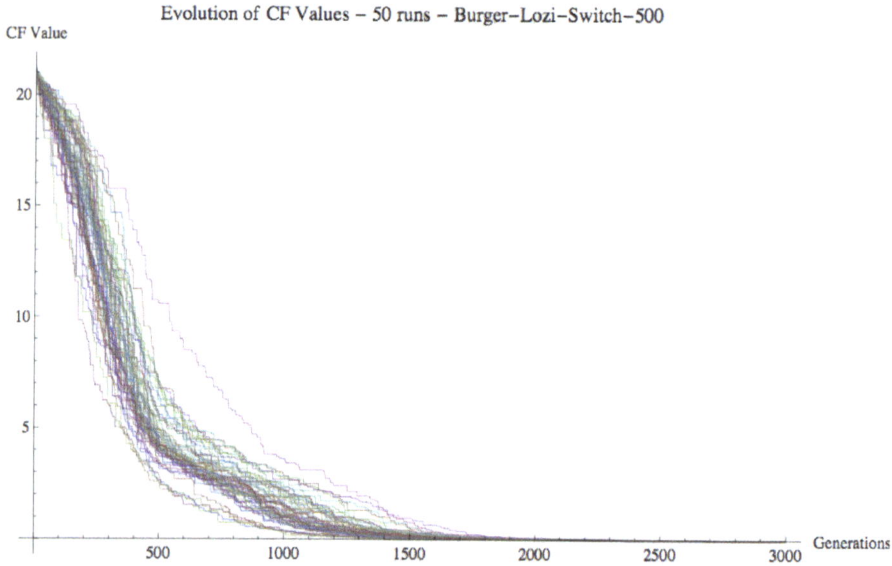


**Fig. 6.** Comparison of the time evolution of CF values for all 50 runs of Multi-ChaosDE version: Burgers-Lozi-Switch-500

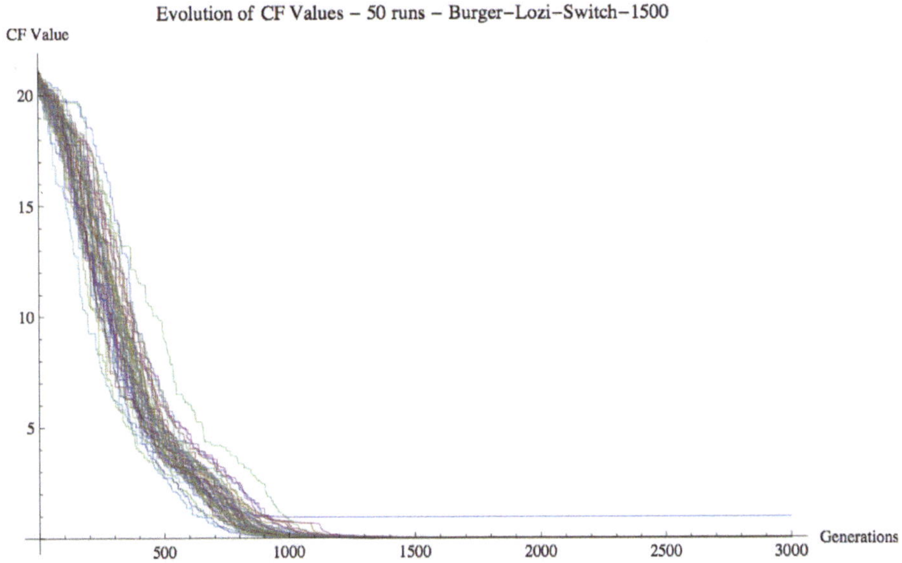


**Fig. 7.** Comparison of the time evolution of CF values for all 50 runs of Multi-ChaosDE version: Burgers –Lozi-Switch-1500

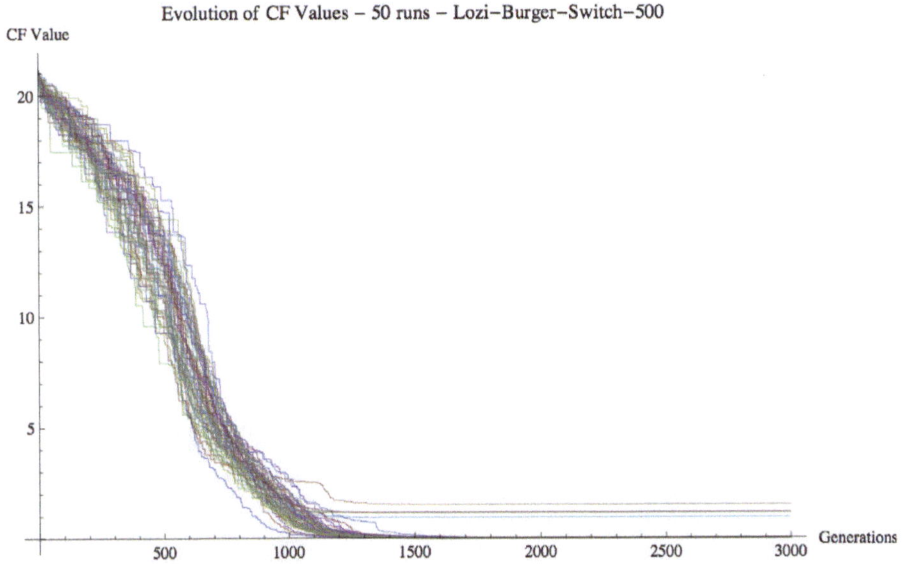


**Fig. 8.** Comparison of the time evolution of CF values for all 50 runs of Multi-ChaosDE version: Lozi-Burger-Switch-500

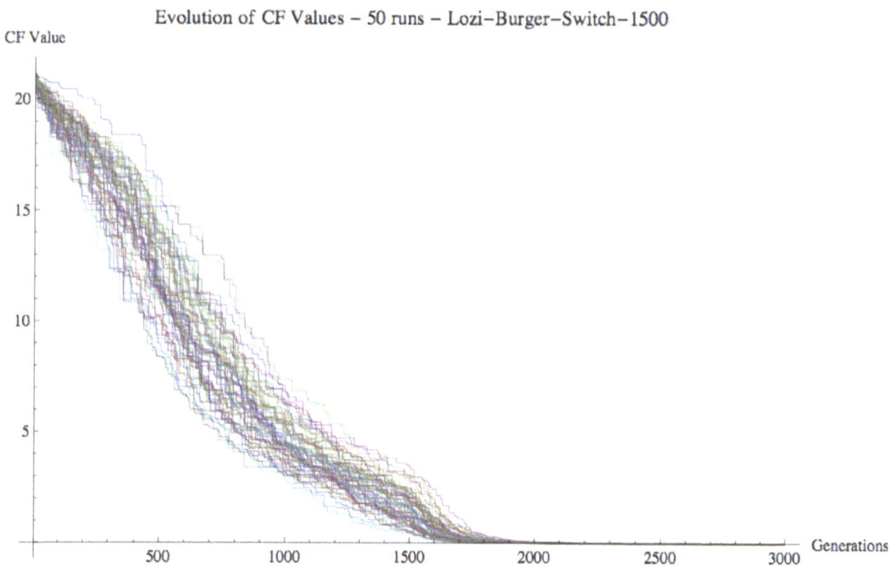


**Fig. 9.** Comparison of the time evolution of CF values for all 50 runs of Multi-ChaosDE version: Lozi- Burgers -Switch-1500

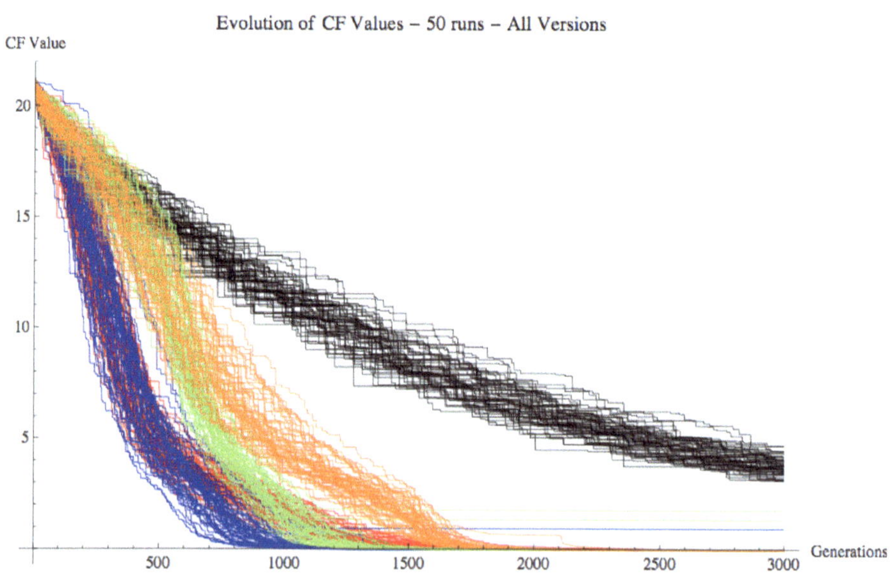


**Fig. 10.** Comparison of the time evolution of CF values for all 50 runs of canonical DE (black) and Multi-ChaosDE versions: Burgers-Lozi-Switch-500 (red), Burgers-Lozi-Switch-1500 (blue), Lozi-Burgers-Switch-500 (green) and Lozi-Burgers-Switch-1500 (orange).

## 7 Conclusion

In this paper, the novel concept of multi-chaos driven DERand1Bin strategy was more deeply analyzed and compared with the canonical DERand1Bin strategy on the selected benchmark function in higher dimension. Based on obtained results, it may be claimed, that the developed Multi-ChaosDE gives considerably better results than other compared heuristics.

The novelty of this research represents the deeper investigation and simulation of the DE performance dependency on switching of the two driving chaotic systems.

For the both *Burgers-Lozi-Switch* versions the progressive Burgers map CPRNG secured the faster approaching towards the global extreme from the very beginning of evolutionary process. The very fast switch over to the Lozi map based CPRNG (*Burgers-Lozi-Switch-500 version*) helped to avoid the Burgers map based CPRNG weak spots, which are the weak overall statistical results, like average CF value and std. dev.; and tendency to stagnation. The aforementioned weak spots of the Burgers map based CPRNG have fully revealed in the case of later alternating of both maps. The initial faster convergence (starting of evolutionary process) and subsequent continuously stable searching process without premature stagnation issues are visible from Fig. 5 (red and green lines).

Through the utilization of *Lozi-Burgers-Switch* versions, the strong progress towards global extreme given by Burgers map CPRNG helped to the evolutionary process driven moderately from the start by mans of Lozi map CPRNG to achieve the best avg. CF and median CF values. The moment of switch (at 500 and 1500 generations) is clearly visible from Fig. 5 (magenta and black lines). From the results, it seems that it is better to keep the Lozi map based CPRNG for more generations to ensure the stable searching process. These versions were able to reach the best individual minimum and median CF values.

Future plans are including the testing of combination of different chaotic systems as well as the adaptive switching and obtaining a large number of results to perform statistical tests.

**Acknowledgements.** This work was supported by Grant Agency of the Czech Republic - GACR P103/13/08195S, partially supported by Grants of SGS No. SP2014/159 and SP2014/170, VŠB - Technical University of Ostrava, Czech Republic, by the Development of human resources in research and development of latest soft computing methods and their application in practice project, reg. no. CZ.1.07/2.3.00/20.0072, further was supported by European Regional Development Fund under the project CEBIA-Tech No. CZ.1.05/2.1.00/03.0089 and by Internal Grant Agency of Tomas Bata University under the project No. IGA/FAI/2014/010.

## References

1. Price, K.V.: An Introduction to Differential Evolution. In: Corne, D., Dorigo, M., Glover, F. (eds.) New Ideas in Optimization, pp. 79–108. McGraw-Hill Ltd. (1999)
2. Qin, A.K., Huang, V.L., Suganthan, P.N.: Differential Evolution Algorithm With Strategy Adaptation for Global Numerical Optimization. IEEE Transactions on Evolutionary Computation 13(2), 398–417 (2009)

3. Mallipeddi, R., Suganthan, P.N., Pan, Q.K., Tasgetiren, M.F.: Differential evolution algorithm with ensemble of parameters and mutation strategies. Applied Soft Computing 11(2), 1679–1696 (2011)
4. Aydin, I., Karakose, M., Akin, E.: Chaotic-based hybrid negative selection algorithm and its applications in fault and anomaly detection. Expert Systems with Applications 37(7), 5285–5294 (2010)
5. Liang, W., Zhang, L., Wang, M.: The chaos differential evolution optimization algorithm and its application to support vector regression machine. Journal of Software 6(7), 1297–1304 (2011)
6. Zhenyu, G., Bo, C., Min, Y., Binggang, C.: Self-Adaptive Chaos Differential Evolution. In: Jiao, L., Wang, L., Gao, X.-B., Liu, J., Wu, F. (eds.) ICNC 2006. LNCS, vol. 4221, pp. 972–975. Springer, Heidelberg (2006)
7. Davendra, D., Zelinka, I., Senkerik, R.: Chaos driven evolutionary algorithms for the task of PID control. Computers & Mathematics with Applications 60(4), 1088–1104 (2010)
8. dos Santos Coelho, L., Mariani, V.C.: A novel chaotic particle swarm optimization approach using Hénon map and implicit filtering local search for economic load dispatch. Chaos, Solitons & Fractals 39(2), 510–518 (2009)
9. Davendra, D., Bialic-Davendra, M., Senkerik, R.: Scheduling the Lot-Streaming Flowshop scheduling problem with setup time with the chaos-induced Enhanced Differential Evolution. In: 2013 IEEE Symposium on Differential Evolution (SDE), April 16-19, pp. 119–126 (2013)
10. Pluhacek, M., Senkerik, R., Davendra, D., Kominkova Oplatkova, Z., Zelinka, I.: On the behavior and performance of chaos driven PSO algorithm with inertia weight. Computers & Mathematics with Applications 66(2), 122–134 (2013)
11. Pluhacek, M., Senkerik, R., Zelinka, I., Davendra, D.: Chaos PSO algorithm driven alternately by two different chaotic maps - An initial study. In: 2013 IEEE Congress on Evolutionary Computation (CEC), June 20-23, pp. 2444–2449 (2013)
12. Gandomi, A.H., Yang, X.S., Talatahari, S., Alavi, A.H.: Firefly algorithm with chaos. Communications in Nonlinear Science and Numerical Simulation 18(1), 89–98 (2013)
13. Senkerik, R., Pluhacek, M., Zelinka, I., Oplatkova, Z.K., Vala, R., Jasek, R.: Performance of Chaos Driven Differential Evolution on Shifted Benchmark Functions Set. In: Herrero, A., et al. (eds.) International Joint Conference SOCO'13-CISIS'13-ICEUTE'13. AISC, vol. 239, pp. 41–50. Springer, Heidelberg (2014)
14. Senkerik, R., Davendra, D., Zelinka, I., Pluhacek, M., Kominkova Oplatkova, Z.: On the Differential Evolution Drivan by Selected Discrete Chaotic Systems: Extended Study. In: 19th International Conference on Soft Computing, MENDEL 2013, pp. 137–144 (2013)
15. Lozi, R.: Engineering of Mathematical Chaotic Circuits. In: Zelinka, I., Chen, G., Rössler, O.E., Snasel, V., Abraham, A. (eds.) Nostradamus 2013: Prediction, Model. & Analysis. AISC, vol. 210, pp. 17–29. Springer, Heidelberg (2013)
16. Price, K.V., Storn, R.M., Lampinen, J.A.: Differential Evolution - A Practical Approach to Global Optimization. Natural Computing Series. Springer, Heidelberg (2005)
17. Sprott, J.C.: Chaos and Time-Series Analysis. Oxford University Press (2003)

# Comparison of Pseudorandom Numbers Generators and Chaotic Numbers Generators used in Differential Evolution

Lenka Skanderova and Adam Řehoř

Department of Computer Science, VSB - Technical university of Ostrava,
17. listopadu 15/2172, 708 33 Ostrava - Poruba, Czech Republic
{lenka.skanderova,adam.rehor.st}@vsb.cz

**Abstract.** Differential evolution is one of the great family of evolutionary algorithms. As well as all evolutionary algorithms differential evolution uses pseudorandom numbers generators in many steps of algorithm. In this paper we will compare pseudorandom numbers generators as Mersenne Twister, Crypto Random, Random number generator in Microsoft .NET System.Random class, Visual Studio 2010, Multiply-with-carry, Xorshift and chaotic numbers generators as Logistic map, Arnold Cat Map and Sinai. The main goal of this paper is compare these pseudorandom numbers generators and chaotic numbers generators from the view of differential evolution convergence's speed to the global minimum.

## 1 Introduction

Evolutionary algorithms (EAs) are based on three basic principles – natural selection, crossing and mutation. They work with the population of individuals. Individuals are created by their parameters and fitness value, which says how this individual is good in the population. At the beginning the population is generated randomly, that means that parameters of individuals are generated in their bounds randomly [1]. For this step EAs need pseudorandom numbers generators as well as in crossing etc.

In [2] authors describe the embedding of Lozi map as the generator of pseudorandom numbers in DE, where Schwefel's function with higher dimensions has been used as the testing function. In [3] authors use three different chaotic attractors – Dissipative standard map, Lozi map and Arnold Cat map as the pseudorandom numbers generators in Particle Swarm Optimization (PSO) algorithm. Two chaotic attractors used as the pseudorandom numbers generators in PSO algorithm – the initial study has been described in [4]. Authors of [5] deal with the question if evolutionary algorithms really require randomness. In [6] authors study DE driven by selected six chaotic systems in the task of reactor geometry optimization. Paper [7] mentions chaos driven DE with Lozi Map in the task of chemical reactor optimization and in [8] authors deal with impact of various chaotic maps on the performance of chaos enhanced PSO Algorithm with inertia weight.

I. Zelinka et al. (eds.), *Nostradamus 2014: Prediction, Modeling and Analysis of Complex Systems*, Advances in Intelligent Systems and Computing 289,
DOI: 10.1007/978-3-319-07401-6_11, 

## 2 Differential Evolution (DE)

DE has been used at first by Ken Price and Rainer Storm in 1995 [1].

*Parameters* Exact principle of DE is described for example in [1]. The quality of DE is influenced by it's control parameters:

- Number of population - $NP$ says how many individuals will be in one population. (Suggestion value is $10D$.)
- Problem dimension - $D$ means the number of cost function arguments.
- Generations - *Generations* set the number of evolution cycles.
- Crossing threshold - $CR$ may have value from 0 to 1. (Suggestion value is 0,8 - 0,9.)
- Mutation constant - $F$ - may have value from 0 to 2. (Suggestion value is 0,3 -0,9.)

In present time, there are many improved algorithms of DE, for example in [9] authors speak about a Taguchi-crossover differential evolution (TCDE), in [10] authors mention a prediction adaptive grouping differential evolution (AGDE) and in [11] a modified differential evolution algorithm (MDE) is described. In this research we will deal with DE called DE/rand/1/bin.

## 3 Pseudorandom Numbers Generators

### 3.1 Mersenne Twister (MT)

Mersenne Twister (MT) has been proposed by Makoto Matsumoto and Takuji Nishimura in 1997. It is mentioned that Mersenne Twister (MT) is a modification of a Twisted Generalized Feedback Shift Register (TGFSR). It has a long period - $2^{19937} - 1$ and a 623 dimensional equidistribution up to 32-bit accuracy [12].

MT is very popular and it is used in many areas of research, for example in [13], where Mersenne Twister hardware implementation for the Monte Carlo Localization Algorithm has been described. In [14], where Makoto Matsumoto is one of the author, variants of MT suitable for graphic processors are mentioned. MT is used in [15] as one of the pseudorandom numbers generators for massively parallel simulations on GPU. In [16] authors mention MT in connection with deterministic parallel random-number generation for dynamic-multithreading platforms.

### 3.2 Microsoft .NET System.Random class, Visual Studio 2010, (VS 2010)

As we can read in [17] pseudorandom numbers are chosen from a finite set of numbers. This pseudorandom numbers generator is based on Donald E. Knuth's subtractive random number generator algorithm, see [18].

Authors mention that this pseudorandom numbers generator starts with a seed value. This seed is time-dependent, because it is the way how to produce different sequences of numbers [17].

### 3.3 Crypto Random

This pseudorandom numbers generator is an improvement of pseudorandom number generator of Microsoft .NET System.Random class in Visual Studio 2010 and it has been used in our research as experimental. Code of this pseudorandom numbers generator is described in [19].

### 3.4 Multiply-with-Carry (MWC) and Xorshift

The author of both pseudorandom numbers generators is G. Marsaglia. It is commonly known that G. Marsaglia is an author of the famous Mother of all random numbers generators, which produces uniformly distributed pseudorandom 32-bits values. Its period is then $2^{250}$.

Big advantage of Xorshift is its speed. It generates numbers very fast, see [20] and [21]. Xorshift produces a sequence of $2^{32}-1$ integers, or sequence of $2^{64}-1$ pairs $x$, $y$ or a sequence of $2^{96}-1$ triples $x$, $y$, $z$ by means of repeated use of a simple computer construction [22].

MWC can generate sequences of random numbers with great periods from $2^{60}$ to $2^{2000000}$ very fast, see [23] and [24].

## 4 Discrete Chaotic Systems as Pseudorandom Numbers Generators in Evolutionary Algorithms

In recent time discrete chaotic systems have been used as the generators of pseudorandom numbers in EAs. In [25] chaotic differential evolution algorithm based on competitive coevolution is described. Other researchers deal with evolutionary synthesis of chaotic systems in [26] and in [27] deal with symbolic regression of deterministic chaos systems using GPA-ES system. In [28] authors describe chaos synchronization problem using Takagi-Sugeno fuzzy observer. Performance of enhanced PSO algorithm with Lozi chaotic map is described in [29]. In [30] authors mentioned chaotic populations in genetic algorithms.

### 4.1 Logistic Map

Logistic map is defined by Eg.1. If $0 \leq a < 1$ the map has a sink at $x = 0$ and every initial condition between 0 and 1 is attracted to this sink. If $1 < a < 3$ the map has a sink at $x = \frac{a-1}{a}$. If $a$ is greater than 3, the fixed point $x = \frac{a-1}{a}$ is unstable. When $a$ grows above $1 + \sqrt{6} \approx 3.45$, the period-two sink also becomes unstable. Many periodic orbits arise as $a$ is increased from 3.45 to 4 [31]. For better illustration see Fig.1.

$$x_{n+1} = ax(1 - x_n) \tag{1}$$

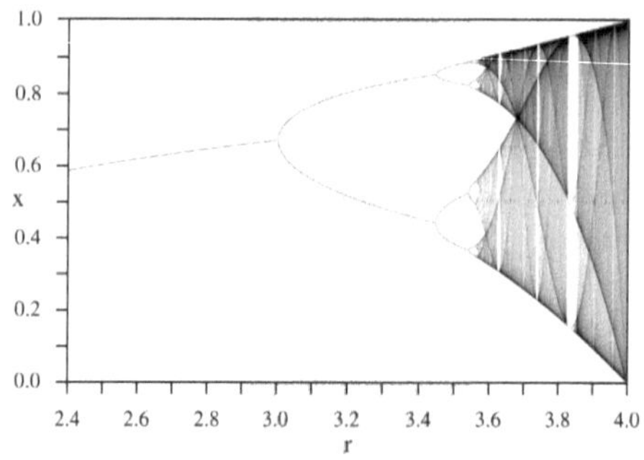


**Fig. 1.** Logistic map, bifurcation diagram

### 4.2 Arnold Cat Map

Arnold Cat Map is defined by Eq.2.

$$\begin{aligned} x_{n+1} &= (x_n + y_n) \bmod 1, \\ y_{n+1} &= (x_n + ky_n) \bmod 1 \end{aligned} \tag{2}$$

In [32] authors study the problem of period distribution of the generalized discrete Arnold Cat Map over the Galois ring Z(2e). Authors of [32] say that one goal of their paper is development of chaotic based Arnold Cat Map for encryption/decryption of DICOM files. In [34] Arnold Cat Map is mentioned in connection with chaos based cryptosystem for still visual data. In [35] Arnold Cat Map is mentioned in connection with secure medical image protection scheme based on chaotic maps and in [36] authors use Arnold Cat Map in connection with a maximum entropy-based chaotic time-variant fragile watermarking scheme for image tampering detection.

### 4.3 Sinai

Sinai is defined by Eq.3.

$$\begin{aligned} x_{n+1} &= (x_n + y_n + \sigma\cos(2\pi y_n)) \bmod 1, \\ y_{n+1} &= (x_n + 2y_n) \bmod 1 \end{aligned} \tag{3}$$

In [37] Kolmogorov-Sinai is mentioned in connection with statistics of Poincare recurrences in local and global approaches, in [38] with entropy production in classical Yang-Mills theory from Glasma initial conditions. In [39] authors connect Kolmogorov-Sinai with relationship between dynamical entropy and energy dissipation far from thermodynamic equilibrium. In [40] correlation properties of exactly solvable chaotic oscillators is described. In [41] Sinai billiard is mentioned.

## 5 Motivation

The main goal of this paper is to compare pseudorandom numbers generators and chaotic numbers generators used in evolutionary algorithm DE together. We have chosen six testing functions and we are interesting in which pseudorandom numbers generator will be better from the view of DE's convergence's speed to the global minimum.

## 6 Experiment Design

For experiments HP Pavilion dv7-6050 with processor Intel Core i7 with frequency 2 GHz, 4 GB RAM and graphic card AMD Radeon HD 6770M and Microsoft Visual Studio 2010 have been used. The experiments have been processed by Mathematica 8 and Gnuplot 4.6. Precise setting of DE is mentioned in Table 1. As testing functions Egg Holder's, Griewangk's, Pathological's, Rana's, Rastrigin's and Schwefel's functions have been chosen. For Egg Holder's, Pathological's and Rana's functions there is not common formula for easy calculation of global minimum. Griewangk's global minimum is $f(x) = 0$ as well as Rastrigin's. Schwefel's global minimum is $f(x) = -418.9829D$ where $D$ means dimension.

The parameter $a$ of logistic map has been set to $a = 4$. The initial value of $x$ has been set to $x_0 = 0.02$. In Arnold Cat Map the initial value of $x$ has been set to $x_0 = 0$ and the initial value of $y$ has been set to $y_0 = \frac{1}{\sqrt{2}}$. Parameter $k$ has been set to $k = 0.1$. In Sinai the initial value of $x$ has been set to $x_0 = 0.5$ and the initial value of $Y$ has been set to $y_0 = 0.5$. The value of parameter $\sigma$ has been set to $\sigma = 0.1$.

**Table 1.** DE setting

| Parameter | Value |
|---|---|
| $NP$ | 50 |
| $D$ | 20 |
| $F$ | 0.4 |
| $CR$ | 0.6 |
| *Generations* | 1500 |

## 7 Results

Results are mentioned in Tables 2, 3, 4, 5, 6 and 7, where we can see minimum, maximum and average reached values for DE, where Egg Holder's, Griewangk's, Pathological's, Rana's, Rastrigin's and Schwefel's functions have been used as testing functions.

**Table 2.** Comparison of pseudorandom numbers generators (PRNGs) and chaotic numbers generators from the view of DE's convergence's speed, where **Egg Holder's function** has been used as cost function. Setting of DE is mentioned in Table 1

| PRNG | Min | Max | Average |
|---|---|---|---|
| Arnold Cat Map | -13243.326 | -6828.035 | -8350.747 |
| Crypto Random | -9876.216 | -7274.433 | -8416.598 |
| Logistic map | -10420.003 | -8143.635 | -9157.682 |
| MT | -10227.992 | -7214.173 | -8389.643 |
| MWC | -8465.763 | -6567.594 | -7266.990 |
| VS 2010 | -15485.638 | -7052.491 | -8800.282 |
| Sinai | -12284.573 | -6804.800 | -8181.811 |
| Xorshift | -10239.755 | -6828.035 | -8350.747 |

**Table 3.** Comparison of pseudorandom numbers generators (PRNGs) and chaotic numbers generators from the view of DE's convergence's speed, where **Griewangk's function** has been used as cost function. Setting of DE is mentioned in Table 1.

| PRNG | Min | Max | Average |
|---|---|---|---|
| Arnold Cat Map | 0.101 | 0.335 | 0.210 |
| Crypto Random | 0.120 | 0.313 | 0.199 |
| Logistic map | 0.111 | 0.304 | 0.195 |
| MT | 0.089 | 0.393 | 0.213 |
| MWC | 0.551 | 0.823 | 0.708 |
| VS 2010 | 0.101 | 0.375 | 0.205 |
| Sinai | 0.140 | 0.332 | 0.228 |
| Xorshift | 0.079 | 0.326 | 0.210 |

**Table 4.** Comparison of pseudorandom numbers generators (PRNGs) and chaotic numbers generators from the view of DE's convergence's speed, where **Pathological's function** has been used as cost function. Setting of DE is mentioned in Table 1.

| PRNG | Min | Max | Average |
|---|---|---|---|
| Arnold Cat Map | 3.943 | 6.471 | 5.663 |
| Crypto Random | 4.821 | 6.278 | 5.744 |
| Logistic map | 3.805 | 6.363 | 5.766 |
| MT | 4.499 | 6.219 | 5.647 |
| MWC | 5.513 | 6.719 | 6.369 |
| VS 2010 | 2.970 | 6.373 | 5.329 |
| Sinai | 3.352 | 6.599 | 5.722 |
| Xorshift | 4.677 | 6.234 | 5.686 |

**Table 5.** Comparison of pseudorandom numbers generators (PRNGs) and chaotic numbers generators from the view of DE's convergence's speed, where **Rana's function** has been used as cost function. Setting of DE is mentioned in Table 1.

| PRNG | Min | Max | Average |
|---|---|---|---|
| Arnold Cat Map | -5908.480 | -4366.865 | -4931.769 |
| Crypto Random | -5808.122 | -4478.561 | -5065.123 |
| Logistic map | -6559.807 | -4732.013 | -5498.817 |
| MT | -5904. 120 | -4608.468 | -5113.732 |
| MWC | -5237.811 | -4139.292 | -4593.440 |
| VS 2010 | -5964.770 | -4514.558 | -5051.146 |
| Sinai | -5993.439 | -4102.931 | -4893.173 |
| Xorshift | -5700.747 | -4447.350 | -5100.393 |

**Table 6.** Comparison of pseudorandom numbers generators (PRNGs) and chaotic numbers generators from the view of DE's convergence's speed, where **Rastrigin's function** has been used as cost function. Setting of DE is mentioned in Table 1.

| PRNG | Min | Max | Average |
|---|---|---|---|
| Arnold Cat Map | 0.100 | 8.812 | 4.296 |
| Crypto Random | 0 | 6.660 | 2.535 |
| Logistic map | 0 | 7.761 | 2.548 |
| MT | 0 | 6.965 | 3.269 |
| MWC | 33.106 | 63.861 | 49.776 |
| VS 2010 | 0 | 8.522 | 3.788 |
| Sinai | 0.048 | 8.039 | 3.632 |
| Xorshift | 0 | 7.683 | 3.194 |

**Table 7.** Comparison of pseudorandom numbers generators (PRNGs) and chaotic numbers generators from the view of DE's convergence's speed, where **Schwefel's function** has been used as cost function. Setting of DE is mentioned in Table 1.

| PRNG | Min | Max | Average |
|---|---|---|---|
| Arnold Cat Map | -8379.658 | -8142.7811 | -8355.970 |
| Crypto Random | -8379.658 | -8261.219 | -8370.183 |
| Logistic map | -8379.658 | -8261.219 | -8378.473 |
| MT | -8379.658 | -8261.219 | -8377.289 |
| MWC | -8379.658 | -8261.219 | -8378.530 |
| VS 2010 | -8379.658 | -8261.219 | -8368.998 |
| Sinai | -8379.658 | -8261.219 | -8363.076 |
| Xorshift | -8379.658 | -8142.781 | -8355.970 |

## 8 Conclusion

From the results mentioned in previous section we can make sme conclusions:

- **Egg Holder's Function:** When we look at the Table 2, we will see that the best results were reached when Logistic map with parameter $a = 4$ had been used as chaotic numbers generator. The speed of DE's convergence to the global minimum is the best in this case. It is clear from the column Average. On the other hand the smallest minimum value was reached when VS 2010 had been used as pseudorandom numbers generator.
- **Griewangk's Function:** From Table 3 we can see that the best results were reached when Logistic map with parameter $a = 4$ had been used as pseudorandom numbers generator. From the column Average we can see that average reached value is the smallest for Logistic map. That means that DE converged to the global minimum faster when Logistic map with parameter $a = 4$ had been used than when other pseudorandom numbers had been used. When we look at the column Min, we will see that minimum reached values are comparable. It is true that the best results were reached when Xorshift had been used. However, DE did not reached the global minimum in any case of pseudorandom numbers generator.
- **Pathological's Function:** Table 4 say that the best results were reached when generator of pseudorandom numbers of VS 2010 had been used. When we look at the other pseudorandom numbers generator we will see, that the results of all pseudorandom numbers generators are comparable for Pathological function. It is interesting that the minimum value of fitness was reached when VS 2010 had been used too.
- **Rana's Function:** DE reached the best results when Logistic map had been used as pseudorandom numbers generator. We can see that the average reached value of fitness is the smallest as well as minimum reached value for the row Logistic map. Results of other pseudorandom numbers generators are comparable.
- **Rastrigin's Function:** We can see in Table 6 that the global minimum was reached when Crypto Random, Logistic map, MT, VS 2010 and Xorshift had been used. When Arnold Cat Map, Sinai and MWC had been used the global minimum was not reached. From the view of DE convergence's speed Crypto Random and Logistic map reached the best results. The average reached values of MT, VS 2010, MT and Xorshift are comparable. The worst values were reached when MWC had been used.
- **Schwefel's Function:** In Table 7 there are results for Schwefel's function. As we can see DE convergence's speed was the best when MWC and Logistic map had been used as the pseudorandom numbers generators. The global minimum was reached for all pseudorandom numbers generators.

From Tables 2, 3, 4, 5, 6 and 7 we can see that the best pseudorandom numbers generator from the view of DE convergence's speed has become Logistic map. DE's convergence's speed was the best almost for all chosen cost functions except

Pathological. On the other hand when we look at the Table 4 we will see that the average reached values are comparable for all pseudorandom numbers generators. When VS 2010 had been used, DE reached the smallest minimum value.

**Acknowledgement.** The following grants are acknowledged for the financial support provided for this research: Grant Agency of the Czech Republic - GACR P103/13/08195S, is partially supported by Grant of SGS No. SP2014/42, VB - Technical University of Ostrava, Czech Republic, by the Development of human resources in research and development of latest soft computing methods and their application in practice project, reg. no. CZ.1.07/2.3.00/20.0072 funded by Operational Programme Education for Competitiveness.

## References

1. Zelinka, I., Celikovsky, S., Richter, H., Chen, G., et al. (eds.): Evolutionary Algorithms and Chaotic Systems. SCI, vol. 267. Springer, Heidelberg (2010)
2. Senkerik, R., et al.: Chaos driven evolutionary algorithm: A new approach for evolutionary optimization. International Journal of Mathematics and Computers in Simulation 7, 363–368 (2013)
3. Pluhacek, M., et al.: On the behavior and performance of chaos driven PSO algorithm with inertia weight. Computers & Mathematics with Applications 66, 122–134 (2013)
4. Pluhacek, M., et al.: Chaos PSO Algorithm Driven Alternately by two Different Chaotic Maps - an Initial Study. In: 2013 IEEE Congress on Evolutionary Computation (CEC), pp. 2444–2449 (2013)
5. Zelinka, I., et al.: Do Evolutionary Algorithms Indeed Require Randomness? In: 2013 IEEE Congress on Evolutionary Computation (CEC), pp. 2283–2289 (2013)
6. Senkerik, R., et al.: Investigation on the Differential Evolution Driven by Selected Six Chaotic Systems in the Task of Reactor Geometry Optimization. In: 2013 IEEE Congress on Evolutionary Computation (CEC), pp. 3087–3094 (2013)
7. Senkerik, R., Davendra, D., Zelinka, I., Pluhacek, M., Kominkova Oplatkova, Z.: Chaos Driven Differential Evolution with Lozi Map in the Task of Chemical Reactor Optimization. In: Rutkowski, L., Korytkowski, M., Scherer, R., Tadeusiewicz, R., Zadeh, L.A., Zurada, J.M. (eds.) ICAISC 2013, Part II. LNCS, vol. 7895, pp. 56–66. Springer, Heidelberg (2013)
8. Pluhacek, M., Senkerik, R., Zelinka, I.: Impact of Various Chaotic Maps on the Performance of Chaos Enhanced PSO Algorithm with Inertia Weight – An Initial Study. In: Zelinka, I., Snasel, V., Rössler, O.E., Abraham, A., Corchado, E.S. (eds.) Nostradamus: Mod. Meth. of Prediction, Modeling. AISC, vol. 192, pp. 153–166. Springer, Heidelberg (2013)
9. Tsai, J.T.: Optimized weights of document keywords for auto-reply accuracy. Neurocomputing 124, 43–56 (2014)
10. Kong, X.Y., et al.: A prediction-based adaptive grouping differential evolution algorithm for constrained numerical optimization. Soft Computing 17, 2293–2309 (2013)
11. Zou, D.X., et al.: A modified differential evolution algorithm for unconstrained optimization problems. Neurocomputing 120, 469–481 (2013)

12. Matsumoto, M., Nishimura, T.: Mersenne Twister: A 623-Dimensionally Equidistributed Uniform Pseudo-Random Number Generator. ACM Transactions on Modeling and Computer Simulation (TOMACS) - Special issue on uniform random number generation 8, 3–30 (1998)
13. Bonato, V., et al.: A Mersenne Twister Hardware Implementation for the Monte Carlo Localization Algorithm. Journal of Signal Processing Systems for Signal Image and Video Technology 70, 75–85 (2013)
14. Saito, M., Matsumoto, M.: Variants of Mersenne Twister Suitable for Graphic Processors. ACM Transactions on Mathematical Software 39 (2013), doi:10.1145/2427023.2427029
15. Manssen, M., et al.: Random number generators for massively parallel simulations on GPU. European Physical Journal-Special Topics 210, 53–71 (2012), doi:10.1140/epjst/e2012-01637-8
16. Leiserson, C.E., et al.: Deterministic Parallel Random-Number Generation for Dynamic-Multithreading Platforms. ACM Sigplan Notices 47, 193–204 (2012), doi:10.1145/2370036.2145841
17. `http://msdn.microsoft.com/en-us/library/system.random%28v=vs.110%29.aspx`
18. Knuth, D.E.: Art of Computer Programming. The: Generating All Trees, History of Combinatorial Generation, vol. 4, Fascicle 4. Pearson Education, Inc. (2006)
19. `http://thinketg.com/how-to-generate-better-random-numbers-in-c-net-2/`
20. Zeng, G., et al.: Improvement of one type Xorshift random number generators. In: Proceedings of the First International Symposium on Data, Privacy, and E-Commerce, pp. 472–474 (2007)
21. Leong, P.H.W., et al.: A comment on the implementation of the Ziggurat method. Journal of Statistical Software 7, 1–4 (2005)
22. Marsaglia, G.: Xorshift RNGs. Journal of Statistical Software (2003)
23. de Oliveira, T., Marranghello, N.: Design of a reconfigurable pseudorandom number generator for use in intelligent systems. Neurocomputing 74(10), 1510–1519 (2011)
24. Marsaglia, G., Tsang, W.W.: The Monty Python method for generating random variables. ACM Transactions on Mathematical Software 24(3), 341–350 (1998)
25. Li, X., et al.: Chaotic Differential Evolution Algorithm Based on Competitive Coevolution and Its Application to Dynamic Optimizzation of Chemical Processes. Intelligent Automation and Soft Computing, 85–98 (February 2013), doi:10.1080/10798587.2013.771437
26. Zelinka, I.: On Evolutionary Synthesis of Chaotic Systems. In: Zelinka, I., Snasel, V., Rössler, O.E., Abraham, A., Corchado, E.S. (eds.) Nostradamus: Mod. Meth. of Prediction, Modeling. AISC, vol. 192, pp. 29–34. Springer, Heidelberg (2013)
27. Brandejsky, T., Zelinka, I.: Specific Behaviour of GPA-ES Evolutionary System Observed in Deterministic Chaos Regression. In: Zelinka, I., Snasel, V., Rössler, O.E., Abraham, A., Corchado, E.S. (eds.) Nostradamus: Mod. Meth. of Prediction, Modeling. AISC, vol. 192, pp. 73–81. Springer, Heidelberg (2013)
28. Chadli, M., Zelinka, I.: Chaos Synchronization Based on Unknown Inputs Takagi-Sugeno Fuzzy Observer. In: Zelinka, I., Snasel, V., Rössler, O.E., Abraham, A., Corchado, E.S. (eds.) Nostradamus: Mod. Meth. of Prediction, Modeling. AISC, vol. 192, pp. 83–92. Springer, Heidelberg (2013)
29. Pluhacek, M., Budikova, V., Senkerik, R., Oplatkova, Z., Zelinka, I.: Extended Initial Study on the Performance of Enhanced PSO Algorithm with Lozi Chaotic Map. In: Zelinka, I., Snasel, V., Rössler, O.E., Abraham, A., Corchado, E.S. (eds.) Nostradamus: Mod. Meth. of Prediction, Modeling. AISC, vol. 192, pp. 167–177. Springer, Heidelberg (2013)

30. Ma, Z.S.: Chaotic populations in genetic algorithms. Applied Soft Computing 12, 2409–2424 (2012), doi:10.1016/j.asoc.2012.03.001
31. Alligood, K., Sauer, T.D., Yorke, J.A.: CHaos - an introduction to dynamical systems. Textbooks in Mathematical Sciences, p. 1197. Springer - Verlag New York, Inc. (1996) ISBN 0- 987-94677-2
32. Chen, F., et al.: Period Distribution of the Generalized Discrete Arnold Cat Map for N=2(e). IEEE Transactions on Information Theory 59, 3249–3255 (2013)
33. Arun Fera, M., Jaganathan, S.: Securing DICOM Format Image Archives Using Improved Chaotic Cat Map Method. In: Meghanathan, N., Nagamalai, D., Chaki, N. (eds.) Advances in Computing & Inform. Technology. AISC, vol. 177, pp. 319–328. Springer, Heidelberg (2012)
34. Taneja, N., et al.: Chaos based cryptosystem for still visual data. Multimedia Tools and Applications 61, 281–298 (2012)
35. Fu, C., et al.: An efficient and secure medical image protection scheme based on chaotic maps. Computers in Biology and Medicine 43, 1000–1010 (2013)
36. Chen, Y.L., et al.: A Maximum Entropy-Based Chaotic Time-Variant Fragile Watermarking Scheme for Image Tampering Detection. Entropy 15, 3170–3185 (2013)
37. Anishchenko, V.S., et al.: Statistics of Poincare recurrences in local and global approaches. Communications in Nonlinear Science and Numerical Simulation 18, 3423–3435 (2013)
38. Iida, H., et al.: Entropy production in classical Yang-Mills theory from glasma initial conditions. Physical Review D 88 (November 12, 2013)
39. Green, J.R., et al.: Relationship between dynamical entropy and energy dissipation far from thermodynamic equilibrium. Proceedings of the National Academy of Sciences of the United States of America 110, 16339–16343 (2013)
40. Blakely, J.N., Corron, N.J.: Correlation properties of exactly solvable chaotic oscillators. Physical Review E 88 (August 12, 2013)
41. Pecora, L.M., et al.: Regularization of Tunneling Rates with Quantum Chaos. International Journal of Bifurcation and Chaos 22 (October 2012)

# Analytic Programming Powered by Chaotic Dynamics

Ivan Zelinka[1,3], Lenka Skanderova[1], Petr Šaloun[1], Roman Senkerik[2], Tran Trong Dao[3], and Duy Vo Hoang[3]

[1] VSB-Technical University of Ostrava,
17. listopadu 15 708 33, Ostrava-Poruba, Czech Republic
ivan.zelinka@vsb.cz

[2] Faculty of Applied Informatics, Tomas Bata University in Zlin, Czech Republic
senkerik@fai.utb.cz

[3] MERLIN, Ton Duc Thang University,
19 Nguyen Huu Tho Str., Dist. 7, Ho Chi Minh City, Vietnam
ivan.zelinka@vsb.cz, {trantrongdao,vohoangduy}@tdt.edu.vn

**Abstract.** In this paper we discuss alternative tool for symbolic regression so called Analytical programming and compare its variants powered by classical random as well as chaotic random-like number generator. Experimental data are used from the previous experiments reported for genetic programming. Selected algorithms are differential evolution, SOMA, particle swarm, simulated annealing and evolutionary strategies. All of them are mutually used in scheme Master-Slave meta-evolution for final complex structure fitting and its parameter estimation.

## 1 Introduction

In [19] is discussed an alternative approach for symbolic structures and solutions synthesis, used here as well as brief well known methods. For example Genetic Programming (GP), [3,4] or Grammatical Evolution (GE), [16], [5]. Generally, there are two well known methods, which can be used for symbolic structures synthesis by means of computers. Another interesting research was carried out by Artificial Immune Systems (AIS) or/and systems, which do not use tree structures like linear GP and other similar algorithm like Multi Expression Programming (MEP), etc. In this chapter, a different method called Analytic Programming (AP) [19], is presented. AP is a grammar free algorithmic superstructure, which can be used by any programming language and also by any arbitrary Evolutionary Algorithm (EA) or another class of numerical optimization method. This paper describes use and results of AP with EA's like Differential Evolution (DE), Self-Organizing Migrating Algorithm (SOMA) on selected GP test examples. All case studies has been carefully prepared and repeated in order to get valid statistical data for proper conclusions.

The initial idea of symbolic regression by means of a computer program was proposed in GP [3,4]. The other approach of GE was developed in [5] and AP in [6]. Another interesting investigation using symbolic regression were carried out in [7] on AIS and Probabilistic Incremental Program Evolution (PIPE), which generates functional programs from an adaptive probability distribution over all possible programs.

I. Zelinka et al. (eds.), *Nostradamus 2014: Prediction, Modeling and Analysis of Complex Systems*, Advances in Intelligent Systems and Computing 289,
DOI: 10.1007/978-3-319-07401-6_12, © Springer International Publishing Switzerland 2014

Yet another new technique is the so called *Transplant Evolution*, see [8], [9] and [10] which is closely associated with the conceptual paradigm of AP, and modified for GE. GE was also extended to include DE by [11]. Generally speaking, it is a process which combines, evaluates and creates more complex structures based on some elementary and noncomplex objects, in an evolutionary way. Such elementary objects are usually simple mathematical operators $(+,-,\times,...)$, simple functions (*sin*, *cos*, *And*, *Not*, ...), user-defined functions (simple commands for robots – MoveLeft, TurnRight, ...), etc. An output of symbolic regression is a more complex "object" (formula, function, command,...), solving a given problem like data fitting of the so-called Sextic and Quintic problem [12,13], randomly synthesized function [13], Boolean problems of parity and symmetry solution (basically logical circuits synthesis) [14,6], or synthesis of quite complex robot control command by [4,15]. Examples mentioned in [19] are just a few samples from numerous repeated experiments done by AP, which are used to demonstrate how complex structures can be produced by symbolic regression in general for different problems, see [19].

## 2 Used Methods

For our experiments described here standard hardware and algorithms has been used. All important information about algorithms used in our experiments are mentioned and referred here.

### 2.1 Evolutionary Algorithms

Comparing to the previous method of genetic programming and grammatical evolution, in this research is used symbolic (in this case AP) regression with evolutionary algorithms like Self-Organizing Migrating Algorithm (SOMA) and Differential Evolution (DE), simulated annealing (SA), Evolutionary strategies (ES) and Particle Swarm (PSO). Together with combinations of all used algorithms has been used approach called Analytic Programming (AP) [19]. AP has been powered by chaotic number generator and compared with the same AP powered by classical pseudorandom number generator. Due to space limits we report here results for DE-SOMA and SOMA-DE test results.

### 2.2 Experiment Design

Our experiments has been set so that analytic programming powered by classical pseudorandom number generator and deterministic chaos generators (see for example [20], [21] and [22]), were used. Based on the fact that deterministic chaos generators were successfully used in the past [20], [21] and [22] for classical evolutionary algorithms, we have selected logistic equation (1), and data series generated by this equation with setting $A = 4$ as a random numbers to replace classical classical pseudorandom number generator in symbolic regression. Algorithms selected for our experiments were SOMA [2] and differential evolution (DERand1Bin) [1], simulated annealing (SA) [24] and

[25], Evolutionary strategies (ES) [23] and Particle Swarm (PSO) [26]. Due to space limits we report here results for DE-SOMA and SOMA-DE test results.

$$x_{n+1} = Ax_n(1 - x_n) \tag{1}$$

The cost function has been defined according to Equation 2 and the main aim of the used evolution was to find formula, that gives the smallest value of Equation 2. To verify more properly the functionality of AP, a set of comparative simulations based on selected examples from Koza's GP, have been done. Two algorithms were used for comparison of AP with GP - DE and SOMA. Simulations were focused on selected examples from [4] and [12], see also Eq. 3 and 4.

$$\sum_{1}^{n} |data_i - synthesized\_data_i| \tag{2}$$

$$x^6 - x^4 + x^2 \tag{3}$$

$$x^5 - 2x^3 + x \tag{4}$$

Based on the studies in [4] and[12], the above mentioned problems have been selected for comparative study. Data (50 equidistantly located points) are generated by means of polynomials $x^6 - x^4 + x^2$ and $x^5 - 2x^3 + x$ in range [-1, 1].

All experiments were done in Mathematica 9, on MacBook Pro, 2.8 GHz Intel Core 2 Duo. Because this is pioneering experiment, we have used one data set and repeat each experiment 50 times. The aim was to compare performance of symbolic regression with classical pseudorandom number generator (Mersenne-twister) and chaotic one. Algorithm setting was based on well known recommended setting for each algorithm, as demonstrated in Tab. 1.

**Table 1.** Algorithms setting

| DE | | SOMA | |
|---|---|---|---|
| NP | 500 | PopSize | 500 |
| Dimensions | 100 | Dimensions | 100 |
| Generations | 500 | Migrations | 10 |
| F | 0.9 | PRT | 0.1 |
| CR | 0.5 | PathLength | 3 |
| | | Step | 0.11 |

## 3 Results

The results of our experiments, based on AP with DE, SOMA, SA, ES and PSO are summarized and visualized on Fig. 1 - 8, and in Tables 2 - 3.

**Table 2.** Sextic problem, better values are bold faced

| DE-SOMA | Logistic eq. | Mersenne Twister | SOMA-DE | Logistic eq. | Mersenne Twister |
|---|---|---|---|---|---|
| Minimum | 0.08469 | **0.03807** | Minimum | 0.02332 | **0.01374** |
| Mean | 0.27846 | **0.24658** | Mean | 0.13580 | **0.08419** |
| Median | 0.25806 | **0.23523** | Median | 0.12108 | **0.06504** |
| Maximum | 0.84326 | **0.53340** | Maximum | 0.34255 | **0.20998** |

**Table 3.** Quintic problem, better values are bold faced

| DE-SOMA | Logistic eq. | Mersenne Twister | SOMA-DE | Logistic eq. | Mersenne Twister |
|---|---|---|---|---|---|
| Minimum | **0.05621** | 0.07492 | Minimum | **0.00870** | 0.01487 |
| Mean | **0.26239** | 0.26832 | Mean | 0.11655 | **0.10726** |
| Median | **0.23946** | 0.24356 | Median | 0.11409 | **0.07113** |
| Maximum | 0.78554 | **0.69073** | Maximum | **0.33879** | 0.39995 |

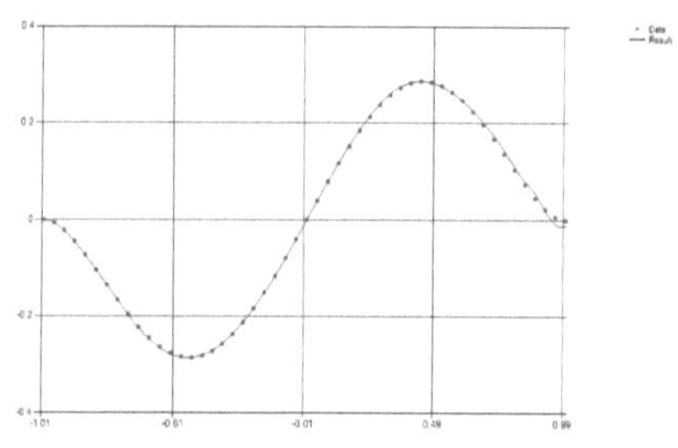

**Fig. 1.** Typical result of evolutionary fitting of quintic problem for DE-SOMA. The generator of the pseudorandom numbers was Logistic equation.

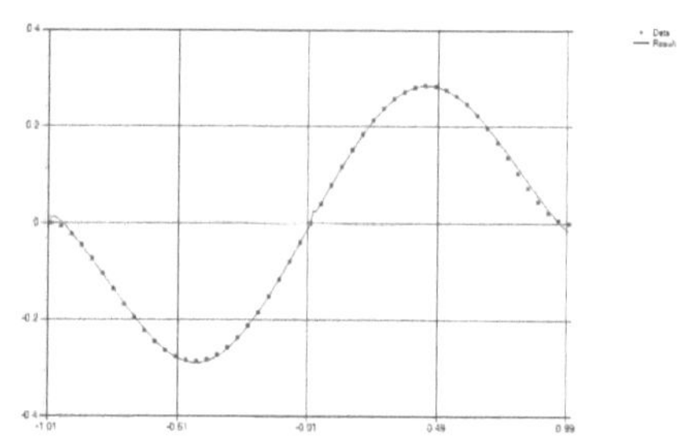

**Fig. 2.** Typical result of evolutionary fitting of quintic problem for DE-SOMA. The generator of the pseudorandom numbers was Mersenne Twister generator.

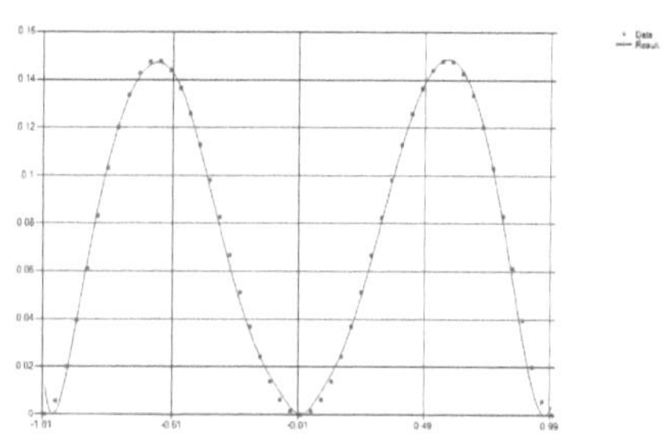

**Fig. 3.** Typical result of evolutionary fitting of sextic problem for DE-SOMA. The generator of the pseudorandom numbers was Logistic equation.

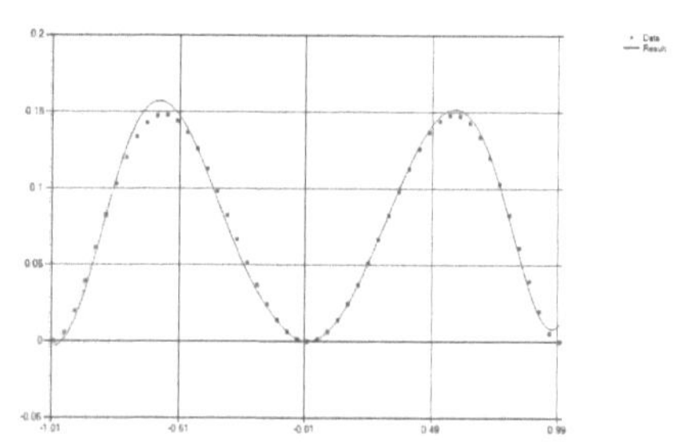

**Fig. 4.** Typical result of evolutionary fitting of sextic problem for DE-SOMA. The generator of the pseudorandom numbers was Mersenne Twister generator.

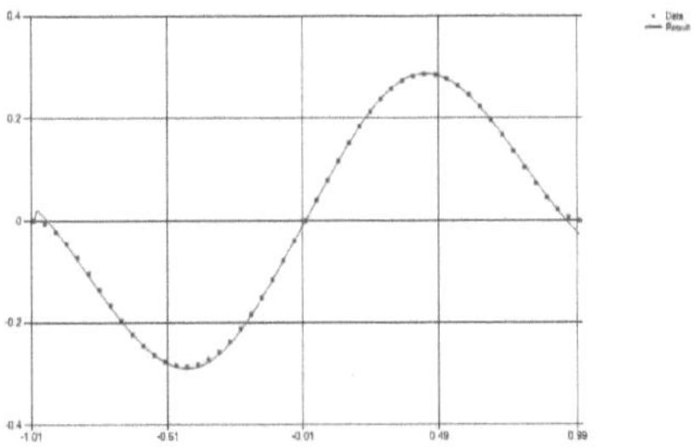

**Fig. 5.** Typical result of evolutionary fitting of quintic problem for SOMA-DE. The generator of the pseudorandom numbers was Logistic equation.

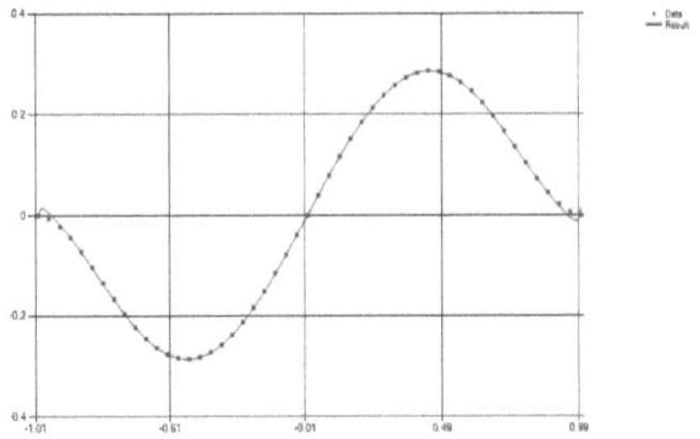

**Fig. 6.** Typical result of evolutionary fitting of quintic problem for SOMA-DE. The generator of the pseudorandom numbers was Mersenne Twister generator.

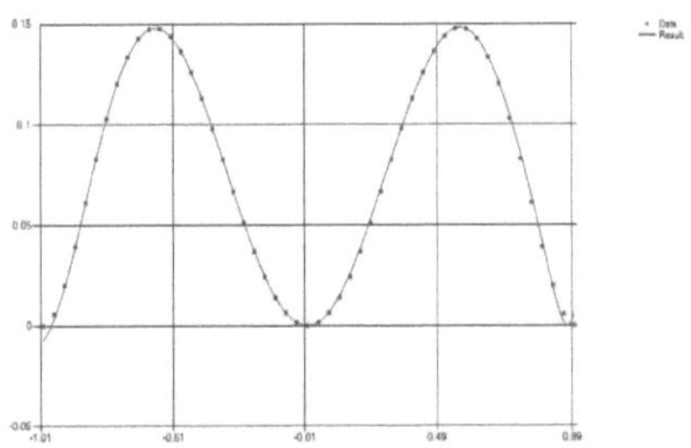

**Fig. 7.** Typical result of evolutionary fitting of sextic problem for SOMA-DE. The generator of the pseudorandom numbers was Logistic equation.

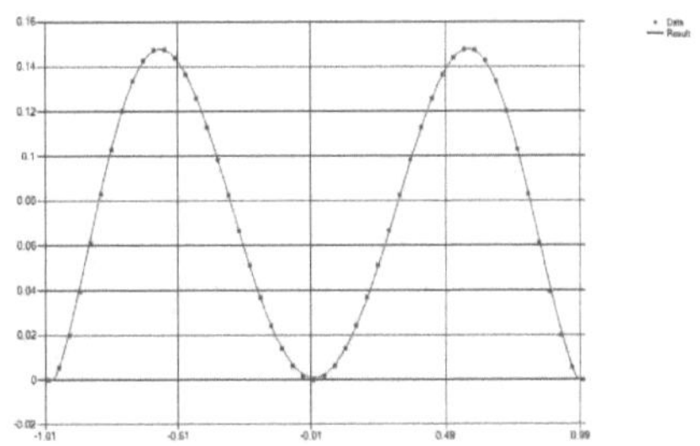

**Fig. 8.** Typical result of evolutionary fitting of sextic problem for SOMA-DE. The generator of the pseudorandom numbers was Mersenne Twister generator.

## 4 Conclusion

The main goal of this research was to test whether it is possible to use AP with deterministic chaos generators instead of pseudorandom one. Two version of AP has been used. The first one was powered by classical pseudorandom number generator and the second one by pseudo-chaotic number generator, in this case by logistic equation. Ap was used with SOMA as well as with DE algorithms. Obtained results (see Figures 1 - 8) shows that AP can be used for that kind of task and both cases, i.e. AP with PRNGS as well as AP with chaos can be used. In Tables 2 - 3 it is visible that mostly Mersenne Twisted generator gives better results, that are slightly better rather than those with logistic equation. A few times vas AP with logistic equation better, see Table 2. It is clear that more experiments under different setting of algorithms and generators is needed. Generators based on chaos can be tuned for different level of chaos and one can get wide spectra of results.

Based on results from [20], [21] and [22], we are going to use chaotic generators with variable level of chaos in order to get more results that show dependance of AP performance on numerical precision of chaotic generators. As was reported in [20], [21] and [22], it is possible to use deterministic chaos generators instead of pseudorandom number and set by control parameter periodicity of obtained data series. It is reported in [20], [21] and [22]. More information about mutual fusion of evolutionary algorithms and deterministic chaos can be found in [17] and [18], while about AP in [19].

**Acknowledgement.** The following two grants are acknowledged for the financial support provided for this research: Grant Agency of the Czech Republic - GACR P103/13/08195S, by the Development of human resources in research and development of latest soft computing methods and their application in practice project, reg. no. CZ.1.07/2.3.00/20.0072 funded by Operational Programme Education for Competitiveness, co-financed by ESF and state budget of the Czech Republic, partially supported by Grant of SGS No. SP2013/159, VŠB - Technical University of Ostrava, Czech Republic, and by European Regional Development Fund under the project CEBIA-Tech No. CZ.1.05/2.1.00/03.0089. Special thanks also belong to the research group MERLIN of Ton Duc Thang University, Ho Chi Minh City, Vietnam.

## References

1. Price, K.: An Introduction to Differential Evolution. In: Corne, D., Dorigo, M., Glover, F. (eds.) New Ideas in Optimization, pp. 79–108. McGraw-Hill, London (1999)
2. Zelinka, I.: SOMA – Self Organizing Migrating Algorithm. In: Babu, B.V., Onwubolu, G. (eds.) New Optimization Techniques in Engineering. STUDFUZZ, vol. 141, pp. 167–218. Springer, Heidelberg (2004)
3. Koza, J.: Genetic Programming: A paradigm for genetically breeding populations of computer programs to solve problems, Stanford University, Computer Science Department, Technical Report, STAN-CS-90-1314 (1990)
4. Koza, J.: Genetic Programming. MIT Press (1998)
5. Ryan, C., Collins, J.J., O'Neill, M.: Grammatical evolution: Evolving programs for an arbitrary language. In: Banzhaf, W., Poli, R., Schoenauer, M., Fogarty, T.C. (eds.) EuroGP 1998. LNCS, vol. 1391, pp. 83–96. Springer, Heidelberg (1998)
6. Zelinka, I., Oplatkova, Z., Nolle, L.: Analytic programming – Symbolic regression by means of arbitrary evolutionary algorithms. Int. J. of Simulation, Systems, Science and Technology 6(9), 44–56 (2005)
7. Johnson, C.: Artificial immune systems programming for symbolic regression. In: Ryan, C., Soule, T., Keijzer, M., Tsang, E.P.K., Poli, R., Costa, E. (eds.) EuroGP 2003. LNCS, vol. 2610, pp. 345–353. Springer, Heidelberg (2003)
8. Weisser, R., Osmera, P.: Two-Level Transplant Evolution for Optimization of General Controllers. In: New Trends in Technologies, Sciyo (2010)
9. Weisser, R., Osmera, P.: Two-level Tranpslant Evolution. In: 17th Zittau Fuzzy Colloquium, Zittau, Germany (2010)
10. Weisser, R., Osmera, P., Matousek, R.: Transplant Evolution with Modified Schema of Differential Evolution: Optimization Structure of Controllers. In: International Conference on Soft Computing MENDEL, Brno, Czech Republic (2010)
11. O'Neill, M., Brabazon, A.: Grammatical Differential Evolution. In: Proceedings of International Conference on Artificial Intelligence, pp. 231–236. CSEA Press (2006)
12. Koza, J., Bennet, F., Andre, D., Keane, M.: Genetic Programming III. Morgan Kaufmann, New York (1999)

13. Zelinka, I., Oplatkova, Z.: Analytic programming – Comparative study. In: Proceedings of Second International Conference on Computational Intelligence, Robotics, and Autonomous Systems, Singapore (2003)
14. Koza, J., Keane, M., Streeter, M.: Evolving inventions, pp. 40–47. Scientific American (2003)
15. Oplatkova, Z., Zelinka, I.: Investigation on artificial ant using analytic programming. In: Proceedings of Genetic and Evolutionary Computation Conference, Seattle, WA, pp. 949–950 (2006)
16. O'Neill, M., Ryan, C.: Grammatical Evolution, Evolutionary Automatic Programming in an Arbitrary Language. Springer, New York (2003)
17. Zelinka, I., Chen, G., Celikovsky, S.: Chaos Synthesis by Means of Evolutionary algorithms. International Journal of Bifurcation and Chaos 18(4), 911–942 (2008) ISSN 0218-1274
18. Zelinka, I., Celikovsky, S., Richter, H., Chen, G. (eds.): Evolutionary Algorithms and Chaotic Systems. SCI, vol. 267. Springer, Heidelberg (2010)
19. Zelinka, I., Davendra, D., Senkerik, R., Jasek, R., Oplatkova, Z.: Analytical Programming - a Novel Approach for Evolutionary Synthesis of Symbolic Structures. In: Kita, E. (ed.) Evolutionary Algorithms. InTech (2011), `http://www.intechopen.com/books/evolutionary-algorithms/analytical-programming-a-novel-approach-for-evolutionary-synthesis-of-symbolic-structures`, doi:10.5772/16166, ISBN: 978-953-307-171-8
20. Zelinka, I., Senkerik, R., Pluhacek, M.: Do Evolutionary Algorithms Indeed Require Randomness? In: IEEE Congress on Evolutionary Computation, Cancun, Mexico, pp. 2283–2289 (2013)
21. Zelinka, I., Chadli, M., Davendra, D., Senkerik, R., Pluhacek, M., Lampinen, J.: Hidden Periodicity - Chaos Dependance on Numerical Precision. In: Zelinka, I., Chen, G., Rössler, O.E., Snasel, V., Abraham, A. (eds.) Nostradamus 2013: Prediction, Model. & Analysis. AISC, vol. 210, pp. 47–59. Springer, Heidelberg (2013)
22. Zelinka, I., Chadli, M., Davendra, D., Senkerik, R., Pluhacek, M., Lampinen, J.: Do Evolutionary Algorithms Indeed Require Random Numbers? Extended Study. In: Zelinka, I., Chen, G., Rössler, O.E., Snasel, V., Abraham, A. (eds.) Nostradamus 2013: Prediction, Model. & Analysis. AISC, vol. 210, pp. 61–75. Springer, Heidelberg (2013)
23. Beyer, H.-G.: Theory of Evolution Strategies. Springer, New York (2001)
24. Cerný, V.: Thermodynamical approach to the traveling salesman problem: An efficient simulation algorithm. J. Opt. Theory Appl. 45(1), 41–51 (1985)
25. Kirkpatrick, S., Gelatt Jr., C.D., Vecchi, M.P.: Optimization by Simulated Annealing. Science 220(4598), 671–680 (1983)
26. Clerc, M.: Particle Swarm Optimization. ISTE Publishing Company (2006)

# A New Approach to Modeling of Bio-inspired Information Diffusion with Ant Colony Optimization in Complex Networks

Reisa Rahmatu Dewi and Tae-Hyong Kim

Department of Computer Engineering at Kumoh National Institute of Technology
1 Yangho-dong, Gumi, 730-701, South Korea
reisadewi@gmail.com, taehyong@kumoh.ac.kr

**Abstract.** This paper proposed a bio-inspired model for information diffusion in complex networks using ant colony optimization. This model introduces selfishness in forwarder nodes and unacquainted nodes. Ant colony optimization important parts are finding shortest path and managing the selfish nodes and disjoined nodes. The proposed approach is simulated in two types of networks: lattice network and scale free network. The simulation results show that the proposed approach has higher performance and higher reachability than epidemic model.

**Keywords:** Bio-inspired, information diffusion, ant colony optimization, complex networks.

## 1 Introduction

Complex networks have been studied in many fields of science over the past decade. A complex network is constructed with many non-identical elements connected by diverse interactions of networks [1],[2]. The Internet and World Wide Web are good examples of complex networks which have led many computer scientists to model and manage the complexity of these two networks. One interesting researche in complex network area is understanding dissemination information and its relation to network characteristic.

In order to describe the complexity of the network, some models were proposed. In 1998, Lattice network was introduced by Watts and Strogatz (WS) as a concept of small-world network [3]. The small-world pattern has been shown to be ubiquitous in many real networks. Lattice network is widely used in many applications, for example, in distributed parallel computation, distributed control and wired circuits [4],[5]. Other model described evolving networks model which presented the dynamics (assembly, evolution) of network by addition or removal of nodes and edges known as complex dynamical network. This network was first introduced by Barabasi and Albert as Scale-free network [6]. The social network like blogs, Facebook and twitter are the examples of complex dynamical networks [7].

I. Zelinka et al. (eds.), *Nostradamus 2014: Prediction, Modeling and Analysis of Complex Systems*, Advances in Intelligent Systems and Computing 289,
DOI: 10.1007/978-3-319-07401-6_13,  

Some recent studies of complex networks discussed about epidemic spreading scenarios which in general focus on deriving epidemic threshold [8] or virus spreading [9]. These works mainly concentrate on the susceptible-infected-susceptible (SIS) and susceptible-infected-removed (SIR) models [10],[11]. SIR model goal is to describe virus spreading and how the vaccination works, in this case, the recovered population either gets vaccinated or immune. This model cannot represent information diffusion because not all the nodes in the network can get infected by the virus. Blocking or immune in SIR model happen when the node already infected by the virus and the rest of the population will not get infected by the virus. Information diffusion often uses SIS model paradigm to infect all susceptible node. SIS model do not take node behavior into account. Information rate for all nodes pair are the same. Epidemic routing of complex network has been studied over various complex network models, e.g. heterogeneous networks, Lattice network and Scale-free network [12], [13]. However, these models cannot properly described how dissemination process of information in the real world. For example, in many technological communication network, each node not only acts as source or target but also forwards information to others [14],[15] or selfish node which only save or ignore the information [16]. In information dissemination of social network e.g. Facebook [17], people can share or forward information to others or choose to save and ignore the information.

The ubiquity of information dissemination in complex network technology has led to following question: *How do social networks mediate the transmission of information? What factors are there affecting the spreading of information in real world?* To answer these questions, this paper introduces the bio-inspired model approach using the ant colony optimization (ACO) [18]. ACO is based on real ant behavior attempting to find a shortest path from nest to food source. While foraging, ants communicate with pheromone to mark the used path and attracts other ants. This cooperative behavior of ants provides a positive feedback. Since each ant moves simultaneously and independently, this decentralized control makes ants adapt to traffic condition and networks. This model can overcome some important factors in dissemination information, i.e. disjoined node will be discovered in short time, forwarding information to all close neighbor and handle selfish node which only save or ignore the information. Simulation of information diffusion using two types of networks, lattice network and scale free network will be presented and discussed under this model.

## 2 Preliminaries and Definitions

### 2.1 Lattice and Scale Free Network Model

**Lattice Network Model.** is defined as an approximation of networks whose nodes are randomly located [12]. Lattice network can also be viewed as a homogeneous network, in which all nodes have approximately the same number of edges. Lattice network characterized by high clustering coefficient, low diameter and same degree distribution. The structure of the lattice network is simple and useful for theoretical analysis. An algorithm to construct Watts-Stogatz Lattice network in Small-world model is presented in Algorithm 1.

---

**Algorithm 1.** Lattice Network in Small-world Model

---

**Require:** start with ring lattice with $N$ nodes
**Ensure:** every node connected to its first $2M$ neighbors
  **while** constructing network **do**
    **if** transition probability $p$ between $(p = 0)$ and $(p = 1)$ **then**
      randomly rewire each edge with probability $p$
      delete duplicate edges
    **else**
      stop randomization
    **end if**
  **end while**

---

**Scale Free Network Model.** is a dynamic network which aim is to reproduce how the network was built and evolved. This network first introduced by Barabasi and Albert in 1998 by considering two main ingredients: growth and preferential attachment.

- Growth: Observation from most networks developed over time by adding new nodes and new links to existing graph structure
- Preferential attachment: expresses the probability of a new or existing node that to be connected to a node which already has a large number of links.

The Scale Free network is the model of the network based on power-law degree distribution. Algorithm 2 is the scheme of Scale-free model.

---

**Algorithm 2.** Scale-Free Network Model

---

**Require:** start with a small number $(m_o)$ of nodes
**Ensure:** preferential attachment with probability $\pi_i$
  **while** each time step **do**
    **if** a new node is introduced $m$ where $m <= m_o$ **then**
      link the new node m to existing nodes in the network
      choose the nodes to which the new node connects based on $\pi_i$
    **else**
      stop linked
    **end if**
  **end while**

---

### 2.2 Information Diffusion Model

Modeling the diffusion of information has been an active research area nowadays and has proven to be a challenging task. There are several types of approach or modeling dissemination of information e.g. probabilistic method [19], rumor spreading [20] and based on epidemic routing technique [21]. However, this approach only considered to spread the information to as many users as possible. The information dissemination process is almost the same with the spreading epidemic, where the final target is to ensure all the nodes have the information. Below are description to understand the basic concept of epidemic routing and the proposed ACO for dissemination information:

**Epidemic Routing Model.** was routing scheme proposed for disruption/delay tolerant network (DTN). The goals of epidemic routing are to maximize message delivery rate, minimize message delivery latency and aggregate system resources consumed in message delivery by set a maximum buffer size in node [22].

However, due to many delivered messages and increasing traffics make network exhausted and cause many dropping packets. To overcome this problem, some epidemic routing enhancements are proposed i.e. prioritized epidemic routing (PREP) by classifying the type of message to reduce the delay and the overhead [23], Oracle based optimal algorithm to lower traffic and contention [24], and opportunistic routing by making replication packet and anti-packet to control the overhead [25].

All these solutions do not take into account the node characteristic in the network where there are disjoint node, forwarding node and selfish node. To overcome this problem the proposed ACO model is introduced.

**Ant Colony Optimization Model.** The basic idea of the ant colony optimization (ACO) is taken from the food searching behavior of real ants. Ants deposit a pheromone on the ground when they are on the way to search for food.The concentration of pheromone on a certain part is an indication of its usage. Ants smell the presence of pheromone and tend to choose paths where pheromone concentration is higher. In ACO, solutions are constructed repetitively by adding solution components to partial solutions stochastically [26]. The ACO meta-heuristic is targeted towards optimization problems that can be solved as shortest path problems on graphs. Ants construct solutions to optimization problems by moves around in a graph and use stigmergy to communicate their experiences.

## 3 The Proposed Technique

The proposed diffusion information model has similarity with epidemic spreading model. In this model, the number of total nodes $N$ is divided into two compartments: Susceptible and Infected. The structure of bio-inspired modeling is shown in Fig. 1.

As shown in Fig. 1, the information diffusion in bio-inspired model can be summarized as follows:

1. Susceptible compartment consist of two groups, unacquainted and selfish nodes. Unacquainted are people who are not informed and had interest to

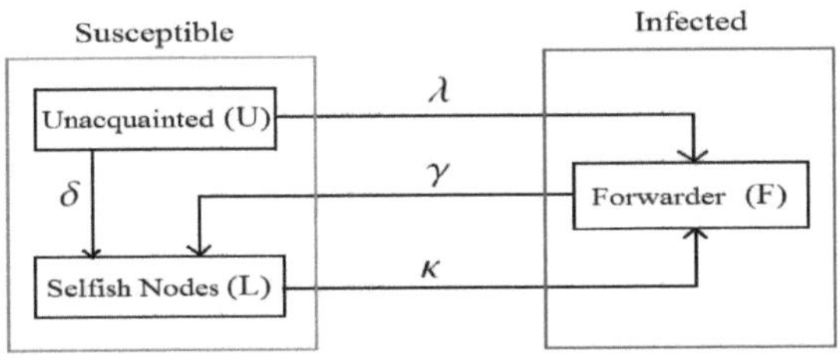


**Fig. 1.** Structure of Bio-Inspired Information Diffusion Model

forward the information. Selfish nodes are people who are decided to save the information and ignore the information.

2. Infected or forwarder are the people who diffuse the information to other contacts.
3. When a forwarder contacts the unacquainted, the unacquainted will become a forwarder with rate $\lambda$.
4. A forwarder can become a selfish node at rate $\gamma$ if he or she does not want to forward the information.
5. The unacquainted can become a selfish node at rate $\delta$ if he or she do not interested at the information.
6. A selfish node can become forwarder at rate $\kappa$ if he or she decided to forward the information.

$S(t)$ define number of nodes at time $t$ that have no acknowledge about the information and $I(t)$ define number of nodes at time $t$ that are forwarding the information. The proposed method is described as follow:

$$\begin{gathered} S(t) + I(t) = N \\ [U(t) + L(t)] + F(t) = N \end{gathered} \tag{1}$$

$U(t)$ denoted the number of nodes unacquainted and $L(t)$ define number of nodes that become selfish nodes and $F(t)$ denoted number of nodes forward the information. This is similar to real life where a person may or may not forward the information when he or she hears it from another person, which is explained by these equation:

$$\frac{\mathrm{d}F(t)}{\mathrm{d}t} = \lambda U(t)F(t) + \kappa L(t)F(t) - \gamma F(t)L(t) \tag{2}$$

$$\frac{\mathrm{d}U(t)}{\mathrm{d}t} = -\lambda F(t)U(t) - \delta U(t)L(t) \tag{3}$$

$\lambda$ is information rate which forwarder can contact the unacquainted and $\gamma$ or saving rate defined the rate of forwarder become a selfish node. In the real world, people are not selfish all the time, but depend on the willingness. From social perspective, a selfish user is willing to help or give information to others with whom he or she has interest or concern, on the other side, selfish due to situation [27]. The selfishness of a node is affected by its neighbor. If a selfish node has selfish neighborhood, the selfishness will become higher, but if a selfish node has cooperative node neighborhood then the selfish node may become cooperative node [28]. Selfish node equation is described as follow:

$$\frac{\mathrm{d}L(t)}{\mathrm{d}t} = \delta L(t)U(t) + \gamma F(t)L(t) - \kappa L(t)F(t) \tag{4}$$

$\delta$ is ignore rate which unacquainted not interested to the information and become selfish node and $\kappa$ define the rate which selfish node decided or willingly forward the information . Information rate usually assumed have the same rate for all pairs by calculate the degree of nodes. However it might not be a good approximation to modeling the information diffusion [29]. To enhance the modeling of information diffusion, information rate by each pair of node will be calculated using ACO approach. To find the information rate or $\lambda$ for each pair of nodes at time $t$ :

$$\lambda = \frac{P_{(i,s)}\bar{k}}{N} \tag{5}$$

Where $\bar{k}$ define mean degree and $P_{(i,s)}$ define the probability of diffuse the information between interested nodes and forwarding nodes.

$$P_{(i,s)} = \frac{[\tau_{is}]^{\alpha}[\eta_{is}]^{\beta}}{\sum_{h\epsilon\Omega}[\tau_{ih}]^{\alpha}[\eta_{ih}]^{\beta}} \tag{6}$$

The probability of diffusing information is taken from the Ant Colonization Optimization (ACO) approach where each ant tries to find a path in the network based on pheromone trail and the minimum weight. $\Omega$ defined as set of nodes which are feasible to be visited from node i, $\alpha$ is parameters weight of the influence trails , $\beta$ is parameters weight of the visibility and $\eta_{is}$ define the attractiveness of move from node i to s which is heuristic value. $\tau_{is}$ define pheromone level laid between node i and s. In this paper, the value of $\alpha$ is set to 1.0 because it is the value of maximum network capacity is reached and short path is discovered faster [30].

The pheromone can distributes the traffic load evenly among the nodes, thus it is capable of discovering disjoined node by marking the intermediate nodes which have been visited. Pheromone values are the results of stigmergy and heuristic values reflect a node local situation. This equation combine between pheromone and heuristic values that resulting in balanced load, because data flows are adaptively spread over multiple paths with a preference of the best paths.

To update the pheromone value:

$$\tau_{is} = (1-\rho)\cdot\tau_{is} + \rho\tau_0 \tag{7}$$

Where $(1-\rho)$ is called the pheromone decrease constant. This value is used to simulate that each ants has to make decision about the next hop over to take and $\tau_0$ is initial pheromone value. Pheromone evaporation is important value to balance between exploitation (e.g. of the route that have good resource) and exploration (e.g. of new alternative routes).

## 4 Evaluation and Simulation Results

This section compares the proposed bio-inspired model with Ant Colony Optimization (ACO) with standard epidemic routing model. The proposed model is simulated using MATLAB. The parameter values that used in the simulation is presented in Table 1.

**Table 1.** Experiment Parameter Value

| Parameter | Value |
|---|---|
| Total number of Nodes | 2500 |
| Total area | 250x250 |
| Mean degree Lattice Network | 4 |
| Mean degree Scale-Free Network | 2-20 |
| Ignore rate | 0.001 |
| Saving rate | 0.001 |

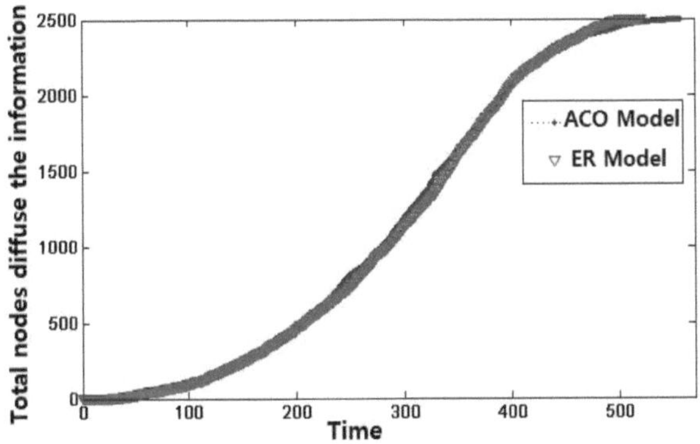


(a)

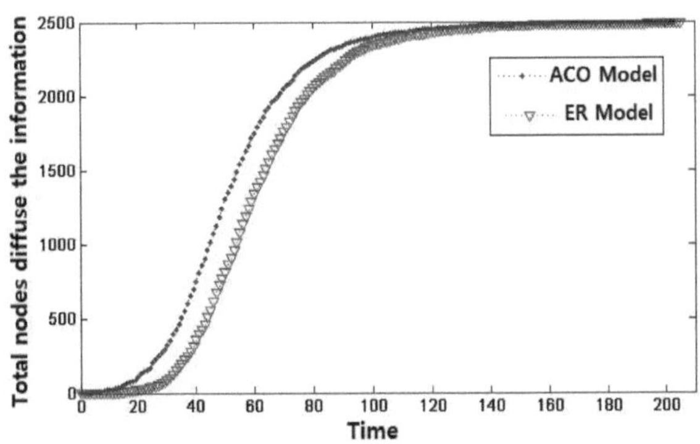


(b)

**Fig. 2.** Simulation result of total nodes and time step in (a) Lattice Network (b) Scale Free Network

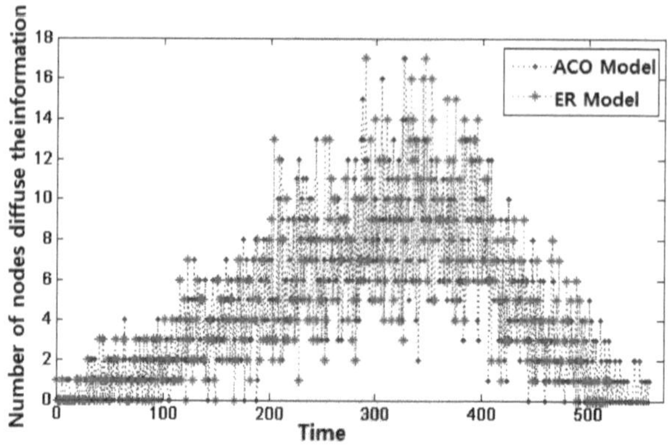


(a)

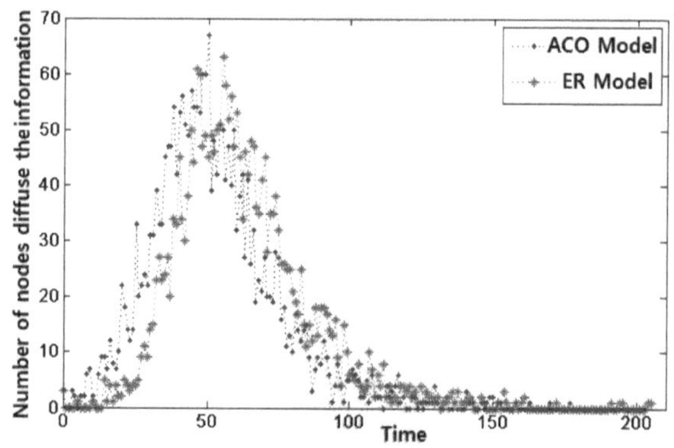


(b)

**Fig. 3.** Number of nodes diffuse the information in (a) Lattice Network (b) Scale Free Network

In the lattice network showed by Fig. 2.a, the ER model has diffusion information time slightly faster than the ACO model, because the information diffusion model in ER does not take into account the possibility of storing or ignoring information, but assumes that each node diffuse the information to the neighboring nodes. In the scale free network (Fig. 2.b), ACO models have diffusion time faster than the ER model because it can find the shortest route. ACO model is checked every disjoint node and update each path in the routing table periodically.

Fig. 3 respectively shows the total number of nodes that diffuse the information in lattice network and scale free network. In Lattice network, the number of nodes diffuse the information is slightly higher in ER model compare to ACO model since all nodes have the same degree. Meanwhile, in Scale Free network, node degree is distributed by power law function. There for, ACO model can adaptively diffuse the information to all nodes faster than ER model.

# 5 Conclusion

This paper introduced the bio-inspired modeling with Ant Colony Optimization (ACO) for information diffusion. The proposed model takes into account selfishness in unacquainted nodes and forwarder nodes. In real world, selfish node behaviors exist when people intend to choose to ignore the information or just save the information by itself.In addition, disjoint node represent people who had small circle or small contacts. Information diffusion for each pair of node is calculated by ACO approach by combining heuristic value and pheromone value. Therefore, the route exploration become faster and balanced because the data spreaded adaptively. It has been shown that proposed model has higher performance and reachability than epidemic model. For the future work, the type of information and calculation of spreading delay will be presented.

# References


1. Newman, M.E.J.: Scientific collaboration networks: I. network construction and fundamental results. Phys. Rev. E 64, 016131 (2001)
2. Wang, X.F., Chen, G.: Complex networks: small-world, scale-free and beyond. IEEE Circuits and Systems Magazine 3(1), 6–20 (2003)
3. Watts, D., Strogatz, S.: Collective dynamics of 'small-world' networks. Nature (393), 440–442 (1998)
4. Milgram, S.: The small world problem. Psychology Today 61, 60–67 (1967)
5. Barrenetxea, G., Berefull-Lozano, B., Vetterli, M.: Lattice networks: capacity limits, optimal routing, and queueing behavior. IEEE/ACM Transactions on Networking 14(3), 492–505 (2006)
6. Barabasi, A.L., Albert, R.: Emergence of scaling in random networks. Science 286(5439), 509–512 (1999)
7. Yook, S., Jeong, H., Barabási, A.L.: Modeling the internet's large-scale topology. PNAS 99, 13382–13386 (2002)
8. Zhang, X., Neglia, G., Kurose, J., Towsley, D.: Performance modeling of epidemic routing. Comput. Netw. 51(10), 2867–2891 (2007)
9. Moore, C., Newman, M.E.J.: Epidemics and percolation in small-world networks. Phys. Rev. E 61, 5678–5682 (2000)
10. Zhou, T., Liu, J.G., Bai, W.J., Chen, G., Wang, B.H.: Behaviors of susceptible-infected epidemics on scale-free networks with identical infectivity. Phys. Rev. E 74, 056109 (2006)
11. Xinli, H.: Threshold dynamics for sir epidemic model in periodic environments. In: 2010 International Conference on Computer Application and System Modeling (ICCASM), vol. 7, pp. V7-41–V7-45 (2010)
12. Wang, Y., Chakrabarti, D., Wang, C., Faloutsos, C.: Epidemic spreading in real networks: an eigenvalue viewpoint. In: Proceedings of the 22nd International Symposium on Reliable Distributed Systems, pp. 25–34 (2003)
13. Pastor-Satorras, R., Vespignani, A.: Epidemic spreading in scale-free networks. Phys. Rev. Lett. 86, 3200–3203 (2001)
14. Ye, W., Heidemann, J.S., Estrin, D.: An energy-efficient mac protocol for wireless sensor networks. In: INFOCOM, vol. 3, pp. 1567–1576 (2002)

15. Ephremides, A.: Energy concerns in wireless networks. IEEE Wireless Communications 9(4), 48–59 (2002)
16. Xu, L., Lin, Z., Ye, A.: Analysis and countermeasure of selfish node problem in mobile ad hoc network. In: 10th International Conference on Computer Supported Cooperative Work in Design, CSCWD 2006, pp. 1–4 (2006)
17. Mohaisen, A., AbuHmed, T., Zhu, T., Mohaisen, M.: Collaboration in social network-based information dissemination. In: 2012 IEEE International Conference on Communications (ICC), pp. 2103–2107 (2012)
18. Dorigo, M., Birattari, M., Stutzle, T.: Ant colony optimization. IEEE Computational Intelligence Magazine 1(4), 28–39 (2006)
19. Passarella, A., Conti, M., Boldrini, C., Dunbar, R.I.: Modelling inter-contact times in social pervasive networks. In: Proceedings of the 14th ACM International Conference on Modeling, Analysis and Simulation of Wireless and Mobile Systems, MSWiM 2011, pp. 333–340. ACM, New York (2011)
20. Karp, R., Schindelhauer, C., Shenker, S., Vocking, B.: Randomized rumor spreading. In: Proceedings of the 41st Annual Symposium on Foundations of Computer Science, pp. 565–574 (2000)
21. Vahdat, A., Becker, D.: Epidemic routing for partially connected ad hoc networks. Technical report (July 2000)
22. Yuan, P., Ma, H.: Impact of infection rate on scaling law of epidemic routing. In: 2012 IEEE Wireless Communications and Networking Conference (WCNC), pp. 2934–2939 (2012)
23. Ramanathan, R., Hansen, R., Basu, P., Rosales-Hain, R., Krishnan, R.: Prioritized epidemic routing for opportunistic networks. In: Proceedings of the 1st International MobiSys Workshop on Mobile Opportunistic Networking, MobiOpp 2007, pp. 62–66. ACM, New York (2007)
24. Spyropoulos, T., Psounis, K., Raghavendra, C.: Single-copy routing in intermittently connected mobile networks. In: 2004 First Annual IEEE Communications Society Conference on Sensor and Ad Hoc Communications and Networks, IEEE SECON 2004, pp. 235–244 (2004)
25. Boldrini, C., Conti, M., Passarella, A.: Modelling data dissemination in opportunistic networks. In: Proceedings of the Third ACM Workshop on Challenged Networks, CHANTS 2008, pp. 89–96. ACM, New York (2008)
26. Dorigo, M., Di Caro, G.: Ant colony optimization: a new meta-heuristic. In: Proceedings of the 1999 Congress on Evolutionary Computation, CEC 1999, vol. 2, pp. 1470–1477 (1999)
27. Li, Q., Zhu, S., Cao, G.: Routing in socially selfish delay tolerant networks. In: 2010 Proceedings IEEE INFOCOM, pp. 1–9 (2010)
28. Li, Y., Su, G., Wu, D., Jin, D., Su, L., Zeng, L.: The impact of node selfishness on multicasting in delay tolerant networks. IEEE Transactions on Vehicular Technology 60(5), 2224–2238 (2011)
29. Sermpezis, P., Spyropoulos, T.: Information diffusion in heterogeneous networks: The configuration model approach. In: 2013 Proceedings IEEE INFOCOM, pp. 3261–3266 (2013)
30. Ling, X., Hu, M.B., Jiang, R., Wang, R., Cao, X.B., Wu, Q.S.: Pheromone routing protocol on a scale-free network. Phys. Rev. E 80, 066110 (2009)

# Scheduling the Flowshop with Zero Intermediate Storage Using Chaotic Discrete Artificial Bee Algorithm

Magdalena Metlická and Donald Davendra

Faculty of Electrical Engineering and Computer Science
VŠB-Technical University of Ostrava, 17. listopadu 15 708 33,
Ostrava-Poruba, Czech Republic
{magdalena.metlicka.st,donald.davendra}@vsb.cz

**Abstract.** This paper analyses the application of the Chaos driven Discrete Artificial Bee Algorithm to the flowshop with zero intermediate storage problem. Nine unique chaos maps are embedded in the Discrete Artificial Bee Algorithm alongside the Mersenne twister and evaluated on the Taillard problem sets for the total flowtime criterion. Based on the obtained results and statistical analysis, it is shown that a number of chaos driven algorithms significantly performed better than the Mersenne Twister variant.

**Keywords:** Artificial Bee Colony, Flowshop with zero intermediate storage, Chaos Maps.

## 1 Introduction

Evolutionary algorithms (EA's) are generally classified by their application of randomness or *stochasticity*, the ability to generate a random event, which in turn, provides the spark of perturbation towards the desired goal. The task of generating this stochasticity is generally in the realm of *pseudorandom number generators (PRNG)*; a structured sequence of mathematical formulation which tries to yield a generally optimal range of distributed numbers within a specified range.

This research evolves around the generation of chaotic sequences which are then used as Chaos Random Number Generators (CPRNG's) in an EA. The chaotic behaviour when observed may seem erratic and somewhat random, however, these systems are deterministic, whose precise description of future behaviour is well known. The proposition is then to reconcile the notion of non-linearity of these systems.

This aperiodic non-repeating behaviour of chaotic systems is the foundation of this research. The objective is then to analyse different chaotic systems, which in this case are the discrete dissipative systems, and to analyse as to which of these improve the application of EA's.

Current literature contains a number of research devoted to certainty, ergodicity and the stochastic property of chaotic systems. Recently, chaotic sequences

I. Zelinka et al. (eds.), *Nostradamus 2014: Prediction, Modeling and Analysis of Complex Systems*, Advances in Intelligent Systems and Computing 289,
DOI: 10.1007/978-3-319-07401-6_14, © Springer International Publishing Switzerland 2014

have been adopted instead of random sequences with improved results. They have been used to enhance the performance of EA's ([1], [2]). They have also been used together with some heuristic optimisation algorithms ([4], [33]) to express optimisation variables. The choice of chaotic sequences is justified theoretically by their unpredictability, i.e. by their spread-spectrum characteristics, non-periodic, complex temporal behaviour, and ergodic properties [18].

A family of enhanced CPRNG's has been developed by [15], where the main imputes is the generation of very long series of PRNG's. The application of using PRNG's in EA's has been debated by [32] and a number of case studies have applied chaos to different problems [21], [22], [25], [6], [18], [16], [31], [4] and [5] amongst others.

This paper looks to expand upon this class of research and to ascertain if chaos can improve the discrete variant of the Artificial Bee Algorithm (ABC) [19]. For comparison, we utilise the Mersenne Twister (MT) as the canonical PRNG in ABC, and compare it with nine different chaotic maps. The following sections describe the ABC algorithm and the utilised chaotic maps. It is followed by the flowshop with zero immediate storage description. The work is concluded by experimentations and conclusion.

## 2 Artificial Bee Colony

Artificial Bee Colony (ABC) algorithm is based on the foraging nature of honey bee swarm, initially modelled by [10] to optimise multi-variable and multi-modal continuous functions. Subsequent experiments by the originator has demonstrated that the performance of the ABC algorithm is competitive to other population-based algorithms with an advantage of employing fewer control parameters ([12], [13] [11]).

The outline of the canonical ABC is quite straightforward. The ABC algorithm classifies the foraging artificial bees into three groups, namely, *employed bees*, *onlookers* and *scouts*. A bee that is currently exploiting a food source is called an *employed bee*. A bee waiting in the hive for making decision to choose a food source is named as an *onlooker*. A bee carrying out a random search for a new food source is called a *scout*. In the ABC algorithm, each solution to the problem under consideration is called a food source and represented by an $D$-dimension real-valued vector, whereas the fitness of the solution corresponds to the *nectar* amount of the associated food resource.

The main steps of the ABC are described as the following:

**Initialisation of Parameters**

The initial parameters are the number of food sources (FS) which is equal to the number of the *employed bees* or *onlooker bees*, the number of trials after which a food source is assumed to be abandoned (*limit*), and a termination criterion. In the basic ABC algorithm, for every food source, there is only one employed bee.

**Initialisation of the Population**

The population can be termed as a list of individuals of *FS* number of randomly generated $D$-dimensional real-valued vectors. Let $X_i = \{x_{i1}, x_{i2}, ..., x_{iD}\}$ be a $i$th food source in the population, where each food source can be generated as:

$$x_{ij} = LB_j + (UB_j - LB_j) \cdot r \tag{1}$$

for $j = 1, 2, \ldots, D$ and $i = 1, 2, \ldots, FS$, where $D$ is the dimension of the problem and $r$ is a real valued random number in the range of [0,1].

**Initialisation of the Bee Phase**

At this stage, each employed bee $x_i$ generates a new food source $x_t$ in the neighbourhood of its present position.

$$x_t = x_{ij} + (x_{ij} - x_{kj}) \cdot r \tag{2}$$

where $k \in \{1, 2, \ldots, FS\} \wedge k \neq i$ and $j \in \{1, 2, \ldots, D\}$ are randomly chosen indices and $r$ is a real valued random number in the range of [-1,1]. The new $x_t$ is evaluated and compared to the current $x_i$ in the populations. if $x_t$ improves on $x_i$, it then replaces the latter in the new population.

**Onlooker Bee Phase**

An onlooker bee evaluates the nectar amount information taken from all the employed bees and selects a food source $x_i$ depending on its probability value $p_i$ given in equation (3).

$$p_i = \frac{f_i}{\sum\limits_{i=1}^{FS} f_i} \tag{3}$$

where $f_i$ is the *nectar* amount of the $i$th food source. Once the food source is obtained, a modification is done on $x_i$ using equation (2). If the new modified food source has a better or equal *nectar* amount then the existing food source, it replaces the incumbent in the population.

**Scout Bee Phase**

If a food source $x_i$ cannot be improved through a predetermined number of iterations (*limit*), the food source is abandoned, and the corresponding employed bee becomes a scout. The scout then produces a new food source randomly as given in equation (1). In the canonical ABC algorithm, at each iteration at most one scout goes outside to search for a new food source.

## 3 Discrete Artificial Bee Algorithm

The discrete variant of the ABC algorithm (DABC) has been developed by [19] to solve the flowshop lot-streaming problem. Subsequently, DABC has been applied to solve the permutative flowshop scheduling problem, with total flowtime minimisation [28], no-idle permutation flowshop scheduling problem with the total tardiness criterion [29] and multi-objective flexible job-shop scheduling problem with maintenance activities [14].

The basic outline of the DABC is now presented.

**Solution Representation**

The permutation based representation constitutes an easy procedure to decode a schedule, which has been widely used in literature for a variety of permutation flow shop scheduling problems [30]. This representation is used for DABC, where each permutation $\pi = \{x_1, x_2, .., x_D\}$, where $D$ is the size of the permutation.

**Employed Bee Phase**

In the DABC, which has a permutation based neighbourhood structure, *insert* and *swap* operators are commonly used to produce neighbouring solutions in the literature [30]. The insert operator of a permutation $\pi$ is defined by removing a job from $\pi$ from its original position $j$ and inserts it into another position $k$ such that $(k \in \{j, j-1\})$.

The *swap* operator produces a neighbour of $\pi$ by interchanging two jobs of $\pi$ in the different positions. To enrich the neighbourhood structure and diversify the population, four neighbouring approaches based on the insert or swap operator are separately utilised to generate neighbouring food sources for the employed bees as follows:

- Performing one insert operation to a sequence $\pi$.
- Performing one swap operation to a sequence $\pi$.
- Performing two insert operations to a sequence $\pi$.
- Performing two swap operations to a sequence $\pi$.

The best strategy is selected using an adaptive mechanism, which is generally problem dependent. In terms of selection, new food source is always accepted if it is better than the current food source.

**Onlooker Bee Phase**

A tournament selection with the size of two is used to select a new food source. In the tournament selection, an onlooker bee selects a food source $x_i$ in such a way that two food sources are picked up randomly from the population, and compared to each other, then the better one is chosen. Onlooker bees use the same method of producing a new neighbouring solution as employed bees. Naturally, if the new food source improves upon the current food source, it replaces it in the population.

**Self Adaptive Strategy**

Both employed bees and onlookers apply a self-adaptive strategy to find neighbouring food sources. The self-adaptive strategy is presented as follows. At the beginning, an initial neighbour list (NL) with a specified length is generated by filling the list one by one randomly from four neighbouring approaches explained before. Then the DABC algorithm is started. During the evolution process, one approach from the NL is taken out and used to generate a new food source for an employed bee or onlooker. If the new food source successfully replaces the current one, this approach will enter into a winning neighbouring list (WNL). Once the NL is empty, it is refilled as follows: 75% of the NL is refilled from the WNL list, and then the rest of 25% is refilled by a random selection from four different approaches. If the WNL is empty, the latest NL is used again. The above process is repeated until a termination criterion is reached. As a result, the proper neighbouring approach can be gradually learned by the algorithm itself to suit the particular problem and the particular phase of search process [19].

**Local Search**

DABC contains embedded local search. In employed bee phase, each bee may perform local search with given probability. If a random number in range [0, 1] is lesser than this probability, fixed count of swap or insert operations are applied to a food source generated by a bee. Local search serves to enhance the exploitation ability of DABC algorithm.

**Scout Bee Phase**

Contrary to ABC, in the DABC scout bee phase, the exhausted food source is replaced by new solution generated from the best solution in the population, upon which at least three insert operations are performed. This way DABC exploits the knowledge of best food source found so far, rather than generating new random one.

## 4 Chaos Induced Discrete Artificial Bee Algorithm

In all the variants of ABC and DABC, very little attention has been paid to the stochasticity of the algorithm. The basic premise is the use of widely available PRNG's. Using the stock DABC algorithm of [19], we have included the most popular Mersenne Twister [17] as the default PRNG.

Alternatively, we have included nine unique chaotic systems as CPRNG for the DABC. These new chaos embedded algorithms (hereafter referred to as variants) can be collectively labelled as **CDABC**. The basic premise of this work is to ascertain if any improvement can be achieved in DABC by using chaotic systems in place of PRNG.

The chaotic map can be utilised in two forms, the first is to generate a large chaotic sequence from *inception* of the map and use it iteratively. The second approach is to use a *random* start position of the algorithm for each experiment. We use the second approach, as to have a unique sample for *each* experiment, and eliminate the need to store large values in memory. This is similar concept to having a *seed* input to a PRNG. The different chaotic systems used as CPRNG's are given in the following section 5.

## 5 Chaos Systems

The most interesting chaotic systems, which can be utilised as CPRNG are discrete dissipative chaotic maps. These maps have the general description of being a linear set of equations, easily formulated, with a fine course grain over the solution landscape. This last attribute allows the parsing of unique values over a period of the chaotic oscillation. In total, nine unique chaotic systems were considered for this experiment. The following sections describe the different systems. All operating parameters were obtained from [26].

**Arnold's Cat Map**

The Arnold's cat map is a two dimensional discrete chaotic map, which is a torus into itself. The equations are given in (4). The parameter of $k = 2.0$.

$$\begin{aligned} X_{n+1} &= X_n + Y_n \cdot (\mathrm{mod}1) \\ Y_{n+1} &= X_n + k \cdot Y_n \cdot (\mathrm{mod}1) \end{aligned} \tag{4}$$

**Burgers Map**

The Burgers map arose from the study of hydrodynamics, where the discretization of coupled differential equations led to a bifurcation effect of the system. The equation is given in (5) and the control parameters are $\alpha = 0.75$ and $\beta = 1.75$.

$$\begin{aligned} X_{n+1} &= (\alpha \cdot X_n) - Y_n^2 \\ Y_{n+1} &= (\beta \cdot Y_n) + (X_n \cdot Y_n) \end{aligned} \tag{5}$$

**Delayed Logistic**

The Delayed Logistic is a two-dimensional map which is a phase shifted one-dimensional logistic equation. The equation is given in (6) and the parameter $\alpha = 2.27$.

$$\begin{aligned} X_{n+1} &= \alpha \cdot X_n \cdot (1 - Y_n) \\ Y_{n+1} &= X_n \end{aligned} \tag{6}$$

**Dissipative Standard Map**

The Dissipative Standard Map is a two-dimensional chaotic system. The equation is given in (7) and the operating parameters are $\beta = 0.1$ and $k = 8.8$.

$$\begin{aligned} X_{n+1} &= X_n + Y_{n-1} \cdot (\mathrm{mod}2\pi) \\ Y_{n+1} &= (\beta \cdot Y_n) + (k \cdot \sin X_n \, (\mathrm{mod}2\pi)) \end{aligned} \tag{7}$$

**Henon Map**

The Henon map is a discrete-time dynamical system, which was introduced as a simplified model of the Poincare map for the Lorenz system. The equation is given in (8) and the control parameters are $\alpha = 1.4$ and $\beta = 0.3$.

$$\begin{aligned} X_{n+1} &= \alpha - X_n^2 + (\beta \cdot Y_n) \\ Y_{n+1} &= X_n \end{aligned} \tag{8}$$

**Ikeda Map**

The Ikeda map is a discrete-time dynamical system derived as a model of light going around across a nonlinear optical resonator. A 2D real example of the Ikeda map is given in equation (9). The operating parameters are $\alpha = 0.75$, $\beta = 1.75$, $\gamma = 1$ and $\mu = 0.9$.

$$\begin{aligned} X_{n+1} &= \gamma + \mu \cdot ((X_n \cdot \cos\phi) - (Y_n \cdot \sin\phi)) \\ Y_{n+1} &= \mu \cdot ((X_n \cdot \sin\phi) + (Y_n \cdot \cos\phi)) \\ \phi &= \beta - \frac{\alpha}{(1+X_n^2+Y_n^2)} \end{aligned} \tag{9}$$

**Lozi Map**

The Lozi map is a simple discrete two-dimensional chaotic map. The equation is given in (10) and the control parameters are $\alpha = 1.7$ and $\beta = 0.5$.

$$\begin{aligned} X_{n+1} &= 1 - (\alpha \cdot |X_n|) + (\beta \cdot Y_n) \\ Y_{n+1} &= X_n \end{aligned} \tag{10}$$

**Sinai Map**

The Sinai map is a simple two-dimensional discrete system similar to the Arnolds Cat map. The equation is given in (11) and the control parameter is $\delta = 0.1$.

$$\begin{aligned} X_{n+1} &= X_n + Y_n + (\delta \cdot \cos 2\pi \cdot Y_n \cdot (\text{mod}1)) \\ Y_{n+1} &= X_n + 2 \cdot Y_n \cdot (\text{mod}1) \end{aligned} \tag{11}$$

**Tinkerbell Map**

The Tinkerbell map is a two-dimensional complex discrete-time dynamical system. The equation is given in (12) and the operating parameters are $\alpha = 0.9$, $\beta =$ -0.6, $\rho = 2$ and $\upsilon = 0.5$.

$$\begin{aligned} X_{n+1} &= X_n^2 - Y_n^2 + (\alpha \cdot X_n) + (\beta \cdot Y_n) \\ Y_{n+1} &= (2 \cdot X_n \cdot Y_n) + (\rho \cdot X_n) + (\upsilon \cdot Y_n) \end{aligned} \tag{12}$$

## 6 Flowshop with Zero Intermediate Storage

One of the most challenging and practical scheduling problem in the flowshop class is the one with no storage or stoppage between machines [20]. Consider a flow shop with zero intermediate storage (FSSZIS) subject to different operating procedures. A job, when it goes through the system, is not allowed to *wait* at any machine. For this process, all subsequent machines have to be *idle*, at the completion of the job on a machine upstream. Therefore, the jobs are ***pulled*** down the line by machines which have become idle. This constraint can be also refereed to as the **no-wait** constraint, and minimising the makespan in such a flow shop is referred to as the

$$Fm\,|nwt\,|C_{\max}$$

Among all types of scheduling problems, FSSZIS owns lots of important applications in different industries such as chemical processing [24], food processing

[9], concrete ware production [8], and pharmaceutical processing [23] amongst others.

For the computational complexity of the FSSZIS scheduling problem, [7] proves that it is strongly NP-complete. Therefore, only small size instances of this flowshop problem can be solved with reasonable computational time by exact algorithms.

The following notations are used to formulate the FSSZIS problem: assume $n$ as number of jobs to be scheduled, $m$ as the number of machines in the flowshop, $t_{i,j}$ as the processing time for the $i^{th}$ job on the $j^{th}$ machine, $d_{i,k}$ as the minimum delay on the first machine between the start of job $i$ and job $k$ due to the no-wait constraint, $[i]$ as the job processed in position $i$ , $C_{[i]}$ as the completion time of the job processed in position. $TFT$ represents the total flow time, i.e. the sum of flow times of all jobs.

The minimum delay time $d_{i,k}$ and completion time $C_{[i]}$ can be calculated as:

$$
\begin{aligned}
d_{i,k} &= t_{i,1} + \max_{2\leq j\leq m} \left( \sum_{p=2}^{j} t_{i,p} - \sum_{p=1}^{j-1} t_{k,p} \right) \\
C_{[i]} &= \sum_{j=1}^{m} t_{[1],j}, \\
C_{[i]} &= \sum_{k=2}^{i} d_{[k-1],[k]} + \sum_{j=1}^{m} t_{[1],j}, \qquad i = 2, 3, \ldots, n.
\end{aligned}
\tag{13}
$$

All jobs are assumed to be available at time zero, the total flow time can then be given as in (14).

$$
\begin{aligned}
&TFT = \sum_{i=2}^{n} \left( \sum_{k=2}^{i} d_{[k-1],[k]} + \sum_{j=1}^{m} t_{[1],j} \right) + \sum_{j=1}^{m} t_{[1],j} = \\
&\sum_{i=2}^{n} \sum_{k=2}^{i} d_{[k-1],[k]} + \sum_{i=1}^{n} \sum_{j=1}^{m} t_{[i],j} = \\
&\sum_{i=2}^{n} (n+1-i) d_{[i-1],[i]} + \sum_{i=1}^{n} \sum_{j=1}^{m} t_{i,j}
\end{aligned}
\tag{14}
$$

where $\sum_{i=1}^{n} \sum_{j=1}^{m} t_{i,j}$ is the sum of the processing time of all jobs in all machines [3].

## 7 Results

The experimentations was conducted on the Taillard data sets [27], which is a set of 12 data classes of different sizes, each of which contains ten unique instances; therefore a total of 120 data instances.

The operating parameters of CDABC are given in Table 1. All parameters were kept constant for all the experimentation, in order not to introduce a bias. All experiments were conducted on the machine having Intel i7-3610QM CPU processor running at 2.3GHz with 8GM of RAM. All codes were written in the C programming language.

**Table 1.** DABC Operating parameters

| Parameter | Value |
|---|---|
| Food Source (FS) | 30 |
| Limit (food source) | 50 |
| Loop (Local Search) | 200 |
| Local search probability ($P_L$) | 0.2 |
| Neighbourhood List (NL) | 20 |
| Winning Neighbourhood List (WNL) | 0.75 x NL |
| Iterations | 100 |

For each instance, fourteen (14) repeated experimentations were conducted by each variant of CDABC in order to obtain statistical variance. Therefore, 1680 individual experiments were conducted by each variant, leading to a sum total of 16800 experimentations for all ten variants.

The *average* results obtained by the 140 experiments of each problem data class are given in Tables 2 and 3. From the results, it can be concluded that the two most promising results are from the Tinkerbell and Delayed Logistic map systems. Tinkerbell has the best average results for the 20x5, 20x10, 100x5, 100x10, 100x20, 200x10, 200x20 and 500x20 data sets. Delayed Logistic obtains the best results for the remaining data sets of 20x20, 50x5, 50x10 and 50x20 data sets. Additionally, it obtains better cumulative average value and standard deviation.

**Table 2.** Summarised results for Mersenne Twister, Arnold Cat, Burgers, Delayed Logistic and Dissipative Maps

| | MT | Arnold Cat | Burgers | Delayed Logistic | Dissipative |
|---|---|---|---|---|---|
| 20x5 | 16211.94 | 16333.94 | 16188.50 | 16183.76 | 16315.27 |
| 20x10 | 23560.44 | 23787.04 | 23525.30 | 23520.45 | 23756.47 |
| 20x20 | 38593.95 | 38857.17 | 38551.00 | **38537.12** | 38842.35 |
| 50x5 | 84813.39 | 86367.90 | 83775.64 | **83207.16** | 85919.88 |
| 50x10 | 119748.21 | 122404.80 | 118449.56 | **117580.99** | 121999.04 |
| 50x20 | 175302.27 | 178829.33 | 173224.56 | **171925.68** | 178313.36 |
| 100x5 | 332715.06 | 344226.62 | 326877.84 | 324506.33 | 341968.45 |
| 100x10 | 458246.87 | 475807.93 | 450901.06 | 447582.99 | 472777.88 |
| 100x20 | 637540.68 | 661497.84 | 628598.67 | 623943.67 | 657132.23 |
| 200x10 | 1821165.26 | 1912745.61 | 1786033.84 | 1770658.18 | 1895025.71 |
| 200x20 | 2463227.25 | 2591044.94 | 2420490.61 | 2398351.19 | 2571131.01 |
| 500x20 | 15572075.59 | 16013990.10 | 15067932.90 | 14815045.71 | 15870479.40 |
| **Average** | 1811933.41 | 1872157.77 | 1761212.46 | **1735920.27** | 1856138.42 |
| **StdDev** | 4403079.05 | 4528691.58 | 4260041.78 | **4188289.04** | 4487960.06 |
| **Time** | 5.76 | 5.60 | 13.99 | 19.58 | 6.47 |

The paired *t-test* experiment is conducted pairwise on all the different variants of CDABC. All the raw results were used for the computations, implying that for each variant, all 1680 results were pairwise compared. The results comprising of the $t$ and $p$ values is given in Table 4. For all the *t-test* comparisons, the $p$ value is compared to a 95% confidence level. In terms of significance, it can be

**Table 3.** Summarised results for Henon, Ikeda, Lozi, Sinai and Tinkerbell Maps

| | Henon | Ikeda | Lozi | Sinai | Tinkerbell |
|---|---|---|---|---|---|
| 20x5 | 16432.11 | 16241.76 | 16233.55 | 16378.60 | **16182.42** |
| 20x10 | 23911.16 | 23641.86 | 23622.57 | 23826.85 | **23516.82** |
| 20x20 | 39085.52 | 38667.38 | 38646.44 | 38959.71 | 38546.31 |
| 50x5 | 86881.14 | 85028.39 | 84597.73 | 86821.97 | 83370.07 |
| 50x10 | 123106.13 | 120354.07 | 119831.39 | 122926.35 | 117754.74 |
| 50x20 | 180033.85 | 175847.83 | 175407.02 | 179771.59 | 172261.97 |
| 100x5 | 347755.44 | 334855.24 | 333891.74 | 346144.03 | **323948.20** |
| 100x10 | 480264.36 | 462990.53 | 460947.43 | 478851.04 | **447040.17** |
| 100x20 | 668043.29 | 645101.83 | 641935.18 | 665638.75 | **622663.85** |
| 200x10 | 1933300.76 | 1847125.13 | 1837010.47 | 1929469.41 | **1764057.10** |
| 200x20 | 2623881.93 | 2504658.01 | 2492349.47 | 2614873.97 | **2391550.34** |
| 500x20 | 16042336.93 | 15575355.41 | 15441420.64 | 16147232.41 | **14890741.87** |
| **Average** | 1880419.38 | 1819155.62 | 1805491.14 | 1887574.56 | 1740969.49 |
| **StdDev** | 4536745.05 | 4404156.64 | 4366160.60 | 4566556.19 | 4209720.31 |
| **Time** | 5.67 | 8.35 | 9.70 | 5.15 | 18.28 |

postulated from the results that all variants of CDABC are significantly different. Therefore, it becomes obvious that the order of the best performing variants is given as Delayed Logistic, Tinkerbell, Burgers, Lozi, Mersenne Twister, Ikeda, Dissipative, Arnold Cat, Sinai and Henon Map.

**Table 4.** Paired t-test results: $t$ and $p$ values

| | MT | | Arnold Cat | | Burgers | | DL | | Disspative | | Henon | | Ikeda | | Lozi | | Sinai | |
|---|---|---|---|---|---|---|---|---|---|---|---|---|---|---|---|---|---|---|
| | $t$ | $p$ | $t$ | $p$ | $t$ | $p$ | $t$ | $p$ | $t$ | $p$ | $t$ | $p$ | $t$ | $p$ | $t$ | $p$ | $t$ | $p$ |
| MT | - | - | - | - | - | - | - | - | - | - | - | - | - | - | - | - | - | - |
| Arnold Cat | 19.58 | **0.00** | - | - | - | - | - | - | - | - | - | - | - | - | - | - | - | - |
| Burgers | 15.2 | **0.00** | 18.1 | **0.00** | - | - | - | - | - | - | - | - | - | - | - | - | - | - |
| DL | 15.42 | **0.00** | 17.61 | **0.00** | 14.69 | **0.00** | - | - | - | - | - | - | - | - | - | - | - | - |
| Dissipative | 20.49 | **0.00** | 11.23 | **0.00** | 11.205 | **0.00** | 17.61 | **0.00** | - | - | - | - | - | - | - | - | - | - |
| Henon | 21.16 | **0.00** | 6.67 | **0.00** | 18.84 | **0.00** | 18.25 | **0.00** | 15.54 | **0.00** | - | - | - | - | - | - | - | - |
| Ikeda | 7.43 | **0.00** | 17.71 | **0.00** | 17.37 | **0.00** | 16.92 | **0.00** | 17.45 | **0.00** | 19.33 | **0.00** | - | - | - | - | - | - |
| Lozi | 5.17 | **0.00** | 17.52 | **0.00** | 17.79 | **0.00** | 17.13 | **0.00** | 17.49 | **0.00** | 18.79 | **0.00** | 11.82 | **0.00** | - | - | - | - |
| Sinai | 19.47 | **0.00** | 9.95 | **0.00** | 18.08 | **0.00** | 17.65 | **0.00** | 14.82 | **0.00** | 4.64 | **0.00** | 17.9 | **0.00** | 17.68 | **0.00** | - | - |
| Tinkerbell | 15.97 | **0.00** | 18.06 | **0.00** | 15.93 | **0.00** | 6.84 | **0.00** | 18.21 | **0.00** | 18.72 | **0.00** | 17.57 | **0.00** | 17.98 | **0.00** | 18.07 | **0.00** |

# 8 Conclusion

The main premise of this research is to experimentally verify the application of chaos map used as CPRNG's in the DABC algorithm. A total of nine different discrete dissipative chaotic maps were embedded in the canonical DABC algorithm and experiment was conducted on the flowshop with zero intermediate storage problem. The Taillard data sets were utilised and a total of 16,800 experiments were conducted by the different variants to validate the hypothesis.

From the obtained results, the top four performing variants were the Delayed Logistic, Tinkerbell, Burgers and Lozi maps. In the paired *t-test* results, all variants were shown to be significantly different. Therefore, we can state that using chaos maps improves the DABC algorithm, and provide significant improvement over Mersenne Twister for the this specific flowshop with zero intermediate storage problem.

**Acknowledgments.** This work was supported by SGS SP2014/170 and Grant Agency of the Czech Republic - GACR P103/13/08195S

## References

1. Alatas, B., Akin, E., Ozer, A.: Chaos embedded particle swarm optimization algorithms. Chaos, Solitons and Fractals 40(4), 1715–1734 (2009)
2. Caponetto, R., Fortuna, L., Fazzino, S., Xibilia, M.: Chaotic sequences to improve the performance of evolutionary algorithms. IEEE Transactions on Evolutionary Computation 7(3), 289–304 (2003)
3. Chang, J.H., Chiu, H.N.: A comprehensive review of lot streaming. International Journal of Production Research 43(8), 1515–1536 (2005)
4. Davendra, D., Zelinka, I., Senkerik, R., Bialic-Davendra, M.: Chaos driven evolutionary algorithm for the traveling salesman problem. In: Davendra, D. (ed.) Traveling Salesman Problem, Theory and Applications, pp. 55–70. InTech Publishing, Croatia (2010)
5. Davendra, D., Senkerik, R., Zelinka, I., Pluhacek, M., Bialic-Davendra, M.: Utilising the chaos-induced discrete self organising migrating algorithm to solve the lot-streaming flowshop scheduling problem with setup time. Soft Computing (2014), doi:10.1007/s00500-014-1219-7
6. Davendra, D., Zelinka, I., Senkerik, R.: Chaos driven evolutionary algorithms for the task of pid control. Computers & Mathematics with Applications 60(4), 1088–1104 (2010)
7. Garey, M., Johnson, D.: Computers and intractability: A guide to the theory of NP-completeness. Freeman, San Francisco (1979)
8. Grabowski, J., Pempera, J.: Sequencing of jobs in some production system. European Journal of Operational Research, 535–550 (2000)
9. Hall, N., Sriskandarayah, C.: A survey of machine scheduling problems with blocking and no-wait in process. Operations Research, 510–525 (1996)
10. Karaboga, D.: An idea based on honey bee swarm for numerical optimization. Technical Report TR06, Computer Engineering Department, Erciyes University, Turkey (2005)
11. Karaboga, D., Akay, B.: A comparative study of artificial bee colony algorithm. Applied Mathematics and Computation 214, 108–132 (2009)
12. Karaboga, D., Basturk, B.: A powerful and efficient algorithm for numerical function optimization: artificial bee colony (abc) algorithm. Journal of Global Optimization 39, 459–471 (2007)
13. Karaboga, D., Basturk, B.: On the performance of artificial bee colony (abc) algorithm. Applied Soft Computing 8, 687–697 (2008)
14. Li, J.Q., Pan, Q.K., Tasgetiren, M.F.: A discrete artificial bee colony algorithm for the multi-objective flexible job-shop scheduling problem with maintenance activities. Applied Mathematical Modelling (2013)
15. Lozi, R.: New enhanced chaotic number generators. Indian Journal of Industrial and Applied Mathematics 1(1), 1–23 (2008)
16. Lu, Y., Zhou, J., Qin, H., Wang, Y., Zhang, Y.: Chaotic differential evolution methods for dynamic economic dispatch with valve-point effects. Engineering Applications of Artificial Intelligence 24(2), 378–387 (2011)
17. Matsumoto, M., Nishimura, T.: Mersenne twister: A 623-dimensionally equidistributed uniform pseudorandom number generator. ACM Transaction on Modeling and Computer Simulation 8(1), 3–30 (1998)

18. Ozer, A.B.: Cide: Chaotically initialized differential evolution. Expert Systems with Applications 37(6), 4632–4641 (2010)
19. Pan, Q.K., Tasgetiren, M.F., Suganthan, P., Chua, T.: A discrete artificial bee colony algorithm for the lot-streaming flow shop scheduling problem. Information Sciences 181, 2455–2468 (2011)
20. Pinedo, M.: Scheduling: theory, algorithms and systems. Prentice Hall, Inc., New Jersey (1995)
21. Pluhacek, M., Senkerik, R., Davendra, D., Kominkova Oplatkova, Z., Zelinka, I.: On the behavior and performance of chaos driven pso algorithm with inertia weight. Computers and Mathematics with Applications 66(2), 122–134 (2013)
22. Pluhacek, M., Senkerik, R., Zelinka, I., Davendra, D.: Chaos pso algorithm driven alternately by two different chaotic maps-an initial study, pp. 2444–2449 (2013)
23. Raaymakers, W., Hoogeveen, J.: Scheduling multipurpose batch process industries with no-wait restrictions by simulated annealing. European Journal of Operational Research, 131–151 (2000)
24. Rajendran, C.: A no-wait flowshop scheduling heuristic to minimize makespan. Journal of the Operational Research Society, 472–478 (1994)
25. Senkerik, R., Pluhacek, M., Davendra, D., Zelinka, I., Kominkova Oplatkova, Z.: Chaos driven evolutionary algorithm: A new approach for evolutionary optimization. International Journal of Mathematics and Computers in Simulation 7(4), 363–368 (2013)
26. Sprott, J.: Chaos and Time-Series Analysis. Oxford University Press, UK (2003)
27. Taillard, E.: Benchmarks for basic scheduling problems. European Journal of Operations Research 64, 278–285 (1993)
28. Tasgetiren, M.F., Pan, Q.K., Suganthan, P., Chen, A.: A discrete artificial bee colony algorithm for the total flowtime minimization in permutation flow shops. Information Sciences 181, 3459–3475 (2011)
29. Tasgetiren, M.F., Pan, Q.K., Suganthan, P., Oner, A.: A discrete artificial bee colony algorithm for the no-idle permutation flowshop scheduling problem with the total tardiness criterion. Applied Mathematical Modelling 37, 6758–6799 (2013)
30. Wang, L.: Shop Scheduling with Genetic Algorithms. Tsinghua Univ. Press, Beijing (2003)
31. Yuan, X., Cao, B., Yang, B., Yuan, Y.: Hydrothermal scheduling using chaotic hybrid differential evolution. Energy Conversion and Management 49(12), 3627–3633 (2008)
32. Zelinka, I., Chadli, M., Davendra, D., Senkerik, R., Pluhacek, M., Lampinen, J.: Do evolutionary algorithms indeed require random numbers? extended study. Advances in Intelligent Systems and Computing 210, 61–75 (2013)
33. Zuo, X., Fan, Y.: A chaos search immune algorithm with its application to neuro-fuzzy controller design. Chaos, Solitons and Fractals 30(1), 94–109 (2006)

# Better and Faster Spectra Analysis Using Analytical Programming on CUDA

Peter Drábik, Petr Šaloun, Ivan Zelinka, and Marie Vraná

VŠB-Technical University of Ostrava, 17. listopadu 15/2172
708 33 Ostrava-Poruba, Czech Republic
{peter.drabik.st,petr.saloun,ivan.zelinka,marie.vrana.st}@vsb.cz

**Abstract.** In this paper we discuss a method useful for spectra analysis – analytical programming and its implementation. Our goal is to create mathematical formulas of emission lines from spectra, which are characteristic for Be stars. One issue in performing this task is symbolic regression, which represents the process in our application, when measured data fit the best represented mathematical formula. In past this was only a human domain; nowadays, there are computer methods, which allow us to do it more or less effectively. A novel method in symbolic regression, compared to genetic programming and grammar evolution, is analytic programming. The aim of this work is to verify the efficiency of the parallel approach of this algorithm, using CUDA architecture.

**Keywords:** analytical programming, spectra analysis, CUDA, evolutionary algorithm, differential evolution, parallel implementation, symbolic regression.

## 1 Introduction

Nowadays, astronomy is one of the scientific disciplines which produce enormous amounts of data per day, which needs to be processed automatically. In our work we focused on the spectra analysis of Be star-candidates. Be stars are hot B-type stars (effective temperature 10,000 to 30,000 K) with luminosity class III to V (i.e. not supergiant stars) whose spectrum has shown at least once an emission line – usually hydrogen in the Balmer line, see Figure 1. Sometimes, other emission lines are visible, for example neutral helium. Even when the spectrum goes back to *normal*, the star remains in the Be star class [9]. Some of them are among the brightest stars in the sky [2].

Non-image data, such as spectra, are also mainly distributed in astronomy in Flexible Image Transport System (fits) format, standardized in 1981 [1]. This format can have the `*.fts`, `*.fits`, `*.fit` extensions and a major feature it has is that images metadata are store in a human readable ASCII header. In our work we processed fits format data files with fv FITS Editor, available from the standard Ubuntu repository.

I. Zelinka et al. (eds.), *Nostradamus 2014: Prediction, Modeling and Analysis of Complex Systems*, Advances in Intelligent Systems and Computing 289,
DOI: 10.1007/978-3-319-07401-6_15, 

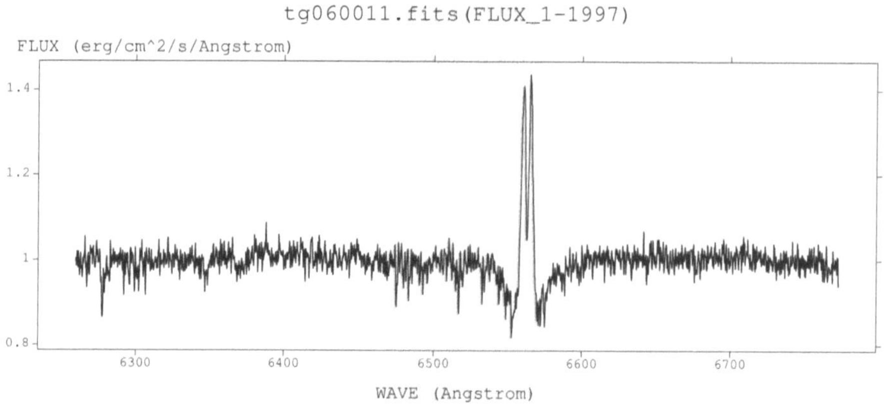


**Fig. 1.** An example of spectra with a characteristic Be star emission.

## 2 Overview of Analytical Programming

Analytical programming, proposed in 2005 [4], was inspired by Hilbert spaces and genetic programming. The principles and general philosophy behind analytical programming (AP) stem from these two methods. Into AP an idea about the evolutionary creation of symbolic solutions is taken from GP while from Hilbert spaces the idea of functional spaces and the building of the resulting function by means of the search process is adopted in AP. The core of AP is based on a set of functions, operators and so-called terminals, which are usually constants or independent variables as well as in GP and GE. The main aim of AP is to synthesize a suitable program which would fit the measured data as well as possible (with the given precision). For this reason, a discrete set handling (DSH) idea [7,8] was adopted in AP [5]. Discrete set handling creates an interface between the Evolutionary Algorithm (EA) and the problem. Therefore, we can use in AP almost any evolutionary algorithm. The individual is represented as a non-numerical value and a numerical value is added to the evolutionary process as an integer index. This index represents an individual from the General Function Set (GFS).

### 2.1 Versions of AP

There are three versions of AP. $AP_{basic}$ is a basic version of AP and uses constants from the terminal set. $AP_{meta}$ – in this version there are constants not defined in the terminal set; in the terminal set there is only one general constant K and every constant K is estimated by a different or the same evolutionary algorithm. We can mention one disadvantage of this version – because of running evolution under evolution, it could be very slow for a large number of steps. There is a big number of evaluations of the fitness function. The last version is $AP_{nf}$ – constants are estimated by non-linear fitting algorithm.

### 2.2 General Function Set

The GFS is a set of all mathematical objects – functions, operators and terminals. GFS consists of functions with different numbers of arguments. GFS is user-defined, so the content may differ. We must split the content of GFS into classes based on the numbers of arguments: $0_{args}$ are terminals, $1_{args}$ $(sin, cos, tan, \ldots)$, $2_{args}$ $(+, -, *, /, \ldots)$, etc. Choosing the right set may have a large influence on the convergence of AP.

### 2.3 Evolutionary Algorithm

AP was designed to be a very robust method and we can use almost any evolutionary algorithm. Individual steps, such as mutation, crossover, etc. are fully handled by the chosen evolutionary algorithm. Operations which make EAs when the algorithm is run do not have any influence on performance. Thus the result performance depends mainly on the correct choice of GFS individuals. For every evolutionary algorithm, the main goal is generally to reduce the fitness value to below a user-defined threshold, or to achieve maximum number of migrations for a Self Organizing Migrating Algorithm (SOMA), or in the case of Differential Evolution (DE) – generations.

### 2.4 Mapping Operators

Mapping is the phase when an individual is transformed into a useful mathematical function. It consist of two parts, DSH and security functions, to exclude the creation of pathological individuals. In the evolutionary algorithm the individual is represented by a vector of indexes and is remapped to mathematical objects from GFS.

### 2.5 Reinforced Evolution

By running evolution, more or less suitable individuals are created, so one very good idea is to include the best individual as the terminal in GFS. The main idea of reinforcement is based on the addition of a just-synthesized and partly successful program in an initial set of terminals [3]. The decision on whether the best value will be added to GFS is ensured by a user-defined threshold value. If it is reached, from that moment it is added in GFS as a terminal, best solution and updated whenever a better solution with lower fitness is found.

### 2.6 Security Procedures

Evolution can result in an expression, which is not mathematically correct, thus we create a pathological individual (without argument). We can prevent this by distributing GFS into classes ordered by the number of arguments. By mapping from evolution space to GFS, we measure the distance to the end of the expression. When the number of arguments is greater than the distance to the end of

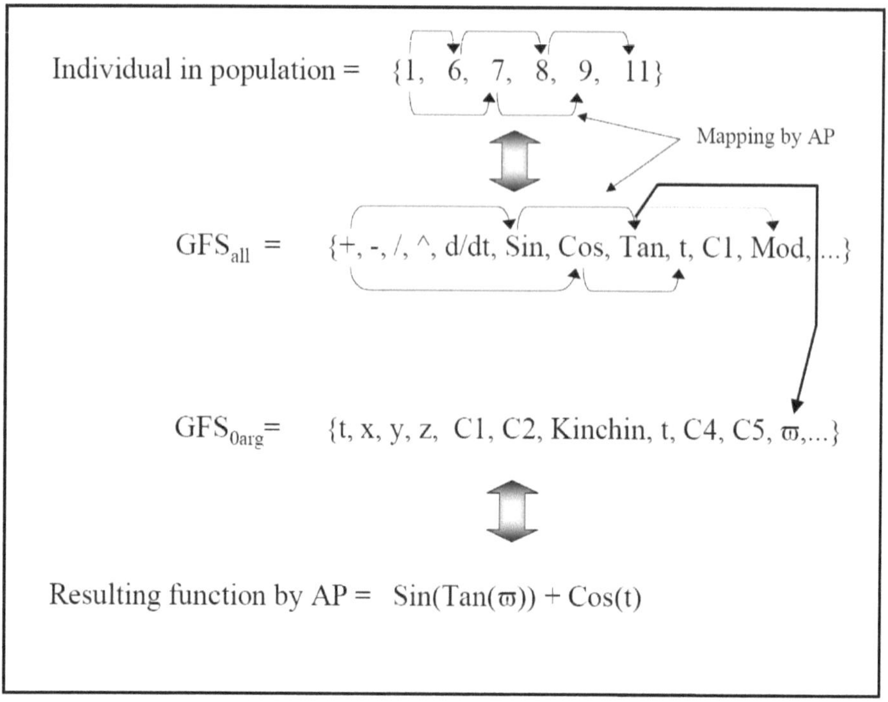


**Fig. 2.** Schema of mapping and security principles. Because of measuring distance to end of expression is *tan* replaced by $\omega$ [4]

the expression, we choose individuals from the lower class. We must also pay attention so as to exclude errors from the fitness function, such as division by zero, functions with an imaginary or real part (if not expected), frozen functions (an extremely long time to get a cost value), etc. [3].

## 3 Parallel Implementation Using CUDA – An Overview

CUDA (Compute Unified Device Architecture) is a platform for parallel computing, developing by NVIDIA. CUDA is SIMT (Single Instruction Multiple Threads). Its main feature is that one instruction is executed by thousands of threads. However, kernels can effectively perform only basic operations. In the CUDA device various types of memory reside, as you can see in Table 1. Note: registers are the fastest memory on the GPU.

The idea of implementing AP on parallel architecture is based on successful parallel implementation of evolutionary algorithms on CUDA, where a significant speeding up was achieved [10,11]. We were inspired to implement a parallel version of AP on CUDA, because of the good results achieved by the evolutionary algorithms when implemented on CUDA, although execution of this number of operations and structures may lead to worse performance results, as expected.

**Table 1.** CUDA Memory Types and Characteristics [13], (u.c. ≡ unless cached)

| Memory | Location | Cached | Access | Scope |
|---|---|---|---|---|
| Register | On-chip | No | Read/Write | One thread |
| Local | On-chip | Yes | Read/Write | One thread |
| Shared | On-chip | N/A | Read/Write | All threads in a block |
| Global | Off-chip (u.c) | Yes | Read/Write | All threads + host |
| Constant | Off-chip (u.c.) | Yes | Read | All threads + host |
| Texture | Off-chip (u.c.) | Yes | Read/Write | All threads + host |

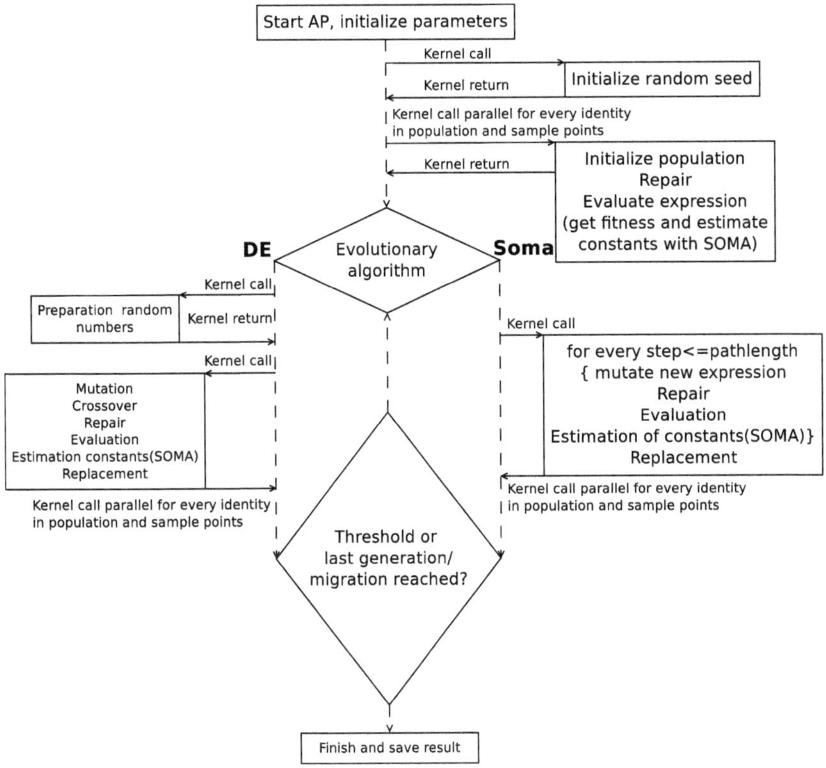


**Fig. 3.** The flowchart of AP implementation

# 4 Testing Parallel Implementation – Methods, Datasets, Results

We cut the spectra from Figure 1 and obtained only the emission line in Figure 4 with 50 equidistant points, to satisfy our requirements. The data on the Be stars spectra come from the archive of the Astronomical Institute of the Academy of Sciences of the Czech Republic[1].

[1] Available from: `http://astropara.projekty.ms.mff.cuni.cz/spectra/newest/`

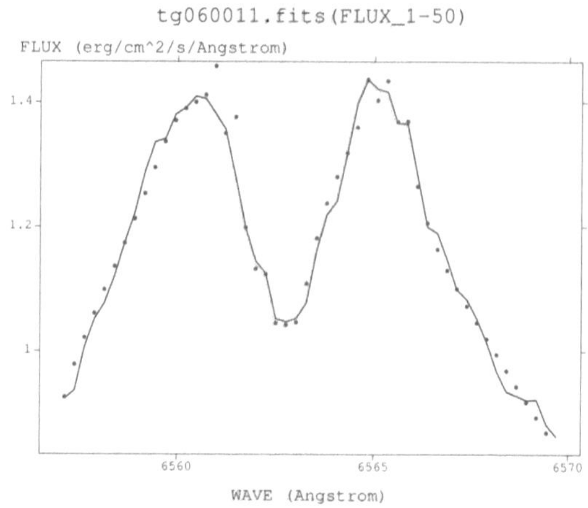


**Fig. 4.** Extracted emission line (blue) of Be star and our best result (red dots) of the SOMA algorithm with deterministic chaos

To by able to carry out a comparison in our tests we used a classical pseudo-random number generator and on the other site we tried to compare it to implemented deterministic chaos, with setting $A = 4$, which equation is as follows:

$$x_{n-1} = Ax_n(1 - x_n)$$

We carried out tests with SOMA and DE as the main evolutionary algorithm of AP. For estimating constants we chose the SOMA algorithm. The measured results are shown in Tables 2 and 3.

The settings for our implementation are shown in Table 2.

**Table 2.** Algorithm settings

| DE | | SOMA | | SOMA constants | |
|---|---|---|---|---|---|
| NP | 500 | PopSize | 500 | PopSize | 10 |
| Dimensions | 40 | Dimensions | 20 | Dimensions | 20 |
| Generations | 100 | Migrations | 10 | Migrations | 5 |
| F | 0.9 | PRT | 0.1 | PRT | 0.1 |
| CR | 0.5 | PathLength | 4 | PathLength | 3 |
| | | Step | 0.21 | Step | 0.25 |

Thanks to the use of deterministic chaos against a pseudo-random number generator (PRNG) we speed up the running time by using simple mathematical operations, which is good for the CUDA kernel. Convergence of the fitness value is also a bit better.

**Table 3.** Minimum, average and maximum fitness achieved from our tests. Each method was tested 10 times.

| | DE PRNG | DE Chaos | SOMA PRNG | SOMA Chaos |
|---|---|---|---|---|
| MIN | 4.02327 | 4.00432 | 0.85327 | 0.83326 |
| AVG | 4.30381 | 4.27877 | 1.41000 | 1.39508 |
| MAX | 4.57460 | 4.51520 | 1.84721 | 1.91367 |

**Table 4.** Duration of execution of AP: with SOMA 95238096 evaluation of the cost function in the main EA, with DE 100000000

| [s] | DE PRNG | DE Chaos | SOMA PRNG | SOMA Chaos |
|---|---|---|---|---|
| MIN | 2688.59 | 1962.53 | 3161.47 | 2761.57 |
| AVG | 2805.84 | 2296.42 | 3305.62 | 2942.40 |
| MAX | 2873.77 | 3037.32 | 3372.46 | 3098.13 |

## 5 Conclusion

We successfully implemented AP on CUDA and obtained relevant results. Both PRNG and deterministic chaos are suitable for running with parallel implementation of AP. The use of deterministic chaos seems to be a bit better. This is given by the fewer and simpler math operations needed to get a value. CUDA is not designed to run as long kernels, as we implemented, but there is a lot of room for future optimization of these kernels. We use lots of registers in the kernel, so performance falls from the ideal. In feature CUDA architecture, kernel under kernel will be probably run, which should bring a large improvement in performance. We want to compare our results with another parallel approach in the near future. On graphics hardware it will be Opencl, and OpenPM as a CPU variant.

**Acknowledgement.** The following grants are acknowledged for the financial support provided for this research: Grant Agency of the Czech Republic – GACR P103/13/08195S, is partially supported by Grant of SGS No. SP2014/159, VŠB–Technical University of Ostrava, Czech Republic, by the Development of human resources in research and development of latest soft computing methods and their application in practice project, reg. no. CZ.1.07/2.3.00/20.0072 funded by Operational Programme Education for Competitiveness, co-financed by ESF and state budget of the Czech Republic.

## References

1. Wells, D.C., Greisen, E.W., Harten, R.H.: FITS: A Flexible Image Transport System. Astronomy and Astrophysics Supplement Series 44, 363–370 (1981)
2. Rivinius, T., Carciofi, A.C., Martayan, C.: Classical Be stars. The Astronomy and Astrophysics Review, 1–86 (2013), doi:10.1007/s00159-013-0069-0

3. Zelinka, I., Davendra, D., Senkerik, R., Jasek, R., Oplatkova, Z.: Analytical Programming – a Novel Approach for Evolutionary Synthesis of Symbolic Structures. In: Kita, E. (ed.) Evolutionary Algorithms. InTech (2011), doi:10.5772/16166 ISBN: 978-953-307-171-8
4. Zelinka, I., Oplatkova, Z., Nolle, L.: Analytic Programming – Symbolic Regression by Means of Arbitrary Evolutionary Algorithms. Special Issue on Inteligent Systems, International Journal of Simulation, Systems, Science and Technology 6(9), 44–56 (2005) ISSN 1473-8031
5. Zelinka, I.: Symbolic regression – an overview, `http://www.mafy.lut.fi/EcmiNL/older/ecmi35/node70.html`
6. Zelinka, I., Oplatkova, Z.: Analytic programming – Comparative Study. In: CIRAS 2003: The second International Conference on Computional Intelligence, Robotics and Autonomous Systems (2003) ISSN 0219-6131
7. Lampinen, J., Zelinka, I.: New Ideas in Optimization – Mechanical Engineering Design Optimization by Differential Evolution, vol. 1, 20 p. McGraw-Hill, London (1999)
8. Zelinka, I.: Artificial Intelligence in The Problems of Global Optimization, 190 p. BEN (2002) (in Czech) ISBN 80-7300-069-5
9. Thizzy: Be stars (2008), `http://www.shelyak.com/contenu.php?id_contenu=30&id_dossier=24`
10. de Veronese, L.P., Krohling, R.A.: Differential evolution algorithm on the GPU with C-CUDA. In: IEEE Congress on Evolutionary Computation, CEC (2010)
11. Kralj, P.: Differential Evolution with parallelised objective functions using CUDA (2013), `http://www.codehunter.eu/media/Kralj_Differential_Evolution_with_parallelised_objective_functions_using_CUDA.pdf`
12. Kromer, P., Platos, J., Snasel, V., Abraham, A.: Many-threaded implementation of differential evolution for the CUDA platform. In: Krasnogor, N. (ed.) Proceedings of the 13th Annual Conference on Genetic and Evolutionary Computation (GECCO 2011), pp. 1595–1602. ACM, New York (2011), `http://doi.acm.org/10.1145/2001576.2001791`, doi:10.1145/2001576.2001791
13. Farber, R.: CUDA Application Design and Development, 1st edn. Morgan Kaufmann Publishers Inc., San Francisco (2011)

# Complex Network Analysis of Discrete Self-organising Migrating Algorithm

Donald Davendra[1], Ivan Zelinka[1], Roman Senkerik[2], and Michal Pluhacek[2]

[1] VŠB - Technical University of Ostrava,
Faculty of Electrical Engineering and Computer Science, 17. listopadu 15, 708 33
Ostrava-Poruba, Czech Republic
donald.davendra@vsb.cz, ivan.zelinka@vsb.cz
[2] Tomas Bata University in Zlin,
Faculty of Applied Informatics, T.G. Masaryka 5555, 760 01 Zlin, Czech Republic
senkerik@fai.utb.cz, pluhacek@fai.utb.cz

**Abstract.** This research analyses the development of a complex network in the swarm based Discrete Self-Organising Migrating Algorithm (DSOMA). The main aim is to evaluate if a complex network is generated in DSOMA, and how the population can be evaluated when the objective is to optimise the flow shop scheduling with blocking problem. The population is evaluated as a complex network over a number of migrations, and different attributes such as adjacency graph, minimal cut, degree centrality, closeness centrality, betweenness centrality, Katz centrality, mean neighbour degree, k-Clique, k-Plan, k-Club, k-Clan and community graph plots are analysed. From the results, it can be concluded that an DSOMA population does behave like a complex network, and therefore can be analysed as such, in order to obtain information about population development.

**Keywords:** Evolutionary algorithm, complex network, flow shop scheduling with blocking.

## 1 Introduction

Complex networks have come to play an important role in the analysis of complex dynamical behaviour. This research looks to expand the application of complex networks to analyse the development of Evolutionary Algorithms (EA's). EA's are used to solve complex engineering problems, using principles of evolution, heredity, mathematical and swarming amongst others. There are two generic components for EA's, a stochastic population and propagation techniques.

Through iteration of the population, the population undergoes transformations and exhibits complex behaviour. As the population is interconnected, complex network behaviour can be generated [2]. Some of the key questions regarding EA's relates to how exactly does the population behave? What impact does the different combination criterion's have on the improvement of the population? What effect does the problem being solved have on the population? Can the

I. Zelinka et al. (eds.), *Nostradamus 2014: Prediction, Modeling and Analysis of Complex Systems*, Advances in Intelligent Systems and Computing 289,
DOI: 10.1007/978-3-319-07401-6_16, © Springer International Publishing Switzerland 2014 

impact of different individuals within the population be measured? Where are the regions of stagnation in the population? [1].

The aim of this research is to analyse and attempt to answer these questions. The tool used for this task is complex network analysis. A complex network is a graph, which has non-trivial features, usually in the domain of real-world graphs. A complex network contain features, which are unique to the problem. It exhibits features such as degree distribution, clustering, community structures etc, which are important markers for EA's [2]. It can be postulated that a population under EA's jurisdiction exhibits such complex network behaviour. This is reinforced by the application of an EA, especially since the network is designed to interconnect, in order for new generation of individuals [3].

Each individual in the population can be regarded as a node in the complex network graph, where its linkage specifies the exchange of information in the population. Additionally, it can be specified that the node is only active for successful transfer of information i.e. if the individual is successful in improving itself. Using this approach, the network size is stagnant. The EA utilised in this research is the Discrete Self-Organising Migrating Algorithm (DSOMA) [4], a discrete variant of Self-Organising Migrating Algorithm (SOMA) algorithm [5]. DSOMA is a swarm based algorithm, where the propagation of the population is based on the sampling of the fitness landscape towards the leader.

The test problem is the permutative flow shop scheduling with blocking constraint (FSSB) problem. The FSSB problem is a *strict-sense* permutative-based combinatorial optimisation problem. A strict-sense problem is one where the encoding is non-over lapping and follows strict ordering. An ordering in this sense can be shown as an example of five jobs in a scheduling problem. In this case, each job is strictly identified and is non-replicating, therefore the encoding can take any permutation based on the basic encoding of $\{1, 2, 3, 4, 5\}$. These types of problems are strongly NP-Hard, and are generally considered some of the most difficult problems to solve.

The paper is organised as follows: section 2 briefly describes DSOMA, and section 3 outlines the FSSB problem. Complex networks and the different attributes tested is described in section 4, which also analyses the generated networks. Finally, the work is concluded in section 5.

## 2 Discrete Self-organising Migrating Algorithm

DSOMA [4] is the discrete version of SOMA [5], developed to solve permutation based combinatorial optimisation problem. The same ideology of the sampling of the space between two individuals of SOMA is retained. Assume that there are two individuals in a search space, where the objective for DSOMA is to transverse from one individual to another, while mapping each discrete space between these two individuals.

The major input of this algorithm is the sampling of the jump sequence between the individuals in the populations, and the procedure of constructing new trial individuals from these sampled jump sequence elements. The overall outline for DSOMA can be given as:

**Table 1.** DSOMA parameters

| Name | Range | Type | Description |
|---|---|---|---|
| $J_{min}$ | (1+) | Control | Min number of jumps |
| Population | 10+ | Control | Num. of individuals |
| Migrations | 10+ | Termination | Number of iterations |

1. **Initial Phase**
   (a) *Population Generation*: An initial number of permutative trial individuals is generated for the initial population.
   (b) *Fitness Evaluation*: Each individual is evaluated for its fitness.
2. **DSOMA**
   (a) *Creating Jump Sequences*: Taking two individuals, a number of possible jump positions is calculated between each corresponding element.
   (b) *Constructing Trial Individuals*: Using the jump positions; a number of trial individuals is generated. Each element is selected from a jump element between the two individuals.
   (c) *Repairment*: The trial individuals are checked for feasibility and those, which contain an incomplete schedule, are repaired.
3. **Selection**
   (a) *New Individual Selection*: The new individuals are evaluated for their fitness and the best new fitness based individual replaces the old individual, if it improves upon its fitness.
4. **Migrations**
   (a) *Iteration*: Iterate the population till a specified migration.

DSOMA requires a number of parameters as given in Table 1. The major addition is the parameter $J_{min}$, which gives the minimum number of jumps (sampling) between two individuals. The SOMA variables PathLength, StepSize and PRT Vector are not initialised as they are dynamically calculated by DSOMA using the adjacent elements between the individuals. The detailed description of DSOMA is given in [4] and [6]. The detailed schematic is given in Figure 1.

## 3 Flow Shop Scheduling with Blocking

Consider $m$ machines in series with *zero* intermediate storage between successive machines, which have to process $n$ jobs. If a given machine finishes the processing of any given job, the job cannot proceed to the next machine while that machine is busy, but must remain on that machine, which therefore remains *idle*. This phenomenon is refereed to as *blocking* [7].

In this paper, only flow shops with zero intermediate storage (FSSB) are considered, since any flow shop with positive (but finite) intermediate storage between machines can be modelled as a flow shop with zero intermediate storage. This is due to the fact that the storage space capable of containing one job may

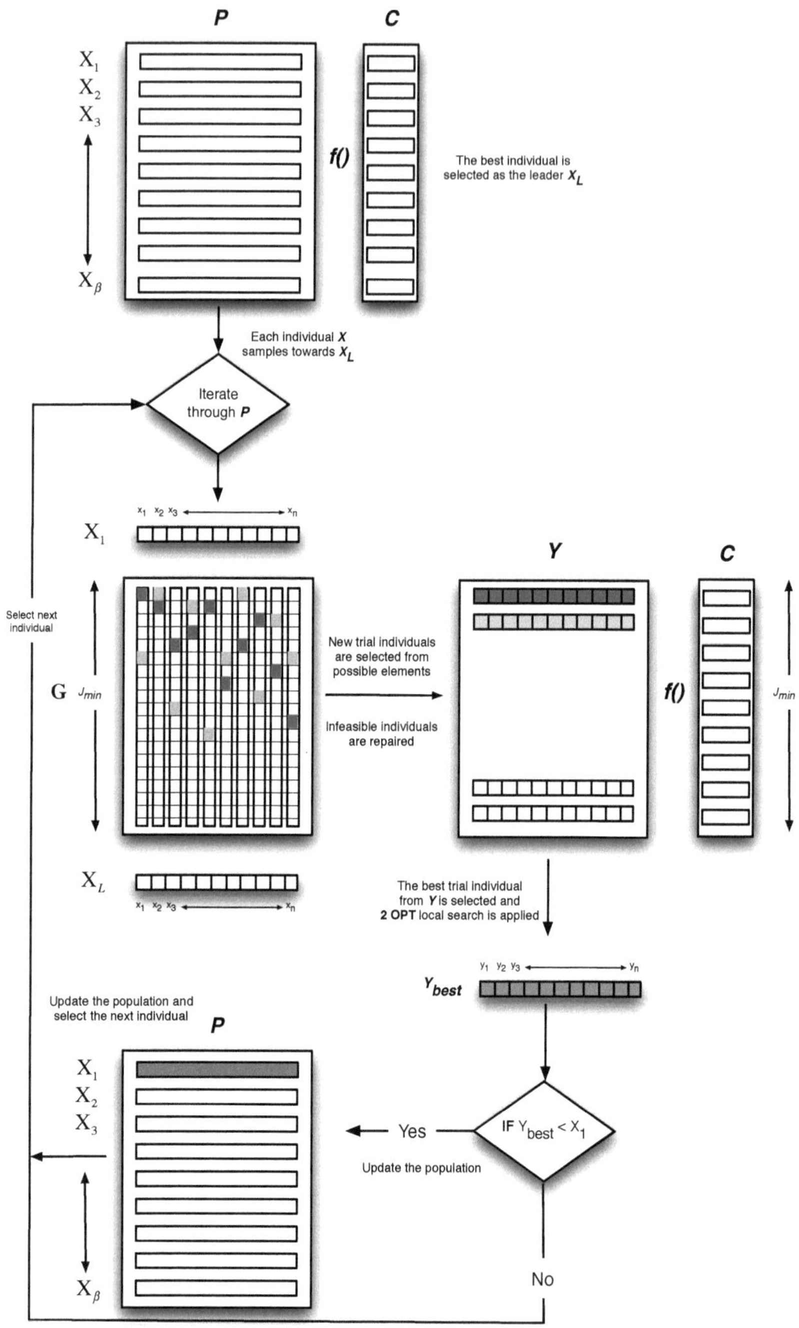


**Fig. 1.** DSOMA Outline

be regarded as a machine on which the processing time of all machines is equal to zero.

Pinedo [7] has defined the problem of minimising the makespan in a flow shop with zero intermediate storages is referred to in what follows as:

$$Fm \left| block \right| C_{\max}$$

Let $D_{i,j}$ denote the time that job $j$ actually departs machine $i$. Clearly $D_{i,j} \geq C_{i,j}$. Equality holds that job $j$ is not blocked. The time job $j$ starts its processing at the first machine is denoted by $D_{0,j}$. The following recursive relationship hold under the job sequence $j_1, \ldots, j_n$:

$$D_{i,j_1} = \sum_{l=1}^{i} p_{l,j_1} \quad i = 1, \ldots, m \tag{1}$$

$$\begin{aligned} D_{i,j_k} = \max\left(D_{i-1,j_k} + p_{i,j_k}, D_{i+1,j_{k-1}}\right) \\ i = 2, \ldots, m \;\; k = 2, \ldots, n \end{aligned} \tag{2}$$

$$D_{m,j_k} = D_{m-1,j_k} + p_{m,j_k} \tag{3}$$

The DSOMA algorithm was coded in the C programming language. The data set used was the first Taillard data set; a 5 machine, 20 job problem [8].

## 4 Complex Network Analysis

Complex network analysis can be accomplished using a number of components. In this paper, we utilise the *Adjacency graph* approach in order to show the linkage between different individuals in the population. The creation of the Adjacency graph is given as in Algorithm 1.

**input** : Adjacency matrix as $n$ x $n$ to 0, where $n$ is PopulationSize
**output**: Adjacency matrix
**for** $i \leftarrow 1$ **to** *Migrations* **do**
  **for** $j \leftarrow 1$ **to** *PopulationSize* **do**
    **for** $k \leftarrow 1$ **to** *IndividualSize* **do**
      *Generate values between* $J_k$ *and* $Leader_k$
    **end**
    *trial* ← *Repair the new trial individuals.*;
    $fitness_{trial} \leftarrow$ evaluate $f_{trial}$;
    **if** $fitness_{trial} < fitness_{individual}$ **then**
      add one to corresponding vector row $j$ and *Leader* vector column in the Adjacency matrix.;
    **end**
  **end**
**end**

**Algorithm 1.** Adjacency matrix generation

Once the Adjacency matrix is created, all non-zero values are reset to one, to have a one-to-one relationship. The following subsections outlines the different properties which were analysed using complex networks. For each property, two graphs are presented, one after first migration and the other after the fifth migration. In retrospect, we aim to show the growth of the complex network within population. We omit the presentation of the final migration due to the heavy density of the complex network.

The visualisation of the experimentation was conducted on Mathematica 9 software, using the social network analysis toolkit [9]. The population size was kept at 100 individuals with 20 migrations.

### 4.1 Adjacency Graph

One of the core measures of *degree*, related to the vertex in a graph is the measure of *adjacency*. Put simply, adjacency is the measure of the number of connections between vertices and edges. Some of the most important attributes in a graph is the number of linkages or degrees, or in the opposite case the total devoid of connections (isolated vertex). The sum of all degrees is the *total* degree of the graph. The adjacency matrix graphs are shown in Figure 2.

From the migrations, it is seen that not all individuals in the population improve in the first migration. From the first migration, seven individuals due not improve and therefore do not contribute to the network. In the fifth migration, two separate leaders are observed in the network, and all individuals have improved in the fitness landscape.

### 4.2 Minimal Cut

A minimum k-cut of a graph is a partition of vertices of a graph into k disjoint subsets with the smallest number of edges between them. From the graphs in Figure 3, we can see that the minimal cut is actually dependent on the number of new leaders been generated in the population. From the analysis of the graphs, it can be seen that the minimal cut increases over the migrations.

### 4.3 Degree Centrality

Generally referred to as the simplest centrality, the degree centrality is defined as the number of edges connected to a specific node. Degree centrality is an important distribution hub in the network as it connects and thereby distributes the most information flowing through the network. The graphs are given in Figure 4.

This is one of the most important features under consideration in the complex network. Using degree centrality, we can actually analyse if *stagnation* or *premature convergence* is occurring the population. Looking at the feature on the graphs, it can be seen multiple nodes are increasing (distinguished by their size), emphasising their prominence in the population, and their effect in generating new leaders.

### 4.4 Closeness Centrality

In a connected graph, having natural distance in the vertices, the distance between nodes is measured by the shortest distance. In other words, the inverse of the sum of all distances of a node in a graph represents its closeness.

Closeness is a key measure as to the *rate* of distribution of information in the graph. The graphs are given in Figure 5. The closeness centrality is actually distributed over the entire network (outer nodes). This actually conveys that all the nodes are contributing equally in the network.

### 4.5 Betweenness Centrality

Betweenness centrality is a measure of a specific vertex within the entire graph. It reflects as the number of connections it is part of as a *shortest* link between different nodes. This is an important measure of the *control* and *management* of information within the graph. Central nodes are seen to play a more fundamental role as shown in Figure 6. Incidentally, the node shown to have the highest betweenness centrality also has the best fitness function, therefore the role of the leader increases over the migrations. Another important note is that once the node for the best fitness changes, as does the betweenness centrality of the system.

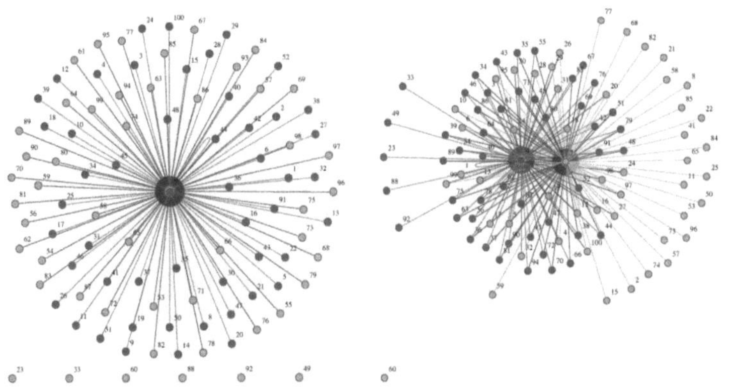

**Fig. 2.** Adjacency Graph

### 4.6 Katz Centrality

Katz centrality is the generalised measure of degree centrality, where it measures the number of all nodes that can be connected through a path with the distances nodes having adverse weighting. The graphs are given in Figure 7. As expected, an early development of the graph can be seen, with equal weighting for all leaders, as given in degree centrality.

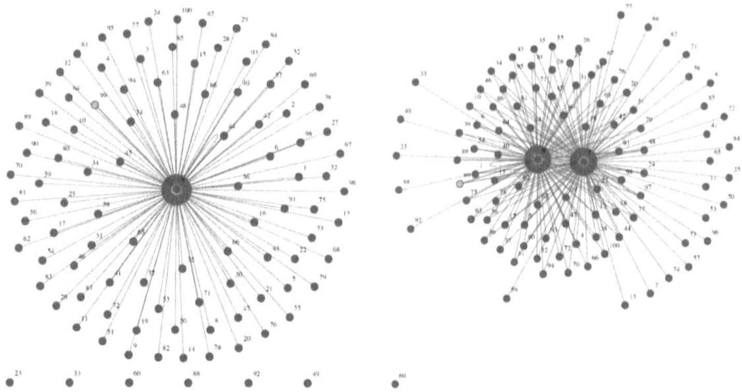

**Fig. 3.** Minimal Cut

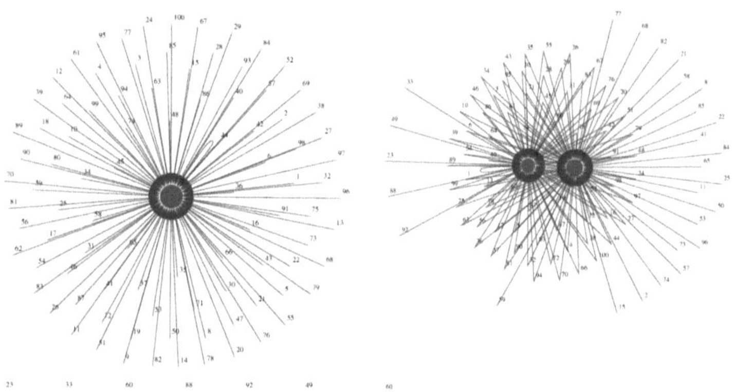

**Fig. 4.** Degree Centrality

### 4.7 Mean Neighbour Degree

The mean neighbour degree is the average neighbour degree of the given vertices in a graph as given in Figure 8. From this specific property, the network is shown to grow more or less proportionally over the migrations, with all nodes playing equal role in its development.

### 4.8 k-Clique

A k-Clique of a graph is a maximal subset $S$ such that geodesic distance between every pair of vertices in the set is less than or equal to k (set to 2 in this case). That means no two nodes in the set can be more than k steps away from each other. The graphs are given in Figure 9. From the graphs, it can be seen that the $S$ increases from one node to two nodes over the migrations. Also, the position of $S$ is centred towards the leader.

### 4.9 k-Plex

A k-Plex is a maximal subgraph with the following property: each vertex of the induced subgraph is connected to at least $n$-k other vertices, where $n$ is the number of vertices in the induced subgraph. Here we have used $k = 2$ and $n = 3$. The graphs are given in Figure 10. The number of k-Plex is kept at three nodes throughout the migrations and again centred on the leader on each migration.

### 4.10 k-Club

A k-Club of a graph is a subset $S$ of the vertex set, which induces a subgraph of diameter $k$ with the following property: each vertex of the induced subgraph is connected to at least $n$-k other vertices, where $n$ is the number of vertices in the induced subgraph. Here we have used $n = 2$. The graphs are given in Figure 11. K-Club is an important attribute used to analyse dependencies of the network, and to isolate non-productive nodes. Therefore, it is seen that the subset $S$ increases over the migrations, implying the growth of the network over the migrations.

### 4.11 k-Clan

A k-Clan of a graph is a *k-clique* in which the subgraph induced by subset $S$ has diameter less than or equal to k. So a subset to be a k-Clan-

1. should be a k-clique.
2. all nodes are connected by a path less than or equal to k.

We have used $k = 5$. The graphs are given in Figure 12 and it can be seen that there is an increase of interconnectivity of the network. In fact all nodes are joined over the migrations, therefore all nodes share or have had an influence in the migration of the network. Therefore, all nodes improve on the leader of the population.

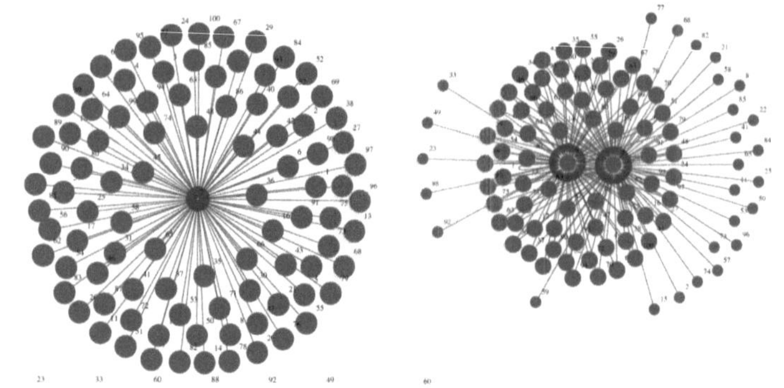

**Fig. 5.** Closeness Centrality

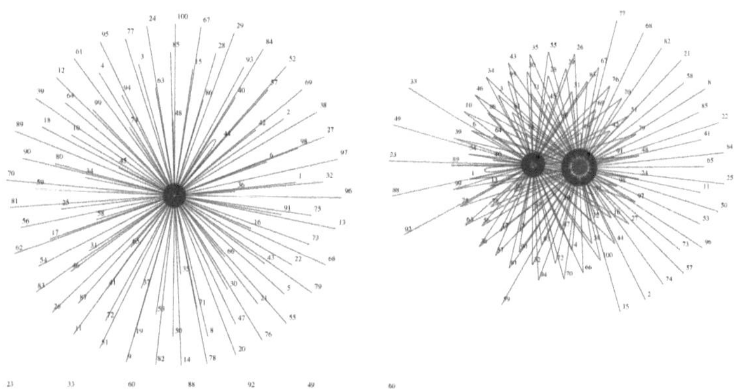

**Fig. 6.** Betweenness Centrality

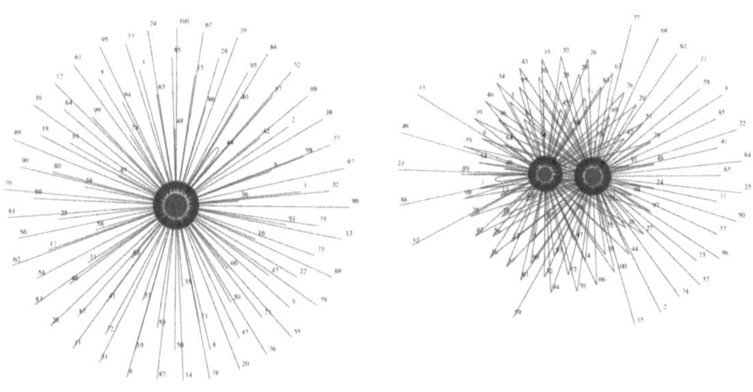

**Fig. 7.** Katz Centrality

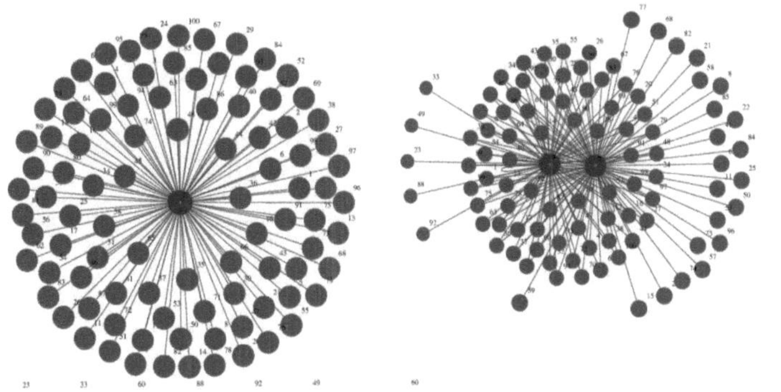

**Fig. 8.** Mean Neighbour Degree

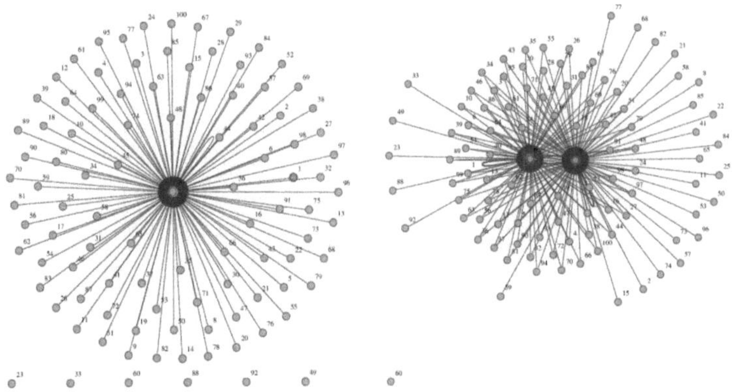

**Fig. 9.** k-Clique

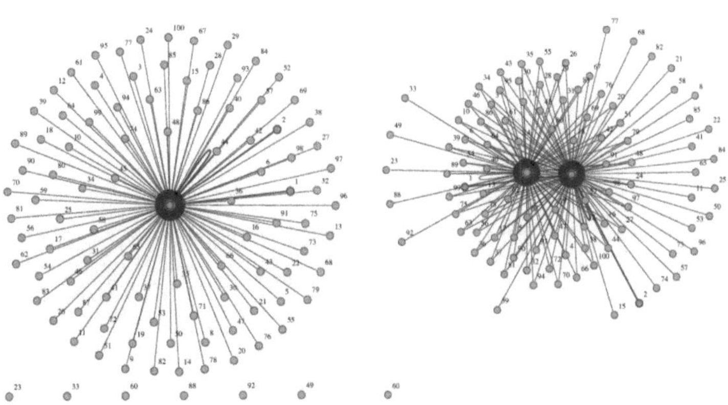

**Fig. 10.** k-Plex

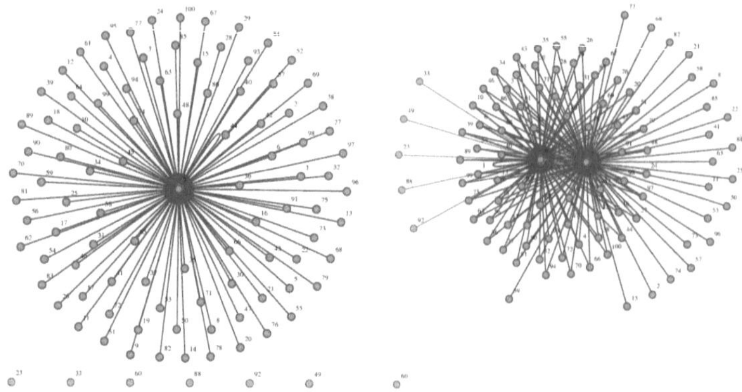

**Fig. 11.** k-Club

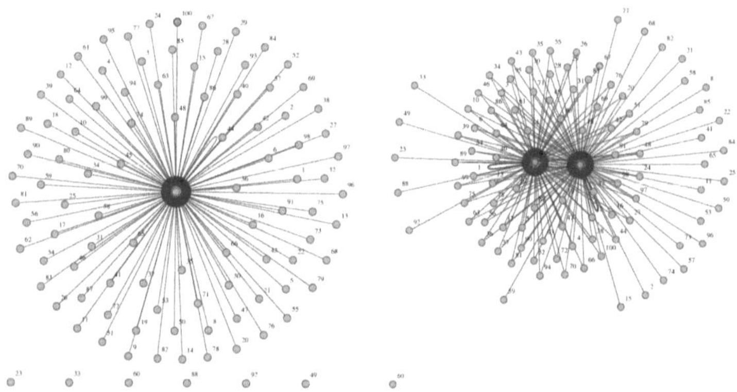

**Fig. 12.** k-Clan

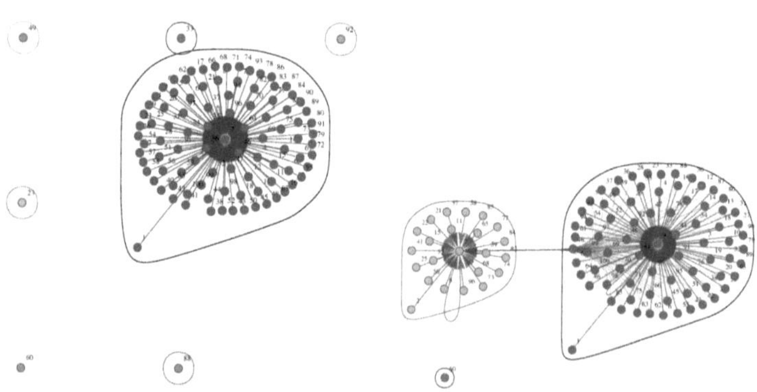

**Fig. 13.** Community Graph Plot

### 4.12 Community Graph Plot

The community graph plot attempts to draw all the vertices grouped into communities, for better representation. In this representation, we display the *hierarchical* mode of graph representation as given in Figure 13. The community plot is simply to decipher. At each migration, nodes are grouped around the leader. As the network becomes more dense over the migrations, the community graph plot increases in terms of connections and decreases in terms of isolated communities. The communities are generally based around the number of leaders.

## 5 Conclusion

From the analysis of the population as an adjacency graph, it becomes clear that the population exhibits complex network behaviour during evaluations. This behaviour is seen as the expansion of the population in terms of improving fitness, clustering due to fitness convergence and community migration due to interaction.

Some of the most notable features is that not all individuals in the population interact at each migration. However, after repeated experimentation, it was clear that at the completion of the migrations, all individuals were connected in the network. Therefore, it is clear that population size has an impact on the selection of new leaders. The different centralities show that a number of hubs, or connection point with more nodes appear according to the number of leaders in the population. In retrospect, it becomes possible to analyse the structure of the problem itself through complex networks, especially the interaction of each individual at different migrations.

The clique analysis gives an overview of the number of sub-connections within the populations. These subsets can give an overview of the density of the population and the depth of the connections at each migration. Generally, all cliques reflect the number of leaders in the populations over the migrations.

Finally, the community plot gives a hierarchical overview of the population. The individuals appear clustered towards the leaders in the populations. The density of the communities is generally based on how many individuals improve upon the leader during the migrations.

It can be seen that the DSOMA population can also be analysed through complex network tools, which give an excellent representation of the interconnections and swarm dynamics, thereby shedding more light on better management on control of DSOMA, with an outline of how many individuals contribute to the development of the populations, throughout the migrations.

**Acknowledgments.** This work was principally supported by the Grant of SGS SP2014/170 and partially by IGA project No. IGA/FAI/2014/010, CEBIA-Tech No. CZ.1.05/2.1.00/03.0089, Grant Agency of the Czech Republic - GACR P103/13/08195S, SGS No. SP2014/42, VB - Technical University of Ostrava, Czech Republic, by the Development of human resources in research and development of latest soft computing methods and their application in practice project, reg. no. CZ.1.07/2.3.00/20.0072 funded by Operational Programme Education for Competitiveness.

## References

1. Davendra, D.: Evolutionary algorithms and the edge of chaos. In: Zelinka, I., Celikovsky, S., Richter, H., Chen, G. (eds.) Evolutionary Algorithms and Chaotic Systems. SCI, vol. 267, pp. 145–161. Springer, Heidelberg (2010)
2. Zelinka, I., Davendra, D., Enkek, R., Jasek, R.: Do Evolutionary Algorithm Dynamics Create Complex Network Structures? Complex Systems 20, 127–140, 0891-2513
3. Zelinka, I., Davendra, D.D., Chadli, M., Senkerik, R., Dao, T.T., Skanderova, L.: Evolutionary Dynamics as The Structure of Complex Networks. In: Zelinka, I., Snasel, V., Abraham, A. (eds.) Handbook of Optimization. ISRL, vol. 38, pp. 215–243. Springer, Heidelberg (2013)
4. Davendra, D., Bialic-Davendra, M.: Scheduling flow shops with blocking using a discrete self-organising migrating algorithm. International Journal of Production Research 51, 2200–2218 (2013)
5. Zelinka, I.: SOMA - Self Organizing Migrating Algorithm. In: Onwubolu, G., Babu, B. (eds.) New Optimization Techniques in Engineering. STUDFUZZ, vol. 141, pp. 167–217. Springer, Heidelberg (2004)
6. Davendra, D., Senkerik, R., Zelinka, I., Pluhacek, M., Bialic-Davendra, M.: Utilising the chaos-induced discrete self organising migrating algorithm to solve the lot-streaming flowshop scheduling problem with setup time. Soft Computing 18, 669–681 (2014)
7. Pinedo, M.: Scheduling: theory, algorithms and systems. Prentice Hall, Inc., New Jersey (1995)
8. Taillard, E.: Benchmarks for basic scheduling problems. European Journal of Operations Research 64, 278–285 (1993)
9. Wolfram Website, `http://www.wolfram.com/`

# GUNT RT 010 Experimental Unit Modelling and Predictive Control Application

Daniel Honc, František Dušek, and Rahul Sharma

Department of Process control, Faculty of Electrical Engineering and Informatics, University of Pardubice, Czech Republic
{daniel.honc,frantisek.dusek}@upce.cz, rahul.sharma@student.upce.cz

**Abstract.** First principle process model of GUNT RT 010 experimental unit is designed for the purpose of various modern control methods laboratory testing and applications. Unknown parameters are estimated from experimental data. Model Predictive Control is applied to the system to verify quality of the model and to demonstrate its use. Two approaches are considered according to process model – transfer function and state-space model. Known future set-point and constraints on input, state and output variable are part of the controller – optimal control actions are calculated by quadratic programming.

## 1 Introduction

Controller design or even simple tuning requires information about the controlled process. Trial-error tuning method by its iteration is getting such information. Critical gain and period, step response characteristics (simple dynamics approximation), transfer functions, state-space models are used often for controller design [1, 2, 3, 4]. More sophisticated controller needs more complex process model and but then special control aims and process features can be considered and solved. Model Predictive Control [5, 6] is a nice example of such a method – control actions are calculated as a discrete-time optimization problem, where the process model is used for the plant behavior prediction and a receding horizon concept is applied to introduce the feedback. Multivariable processes, non-minimum phase or processes with dead-times can be easily controlled and knowledge of the future set-point or process constraints can be directly used in the design procedure. This can lead to a better control quality which is unreachable by the conventional controllers. Drawback is that the solution is quite complex – design, debugging, application, maintenance, modification – experts are needed for such tasks.

Simulations and laboratory experiments are essential in education and training phase [7, 8]. GUNT RT 010 Level Control Training System is very interesting laboratory process [9, 10]. It is shipped together with experiments, software, and simulations. Learning objectives of the experiments are: investigation of a controlled system without feedback, open loop control response, closed loop system with different controllers and parameters. However, the possibilities of applying user algorithms are li-

I. Zelinka et al. (eds.), *Nostradamus 2014: Prediction, Modeling and Analysis of Complex Systems*, Advances in Intelligent Systems and Computing 289,
DOI: 10.1007/978-3-319-07401-6_17, © Springer International Publishing Switzerland 2014

mited. Because commercial acquisition card LabJack U12 [11] with available dynamic linked library is used by the training system there is a way how to communicate with the process from another environment for example from MATLAB or LabView [12].

The outline of the paper is as follows. GUNT Level Control Training System is presented in section 2. Nonlinear first principle process model and its linearization form are derived in section 3. Unknown parameters are estimated in section 4. Two model predictive controllers are applied in section 5. Conclusions are given in section 6.

## 2 GUNT Level Control Training System

Experimental set-up [10] (see Fig. 1) is mounted on the housing with electronics 3. Transparent level-controlled tank 1 is fed from the storage tank 4 with the speed-controlled pump 2. Liquid level is measured using the tensometric pressure sensor. Electromagnetic proportional valve 5 in the tank outlet can serve as a disturbance variable or as a second manipulated variable.

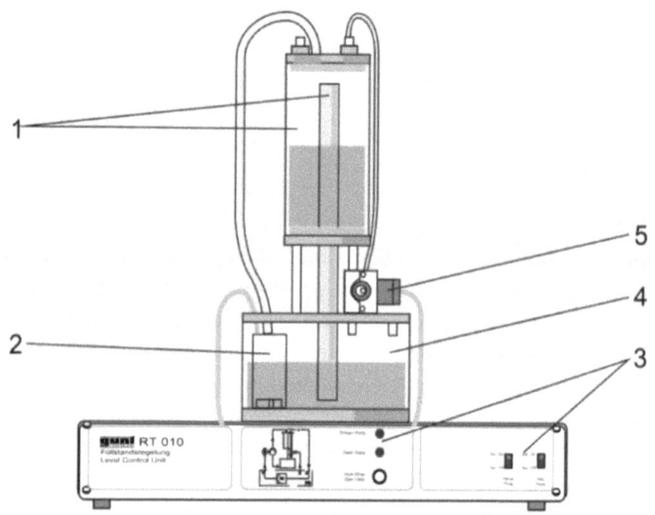


**Fig. 1.** Scheme of training system RT 010

Technical Data:

- Level-controlled tank: capacity 1.2 l
- Storage tank: capacity 3.7 l
- Pump ELEGANT: 12 V, 10-18 W, max. flow rate 8 l/min, 60 kPa
- Electromagnetic valve Burkert 2835: 24 V, 16 W
- Tensometric pressure sensor: 0 – 3 kPa (0 – 300 mm of water level)

Internal wiring with LabJack U12 acquisition card [11] (see Fig. 2):

- Analog input AI0 - pressure sensor output $y$ (0.5 – 2.6 V)
- Analog output AO0 - pump input $u_c$ (0 – 5 V) , 0 off / 5 on
- Analog output AO1 - valve input $u_v$ (0 – 5 V), 0 closed / 5 opened
- Digital input IO2 - pump manual switch output (on/off)
- Digital input IO3 - valve manual button output (on/off)

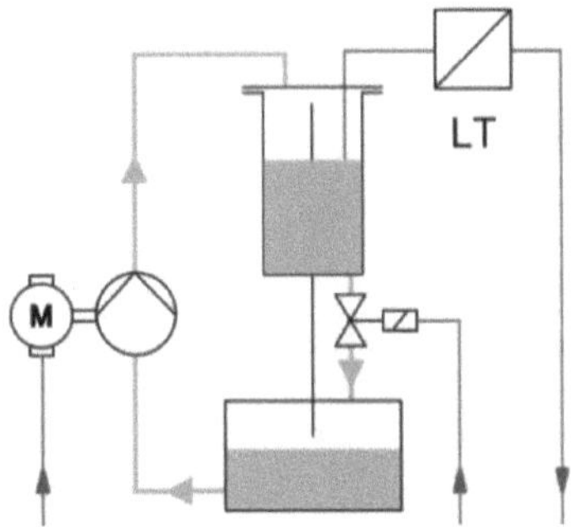


**Fig. 2.** Signal scheme

# 3 Process Model

## 3.1 Nonlinear Process Model

Nonlinear first-principle process model of the GUNT Level Control Training System is derived from mathematical-physical analysis by application of law of energy and mass conservation. Model has two inputs – voltages for the pump and valve and one output – signal from pressure sensor corresponding to water level in level-controlled tank, construction parameters and 6 unknown parameters.

Pump model is described with static characteristic

$$Q_{in} = a(u_c - \tilde{u}_c)^b \tag{1}$$

where $Q_{in}$ is tank input flow rate, $m^3/s$,
$u_c$ is pump voltage, V,
$\tilde{u}_c$ is voltage when the water begins to flow into tank, V and
$a, b$ are pump parameters.

Tank model is derived from law of mass conservation

$$Q_{in} = Q_{out} + S\frac{dh}{dt} \tag{2}$$

where $Q_{in}$ is tank input flow rate, $m^3/s$,
$Q_{out}$ is tank output flow rate, $m^3/s$ and
$S$ is tank cross-section, $m^2$.

Valve model is described with static characteristic – derived from law of energy conservation – conversion of potential energy into kinetic energy

$$Q_{out} = e\left(u_v - \tilde{u}_v\right)^f \sqrt{h} \tag{3}$$

where $Q_{\text{out}}$ is tank output flow rate, m$^3$/s,
$u_v$ is valve voltage, V,
$\tilde{u}_v$ is voltage when the valve begins to open, V,
$h$ is water level, m and
$e, f$ are valve parameters.

Pressure sensor model is described with static characteristic

$$y = c\,h + d \tag{4}$$

where $y$ is pressure sensor voltage, V,
$h$ is water level, m and
$c, d$ are pressure sensor parameters.

### 3.2 Process Model Linearization

Nonlinear model is linearized and rewritten into deviation form for the purpose of linear controller design.

Nonlinear static characteristic of the pump is approximated with a line with slope equal to a derivative in linearization point given by pump voltage $u_{c0}$

$$\overline{Q}_{in} = \left.\frac{dQ_{in}}{du_c}\right|_{u_{c0}} \cdot \overline{u}_c = ab(u_{c0} - \tilde{u}_c)^{b-1}\,\overline{u}_c \tag{5}$$

Variables signed with stripe are deviations of original variables in linearization point [$u_{c0}$, $u_{v0}$, $Q_0$, $h_0$, $y_0$]. For example

$$\overline{Q}_{in} = Q_{in} - Q_0 \tag{6}$$

where $Q_0$ is flow rate in linearization point, m$^3$/s.

Tank model is linear so it is only rewritten into deviation form

$$\overline{Q}_{in} = \overline{Q}_{out} + S\frac{d\overline{h}}{dt} \tag{7}$$

Valve static characteristic is linearized in a similar manner as the pump static characteristic

$$\overline{Q}_{out} = \left.\frac{dQ_{out}}{du_v}\right|_{u_{v0},h_0} \cdot \overline{u}_v + \left.\frac{dQ_{out}}{dh}\right|_{u_{v0},h_0} \cdot \overline{h} = e\,f\left(u_{v0} - \tilde{u}_v\right)^{f-1}\sqrt{h_0}\,\overline{u}_v + \frac{e\left(u_{v0} - \tilde{u}_v\right)^f}{2\sqrt{h_0}}\overline{h} \tag{8}$$

Pressure sensor static characteristic is linear, so it is only rewritten into deviation form

$$\bar{y} = c \cdot \bar{h} \tag{9}$$

### 3.3 Linear Transfer Function Model

Tank input flow rate $\overline{Q}_{in}$ from Eq. (5) and output low rate $\overline{Q}_{out}$ from Eq. (8) are substituted into Eq. (7)

$$a\,b\,(u_{c0} - \tilde{u}_c)^{b-1}\,\bar{u}_c = e\,f\,(u_{v0} - \tilde{u}_v)^{f-1}\sqrt{h_0}\,\bar{u}_v + \frac{e\,(u_{v0} - \tilde{u}_v)^f}{2\sqrt{h_0}}\bar{h} + S\frac{d\bar{h}}{dt} \tag{10}$$

Eq. (10) is modified so that time constant $T$, gain $Z_c$ of the pump and gain $Z_v$ of the valve to water level can be expressed

$$\begin{gathered} T\frac{d\bar{h}}{dt} + \bar{h} = Z_c\,\bar{u}_c + Z_v\,\bar{u}_v \\ T = \frac{2S\sqrt{h_0}}{e\,(u_{v0} - \tilde{u}_v)^f},\; Z_c = \frac{2ab(u_{c0} - \tilde{u}_c)^{b-1}\sqrt{h_0}}{e\,(u_{v0} - \tilde{u}_v)^f},\; Z_v = \frac{-2\,fh_0}{u_{v0} - \tilde{u}_v} \end{gathered} \tag{11}$$

If gain for pressure sensor voltage $y$ is needed, gains must be multiplied by constant $c$ and the resulting model is

$$T\frac{d\bar{y}}{dt} + \bar{y} = cZ_c\,\bar{u}_c + cZ_v\,\bar{u}_v \tag{12}$$

### 3.4 Linear State Space Model

State-space model is as follows (water level is state variable)

$$\begin{gathered} \frac{d\bar{h}}{dt} = -\frac{1}{T}\bar{h} + \begin{bmatrix} \frac{Z_c}{T} & \frac{Z_v}{T} \end{bmatrix}\begin{bmatrix} \bar{u}_c \\ \bar{u}_v \end{bmatrix} \\ \bar{y} = c\bar{h} \end{gathered} \tag{13}$$

## 4 Unknown Parameters Estimation

Construction parameters of GUNT Level Control Training System are summarized in Tab. 1.

**Table 1.** Construction parameters

| *Construction parameter* | *Value* |
|---|---|
| Tank diameter | 0.096 m |
| Overflow pipe diameter | 0.025 m |
| Pressure sensor pipe diameter | 0.016 m |
| Distance between tank bottom and valve | 0.083 m |
| Diameter of pipe between tank and valve | 0.02 m |
| Minimum measurable water level (origin of water level scale) | 0.089 m |
| Maximum water level (overflow throat distance from valve) | 0.266 m |

Pump, valve and pressure sensor unknown parameters are estimated from experimental data – steady-states values of flow rate, water level and pressure sensor voltage are measured for different pump and valve voltages and parameters are estimated by numerical optimization. Pump, valve and pressure sensor estimated parameters are summarized in Tab. 2. Static characteristics are shown in Fig. 4-6 (measured data are plotted by crosses).

**Table 2.** Estimated parameters

| *Estimated parameter* | *Value* |
|---|---|
| Pump parameter a | $4.77\times10^{-5}$ |
| Pump parameter b | 0.535 |
| Voltage when the water begins to flow into tank $\tilde{u}_c$ | 1.75 V |
| Valve parameter e | $5.13\times10^{-5}$ |
| Valve parameter f | 0.714 |
| Voltage when the valve begins to open $\tilde{u}_v$ | 0.4 V |
| Pressure sensor parameter c | 12.6 |
| Pressure sensor parameter d | -0.63 |

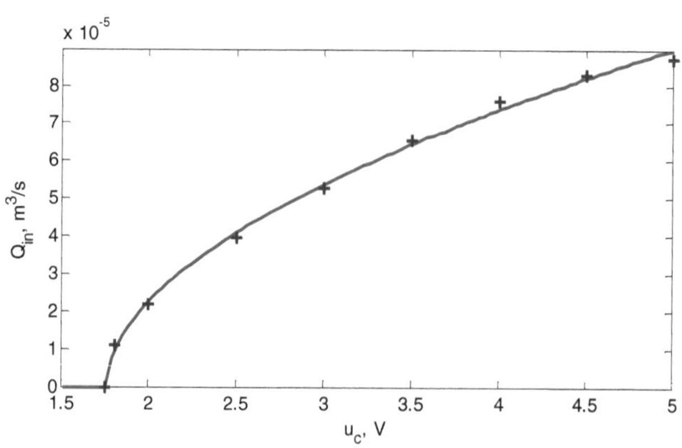


**Fig. 3.** Pump static characteristic

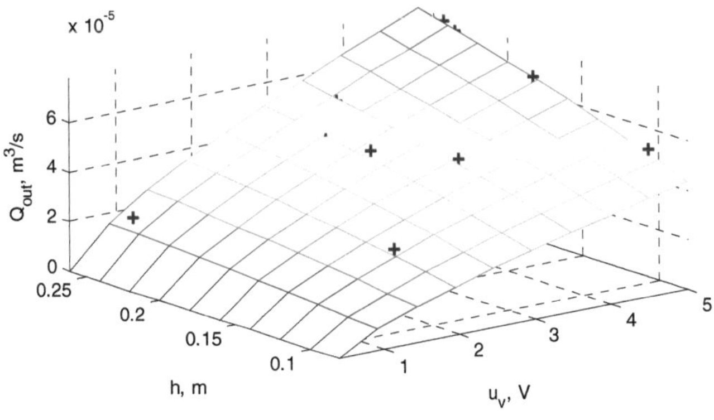


**Fig. 4.** Valve static characteristic

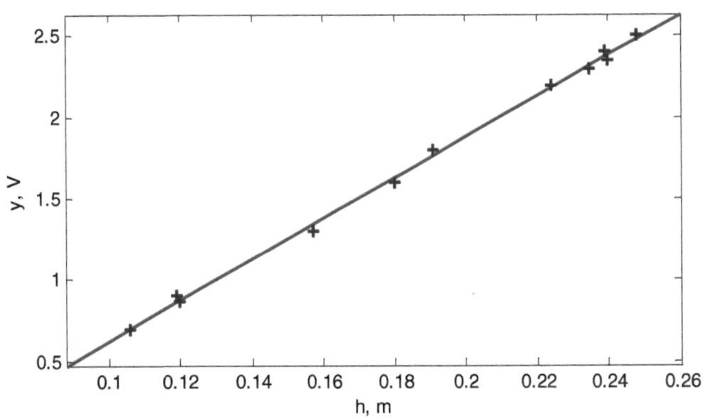


**Fig. 5.** Pressure sensor static characteristic

# 5 Constrained Model Predictive Control

Two Model Predictive Controllers are applied to GUNT Level Control Training System – one predictive controller with transfer function and another one with state-space process model [5, 6]. Incremental forms of the models are used in the controllers. Filtration polynomial is considered in the transfer function case. No observer is needed by the state-space version because water level can be calculated from pressure sensor voltage. In both cases constraints are considered and we assume that the future course of the set-point is known. Linear models are used in the predictive controllers and nonlinear model for the control response simulations. Linearization point - steady-state flow rate, water level and pressure sensor voltage are: $u_{c0} = 2.5$ V, $u_{v0} = 3$ V, $Q_0 = 4.09\times10^{-5}$ m$^3$/s, $h_0 = 0.074$ m and $y_0 = 1.42$ V. Time constant and gains of the linearized model are: $T = 52$ s, $Z_c = 0.23$ and $Z_v = -0.089$. Parameters of all controllers are the same and listed in Tab. 3. Control experiment starts from the calculated steady-state input and output of nonlinear model. Set-point is increased in 60 s stepwise and decreased in 120 s back to its initial value. Simulation response of the nonlinear model is plotted with the dotted line and the solid line is used for the experimental data. Response of the controller with the transfer function is in Fig. 6 and with the state-space model is in Fig. 7.

**Table 3.** Model Predictive Controller parameters

| *Controller parameter* | *Value* |
|---|---|
| Sample time | 0.5 s |
| Start of costing horizon | 1 |
| End of costing horizon | 40 |
| End of control horizon | 40 |
| Weighting coefficient for control error | 1 |
| Weighting coefficient for control increments | 0.1 |

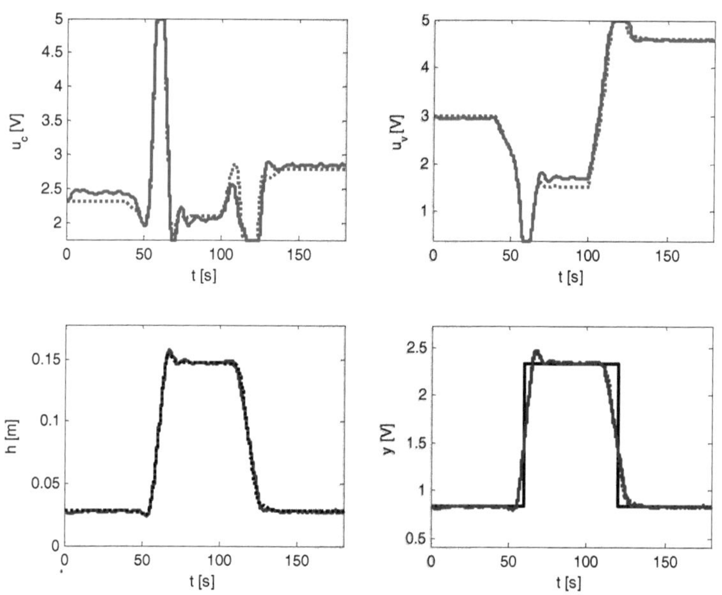


**Fig. 6.** Control response of controller with transfer function

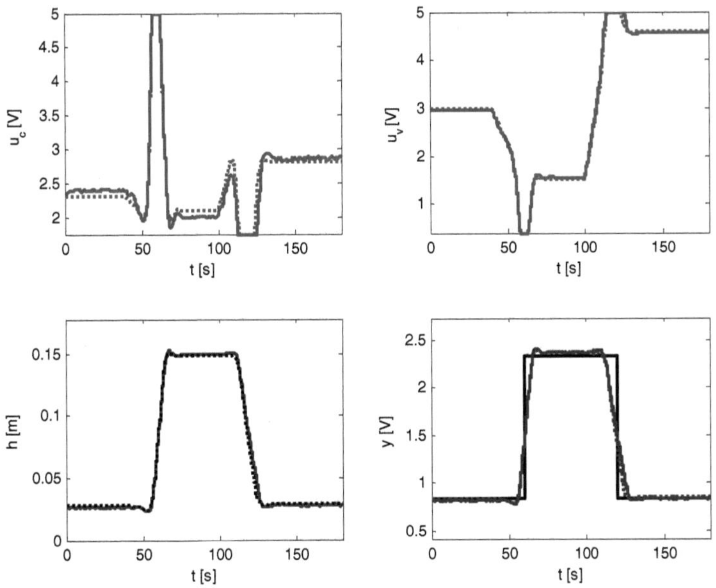


**Fig. 7.** Control response of controller with state-space model

# 6 Conclusion

First principle nonlinear model describes behavior of the real system very well - simulated and experimental responses are very similar (practically identical). The model has only 8 unknown parameters and all of them can be easily estimated from measured steady-state data. Controller design is not covered in this paper. The main objective of this paper was to validate the nonlinear model quality. We have two manipulated variables and only one controlled variable. Using of terminal constraint will force the controller output to follow optimal steady state solution. Now which combination will be at the end of the control experiment depends on used weightings and constraints. Simulation with the nonlinear model will help to implement the controller design in real world applications. Typical noise and disturbances for the process can be added so the simulation is even more realistic and can serve as a benchmark system.

**Acknowledgments.** This research was supported by Institutional support of The Ministry of Education, Youth and Sports of the Czech Republic and by the project "Support for internships and professional activities by innovation of tertiary education at DFJP and FEI University of Pardubice, Reg. No: CZ.1.07/2.4.00/17.0107" in Modern control methods team - development and application of predictive control methods using artificial intelligence.

## References


1. Dorf, R.C., Bishop, R.H.: Modern control systems, 12th edn. Prentice Hall (2011)
2. Nise, N.S.: Control systems engineering, 6th edn., Hoboken, NJ (2011)
3. Ogata, K.: Modern control engineering, 5th edn. Prentice Hall (2010)
4. Goodwin, G.C., Graebe, S.F., Salgado, M.E.: Control System Design. Prentice Hall (2001)
5. Camacho, E.F., Bordons, C.: Model predictive control. Springer, New York (2004)
6. Rossiter, J.A.: Model-based predictive control: a practical approach. CRC Press, Boca Raton (2003)
7. Åström, K.J., Lundh, M.: Lund control program combines theory with hands-on experience. IEEE Contr. Syst. Ag. 12(3), 22–30 (1992)
8. Honc, D., Dušek, F.: Novel multivariable laboratory plant. In: 26th European Conference on Modelling and Simulation, pp. 468-473. ECMS (2012)
9. G.U.N.T. - Equipment for engineering education - Home, `http://www.gunt.de/static/s1_1.php`
10. G.U.N.T. - Equipment for engineering education - RT 010, `http://www.gunt.de/static/s3411_1.php?p1=&p2=&pN=`
11. U12 | LabJack, `http://labjack.com/u12`
12. Honc, D., Dušek, F.: MATLAB/Simulink Support for GUNT Control Units. In: 19th International Conference on Process Control, pp. 534–539. Slovak University of Technology, Bratislava (2013)

# Primary Method of Quadratic Programming in Multivariable Predictive Control with Constraints

Tomáš Barot and Marek Kubalčík

Department of Process Control, Tomas Bata University in Zlín,
Faculty of Applied Informatics, nám. T. G. Masaryka 5555, 76001 Zlín, Czech Republic
{barot,kubalcik}@fai.utb.cz

**Abstract.** General Predictive Control (GPC) is a modern method for process control which is appropriate for many characters of processes. In this paper there is proposed possibility of optimization, which is performed in each sampling period, in the GPC algorithm. Lower time of calculations is in general important for GPC control of multivariable systems with many constraints. An improvement of a primary method of quadratic programming task is proposed in this paper. Computational time can be reduced by changes in details of the optimization method. Time reserves are analyzed in a case of a nonlinear constrained problem which is represented by the Active Set Method. The improved method is presented and results are discussed in simulations.

## 1 Introduction

General predictive control [1]-[2] is an actually researched area of process control, which is suitable for control of wide spectrum of processes. The multivariable variant [3] of this method contains an expression of a controlled system [4] in form of a mathematical model, which has two possible descriptions. A state-space representation [5] is used for an easier form of predictive equations than by using a matrix transfer function. The controller consists of two subsystems - a predictor and an optimizer [6]. A receding horizon principle of GPC [7] gives a control law in a more complicated mathematical form than in case of other common methods of control synthesis [8]. Resulting formulas for increments of manipulated variables are determined from the cooperating of the matrix prediction equation and an optimization expression which is a minimization of an argument of an appropriate cost function. The first part of the results is used in the next time-period in the receding horizon concept.

The quadratic programming [9] is a commonly used approach for optimization principle in GPC because it enables including constraints [10] in a controller synthesis. The cost function has a form of a quadratic function with a positively defined Hessian.[11] The restrictions for signals are defined as constraints defined by matrix inequalities in the GPC. There are two main classical approaches for solving of this optimization task. The first way - a primary method can be seemed more understandable in comparison to a more difficult dual method [9]. However, a computational time of the dual method is significantly lower than the computational time of the primary method. But its principle is theoretically difficult.

I. Zelinka et al. (eds.), *Nostradamus 2014: Prediction, Modeling and Analysis of Complex Systems*, Advances in Intelligent Systems and Computing 289,
DOI: 10.1007/978-3-319-07401-6_18, © Springer International Publishing Switzerland 2014

The primary method - Active Set Method (ASM) [9] is based on a combinatorial strategy. An inequality task is converted to the sequence of partial problems with activating of a determined amount of conditions, which are considered as equalities. Solving of the matrix equations is then more logical than in case of the dual method.

The one dimensional systems are easy to solve by ASM; however, the problems are a time demanding enumeration of equations in case of multivariable systems and more comprehensive set of constraints. In [6] it is described an example of GPC control of one dimensional system with constraints using a dual method with appropriate reduction [6, p. 67]; however, this formula is not commonly included in ASM. Here, it will be improved possibilities for its utilization in a multidimensional-optimization problem. In this paper there is applied the presented improvement of the ASM approach. The ASM method is suitable for control of multivariable systems with many constraints as can be seen from the results of simulations.

## 2 State-Space Model of Multivariable Controlled System

The controlled system can be expressed by many types of mathematical description in GPC control. The state-space modeling [5] is used in this paper. The advantage of this expression is a similar form for one and more dimensional case of a system; the equations in GPC algorithm have the compact and more suitable form. The multivariable linear discrete dynamical system in state-space (Fig. 1) is defined as (1). The system has $m$ inputs, $l$ outputs and $n$ state variables. [5]

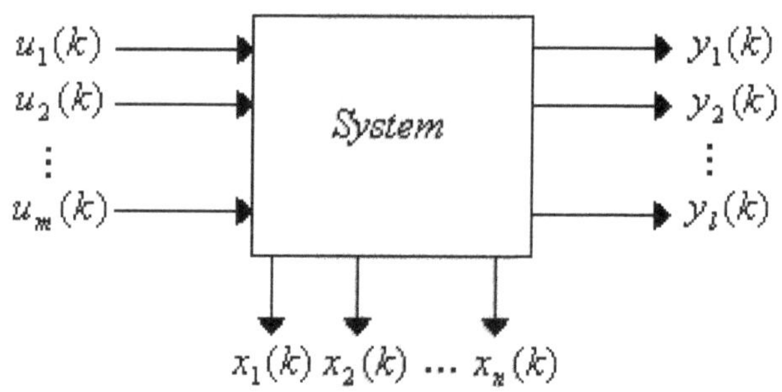


**Fig. 1.** Multivariable Linear Discrete Dynamical System in State-space

$$\begin{aligned}
&\mathbf{x}(k+1) = \mathbf{A}.\mathbf{x}(k) + \mathbf{B}.\mathbf{u}(k) : \mathbf{A} \in \Re^{n,n}, \mathbf{B} \in \Re^{n,m}, \\
&\mathbf{y}(k) = \mathbf{C}.\mathbf{x}(k) + \mathbf{D}.\mathbf{u}(k) : \mathbf{C} \in \Re^{l,n}, \mathbf{D} \in \Re^{l,m}, \\
&\mathbf{x}(k) = \begin{bmatrix} x_1(k) & x_2(k) & \cdots & x_n(k) \end{bmatrix}^T, \\
&\mathbf{y}(k) = \begin{bmatrix} y_1(k) & y_2(k) & \cdots & y_l(k) \end{bmatrix}^T, \\
&\mathbf{u}(k) = \begin{bmatrix} u_1(k) & u_2(k) & \cdots & u_m(k) \end{bmatrix}^T
\end{aligned} \tag{1}$$

# 3 Multivariable Predictive Control in State-Space

Predictive control is based on a receding horizon strategy [7], which is used in a feedback control. Algorithm procedures are provided in each sampling period of a discrete control. The future outputs are computed using equations of predictor subsystem and the results are given to the optimizer for control law determination. Structure of predictive controller is displayed in Fig. 2. Variables are vectors of signal-values. A reference signal **w** and the horizon parameters $N_1$, $N_u$, $N_2$ are the initial conditions of a predictive control, where it is determined a vector of increments of manipulated variable **du**. Other signals correspond with the state-space model (1).

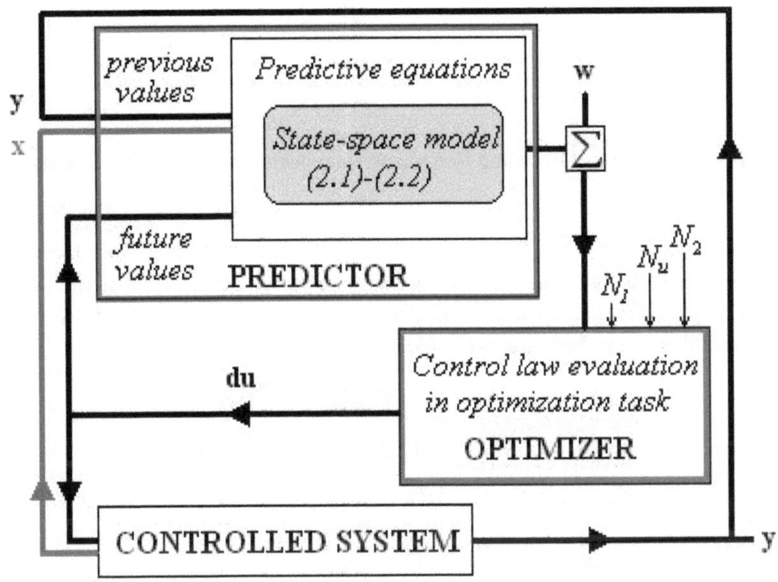


**Fig. 2.** Scheme of Multivariable Predictive Control in State-Space

## 3.1 Predictor of GPC Controller

By the mathematical model (1) are determined equations for the future output variables **y** in a multidimensional case. All steps are based on a receding horizon strategy with horizon parameters ($N_1$ – minimum, $N_u$ control and $N_2$ – maximum horizon), as can be seen in Fig. 3. The final equation will contain the unknown values of **du**, which are determined in the optimizer using a quadratic programming task. Then it will be expressed the control law for GPC of multivariable controlled system. [6]

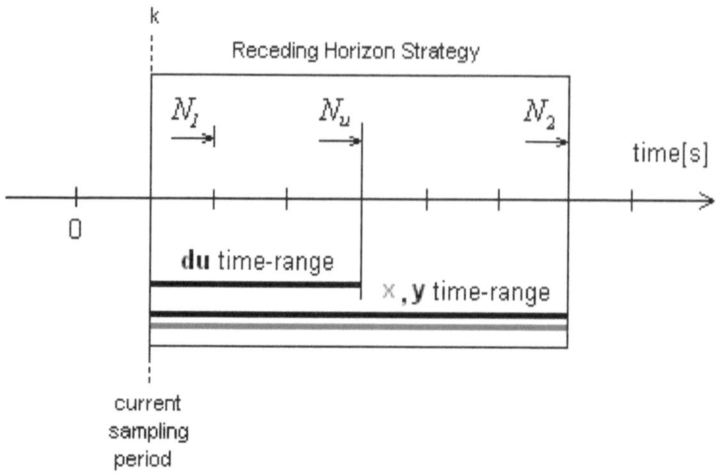


**Fig. 3.** Principal of Receding Horizon Strategy with Time Ranges for Signals

The mathematical model (1) consists of a vector of manipulated variable **u.** An integrator is needed to be implemented in the state-space description of the controlled system. The previous expression (1) is converted to an alternative form of the vector of state variables as can be seen in (2)-(3). Where **I** is an identity matrix.

$$\begin{aligned}\mathbf{x}(k+1) &= \mathbf{A}.\mathbf{x}(k)+\mathbf{B}.(\mathbf{du}(k)+\mathbf{u}(k-1)),\\ \mathbf{y}(k) &= \mathbf{C}.\mathbf{x}(k)+\mathbf{D}.(\mathbf{du}(k)+\mathbf{u}(k-1)),\\ \mathbf{du}(k) &= \mathbf{u}(k)-\mathbf{u}(k-1)\end{aligned} \tag{2}$$

$$\begin{aligned}\begin{bmatrix}\mathbf{x}(k+1)\\ \mathbf{u}(k)\end{bmatrix} &= \begin{bmatrix}\mathbf{A} & \mathbf{B}\\ \mathbf{0} & \mathbf{I}\end{bmatrix}.\begin{bmatrix}\mathbf{x}(k)\\ \mathbf{u}(k-1)\end{bmatrix}+\begin{bmatrix}\mathbf{B}\\ \mathbf{I}\end{bmatrix}.\mathbf{du}(k),\\ \mathbf{y}(k) &= \begin{bmatrix}\mathbf{C} & \mathbf{0}\end{bmatrix}.\begin{bmatrix}\mathbf{x}(k)\\ \mathbf{u}(k-1)\end{bmatrix}\end{aligned} \tag{3}$$

The form of the output equation is converted to (4).

$$\mathbf{y}(k) = \underbrace{\begin{bmatrix}\mathbf{C} & \mathbf{0}\end{bmatrix}}_{\tilde{\mathbf{C}}}.\underbrace{\begin{bmatrix}\mathbf{A} & \mathbf{B}\\ \mathbf{0} & \mathbf{I}\end{bmatrix}}_{\tilde{\mathbf{A}}}^{k}\begin{bmatrix}\mathbf{x}(0)\\ \mathbf{0}\end{bmatrix}+\sum_{r=0}^{k-1}\begin{bmatrix}\mathbf{C} & \mathbf{0}\end{bmatrix}.\begin{bmatrix}\mathbf{A} & \mathbf{B}\\ \mathbf{0} & \mathbf{I}\end{bmatrix}^{k-r-1}\underbrace{\begin{bmatrix}\mathbf{B}\\ \mathbf{I}\end{bmatrix}}_{\tilde{\mathbf{B}}}.\mathbf{du}(r) \tag{4}$$

For future outputs it is in general determined the recursive matrix equation [3]:

$$\begin{bmatrix} \mathbf{y}(k+1) \\ \mathbf{y}(k+2) \\ \vdots \\ \mathbf{y}(k+N_2) \end{bmatrix} = \mathbf{P}.\begin{bmatrix} \mathbf{x}(k) \\ \mathbf{u}(k-1) \end{bmatrix} + \mathbf{G}.\begin{bmatrix} \mathbf{du}(k) \\ \mathbf{du}(k+1) \\ \vdots \\ \mathbf{du}(k+N_u-1) \end{bmatrix} \tag{5}$$

Where matrices **P** and **G** are defined in (6) by new symbols from (4).

$$\mathbf{P} = \begin{bmatrix} \tilde{\mathbf{C}}.\tilde{\mathbf{A}} \\ \tilde{\mathbf{C}}.\tilde{\mathbf{A}}^2 \\ \vdots \\ \tilde{\mathbf{C}}.\tilde{\mathbf{A}}^{N_2} \end{bmatrix}, \mathbf{G} = \begin{bmatrix} \tilde{\mathbf{C}}.\tilde{\mathbf{B}} & \mathbf{0} & \cdots & \mathbf{0} \\ \tilde{\mathbf{C}}.\tilde{\mathbf{A}}\tilde{\mathbf{B}} & \tilde{\mathbf{C}}.\tilde{\mathbf{B}} & \cdots & \mathbf{0} \\ \vdots & \vdots & \ddots & \vdots \\ \tilde{\mathbf{C}}.\tilde{\mathbf{A}}^{N_2-1}\tilde{\mathbf{B}} & \tilde{\mathbf{C}}.\tilde{\mathbf{A}}^{N_2-2}\tilde{\mathbf{B}} & \cdots & \tilde{\mathbf{C}}.\tilde{\mathbf{B}} \end{bmatrix} \tag{6}$$

### 3.2 Optimizer of GPC Controller

The prediction equations in matrix form (5) are prepared for predictor; however, variables **du** should be evaluated. A way for expression of these variables is an optimization task, which contains classical requirements for a feedback control just in a cost function. The cost function has a quadratic form (7). [3]

$$J = \frac{1}{2}\mathbf{du}^T.\mathbf{H}.\mathbf{du} + \boldsymbol{b}^T.\mathbf{du} : \mathbf{du} \in \Re^{m,n} \tag{7}$$

In predictive control it is needed to minimize a cost function (7) using an argument **du**, when it is achieved an enumeration of a control law in a current sampling period.

Where matrices **H** and **b** are defined as follow:

$$\mathbf{H} = 2.(\mathbf{G}^T.\mathbf{G} + \mathbf{I}^{mNu,mNu}) \tag{8}$$

$$\mathbf{b} = 2.(\mathbf{P}.[\mathbf{x}(k), \mathbf{u}(k-1)]^T - [\mathbf{w}(k+N_1), ..., \mathbf{w}(k+N_2)]^T)^T.\mathbf{G} \tag{9}$$

The included constraints are considered in a form of matrix inequalities (10).

$$\min\{J\}, \mathbf{M}.\mathbf{du} \le \mathbf{K} \tag{10}$$

Constraints can restrict ranges of variables – a minimum or maximum values of signals or their increments. Definitions of matrices **M** and **K** are in Table 1. In case of more than one constraints it is possible to expand a definition with the appropriate rows of a new constraint. Where **E** is a matrix with unit elements. [6]

**Table 1.** Definition of Matrices for Constraints Conditions

| *type* | **M** | **K** |
|---|---|---|
| $u_{\min}$ | $\begin{bmatrix} -\mathbf{E}^{m,m} & \mathbf{0} & \mathbf{0} \\ -\mathbf{E}^{m,m} & -\mathbf{E}^{m,m} & \ddots \\ -\mathbf{E}^{m,m} & -\mathbf{E}^{m,m} & -\mathbf{E}^{m,m} \end{bmatrix}$ | $\begin{bmatrix} \mathbf{u}(k-1)+u_{\min}\mathbf{I}^{m,1} \\ \mathbf{u}(k-1)+u_{\min}\mathbf{I}^{m,1} \\ \vdots \\ \mathbf{u}(k-1)+u_{\min}\mathbf{I}^{m,1} \end{bmatrix}$ |
| $u_{\max}$ | $\begin{bmatrix} \mathbf{E}^{m,m} & \mathbf{0} & \mathbf{0} \\ \mathbf{E}^{m,m} & \mathbf{E}^{m,m} & \ddots \\ \mathbf{E}^{m,m} & \mathbf{E}^{m,m} & \mathbf{E}^{m,m} \end{bmatrix}$ | $\begin{bmatrix} -\mathbf{u}(k-1)+u_{\max}\mathbf{I}^{m,1} \\ -\mathbf{u}(k-1)+u_{\max}\mathbf{I}^{m,1} \\ \vdots \\ -\mathbf{u}(k-1)+u_{\max}\mathbf{I}^{m,1} \end{bmatrix}$ |
| $y_{\min}$ | $-\mathbf{G}$ | $-\mathbf{E}^{N_2-N_1+1,1}.y_{\min}+\mathbf{P}.\begin{bmatrix} \mathbf{x}(k) \\ \mathbf{u}(k-1) \end{bmatrix}$ |
| $y_{\max}$ | $\mathbf{G}$ | $\mathbf{E}^{N_2-N_1+1,1}.y_{\max}+\mathbf{P}.\begin{bmatrix} \mathbf{x}(k) \\ \mathbf{u}(k-1) \end{bmatrix}$ |
| $du_{\min}$ | $\begin{bmatrix} -\mathbf{I}^{m,m} & \mathbf{0} & \mathbf{0} \\ \mathbf{0} & -\mathbf{I}^{m,m} & \vdots \\ \mathbf{0} & \mathbf{0} & \ddots \end{bmatrix}$ | $\begin{bmatrix} -du_{\min}\mathbf{I}^{m,1} \\ -du_{\min}\mathbf{I}^{m,1} \\ \vdots \\ -du_{\min}\mathbf{I}^{m,1} \end{bmatrix}$ |
| $du_{\max}$ | $\begin{bmatrix} \mathbf{I}^{m,m} & \mathbf{0} & \mathbf{0} \\ \mathbf{0} & \mathbf{I}^{m,m} & \vdots \\ \mathbf{0} & \mathbf{0} & \ddots \end{bmatrix}$ | $\begin{bmatrix} du_{\max}\mathbf{I}^{m,1} \\ du_{\max}\mathbf{I}^{m,1} \\ \vdots \\ du_{\max}\mathbf{I}^{m,1} \end{bmatrix}$ |

## 4 Quadratic Programming Task

It is needed to solve the optimization problem (10) as a quadratic programming task. The most common used application of GPC control is implemented by methods from category of dual methods, e.g. the command *quadprog* in MathWorks: MATLAB software.

For own numerical solution of quadratic programming task a dual method is suitable; however, its explanation is not so clear as in case of a primary method, e.g. Active Set Method. If the Active Set Method were used, the application of it would not be solvable in real time. Its principals lead from the combinatorial step sequence. In the first part, this method is explained itself, but its application for multivariable systems with many constraints is not possible.

## 4.1 Active Set Method

The quadratic programming problem is based on minimization of the cost function by its argument with constraints on this variable. All constraints have a form of inequalities. The Active Set Method considers any constraints as equalities "active" without any selected (non-active). They are determined by testing on value of Lagrange multipliers [6] and then they are reduced from the whole evaluation.

At first, a new matrix $\mathbf{M}_a$ is set in the same way as the matrix $\mathbf{M}$. The same rule is applied for the matrix $\mathbf{K}$, which becomes a new matrix $\mathbf{K}_a$. This is an initial part of the algorithm. Variables from task (7) with these matrices are substituted into (11). If any $i$-element of the vector $\boldsymbol{\lambda}$ is negative, than it is removed each corresponding $i$-constraint from the rows of matrices $\mathbf{M}_a$ and $\mathbf{K}_a$ for the next iteration. The cycles of first part ends when all elements of vector $\boldsymbol{\lambda}$ are positive. In the worst case, it remains only one constraint.

$$\boldsymbol{\lambda} = -(\mathbf{M}_a.\mathbf{H}^{-1}.\mathbf{M}_a^T)^{-1}(\mathbf{K}_a + \mathbf{M}_a\mathbf{H}^{-1}\mathbf{b}) \tag{11}$$

If all constraints give the vector $\boldsymbol{\lambda}$ with all positive elements, than follows a second part of the method. There are tested all positive or "active" constraints. At first are included all constraints and then are included just the combinations of them in a different order - with an aim to combine all variants of them in matrices $\mathbf{M}_a$ and $\mathbf{K}_a$. For each case it is determined a potential solution in (12). The most appropriate minimum is verified using substitution into the cost function. The achieved minimum gives the least value of the cost function $J$.

$$\mathbf{du} = -\mathbf{H}^{-1}(\mathbf{b} + \mathbf{M}_a^T.\boldsymbol{\lambda}) \tag{12}$$

## 4.2 Improving of Active Set Method

In the previous chapter was described the common presented approach for a quadratic programming problem. For many constraints and many inputs and outputs of controlled system, the Active Set Method does not achieve the solution in real time. There is proposed an additional part for ASM algorithm, which can reduce a computational time. The main strategy is an evaluating of minimum without any constraints and testing for all constraints (13), known in case of dual methods [6]. These cases are possible in GPC control. If the result is correct, than it is reduced a next time-consuming part of the algorithm. The important advantage of the improvement brings time-saving and makes this understandable method accessible for a practice.

$$\{\mathbf{du} = -\mathbf{H}^{-1}\mathbf{b}; \mathbf{M}(-\mathbf{H}^{-1}\mathbf{b}) \le \mathbf{K}\} \tag{13}$$

## 5 Results

The proposed approaches were implemented in MathWorks: MATLAB software in a form of scripts, which are able to analyse a computational time during the multivariable predictive control by a predictor and optimizer activities. For the multivariable system (14) the simulations of GPC control were run with setting of constraints on variables and parameters for a feedback control: $u_{min}$=0, $u_{max}$= 0.9, $y_{min}$ =0, $y_{max}$ =0.5, $du_{min}$ =0.01, $du_{max}$ =0.5, $N_1$=1, $N_u$=35, $N_2$=38. References signals have form: $w_{min}$=0.25, $w_{max}$=0.5. Sampling period was chosen as 1 [s].

$$\begin{gathered}\mathbf{x}(k+1) = \begin{bmatrix} 0.583 & 0.022 & 1 & 0 \\ -0.017 & 0.456 & 0 & 1 \\ -0.175 & -0.179 & 0 & 0 \\ 0.089 & 0.083 & 0 & 0 \end{bmatrix} . \mathbf{x}(k) + \begin{bmatrix} -0.004 & 0.148 \\ 0.278 & -0.037 \\ 0.096 & 0.219 \\ 0.311 & -0.349 \end{bmatrix} . \mathbf{u}(k) \\ \mathbf{y}(k) = \begin{bmatrix} 1 & 0 & 0 & 0 \\ 0 & 1 & 0 & 0 \end{bmatrix} . \mathbf{x}(k) + \begin{bmatrix} 0 & 0 \\ 0 & 0 \end{bmatrix} . \mathbf{u}(k), \ \mathbf{x}(0) = \begin{bmatrix} 0 & 0 & 0 & 0 \end{bmatrix}^T\end{gathered} \tag{14}$$

As can be seen in Fig. 4, the successful results were achieved using the proposed improvement of Active Set Method. The number of rows in matrices **M** and **K** was in this case 432. The average time of the optimization part was approximately 0.438 [s] in the sampling period.

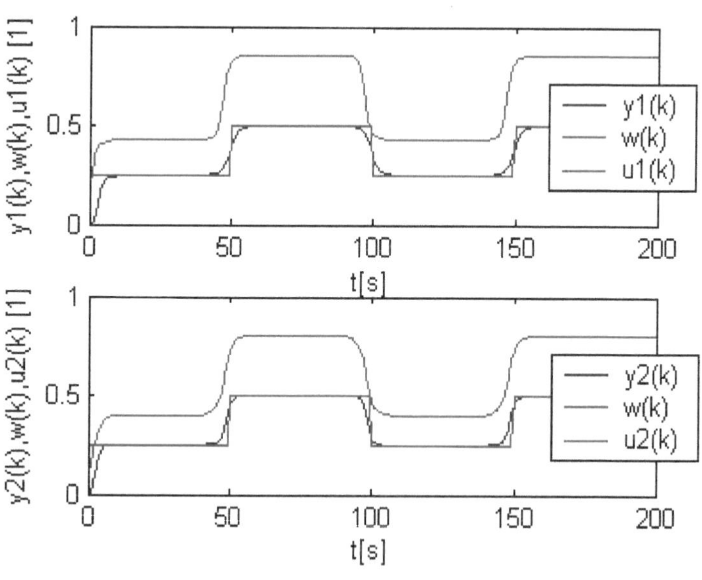


**Fig. 4.** Results of Proposed Method

## 6 Conclusion

The improved Active Set Method was used for a predictive control of a multivariable system. There were included constraints for penalisation of variable ranges. It was possible to control this multi-input multi-output system with many constraints by the proposed method with an average computation time 0.438 [s]. Without including of improvement it will not be possible to control in real time in a case of multidimensional optimization problem with many restrictions. The aim of this paper was to make this method available, because its general principals are easier to explain, than it is by using the dual approaches of quadratic problem optimization. The improvement is suitable for a multidimensional problem in each sampling period in a predictive control, where the least computational time is in general required.

**Acknowledgment.** The article was realized with financial support of IGA, Tomas Bata University in Zlín, Faculty of Applied Informatics number IGA/FAI/2014/002.

## References

1. Camacho, E.F., Bordons, C.: Model Predictive Control. Springer, London (2007)
2. Rawlings, J.B., Mayne, D.Q.: Model Predictive Control Theory and Design. Nob Hill Pub. (2009)
3. Kubalčík, M., Bobál, V.: Computation of predictions in multivariable predictive control. In: Proceedings of the 13th WSEAS International Conference on Automatic Control, Modelling & Simulation, ACMOS 2011 (2011)
4. Bobal, V., Kubalcik, M., Dostal, P., Matejicek, J.: Adaptive predictive control of time-delay systems. Computers & Mathematics with Applications 66(2), 165–176 (2013)
5. Kučera, V.: Analysis and Design of Discrete Linear Control Systems. Nakladatelství Československé akademie věd, Praha (1991)
6. Wang, L.: Model Predictive Control System Design and Implementation Using MATLAB. Springer-Verlag Limited, London (2009)
7. Kwon, W.H.: Receding horizon control: model predictive control for state models. Springer, London (2005)
8. Corriou, J.-P.: Process control: theory and applications. Springer, London (2004)
9. Lee, G.M., Tam, N.N., Yen, N.D.: Quadratic Programming and Affine Variational Inequalities: A Qualitative Study. Springer (2005)
10. Luenberger, D.G., Ye, Y.: Linear and nonlinear programming, 3rd edn. Springer, New York (2008)
11. Dostál, Z.: Optimal Quadratic Programming Algorithms: With Applications to Variational Inequalities. Springer, New York (2009)

# Control of Concentration inside CSTR Using Nonlinear Adaptive Controller

Jiri Vojtesek and Petr Dostál

Tomas Bata University in Zlin, Faculty of Applied Informatics
Nam. T.G.Masaryka 5555, 760 01 Zlin, Czech Republic
{vojtesek,dostalp}@fai.utb.cz
http://www.utb.cz/fai

**Abstract.** An adaptive nonlinear control is modification of the classic adaptive control where the controller is divided into the dynamic linear part and the static nonlinear part. The dynamic linear part is constructed with the use of polynomial synthesis together with the pole-placement method and the spectral factorization. The static nonlinear part uses static analysis of the controlled plant for introducing the mathematical nonlinear description of the relation between the controlled output and the change of the control input. In this case, the output response could be tuned by the change of the closed-loop pole. The verification of the proposed control strategy was made by simulations on the mathematical model of CSTR with cooling in the jacket as a typical nonlinear system.

**Keywords:** Adaptive Nonlinear Control, CSTR, Mathematical Model, Simulation, Recursive Identification.

## 1 Introduction

The control of the chemical processes in the industry is always challenging because of the nonlinearity of the major group of systems. The continuous stirred-tank reactor (CSTR) is one of the most common used types of chemical reactors because of easily controllability [1].

The adaptive control [2] is a control technique with good theoretical background and also practical implementations. It uses idea of the living organisms that adopts their behavior to the actual environmental conditions. There are also various adaptation techniques and variations described for example in [3].

The control method used here is based on the combination of the adaptive control and nonlinear control. Theory of nonlinear control (NC) can be found for example in [4] and [5]. The nonlinear adaptive controller is divided via Wieners model [6] into two parts the dynamic linear part (DLP) and the static nonlinear part (SNP). The DLP uses polynomial synthesis [7] with pole-placement method and spectral factorization and all these methods satisfy basic control requirements such as a disturbance attenuation, a stability and a reference signal tracking. The second, nonlinear, part uses measurements of the steady-state behavior of the system for mathematical description of the dependence between the controlled output variable and the control input variable.

I. Zelinka et al. (eds.), *Nostradamus 2014: Prediction, Modeling and Analysis of Complex Systems*, Advances in Intelligent Systems and Computing 289,
DOI: 10.1007/978-3-319-07401-6_19, © Springer International Publishing Switzerland 2014

The controlled system, CSTR, with originally nonlinear behavior could be math-ematically described for the control purposes by the External Linear Model (ELM) [8], parameters of which could vary because of the nonlinearity of the system. This problem could be overcome with the use of recursive identification which recomputes parameters of the ELM according to the actual state and the behavior of the system. There were used delta ($\delta$-) model [8] as a special type of discrete-time models parameters of which approaches to the continuous ones for the small sampling period as it is proofed for example in [9].

The proposed control strategy was verified by simulations on the mathematical model of CSTR with cooling in the jacket [10]. This mathematical model was studied also in [11] and pure adaptive controller was applied in [12]. All simulations were done in the mathematical software Matlab, version 7.0.1.

## 2 Controlled Plant

The system under the consideration is a Continuous Stirred-Tank Reactor (CSTR) with the so called *Van der Vusse reaction* $A \rightarrow B \rightarrow C, 2A \rightarrow D$ inside and cooling jacket.

If we introduce common simplifications like the perfect mixture of the reactant, all densities, transfer coefficients, heat capacities and the volume of the reactant are constant throughout the reaction, the mathematical model developed with the use of material and heat balances inside has form of the set of Ordinary Differential Equations (ODEs) [10]

$$\begin{aligned}
\frac{dc_A}{dt} &= \frac{q_r}{V_r}\left(c_{A0} - c_A\right) - k_1 c_A - k_3 c_A^2 \\
\frac{dc_B}{dt} &= -\frac{q_r}{V_r} c_B + k_1 c_A - k_2 c_B \\
\frac{dT_r}{dt} &= \frac{q_r}{V_r}\left(T_{r0} - T_r\right) - \frac{h_r}{\rho_r c_{pr}} + \frac{A_r U}{V_r \rho_r c_{pr}}\left(T_c - T_r\right) \\
\frac{dT_c}{dt} &= \frac{1}{m_c c_{pc}}\left(Q_c + A_r U\left(T_r - T_c\right)\right)
\end{aligned} \tag{1}$$

where $t$ in is the time, $c$ are concentrations, $T$ represents temperatures, $c_p$ is used for specific heat capacities, $q_r$ means the volumetric flow rate of the reactant, $Q_c$ is the heat removal of the cooling liquid, $V_r$ is volume of the reactant, $\rho$ stands for densities, $A_r$ is the heat exchange surface and $U$ is the heat transfer coefficient. Indexes $(\cdot)_A$ and $(\cdot)_B$ belong to compounds A and B, respectively, $(\cdot)_r$ denotes the reactant mixture, $(\cdot)_c$ cooling liquid and $(\cdot)_0$ are feed (inlet) values.

The variable $h_r$ and $k_{1-3}$ in (1) denotes the reaction heat and reaction rates which are computed from

$$\begin{aligned}
h_r &= h_1 \cdot k_1 \cdot c_A + h_2 \cdot k_2 \cdot c_B + h_3 \cdot k_3 \cdot c_A^2 \\
k_j\left(T_r\right) &= k_{0j} \cdot \exp\left(\frac{-E_j}{RT_r}\right), \text{ for } j = 1, 2, 3
\end{aligned} \tag{2}$$

where $h_i$ stands for reaction enthalpies. Reaction rates $k_{1-3}$ in the second equation are nonlinear functions of the reactants temperature computed via *Arrhenius law* with $k_{0j}$ as rate constants, $E_j$ are activation energies and $R$ means gas constant.

Equations (1) together with (2) construct the *mathematical model of the plant* used later for simulation studies. Due to simplifications introduced above we can say, that this type of reactor is a nonlinear lumped-parameters system. We have four state variables $c_A$, $c_B$, $T_r$ and $T_c$ and four input variables the volumetric flow rate of the reactant, $q_r$, the heat removal of the coolant, $Q_c$, the input concentration $c_{A0}$ and input temperature of the reactant, $T_{r0}$. The fixed values of the reactor are shown in [10].

## 3 Nonlinear Adaptive Control

The control strategy here is based on the factorization of controller into the static nonlinear part (SNP) and the dynamic linear part (DLP) see Fig. 1. This control scheme configuration is called a *Wiener system.*

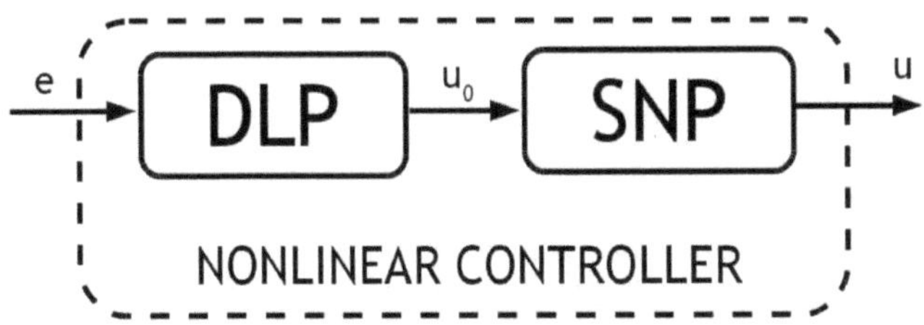


**Fig. 1.** The scheme of the nonlinear controller

As it written in the previous part, there are theoretically four input and four output variables. In this case, the change of the output concentration, $c_B$, from its steady-state value, $c_B^s$, was controlled with the change of the volumetric flow rate of the reactant, $q_r$, from the working point, $q_r^s$, i.e.

$$\begin{array}{l} u(t) = \Delta q_r = q_r(t) - q_r^s \; \left[m^3 \cdot min^{-1}\right] \\ y(t) = \Delta c_B = c_B(t) - c_B^s \; \left[kmol \cdot m^{-3}\right] \end{array} \quad (3)$$

The dynamic part DLP in Fig. 1 represents linear dynamic relation between the tracking error $e(t)$ and the input to the nonlinear static part $u_0(t) = \Delta c_{Bw}(t)$ which is difference between the concentration of the product, $c_B(t)$, and its desired value. The second static nonlinear part then describes the relation between $u_0(t)$ and corresponding change of the input volumetric flow rate of the reactant $\Delta q_r(t)$.

The schematic representation of the control system can be found in Figure 2.

### 3.1 Static Nonlinear Part

The nonlinear part uses properties of the system in the steady-state, i.e. for time $t \to \infty$ which means that the set of ODE (1) is transformed into the set of

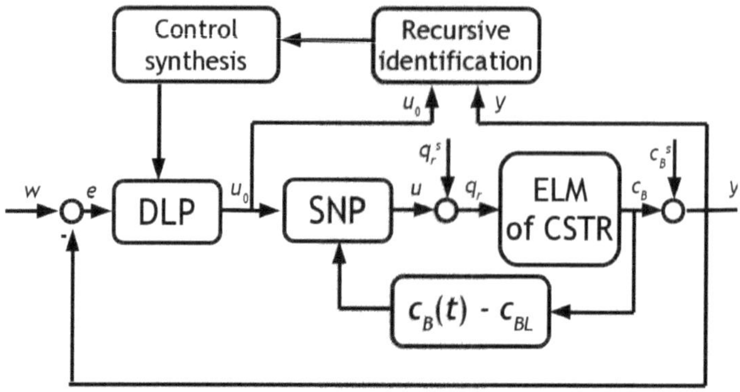


**Fig. 2.** Control Scheme

algebraic nonlinear equations because derivations with respect to time are equal to zero in the steady-state.

There was done the steady-state characteristic for the volumetric flow rate of the reactant, $q_r$, from the range $q_r =< 0.001; 0.04 >$ $m^3.min^{-1}$ and results for the steady-state values of the products concentration, $c_B^s$, are shown in Fig. 3 (left graph). The operation of the controller was chosen in the interval where $q_{r,min} = 0.0055\ m^3.min^{-1}$ and $q_{r,max} = 0.03\ m^3.min^{-1}$. Working point of the system was chosen in the middle of this interval and includes also the nonlinearity of the system. This point is defined by the volumetric flow rate $q_r^s = 0.015\ m^3.min^{-1}$ and heat removal of the coolant $Q_c^s = -18.56\ kJ.min^-1$. The steady-state value of the controlled concentration is in this point $c_B^s = 0.442\ kmol.m^{-3}$.

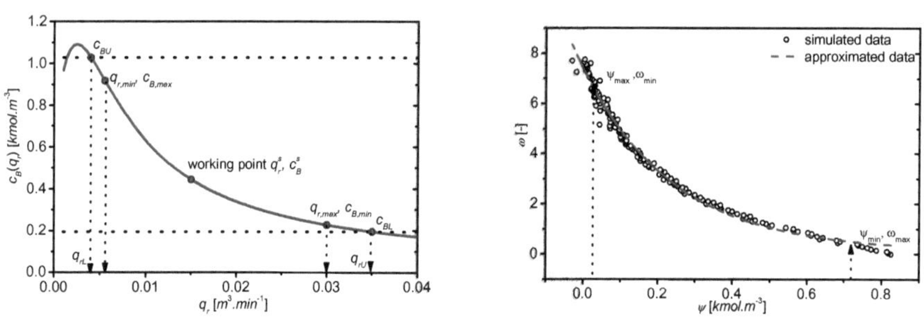


**Fig. 3.** The steady-state characteristic (left graph) and simulated and approximated characteristics $\omega = f(\psi)$ (right graph)

Due to later approximation and better unification of the variables, the new x and y variables $\omega$ and $\psi$ are introduced and

$$\omega = \frac{q_r^s - q_{rL}}{q_{rL}}\ [-];\ \psi = c_B^s - c_{BL}\ \left[kmol \cdot m^{-3}\right] \qquad (4)$$

where $q_{rL}$ is lower bound from the interval and $c_{BL}$ is corresponding products concentration from the upper bound $q_{rU}$ see Fig 3. It is recommended to choose this interval slightly longer than those in $q_{r,min} \leq q_r(t) \leq q_{r,max}$ which means in this case that lower and upper bounds of the input variable and equivalent values of the concentrations are

$$\begin{aligned} q_{rL} &= 0.004\, m^3 \cdot min^{-1} \qquad & c_{BL} &= 0.1953\, kmol.m^{-3} \\ q_{rU} &= 0.035\, m^3 \cdot min^{-1} \qquad & c_{BU} &= 1.0274\, kmol.m^{-3} \end{aligned} \tag{5}$$

It is common, that the measured data on the real system are affected by the measurement errors. To emulate these errors, the random white-noise error on the output variable is introduced here. Later computations use the inverse of these coordinates for computing of the action value. This inverse is shown in Fig. 3 (right graph) and this nonlinear course could be replaced by the function from the ring of exponential, polynomial etc. functions.

The exponential function in the general form

$$\omega = f(\psi) = a \cdot e^{-b \cdot \psi} + c \tag{6}$$

was used in this case. The course of this approximation is shown in Fig. 3 (red dashed line on the right graph) with the identified values of constants $a = 7.1601$, $b = 4.1806$ and $c = 0.1707$.

The difference of the input volumetric flow rate of the coolant is from (3) $u(t) = \Delta q_r(t)$ and the nonlinear part can be then computed from

$$u(t) = \Delta q_r(t) = q_{rL} \left( \frac{d\omega}{d\psi} \right)_{\psi(c_B)} u_0(t) \tag{7}$$

and the derivative $d\omega/d\psi$ in the previous equation is computed for each measured products concentration, $c_B$, from the derivative of the function $f$ in (6), i.e.

$$\frac{d\omega}{d\psi} = -29.9335 \cdot e^{-4.1806 \cdot \psi} \tag{8}$$

The course of this function is shown in following Fig. 4 left graph.

### 3.2 External Linear Model of CSTR

The dynamic behavior of the controlled system, in our case CSTR, together with the SNP derived above is observed for the step responses of the input $u_0$ see Fig. 2. There were done five changes $u_0$ for the working point $q_r^s = 0.015\, m^3.min^{-1}$ and $Q_c^s = -18.56\, kJ.min^{-1}$ and results are shown in Fig. 4 right graph.

The gain of the system SNP+CSTR is computed as

$$g_s = \lim_{t \to \infty} \frac{y(t)}{u_0} \tag{9}$$

and the values of gs are shown also in Fig. 4 right graph.

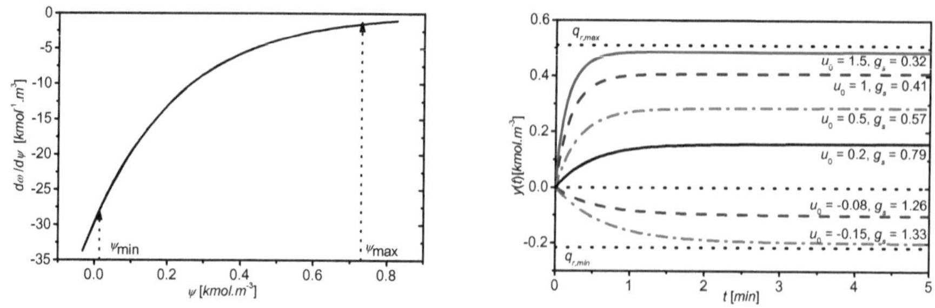


**Fig. 4.** The course of the derivative of $d\psi/d\omega$ (left graph) and results of dynamic analyses for the changes of input $u_0$ (right graph)

Although the system has nonlinear behavior, presented output dynamic responses could be described by the first order continuous-time transfer function

$$G(s) = \frac{Y(s)}{U(s)} = \frac{b(s)}{a(s)} = \frac{b_0}{s + a_0} \tag{10}$$

with $s$ as a complex variable a polynomials $a(s)$ and $b(s)$ comes from identification. This transfer function could be then in the form of the differential equation

$$\dot{y}(t) + a_0 y(t) = b_0 u(t) \tag{11}$$

### 3.3 Identification of the ELM

The online identification of the continuous-time ELM (10) is not very simple. On the other hand, $\delta$-identification models belong to the class of discrete models but their parameters are close to the continuous ones for very small sampling period.

The delta model introduces a new complex variable $\gamma$ as an alternative to complex variables $s$ in continuous-time and $z$ in discrete-time. A so called forward $\delta$-model for $\beta = 0$ was used here with the $\gamma$ operator $\gamma = (z-1)/T_v$ where $T_v$ is a sampling period and $z$ is a discrete-time complex variable.

The continuous model (10) is then rewritten to the form

$$a^{\delta}(\delta)\, y(t') = b^{\delta}(\delta)\, u(t') \tag{12}$$

where polynomials $a^{\delta}(\delta)$ and $b^{\delta}(\delta)$ are discrete polynomials and their coefficients are different from those of the CT model $a(s)$ and $b(s)$ in (10). Time $t'$ denotes discrete time.

The eq. (11) could be then with the substitution and simplifications rewritten to

$$y_{\delta}(k) = -a_0^{\delta} y_{\delta}(k-1) + b_0^{\delta} u_{\delta}(k-1) \tag{13}$$

where new, recomputed, values of input and output variables are

$$y_\delta(k) = \frac{y(k)-y(k-1)}{T_v};\; y_\delta(k-1) = y(k-1);\; u_\delta(k-1) = u(k-1) \quad (14)$$

The regression vector, $\phi_\delta$ , and vector of parameters, $\theta_\delta$, used for identification are then

$$\varphi_\delta(k-1) = \left[-y_\delta(k-1), u_\delta(k-1)\right]^T; \quad \theta_\delta(k) = \left[a_1^\delta, a_0^\delta, b_1^\delta, b_0^\delta\right]^T \quad (15)$$

and the differential equation (13) could be rewritten to the vector form:

$$y_\delta(k) = \theta_\delta^T(k) \cdot \varphi_\delta(k-1) + e(k) \quad (16)$$

where $e(k)$ is a general random immeasurable component. The task of the recursive identification is to find unknown vector of parameters, $\theta_\delta$, from the measured data vector $\phi_\delta$. The simple Recursive Least-Squares (RLS) method was used in this work. This method together with exponential and directional forgetting modifications produces sufficient results as it was proofed by the previous experiments.

### 3.4 Dynamic Linear Part

The last part from Fig. 2 which has not been discussed is the dynamic linear part (DLP). The feedback controller is designed with the use of polynomial approach [7]. The transfer function of the controller has general polynomial form:

$$\tilde{Q}(s) = \frac{q(s)}{s \cdot \tilde{p}(s)} \quad (17)$$

where parameters of polynomials and $q(s)$ are computed from Diophantine equation:

$$a(s) \cdot s \cdot \tilde{p}(s) + b(s) \cdot q(s) = d(s) \quad (18)$$

On the other hand, polynomials of the ELM $a(s)$ and $b(s)$ in (18) are known from the recursive identification and we expect, that parameters of this polynomial are very close to parameters of correspondent $\delta$ polynomials $a^\delta(\delta)$ and $b^\delta(\delta)$ in (12). The polynomial $d(s)$ on the right side of (18) is an optional stable polynomial and the degree of this polynomial is $\deg d(s) = \deg a(s) + \deg \tilde{p}(s) + 1$ . Roots of this polynomial are called *poles of the closed-loop* and their position affects quality of the control.

There are several ways how to construct this optional polynomial e.g. the Pole-placement method, LQ approach etc. The choice here combines the Pole-placement method with spectral factorization of the identified polynomial $a(s)$. The polynomial $d(s)$ has then two parts

$$d(s) = n(s) \cdot (s+\alpha)^2 \quad (19)$$

where $\alpha > 0$ is an optional coefficient reflecting closed-loop poles and stable polynomial $n(s)$ is obtained from the spectral factorization of the polynomial $a(s)$ - $n^{\star}(s) \cdot n(s) = a^{\star}(s) \cdot a(s)$, which is known from the recursive identification. The transfer function of the controller (17) is for this concrete ELM (10):

$$\tilde{Q}(s) = \frac{q(s)}{s \cdot \tilde{p}(s)} = \frac{q_1 s + q_0}{p_0 \cdot s} \tag{20}$$

and parameters $q_1$, $q_0$ and $p_0$ are computed from Diophantine equation (18).

The control synthesis presented above is derived in the continuous-time, but identification and recomputation of the controllers parameters runs in discrete time ($\delta$-models). That is why we call this controller hybrid adaptive controller.

## 4 Simulation Experiment

The goal of this last chapter is to verify proposed nonlinear adaptive controller by simulations on the mathematical model (1) of the CSTR. The simulations were done for three values of the $\alpha$ from (19) which could be understood as a tuning parameter. The sampling period was $T_v = 0.1\ min$, the simulation time was $75 min$ and 5 step changes of the reference signal $w(t)$ were done during this time.

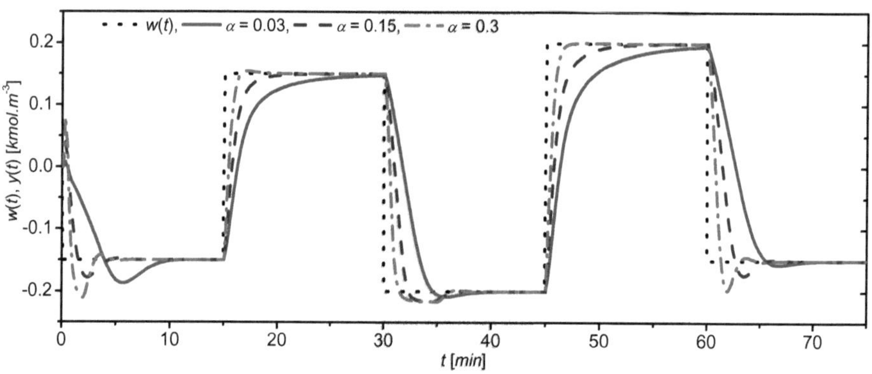


**Fig. 5.** The course of the output variable $y(t)$ and the reference signal $w(t)$ for various values of the tuning parameter $\alpha$

Fig. 5 shows courses of the output variable, $y(t)$, for various $\alpha = 0.03, 0.15$ and 0.3. It is clear, that the increasing value of this parameter results in the quicker output response but overshoots especially for the negative step changes. On the other hand, Fig. 6 shows the course of the input variable $u_0$ (left graph) as an output from the DLP which is also input to the SNP. The graph on the right size is the course of the volumetric flow rate qr as an output from the SNP and the input to the mathematical model of CSTR see schematic representation in Fig. 2. We can say that decreasing value of the parameter $\alpha$ results in smoother course of both input variables.

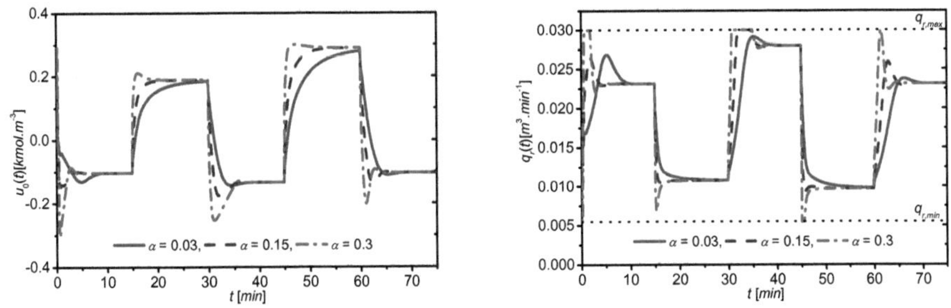


**Fig. 6.** Outputs from the LDP $u_0$ or various values of $\alpha$ (left graph) and the courses of the computed input variable, $q_r$, to the ELM for various $\alpha$. (right graph)

## 5 Conclusion

The paper shows one possible way how to control CSTR as a typical member of nonlinear system with lumped parameters. The nonlinear adaptive control strategy is based on the Wiener system where the controller is divided into the dynamic linear part and the static nonlinear part. The dynamic linear part is based on the adaptive theory where the controlled nonlinear system is simplified to the linear system described by the external linear model parameters of which are identified recursively during the control. The control synthesis employs polynomial theory together with the pole-placement method and spectral factorization. The static nonlinear part then uses simulated or measured steady-state characteristics of the mathematical model to describe the relation between controlled concentration of the product and the change of the reactants volumetric flow rate as an input variable. Simulations have shown that the system could be tuned by the choice of the pole $\alpha$ and an increasing value of this parameter results in the quicker course of both input and output variables but with one negative feature overshoot, which is evident in the negative step changes. Although the system has nonlinear behavior, proposed control strategy cope with it well and it could be used also for similar types of system.

## References

1. Ingham, J., Dunn, I.J., Heinzle, E., Prenosil, J.E.: Chemical Engineering Dynamics. An Introduction to Modeling and Computer Simulation, 2nd Completely revised edn. VCH Verlagsgesellshaft, Weinheim (2000)
2. Astrom, K.J., Wittenmark, B.: Adaptive Control. Addison Wesley, Reading (1989) ISBN 0-201-09720-6
3. Bobal, V., Böhm, J., Fessl, J., Machacek, J.: Digital Self-tuning Controllers: Algorithms. Implementation and Applications. Advanced Textbooks in Control and Signal Processing. Springer-Verlag London Limited (2005) ISBN 1-85233-980-2
4. Astolfi, A., Karagiannis, D., Ortega, R.: Nonlinear and adaptive control with applications. Springer, London (2008)

5. Vincent, T.L., Grantham, W.J.: Nonlinear and optimal control systems. John Wiley & Sons, New York (1997) ISBN 0471042358
6. Nakamura, M., Sugi, T., Goto, S.: Nonlinear separation model and control for a complex process realized by conventional PID controller hardware. In: Proceedings of the 4th Asian Control Conference, Singapore, pp. 274–279 (2002)
7. Kucera, V.: Diophantine equations in control – A survey. Automatica 1993, 1361–1375 (1993)
8. Middleton, H., Goodwin, G.C.: Digital Control and Estimation - A Unified Approach. Prentice Hall, Englewood Cliffs (2004) ISBN 0-13-211798-3
9. Stericker, D.L., Sinha, N.K.: Identification of continuous-time systems from samples of input-output data using the $\delta$-operator. Control-Theory and Advanced Technology 9, 113–125 (1993)
10. Chen, H., Kremling, A., Allgöwer, F.: Nonlinear Predictive Control of a Benchmark CSTR. In: Proceedings of 3rd European Control Conference, Rome, Italy (1995)
11. Vojtesek, J., Dostal, P., Haber, R.: Simulation and Control of a Continuous Stirred Tank Reactor. In: Proc. of Sixth Portuguese Conference on Automatic Control, CONTROLO 2004, Faro, Portugal, pp. 315–320 (2004)
12. Vojtesek, J., Dostal, P.: From steady-state and dynamic analysis to adaptive control of the CSTR reactor. In: Proc. of 19th European Conference on Modelling and Simulation, ESM 2005, Riga, Latvia, pp. 591–598 (2005)

# LMI Approach of Constrained Fuzzy Model Predictive Control of DC-DC Boost Converter

S. Bououden[1], M. Chadli[2], and Ivan Zelinka[3]

[1] Faculty of sciences and technology, University of Abbes Laghrour Khenchela, and Laboratory of automatic and robotic, University Constantine1, Algeria
ss.bououden@gmail.com
[2] University of Picardie Jules Verne, MIS (EA 4029)33 rue Saint-Leu, 80039 Amiens, France
mchadli@u-picardie.fr
[3] Faculty of Electrical Engineering and Computer Science VŠB-TUO 17. listopadu 15, 708 33 Ostrava-Poruba, Czech Republic
ivan.zelinka@vsb.cz

**Abstract.** In this paper, we propose a fuzzy model predictive control (FMPC) using Linear matrix inequalities (LMIs) approach for the voltage tracking control of a DC-DC Boost converter. A mathematical model is required to synthesis this controller, the typically used model is the averaged model, which describes the converter behavior on the operating point. Boost converter has a nonlinear dynamic behavior; the Takagi–Sugeno (T–S) fuzzy model is used to represent the state-space model of nonlinear system where the consequent part of the fuzzy rule is replaced by linear systems. Based on this model, we formulate and solve a constrained optimal control problem using linear matrix inequalities approach.

**Keywords:** Predictive controller, non-linear systems, Boost converter, averaged model, LMI approach.

## 1 Introduction

Power converters are used extensively in most of the power supply systems such as personal computers, laptops, aircrafts and electronic equipment. A DC-DC converter is a switching circuit, which transform a certain electrical voltage to another level of voltage, this is obtained by switches operating (open or closed) at high frequencies; the control objective of such devices is to maintain regulation of the output voltage at the desired value [1],[2]. DC-DC converters are nonlinear systems in essence [3], different control techniques ranging from linear control based on linearized model [10] to passivity based control [11] and sliding mode [12] have been used in recent years in the control of switching converters. These works are usually based on a small signal model using the method of averaging of the state space; the model obtained by these methods is useful only for small variations around a specific operating point, will result in poor dynamic performance.

I. Zelinka et al. (eds.), *Nostradamus 2014: Prediction, Modeling and Analysis of Complex Systems*, Advances in Intelligent Systems and Computing 289,
DOI: 10.1007/978-3-319-07401-6_20, 

Recently, the approach of T-S Fuzzy Modeling reflected much interest for DC-DC converters. The fuzzy model proposed by Takagi and Sugeno [13] is described by fuzzy IF-THEN rules which represent local linear input-output relations of a nonlinear system. The main feature of T-S fuzzy model is to express the local dynamics of each fuzzy implication by a linear system model. The passage of the nonlinear model T-S model is not unique, and the number of sub-models increases exponentially according to the number of nonlinearities considered. The curse of the number of rules makes controller design difficult. To loosen the curse, we proposed a fuzzy model based on the so-called sector nonlinearity concept [14]. The method by transformation in nonlinear sector is based on bounded function and gives a minimum number of local models.

In DC-DC converters, the duty ratio is bounded in the interval [0,1]. The classical control techniques applied to power converters do not take into account neither state nor input constraints, and thus violations of component specifications regularly occur, in practice we can apply model predictive control [2].

Model predictive control (MPC) [4], also known as receding horizon control or moving horizon control is one of the most successful modern control methodologies that offer good solutions, already successfully implemented in industry, for the regulation of constrained linear or nonlinear systems. MPC has ability to handle hard constraints on states/outputs and inputs [5-7]. The fuzzy model predictive control law can be easily obtained by solving a convex optimization problem subject to several linear matrix inequalities [18, 19].

The paper is organized as follows: Section 2 gives the averaged model of basic PWM boost converter. State-space model would be discretized and linearized around equilibrium point of the converter. Section 3 presents the T-S fuzzy model of the DC-DC converter. Section 4 is devoted to the controller theoretical design. In Section 5 we reformulate the predictive control subject to constraints as a LMI-based optimization problem. Section 6 present simulation results illustrating the performance of the proposed control approach. Finally, Section 7 concludes the paper.

# 2 Averaged Model of Basic PWM Boost Converter

## 2.1 State-Space of the Boost Converter

This section shows the state-space averaged model of the boost converter, which includes the PWM (Pulse Width Modulation) and has as control input the duty ratio $d$. The DC-DC converter can operate in both continuous conduction mode (CCM) and discontinuous conduction mode (DCM) .We assume that the converter operates in continuous conduction mode (CCM) and that the inductor current is always larger than zero

Figure 1 shows the schematic circuit diagram of a DC–DC boost converter and the relevant control signals.

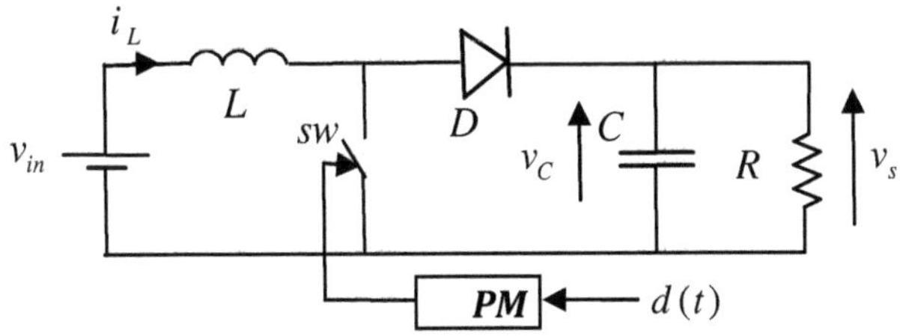


**Fig. 1.** Schematic of the boost converter

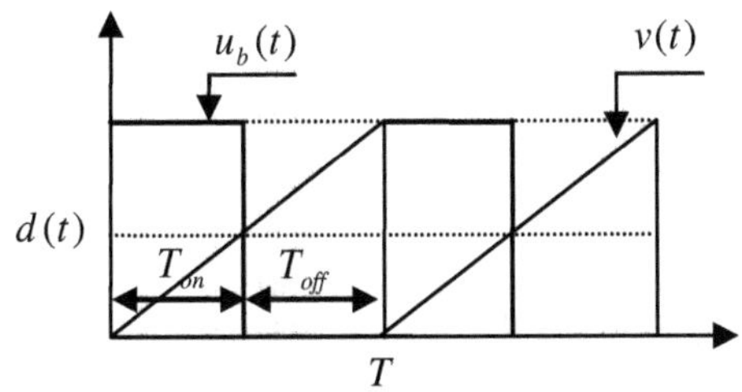


**Fig. 2.** Waveforms of the PWM process

In Fig.1, $v_s$ is the output voltage and $v_{in}$ the line voltage. The output voltage must be kept at a given constant $v_{ref}$ value, the diode $D$ is on inverse polarization, $R$ models the converter load, while $C$ and $L$ represent, respectively, capacitor and inductor values, the switch $sw$ was a power transistor controlled by a binary signal $u_b$.

The binary signal that triggers on and off the switches is controlled by a fixed-frequency pulse width modulation (PWM) circuit (Fig. 2). The constant switching frequency is $1/T_s$, with $T_s$ the switching period is given by the sum of $T_{on}$ (when $u_b = 1$) and $T_{off}$ (when $u_b = 0$) and the ratio

$T_{on}/(T_{on} + T_{off})$ is the duty cycled $d(t)$. Duty cycle is compared with a sawtooth signal $v(t)$ of amplitude equals to 1. Consequently $0 \leq d \leq 1$.

So, the studied converter has two working topologies corresponding to its switch states. The first topology (Fig. 3-a) corresponds to the on state of the switch $(d = 0)$ during a period fraction $dT_s$, the system can be presented by the following equations:

$$\frac{di_L}{dt} = \frac{1}{L} v_{in} \tag{1}$$

$$\frac{dv_C}{dt} = -\frac{1}{CR} v_C \tag{2}$$

$$v_s = v_C \tag{3}$$

The second topology, presented by the (Fig. 3-b) corresponds to the off state of the switch $(d = 1)$ during the rest of the sampling period $d'T_s$, for this case, the diode conducts, and the dynamical equations of the converter are expressed as

$$\frac{di_L}{dt} = \frac{1}{L}v_{in} - \frac{1}{L}v_C \tag{4}$$

$$\frac{dv_C}{dt} = \frac{1}{C}i_L - \frac{1}{CR}v_C \tag{5}$$

$$v_s = v_C \tag{6}$$

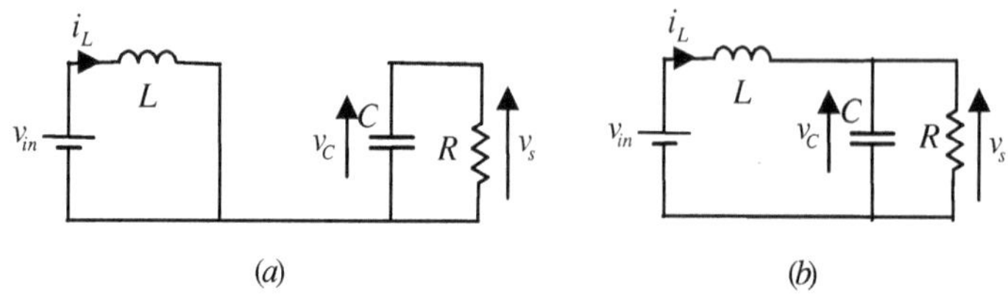


**Fig. 3.** Equivalent circuit of the boost converter(a) switch on. (b) switch off

Choosing the state vector as $x = [i_L, v_C]^T$ where $v_C$ is the capacitor voltage; $i_L$ is the inductor current, the general equation that governs the operation of the boost converter is:

$$\begin{cases} \dot{x} = A_i x + B_i \\ y = C_i x \end{cases} \tag{7}$$

Where: $i = 1$ to the first configuration described in Figure 2-(a) and: $i = 2$ the second configuration shown in Figure 2-(b).

Where:

$$A_1 = \begin{bmatrix} 0 & -1/L \\ 1/C & -1/RC \end{bmatrix},\quad B_1 = \begin{bmatrix} v_{in}/L \\ 0 \end{bmatrix},\quad C_1 = [0 \quad 1]$$

$$A_2 = \begin{bmatrix} 0 & -1/L \\ 1/C & -1/RC \end{bmatrix},\quad B_2 = \begin{bmatrix} 0 \\ 0 \end{bmatrix},\quad C_2 = [0 \quad 1]$$

So the combination of state-space representation of mode 1 (on mode) and mode 2 (off mode) induces the following linear state space representation:

$$\begin{cases} \dot{x} = Ax + Bu \\ y = Cx \end{cases} \tag{8}$$

With: $A = dA_1 + d'A_2; B = dB_1 + d'B_2; C = dC_1 + d'C_2$

And: $d' = 1 - d;\ u = d$

This gives:

$$A = \begin{bmatrix} 0 & -(1-d)/L \\ (1-d)/C & -1/RC \end{bmatrix},\quad B = \begin{bmatrix} v_{in}/L \\ 0 \end{bmatrix},\quad C = [0 \quad 1]$$

**Discrete-Time State-Space Model**

The MPC is to be fed with a discrete time model which is easily obtained from the continuous model assuming that the switching period $T_S$ is much smaller than the time constants associated with the circuit, the following discrete-time model is obtained using the forward Euler approximation, from the continuous time version (8):

$$x(k+1) = (I + T_s A)x(k) + T_s B d(k) \tag{9}$$

Using this assumption, the discrete-time state-space model of the boost converter can be written as:

$$\begin{bmatrix} x_1(k+1) \\ x_2(k+1) \end{bmatrix} = \begin{bmatrix} 1 & -(1-d(k))T_s/L \\ (1-d(k))T_s/C & 1-(T_s/RC) \end{bmatrix} \begin{bmatrix} x_1(k) \\ x_2(k) \end{bmatrix} + \begin{bmatrix} T_s v_{in}(k)/L \\ 0 \end{bmatrix} d(k) \tag{10}$$

The system in (10) is a nonlinear system since there are products of two input signals. It is therefore mandatory to make some sort of linearization at some point [15].

We can rewrite the non linear model of boost converter as follows:

$$\begin{cases} \dot{x} = f(x,u) \\ y = g(x,u) \end{cases} \tag{11}$$

Considering that the system variables consist of the following two components:

$$x = \Delta x + x_0 \tag{12}$$

$$d = \Delta d + d_0 \tag{13}$$

Where $x_0$ and $d_0$ represent the equilibrium values and $\Delta$x and $\Delta$d are the perturbed values of state and duty cycle, where the superscript $\Delta$ represents incremental variations over the equilibrium point.

Assume that the operating point is an equilibrium point which must satisfy

$$\dot{x}_0 = f(x_0, u_0) = 0 \tag{14}$$

$$\dot{x} = \Delta\dot{x} + \dot{x}_0 = f([\Delta x + x_0], [\Delta u + u_0]) \tag{15}$$

The linearized equations are:

$$\begin{cases} \Delta x(k+1) = A'\Delta x(k) + B'\Delta d(k) \\ \Delta y(k) = C'\Delta x(k) \end{cases} \tag{16}$$

Where the matrices $A'$, $B'$ and $C'$ are given by:

$$A' = \left.\frac{\delta f}{\delta x}\right|_{\substack{x=x_0 \\ d=d_0}} ; \ B' = \left.\frac{\delta f}{\delta d}\right|_{\substack{x=x_0 \\ d=d_0}} ; \ C' = \left.\frac{\delta g}{\delta x}\right|_{\substack{x=x_0 \\ d=d_0}}$$

Finally we obtain the following system description:

$$\begin{bmatrix} \Delta\dot{x}_1(k+1) \\ \Delta\dot{x}_2(k+1) \end{bmatrix} = \begin{bmatrix} 1 & -(1-d_0)T/L \\ (1-d_0)T/C & 1-T/RC \end{bmatrix} \begin{bmatrix} \Delta x_1(k) \\ \Delta x_2(k) \end{bmatrix} + \begin{bmatrix} (v_{in}+x_{02})T/L \\ -x_{01}T/C \end{bmatrix} \Delta d(k) \tag{17}$$

## 3 T-S Fuzzy Model of Boost Converter

The system (10) can be represented by linear subsystems called Takagi-Sugeno. The main feature of T-S fuzzy model is to express nonlinear dynamics by means of a set of IF-THEN rules. The consequent parts of the rules are local linear systems obtained from specific information about the original nonlinear plant.

The $i^{th}$ rule of fuzzy model for the nonlinear discrete-time systems has the following form:

$R_i$: If $z_1(t)$ is about $M_{1i}$ and $z_q(t)$ is about $M_{qi}$

$$\text{then}\begin{cases}\dot{x}(t) = A_i x(t) + B_i u(t) \\ y(t) = C_i x(t)\end{cases} \quad \text{for } i = 1, \dots, r \tag{18}$$

in which $M_{ji}$ is the fuzzy set of $z_i(k)$ in rule $R_i$, $r$ is the number of IF-THEN fuzzy rules and $z_i(t)$ are the decision variable assumed measurable, $x(t) \in R^n$ is the system state vector, $y(t) \in R^p$ is the measurable output,

The global dynamic system is inferred as follows:

$$\dot{x}(t) = \sum_{i=1}^{r} \mu_i(z(t))(A_i x(t) + B_i u(t)) \tag{19}$$

$$y(t) = \sum_{i=1}^{r} \mu_i(z(t))(C_i x(t)) \tag{20}$$

Where

$$\mu_i = \frac{\prod_{j=1}^{q} M_{ji}\,(z_i(t))}{\sum_{i=1}^{r}(\prod_{j=1}^{q} M_{ji}\,(z_i(t)))} \tag{21}$$

$M_{ji}\,(z_i(t))$ is the grade of membership of $z_i(t)$ in $M_{ji}$ , The normalized activation function $\mu_i(z(t))$ in relation with the $i^{th}$ sub model is such that

$$\sum_{i=1}^{r} \mu_i(z(t)) = 1, 0 \le \mu_i(z(t)) \le 1 \ \forall\, i = 1, \dots, r \tag{22}$$

A nonlinear system may also be represented by sectors [16].The following Lemma will be used in the sequel of the paper.

***Lemma:*** *Let $f(x(t)): R \to R$ is a bounded function $(\alpha \le f(x(t)) \le \beta)$, it always exist tow functions,$\eta_1(x(t))$ and $\eta_2(x(t))$ and two scalars $\alpha$ and $\beta$ such that:*

$$f(x(t)) = \alpha \times \eta_1(x(t)) + \beta \times \eta_2(x(t)) \tag{23}$$

*With* $\eta_1(x(t)) + \eta_2(x(t)) = 1$

$$\eta_1(x(t)) \ge 0\,, \eta_2(x(t)) \ge 0$$

*And:* $\alpha = max\,(f(x(t)))$, $\beta = min\,(f(x(t))$

$$\eta_1(x(t)) = \frac{f(x(t)) - \beta}{\alpha - \beta} \qquad \eta_2(x(t)) = \frac{\alpha - f(x(t))}{\alpha - \beta}$$

We can rewrite the non linear model of boost converter as follows:

$$\begin{bmatrix} x_1(k+1) \\ x_2(k+1) \end{bmatrix} = \begin{bmatrix} 1 & T_s/L \\ -T_s/C & 1-(T_s/RC) \end{bmatrix} \begin{bmatrix} x_1(k) \\ x_2(k) \end{bmatrix} + \begin{bmatrix} T_s v_C(k)/L \\ -T_s i_L(k)/C \end{bmatrix} d(k) \tag{24}$$

The obtained T-S fuzzy model represents exactly the nonlinear system for $x(k) \in R^n$ with $2^l$ locals models where $l$ represents the number of local models.

Assuming that, $I_{min} \le i_L(k) \le I_{max}$ and $V_{min} \le v_C(k) \le V_{max}$ the system described by (24) is modeled with a four rules T-S fuzzy system as follows:

$R_1$: If $i_L(k)$ is $I_{min}$ and $v_C(k)$ is $V_{min}$
then: $x(k+1) = A_1 x(k) + B_1 d(k)$

$R_2$: If $i_L(k)$ is $I_{max}$ and $v_C(k)$ is $V_{min}$
then: $x(k+1) = A_2 x(k) + B_2 d(k)$

$R_3$: If $i_L(k)$ is $I_{min}$ and $v_C(k)$ is $V_{max}$
then: $x(k+1) = A_3 x(k) + B_3 d(k)$

$R_4$: If $i_L(k)$ is $I_{max}$ and $v_C(k)$ is $V_{max}$
then: $x(k+1) = A_4 x(k) + B_4 d(k)$

The overall model of T-S fuzzy system can be given by the following:

$$x(k+1) = Ax(k) + Bu(k) \tag{25}$$

With

$$B = \sum_{i=1}^{4} \mu_i(x(k)) B_i \tag{26}$$

and

$$A = A_1 = A_2 = A_3 = A_4 = \begin{bmatrix} 1 & -T_s/L \\ T_s/C & 1-(T_s/RC) \end{bmatrix}$$

$$B_1 = \begin{bmatrix} T_s V_{min}/L \\ -T_s I_{min}/C \end{bmatrix}, \qquad B_2 = \begin{bmatrix} T_s V_{min}/L \\ -T_s I_{max}/C \end{bmatrix},$$

$$B_3 = \begin{bmatrix} T_s V_{man}/L \\ -T_s I_{min}/C \end{bmatrix}, \qquad B_4 = \begin{bmatrix} T_s V_{max}/L \\ -T_s I_{max}/C \end{bmatrix}$$

The membership function is such that:

$$\mu_1 = \frac{-i_L(k) + I_{max}}{I_{max} - I_{min}} \qquad \mu_2 = \frac{i_L(k) - I_{min}}{I_{max} - I_{min}}$$

$$\mu_3 = \frac{-v_C(k) + V_{max}}{V_{max} - V_{min}} \qquad \mu_4 = \frac{v_C(k) - V_{min}}{V_{max} - V_{min}}$$

Figure 4 shows the membership functions of the fuzzy model

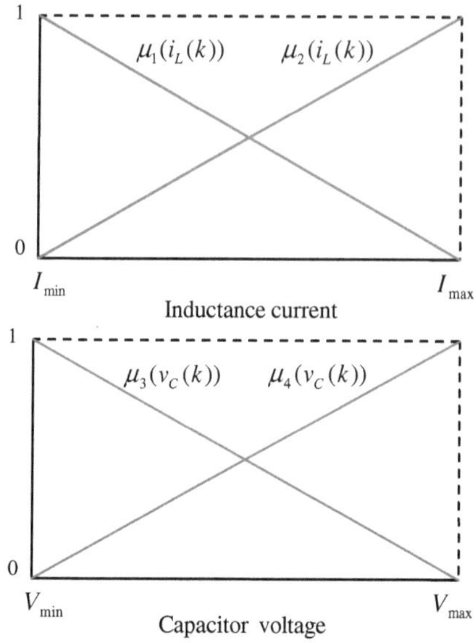


**Fig. 4.** Membership functions of the T-S fuzzy model

# 4 Model-Based Predictive Control

The basic model to calculate the predictions is discrete state model given by [9]:

$$\begin{cases} x(k+1) = Ax(k) + Bu(k) & x(k) \in R^n; u(k) \in R^l \\ y(k) = Cx(k) & y(k) \in R^m \end{cases} \tag{27}$$

$J$ the cost function to be minimized at each sampling period penalizes deviations from the predicted output $y(k+i|k)$ of a reference trajectory $r(k+i|k)$, and variations of the control vector $\Delta u(k) = u(k) - u(k-1)$, it is often given by the quadratic form

$$J(k) = \sum_{i=H_w}^{H_p} \left\| \hat{y}(k+i|k) - r(k+i|k) \right\|_{Q(i)}^2 + \sum_{i=0}^{H_u-1} \left\| \Delta\hat{u}(k+1|k) \right\|_{R(i)}^2 \tag{28}$$

Where $H_u$ is the control horizon, $H_w$ is first point of the prediction horizon, $H_p$ is last point of the prediction horizon, $H_u < H_p$, and $\Delta u(k+i|k) = 0$ if $i < H_u$. $R(i)$, $Q(i)$ are suitable weighting matrixes.

The cost function (28) can be rewritten as [13]:

$$\begin{aligned} J(U_n) = J_{min} &+ 2\left[\left(\Gamma + \Xi \overline{U}_{k-1} - Y_{ref}^n\right)^T Q\Lambda\right]\Delta U_n \\ &+ \Delta U_n^T \left[\Lambda^T Q \Lambda + R + S\right]\Delta U_n \end{aligned} \tag{29}$$

where

$$J_{\min} = Y_{ref}^T Q Y_{ref} + \Gamma^T Q \Gamma - 2Y_{ref}^T Q \Gamma + \overline{U}_{k-1}^T S \overline{U}_{k-1}^T + \overline{U}_{k-1}^T R \overline{U}_{k-1}^T$$

The fuzzy model (27) is used to predict the process output, subject to level and rate constraints on the inputs and outputs

$$\begin{bmatrix} I \\ -I \\ L \\ L \\ \Lambda \\ -\Lambda \end{bmatrix} \Delta U_n \leq \begin{bmatrix} \tilde{U}_{\max} \\ -\tilde{U}_{\min} \\ U_{\max} \\ -U_{\min} \\ Y_{\max} - \Gamma \\ -Y_{\max} + \Gamma \end{bmatrix} \tag{30}$$

Where the predicted output is given by:

$$\hat{Y} = \Gamma + \Lambda \Delta U_n \tag{31}$$

And $\hat{Y} \in R^{H_p n_0}$, $\Gamma \in R^{H_p n_0}$, $\Lambda \in R^{H_p n_0 \times H_u n_i}$ and $\Delta U_n \in R^{H_u n_i}$, $n_0$ and $n_i$ are the number of system outputs and system inputs, respectively. $\Gamma$ is called the free output response as it does not depend on the control sequence, and $\Lambda \Delta U_n$ is called the force output response as it depends on the selected control sequence.

## 5 Transformation in LMI Form

An optimization LMI problem requires that the initial problem is restructured so as to involve a linear cost function and constraints of type strict inequality, for this the control objective imposes the following steps. In general, the minimization of a convex quadratic function $\Im(U_n)$ can result in an equivalent manner by the strategy of minimizing follows:

minimize $\gamma$ and find an $\Delta U$ admissible that satisfied

$$\Im(U_n) < \gamma \tag{32}$$

This inequality becomes linear using Schur lemma.

$Min \;\; \gamma \in R^+$

Subject to

$$\begin{bmatrix} 2\left|\left(\Gamma + \Xi \overline{U}_{k-1} - Y_{ref}^n\right)^T Q \Lambda\right| \Delta U_n + J_{\min} - \gamma & \Delta U_n^T \\ \Delta U_n & -\left[\Lambda^T Q \Lambda + R + S\right]^{-1} \end{bmatrix} < 0$$

And $-\left[\Lambda^T Q \Lambda + R + S\right]^{-1} < 0$ (33)

$$\begin{cases} L(x) = 2\left[\left(\Gamma + \Xi \overline{U}_{k-1} - Y_{ref}^{n}\right)^{T} Q\Lambda\right]\Delta U_{n} + J_{\min} - \gamma \\ M(x) = -\left[\Lambda^{T} Q\Lambda + R + S\right]^{-1} \\ W(x) = \Delta U_{n}^{T} \end{cases}$$

$L(x)$, $M(x)$, $W(x)$ are the Schur matrix.

Finally, the second inequality of (33) is always verified by the Hessian structure in the MPC case. The previous constraints (30) must be presented in a diagonal form defining thus a matrix space convex and symmetric. At the end of the three previous steps, the problem in the final form used by LMI solvers is therefore:

$$Min \ \gamma \in R^{+}$$

Subject to

$$\begin{bmatrix} 2\left[\left(\Gamma + \Xi \overline{U}_{k-1} - Y_{ref}^{n}\right)^{T} Q\Lambda\right]\Delta U_{n} + J_{\min} - \gamma & \Delta U_{n}^{T} \\ \Delta U_{n} & -\left[\Lambda^{T} Q\Lambda + R + S\right]^{-1} \end{bmatrix} < 0$$

$$\begin{aligned} & diag\left(I\Delta U_{n} - \tilde{U}_{\max}\right) \le 0 \\ & diag\left(- I\Delta U_{n} + \tilde{U}_{\min}\right) \le 0 \\ & diag\left(L\Delta U_{n} - U_{\max}\right) \le 0 \\ & diag\left(- L\Delta U_{n} + U_{\min}\right) \le 0 \\ & diag\left(\Lambda\Delta U_{n} - Y_{\max} + \Gamma\right) \le 0 \\ & diag\left(\Lambda\Delta U_{n} - Y_{\max} + \Gamma\right) \le 0 \end{aligned} \qquad (17)$$

## 6 Simulations and Results

The performances of the proposed control design are illustrated through simulations. The parameters nominal values of the boost converter are illustrated in Table 1.

**Table 1.**

| Parameters | Description | Numerical value |
|---|---|---|
| $L$ | The inductance | $200\mu H$ |
| $C$ | The capacitance | $200\mu F$ |
| $R$ | The load resistance | 10Ω |
| $V_{in}$ | Input voltage | 10V |
| $T$ | The sampling | $0.65ms$ |
| $fs$ | The frequency | 200μHz |
| $I_{min}$ | Minimum current | 0 |
| $I_{max}$ | Maximum current | 3 |
| $V_{min}$ | Minimum voltage | 0 |
| $V_{max}$ | Maximum voltage | 2.2 |

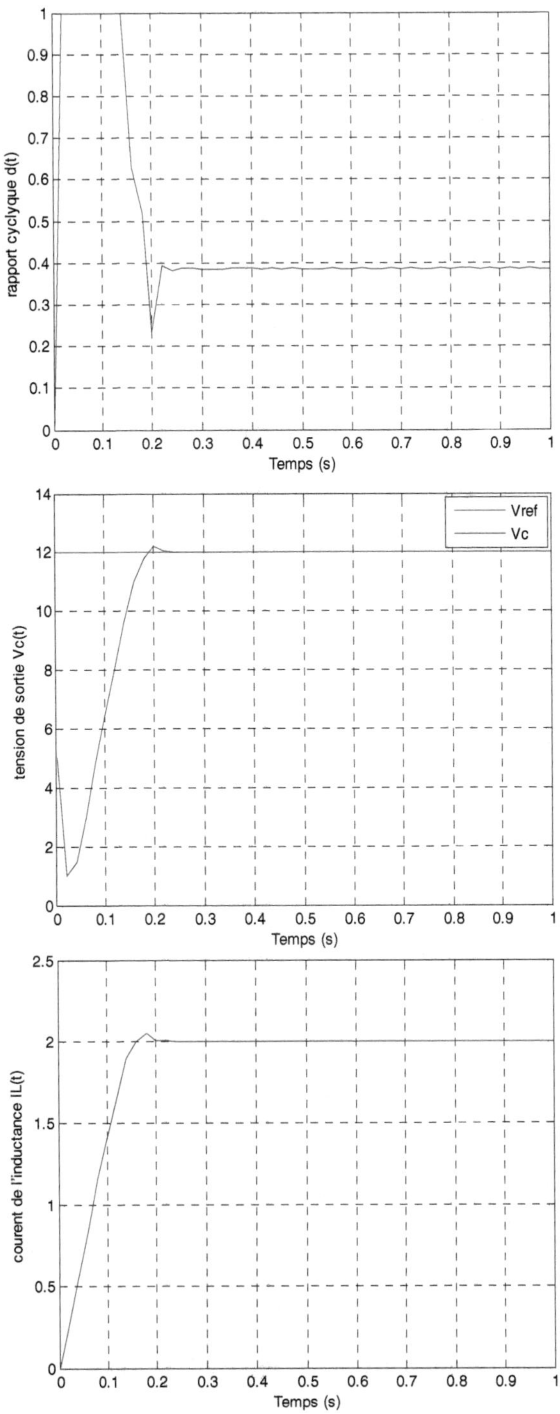


**Fig. 5.** The simulation results of the boost converter

Let : $0 \leq u \leq 1$, $0 \leq y \leq 2.2$

The value of the weighting matrices for the different predictive methods are:

$Q$=eye($H_p$); $S$=0.5*eye(($H_u$+1)*$ni$); $R$=0.1*eye(($H_u$+1)*$ni$);

Figure 5 illustrates simulation of the boost converter in the reference voltage12$V$. The waveforms depicted in the Fig 5 are the duty-cycle $d$ capacitor voltage $v_C$ and inductor current $i_L$. point that the output voltage settle to their desired value with less overshoot and the settling time of the circuit as clearly seen in the graph is $t = 0.1$ seconds. It is noted that with MPC-LMI can ensure a limitation on the control signal and maintain the output voltage at its desired value by using a duty ratio in the range [0,1]. The best performances are obtained for $H_p = 7, H_u = 3$

## 7 Conclusion

In this paper, we have presented a MPC-LMI controller for the boost converter operating in CCM mode. At first we used the fuzzy model based on sector nonlinearity concept, to approximate the nonlinear model, this approximation allows us to obtain the exact model of the system. The fuzzy model obtained was then used to design a MPC controller that ensures the output voltage regulation with good performances. A Linear Matrix Inequalities (LMIs) formulation of the MPC problem permits to obtain a controller optimizing.

## References


[1] Guldemir, H.: Modeling and Sliding Mode Control of Dc-Dc Buck-Boost Converter. In: 6th International Advanced Technologies Symposium (IATS 2011), Elazığ, Turkey, May 16-18 (2011)

[2] Ahmad, A., Zhiliu, K., Kinoshita, H.: High Performance Algorithms for the Control of Buck DC-DC Converters. International Journal of Engineering Science and Technology 2(10), 5799–5812 (2010)

[3] Inglés, J.V., Garcés, P., Leyva, R.: Robust LMI Control of a Buck-Boost Converter with Low Ripple Propagation. Preprints of the 2012 20th Mediterranean Conference on Control & Automation (MED), Barcelona, Spain, July 3-6 (2012)

[4] García, C.E., Prett, D.M., Morari, M.: Model predictive control: Theory and practice, a survey. Automatica 25, 335–348 (1989)

[5] Tatjewski, P., Nczuk, M.: Soft Computing in Model–Based Predictive Control. Int. J. Appl. Math. Comput. Sci. 16(1), 7–26 (2006)

[6] Bououden, S., Chadli, M., Filali, S., El Hajjaji, A.: Fuzzy Model Based Multivariable Predictive Control of a Variable Speed Wind Turbine: LMI approach. Renewable Energy 37(1), 434–439 (2012)

[7] Carneiro, G.L., Galvão, R.K.H.: Model Based Predictive Control of an Aircraft with Actuator Failure in a Terrain Following Task. In: 3rd CTA-DLR Workshop on Data Analysis & Flight Control, S. J. Campos, SP, Brazil, September 14-16 (2009)

[8] Hichem, B., Faouzi, M.: Fast Nonlinear Model Predictive Control using Second Order Valera Models Based Multi-agent Approach. Institut Supérieur Des Etudes Technologiques de SFAX Ecole Nationale d'ingénieur de Monastir Tunisia
[9] Lazar, M., Roset, B.J.P., Heemels, W.P.M.H., Nijmeijer, H., van den Bosch, P.P.J.: Input-To-State Stabilizing Sub-Optimal Nonlinear MPC Algorithms with an Application to DC-DC Converters. In: IFAC 2006 (2006)
[10] Leung, F.H.F., Tam, P.K.S., Li, C.K.: The Control of Switching DC-DC Converters – A General LQR Problem. IEEE Transactions on Industrial Electronics 38, 65–71 (1991)
[11] Sira-Ramirez, H., Perez-Moreno, R.A., Ortega, R., Garcia-Esteban, M.: Passivity-based controllers for the stabilization of DC-to-DC power converters. Automatica 33, 499–513 (1997)
[12] Sira-Ramirez, H.: On the Generalized PI Sliding Mode Control of DC-to-DC Power Converters: A Tutorial. International Journal of Control 76, 1018–1033 (2003)
[13] Takagi, T., Sugeno, M.: Fuzzy identification of systems and its applications to modeling and control. IEEE Transactions on Systems, Man, and Cybernetics 15, 116–132 (1985)
[14] Tanaka, K., Wang, H.O.: Fuzzy Control Systems Design and Analysis: A Linear Matrix Inequality Approach. Wiley, New York (2001)
[15] Mahabir, K., Verghese, G., Thottuvelil, J., Heyman, A.: Linear averaged and sampled data models for large signal control of high power factor converter. In: Proceedings of IEEE Power Electronics Specialists Conference, San Antonio, Texas, pp. 372–381 (1990)
[16] Bououden, S., Chadli, M., Filali, S., El Hajjaji, A.: Fuzzy Model Based Multivariable Predictive Control of a Variable Speed Wind Turbine: LMI approach. Renewable Energy 37(1), 434–439 (2012)
[17] Bououden, S., Benelmir, O., Ziani, S., Filali, S.: A new adaptive fuzzy model and output terminal constraints in predictive control. International Journal of Information and Systems Sciences 3(1), 25–35 (2007)
[18] Miller, A., Muljadi, E., Zinger, D.S.: A variable speed wind turbine power control, Energy Conversion. IEEE Transaction on Power Electronics 12(2), 181–186 (1997)
[19] Nakamura, T., Morimoto, S., Sanada, M., Takeda, Y.: Optimum control of IPMSG for wind generation system. In: Power Conversion Conference, vol. 3, pp. 1435–1440 (2002)

# A Note about Robust Stabilization of Chaotic Hénon System Using Grammatical Evolution

Radomil Matousek, Ladislav Dobrovsky,
Petr Minar, and Katerina Mouralova

Department of Applied Computer Science, Brno University of Technology
Technicka 2, 61669 Brno, Czech Republic
matousek@fme.vutbr.cz

**Abstract.** The paper deals with robust stabilization of a well-known deterministic discrete chaotic system denoted as Hénon map. By means of proper utilization of metaheuristic optimization tool, the Grammatical Evolution (GE) can synthesise a new robust control law. As a model of deterministic chaotic system the two-dimensional Hénon map with original definition was used. The Hénon map is an iterated discrete-time system which exhibits chaotic behaviour in two-dimension. Stabilization for the period-2 orbits of the two-dimensional Hénon map is presented. The chaotic system stabilization is based on a time-delay auto-synchronization with its own synthesized control law. This synthesized chaotic controller utilizes own design of advanced GE algorithm with two-level optimization procedures and a proper objective function. The original objective function design considers a low sensitivity dependence on initial conditions and also proper time for stabilisation of the control process. All computing experiments are performed using Matlab/Simulink environment where the double precision floating point arithmetic was used.

**Keywords:** Hénon map, Chaos Control, Robust stabilization, Metaheuristic optimization, Grammatical evolution.

## 1 Introduction

The *chaos theory* studies the behaviour of *dynamical systems* that are highly sensitive to the initial conditions (this is a paradigm popularly referred to as the butterfly effect) [3], [4]. Small differences in the initial conditions produce widely diverging outcomes for such dynamical systems, and a long-term prediction is impossible in general. Chaos theory currently penetrates natural and engineering sciences.

During the last 20 years from its beginning this theory has changed most of things we have perceived as definitive, right, and normal, under influence of our traditions, customs, and history. There are two fundamental characteristics of common chaotic systems, which can be illustrated by the Hénon system very well. The first one is called *the sensitive dependence on the initial conditions*. This causes systems that have the same values of control parameters but slightly different initial conditions to diverge critically during their iteration in time. The second one is called *ergodicity*. The ergodicity means that a large set of identical systems, which differ only in their initial conditions, will be

I. Zelinka et al. (eds.), *Nostradamus 2014: Prediction, Modeling and Analysis of Complex Systems*, Advances in Intelligent Systems and Computing 289,
DOI: 10.1007/978-3-319-07401-6_21, 

distributed after a given time on the attractor exactly in the same way as the series of iterations of one single system.

Deterministic chaos is typical for open non-linear dynamic macro systems due to which, because of the unstable starting condition or some important system's parameters (control parameters), it comes to unpredictability of its behaviour. Generally if the system is markedly diverting from its steady state it will reach a boundary of its stability. This is characterized by the so-called *bifurcation point*. Primary division is a primary bifurcation, it is generally characterized by the period of boundary cycle. Further from the steady state the number of oscillation frequencies increases and the cascades of bifurcations occur. System is getting to chaos state.

In case of the deterministic chaos it is that state of a system in which it neither converges to a *fixed-point* nor settles into a *periodic oscillation*. Such irregular dynamics of the state is not due to external "noise" or due to an infinite-dimensional state-space. Such behaviour is possible only in systems that are non-linear in nature. It can be seen in mathematical models such as systems of non-linear difference equations or differential equations.

During the past twenty years, utilization of new intelligent methods in engineering and computing has attracted the attention of researchers worldwide. The most current methods are frequently based on soft computing, which represent a class of special methods (algorithms) and belong to the *artificial intelligence* (AI) paradigm. The most popular of these methods are evolutionary algorithms, genetic programming, artificial neural networks, fuzzy logic, etc. Presently, evolutionary algorithms are known as a powerful kind of tools with a possibility to solve any difficult and complex optimization problem. In case of the paper the advance Grammatical Evolution (GE) is used [33]. In the relation to the idea of Genetic Programming (GP) [34] or Analytic Programming (AP) [19] the objective is to find an executable program or a common model (structure and variables), that will achieve a good fitness value for a given objective function. Just an objective function design is very important for the whole optimization process.

The interest in the interconnection between *evolutionary techniques and control of chaotic systems* is spread daily. First steps were done in (Richter et al. 2000) [9], (Zelinka et al. 2008) [19], [20], (Senkerik, Oplatkova et al.,2006; 2010) [22], [23], [26], where the control law was based on the discrete Pyragas method ETDAS (Extended delay feedback control, Pyragas, 1995) [7], [8], [26], [25] etc. These papers were concerned to tune several parameters inside the control technique for the chaotic system and also concerned to design its own control law by special evolutionary computing techniques.

## 2 Hénon Map

The Hénon map has been proposed by the French astronomer and mathematician Michel Hénon [1] as a simplified model of the Poincare map that arises from a solution of the Lorenz equations [2]. As we mentioned above the Hénon map presents a simple two-dimensional invertible iterated map with quadratic non-linearity and chaotic solutions called *strange attractor*. It is important to note that it is a link between the chaos and the fractals. The Hénon map is a discrete-time dynamic system which takes a point $(x_n, y_n)$ in the plane and maps it to a new point by the recurrence relations (1).

$$\begin{aligned} x_{n+1} &= 1 - ax_n^2 + y_n \\ y_{n+1} &= bx_n \end{aligned} \tag{1}$$

The strange attractor (Hénon map) depends on two parameters, $a$ and $b$, which for the classical Hénon map [1] have values of $a = 1.4$ and $b = 0.3$. The Hénon map can be written in terms of a single variable with two time delays [37]. In this case the second equation in (1) is written as $y_n = x_{n-1}$ and the map is rewritten in forms of a single variable with two delays $x_{n+1} = 1 - ax_n^2 + bx_{n-1}$. There is also the second common definition of the Hénon map given by formula (2). This definition is not equal as the original one (1) but similar. We can derive the formula (2) from (1) by substitutions: $X_n = (1/a)x_n$ and $Y_n = (b/a)y_n$. Hénon map (2) with given domain of definition has much bigger region of convergence to chaotic attractor than the original definition (1).

$$\begin{aligned} X_{n+1} &= a - X_n^2 + bY_n \\ Y_{n+1} &= X_n \end{aligned} \tag{2}$$

Many researchers [19], [22], [23] [17] working in the field of the chaotic control probably use the formula (2) for this convenient feature. Unfortunately, many researchers wrote that definition (1) is the original definition by Hénon [1]. In this paper the original definition of Hénon map (1) was used. The strange attractor with its basin of attraction (blue color) are presented in Fig. 1.

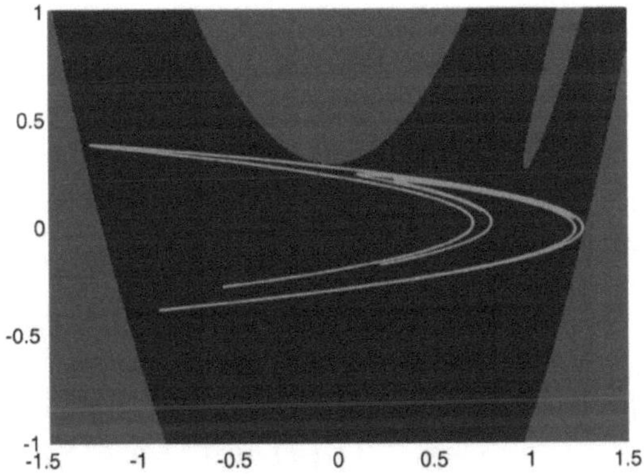


**Fig. 1.** Hénon attractor in $x-y$ plane for $a = 1.4$ and $b = 0.3$: points in blue area form the Hénon attractor, points in red area diverge to negative infinity.

There are two saddle fixed points which for the $a = 1.4$ and $b = 0.3$ have coordinates $(x_A = +0.63135, y_A = +0.18941)$ and $(x_U = -1.13135, y_U = -0.33941)$. The point $(x_A, y_A)$ belongs to the *attractor of the map*, while the point $(x_U, y_U)$ is an *unstable fixed point*.

## 3 Control Design and Optimization

In this section we briefly describe optimization methods and an objective function design used for the design of optimal parametrization of the control model of the given chaotic system. A control of chaos means the stabilization using small system perturbations of unstable periodic orbits. The perturbation has to be tiny to avoid significant modification of the system's natural dynamics. The goal of our optimization task is to present the robust stabilization of the system (1) in the $p-2$ orbital. In our case points $(x_{TS_1} = 0.63, x_{TS_2} = 0.37)$ were used. The model was performed by original objective function based on the common formula (3), where $TS$ index denotes a *target state* (set point in automatic control terminology), $x_n$ means an *actual state* given by model (1) and $N$ means a maximum number of iteration steps of the given computer simulation. The formula minimizes the integral of the time-weighted absolute error (ITAE) in discrete form, where the time is substituted by steps and integral by the summation.

$$ITAE = \sum_{n=1}^{N} n\,|x_{TS} - x_n|, \tag{3}$$

The objective function (3) includes a penalisation of target state error (i.e. actual state value deviation from fixed point in our case) and penalisation in relation of the iteration step. Of course, there are many variants of (3) as in (4), where a sense of robustness by parameters $r, s$ is evident [31].

$$ITAE^{rs} = \sum_{n=1}^{N} n^r (|x_{TS} - x_n|)^s \tag{4}$$

Equation (3) was used as a base for robust objective function design. In case of two fixed point stabilisation the *ITAE* criterion has to be split up into even and odd part by orbit for two fixed points target state $TS_1$ and $TS_2$ (5).

$$ITAE_{p2} = ITAE^{odd} + ITAE^{even}, \text{where} \begin{cases} ITAE^{odd} = n \sum\limits_{n=1,3,5,\ldots}^{N} |x_{TS_1} - x_n| \\ ITAE^{even} = n \sum\limits_{n=2,4,6,\ldots}^{N} |x_{TS_2} - x_n| \end{cases} \tag{5}$$

In the end, we use the union formula (6), which implemented a parallel influence of the given set of initial solution $x_1, y_1$ for the final solution - the robust control design in aspect of sensitivity to initial conditions. The minimization problem given by multi objective function (6) was solved.

$$mITAE = \sum_{k=1}^{K} ITAE_{p2}(x_1(k)) \quad (x_1(k) = rand([min, max]), \forall k \in 1 \ldots K) \qquad (6)$$

The multi objective function *mITAE* covered *K* solutions obtained on the base of *K* random initialisation of $(x_1, y_1)$. The optimization of the control law utilized two algorithms GE (Grammatical Evolution) and NM (Nelder-Mead):

- To choice randomly *K* initial points $(x_1, y_1)$ from given interval $[-1, 1]$.
- Using GE algorithm to search and optimize a control law for the Hénon map based chaotic system according to *mITAE*.
- Using NM algorithm to re-tune the model of control law.

### 3.1 Nelder-Mead Algorithm

The Nelder-Mead algorithm (NM) or the so call non-linnear simplex search algorithm, originally published in 1965 (Nelder and Mead, 1965), is one of the best known and successful algorithms for the multidimensional unconstrained optimization without derivatives. In our case we use this algorithm as a base solver which can be substituted in the future. The Nelder-Mead method is simplex-based. A simplex *S* is defined as the convex hull of $n+1$ vertices $x_0, \ldots, x_n$. Nelder-Mead algorithm generates a new test position of simplex by extrapolating the behaviour of the objective function measured at each test point arranged as the simplex. The algorithm then chooses to replace one of these test points with the new test point and so the technique progresses. Our implementation of the NM used the Matlab function `fminsearch`.

### 3.2 Grammatical Evolution

Grammatical evolution (GE) is one of the newest approaches in evolution algorithms. GE has been designed by Conor Ryan, JJ Collins and Michael O'Neill in 1998 [33]. Practically from optimization point of view it has similar options as a well-known GP or a less known AP. In the GE the programs (or models) are generated in languages described by Backus-Naur form (BNF) of context-free grammars. The GE paradigm is a principle of coding/decoding of known GA [32] using BNF, i.e. a genotype is interpreted in the same way as in GA and decoded a phenotype according to the rules of GE. A choice of terminals and non-terminals of grammar and rewriting rules understandingly changes options of the result model construction. In additional, the resulting models have been generated in Matlab/Simulink environment. The `memory` block interpreted a delay of sequence's members. According presented aim the following grammar has been chosen in listing 1.

**Listing 1.** The grammar for automatic generating control law (disturbance series $f$)

```
<start> ::= <expr>

<expr> ::= ( <expr> <op> <expr> )
         | getXat(x, n, _GE_intConstant_1_<0;3> )
         | _GE_realConstant_1_<0.05;0.95>

<op> ::= +
       | -
       | .*
       | ./
```

# 4 Experimental Results and Conclusion

As described in above Sect. 3 the meta-evolutionary approach means the use of one main evolutionary algorithm (GE) for the first step of an optimization design and the second algorithm for the model parameters estimation. The first optimization technique is used for model design, i.e. it produce its own control sequences $f$ which stabilised the given discrete chaotic system (1). An automatic model design for $f$ sequence generating is based on GE using grammar by Listing 1. The next step is a parameter tuning of $f$. This second level of optimization produces final form of $f$. On the base of practical experiments, we can say that in case of proper GE settings and grammar (BNF) this process only slightly improves the solution $f$.

The system stabilisation of (1) means the use of additional (input) vector $f$ which on disturbance principle stabilizes the progress into given fixed points ($x_{TS_1} = 0.63, x_{TS_2} = 0.37$) (p-2 orbital). The proposal objective function (6) uses 300 iterations ($N = 300$) and 100 random initialize points $x_1$ ($K = 100$) for the robust design of disturbance sequence $f$. An example of simulation results - stabilisation of Hénon chaotic system by means of control law $f_1$ given by Table 1 is presented in the Fig. 2. The multiple ITAE objective function (6) was used simultaneously only for 100 initialise points of $x_1$ ($x_1 = rand(x_{min}, x_{max})$) due to the computing time. Nevertheless, the test of robustness was realized by means of 16384 initial random settings of $x_1$ for the objectivity.

**Table 1.** The results of the optimization process (evolutionary synthesized chaotic system using GE and NM algorithms) of chaos stabilization at p-2 orbit

| Model | The formula of the control law |
|---|---|
| $f_1$ | $x_{n-1} * 0.7000000134726454 - 0.62634000194746675$ |
| $f_2$ | $0.1671485837 - 0.6808108696 * x_{n-2} - 0.6808108696 * x_n$ |
| $f_3$ | $(3.470971 * x_{n-1})/(0.803922 * x_{n-4} - 0.803922 * x_{n-5} + 1.465975/x_{n-7} - 29.200063)$ |
| $f_4$ | $3.4709719638/(x_{n-1} * (1.526849450954374/x_{n-7} - 29.200063240139128))$ |
| $f_5$ | $-3.4709719638/(x_{n-1} * (0.568627 * x_{n-7} - 1.5720859100/x_{n-7} + 28.87672657501012))$ |

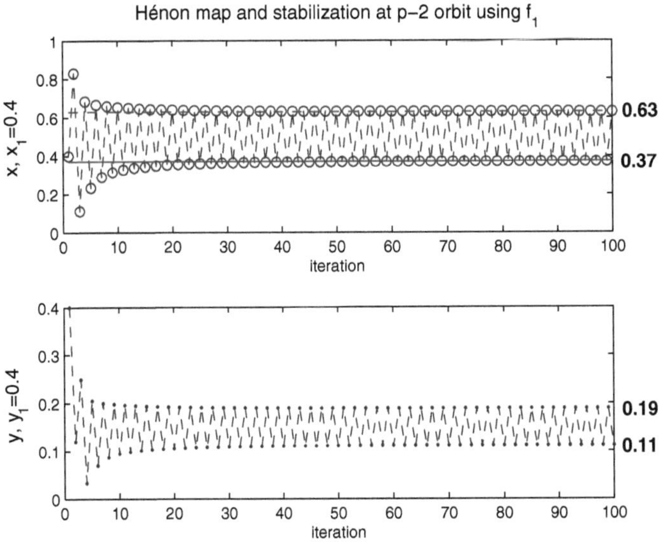


**Fig. 2.** Examples of simulation results - stabilisation of chaotic systems at $p-2$ orbit by robust control laws $f_1$. Set points (fixed points) are $x_{TS_1} = 0.63$ and $x_{TS_2} = 0.37$ (induce points $y_{TS_1} = 0.19$ and $y_{TS_2} = 0.11$), initial values are $x_1 = 0.4$ and $y_1 = 0.4$.

The results for $p-2$ orbit stabilisation are displayed in Table 1 and Table 2. The robust features of the models (disturbance series denoted as $f_1 \ldots f_5$) are presented in Table 2, where the ratio of success for the 16384 random selected initial points $x_1$ and $y_1$ can be seen. There are also average objective function values for the 300 iterations (the less is better) in Table 2. The best presented result $f_1$ and other $f_2 \ldots f_5$ are different in the possibility to reach fixed points at $p-2$ orbit. The dependences of the models on the initial solutions and the mutual comparison is presented in Fig. 3. The figure shows regions of initial points $[x_1, y_1]$ where the convergence to the target points (stabilization of chaotic system) were reached. This measure of dependence is expressed for given models as percentage of success in Table 2.

**Table 2.** The features of the solutions for $p-2$ orbits stabilisation. (GE+NM design, $mIKAE$)

| Model | Delay elements | Success ratio | % | $mIKAE$ |
|---|---|---|---|---|
| $f_1$ | $x_{n-1}$ | 11371/16384 | 69.40 | 11 |
| $f_2$ | $x_n, x_{x-2}$ | 9570/16384 | 58.41 | 354 |
| $f_3$ | $x_{n-1}, x_{n-4}, x_{n-5}, x_{n-7}$ | 4945/16384 | 30.18 | 363 |
| $f_4$ | $x_{n-1}, x_{n-7}$ | 4829/16384 | 29.47 | 466 |
| $f_5$ | $x_{n-1}, x_{n-7}$ | 4691/16384 | 28.63 | 485 |

It is necessary to emphasize that the presented solution based on $mITAE$ is robust considering to $K$ initial conditions.

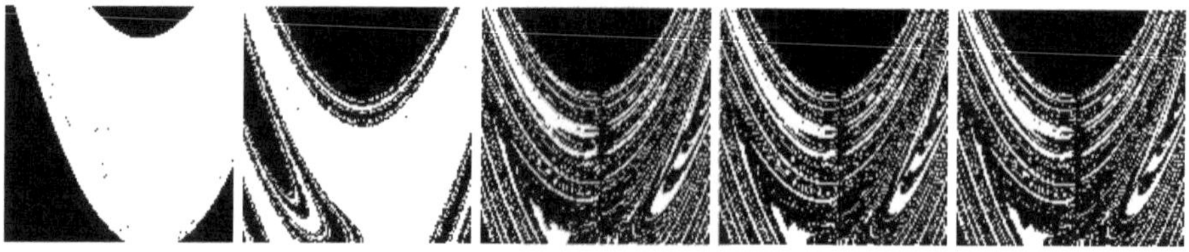

**Fig. 3.** The examples of simulation results - white regions indicate where stabilisation of the chaotic systems at $p-2$ orbit are reached. The figures present models $f_1 \dots f_5$ (from left to right).

The research deals with a synthesis of the control laws by means of GE and NM algorithms for given points (one fixed point is 0.63, one selected point is 0.37) for stabilisation of Hénon chaotic system. The presented results reinforce the argument that the global solutions with low sensitivity to initial conditions can be reached. The way for the complete robustness can be achieved by fusion of a proper variant of control law models $f$. Important question and also future work, will be the use of HPC CUDA implementation of HC12 [29] where we expect a significantly better result. An improved GE result will be expected in case of a new designed BNF grammar base. The last Fig. 4 presents robustness solution given by model $f_1$. There are 144 initial states uniformly distributed in region $[-1,1] \times [-1,1]$ and the given solutions for the first 40 iterations.

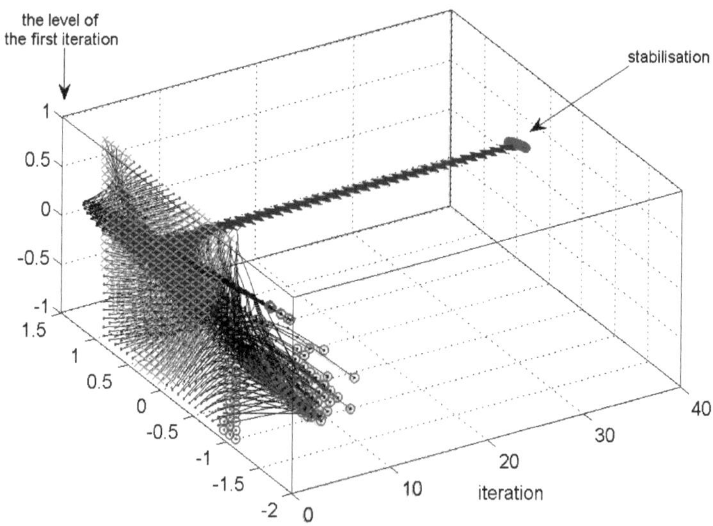


**Fig. 4.** The example of robustness solution given by model $f_1$. The blue color is stabilisation, the black color is divergence. There are 144 initial states uniformly distributed in region $[-1,1] \times [-1,1]$ denoted as the level of the first iteration.

**Acknowledgement.** This work was supported by grants IGA No. FSI-S-14-2533: Applied Computer Science and Control Design and TACR No. TA02021449: Intelligent Alarm System for Nuclear Power Plant Technology. Authors also thank to friends in needs namely Ivan, Roman and Zuzka for convenient discussion about chaos control.

## References

1. Hénon, M.: A two dimensional mapping with a strange attractor. Commun. Math. Phys. 50, 69–77 (1976)
2. Benedicks, M., Carleson, L.: The dynamics of the Hénon maps. Ann. Math. 133, 1–25 (1991)
3. Lozi, R.: Un attracteur étrange du type attracteur de Hé non. Journal de Physique. Colloque C5, Supplément au 39(8), 9–10 (1978)
4. May, R.M.: Simple mathematical models with very complicated dynamics. Nature 261(5560), 459–467 (1976), doi:10.1038/s261459a0
5. Ott, E., Grebogi, C., Yorke, J.A.: Controlling chaos. Phys. Rev. Lett. 64, 1196–1199 (1990)
6. Ott, E., Grebogi, C., Yorke, J.A.: Controlling chaotic dynamical systems. In: Campbell, D.K. (ed.) Chaos, Amer. Inst. of Phys., New York, pp. 153–172 (1990)
7. Pyragas, K.: Continuous control of chaos by self-controlling feedback. Phys. Lett. A. 170, 421–428 (1992)
8. Pyragas, K.: Control of chaos via extended delay feedback. Phys. Lett. A 206, 323–330 (1995)
9. Richter, H., Reinschke, K.J.: Optimization of local control of chaos by an evolutionary algorithm. Physica D 144, 309–334 (2000)
10. Richter, H., Reinschke, K.J.: Local control of chaotic systems: a Lyapunov approach. Int. J. Bifurc. Chaos 8, 1565–1573 (1998)
11. Chen, G.: Control and Synchronization of Chaotic Systems (bibliography). ECE Dept, Univ of Houston, `http://www.ee.cityu.edu.hk/~gchen/chaos-bio.html` (cited: March 13, 2013)
12. Fradkov, A.L.: Chaos Control Bibliography, Russian Systems and Control Archive, RUSYCON (1997-2000), `http://www.rusycon.ru/chaos-control.html` (cited: March 13, 2013)
13. Ramaswamy, R., Sinha, R., Gupte, N.: Targeting chaos through adaptive control. Phys. Rev. Lett. 57(3), 2507–2510 (1998)
14. Starrett, J.: Time-optimal chaos control by center manifold targeting. Phys. Rev. Lett. 66(4), 6206–6211 (2002)
15. Bollt, E.J., Kostelich, E.J.: Optimal targeting of chaos. Physics Letters A 245, 399–406 (1998)
16. Iplikci, S., Denizhan, Y.: An improved neural network based targeting method for chaotic dynamics. Chaos, Solutions & Fractals 17(2-3), 523–529 (2003)
17. Wang, T., Wang, X., Wang, M.: Chaotic control of Hénon map with feedback and nonfeedback methods. Commun. Nonlinear Sci. Numer. Simulat. 16, 3367–3374 (2011)
18. Zelinka, I.: SOMA–self organizing migrating algorithm. In: Babu, B.V., Onwubolu, G. (eds.) New Optimization Techniques in Engineering. ch. 7, vol. 33. Springer-Verlag (2004)
19. Zelinka, I., Guanrong, C., Celikovsky, S.: Chaos synthesis by means of evolutionary algorithms. International Journal of Bifurcation and Chaos 18(4), 911–942 (2008)
20. Zelinka, I., Senkerik, R., Navratil, E.: Investigation on evolutionary optimitazion of chaos control. Chaos, Solitons & Fractals 40, 111–129 (2009)

21. Lampinen, J., Zelinka, I.: Mechanical engineering design optimization by differential evolution. In: Corne, D., Dorigo, M., Glover, F. (eds.) New Ideas in Optimization, pp. 127–146. McGraw-Hill (1999); 007-709506-5
22. Senkerik, R., Zelinka, I., Oplatkova, Z.: Optimal control of evolutionary synthesized chaotic system. In: Matousek, R. (ed.) 15th International Conference on Soft Computing MENDEL 2009, pp. 220–227 (2009) ISSN:1803-3814, ISBN: 978-80-214-3884-2
23. Senkerik, R., Zelinka, I., Davendra, D., Oplatkova, Z.: Utilization of SOMA and differential evolution for robust stabilization of chaotic Logistic equation. Computers & Mathematics with Applications 60(4), 1026–1037 (2010)
24. Senkerik, R., Oplatkova, Z., Zelinka, I., Davendra, D.: Evolutionary chaos controller synthesis for stabilizing chaotic Henon maps. Complex Systems 20(3), 205–214 (2012); 0891-2513
25. Senkerik, R., Oplatkova, Z., Zelinka, I., Davendra, D.: Synthesis of feedback controller for three selected chaotic systems by means of evolutionary techniques: analytic programming. Mathematical and Computer Modelling 57(1-2), 57–67 (2013); 0895-7177
26. Kominkova-Oplatkova, Z., Senkerik, R., Zelinka, I., Pluhacek, M.: Analytic programming in the task of evolutionary synthesis of a controller for high order oscillations stabilization of discrete chaotic systems. Computers & Mathematics with Applications (2013) (available online March 5, 2013)
27. Matousek, R.: GAHC: Improved GA with HC mutation. In: WCECS 2007, San Francisco, pp. 915–920 (2007)
28. Matousek, R.: GAHC: Hybrid Genetic Algorithm. In: Advances in Computational Algorithms and Data Analysis. LNEE, vol. 14, pp. 549–562 (2009)
29. Matousek, R.: HC12: The Principle of CUDA Implementation. In: 16th International Conference on Soft Computing, MENDEL 2010, Brno, pp. 303–308 (2010)
30. Matousek, R., Zampachova, E.: Promising GAHC and HC12 algorithms in global optimization tasks. Journal Optimization Methods & Software 26(3), 405–419 (2011)
31. Matousek, R., Minar, P.: Stabilization of chaotic logistic equation using HC12 and grammatical evolution. In: Zelinka, I., Chen, G., Rössler, O.E., Snasel, V., Abraham, A. (eds.) Nostradamus 2013: Prediction, Model. & Analysis. AISC, vol. 210, pp. 137–146. Springer, Heidelberg (2013)
32. Goldberg, D.E.: Genetic Algorithms in Search, Optimization and Machine Learning. Addison-Wesley Longman Publishing Co. Inc., Boston (1989)
33. O'Neill, M., Ryan, C.: Grammatical Evolution: Evolutionary Automatic Programming in an Arbitrary Language. Kluwer Academic Publishers (2003)
34. Koza, J.R.: Genetic Programming: On the Programming of Computers by Means of Natural Selection. The MIT Press (1992)
35. Doerner, R., Hübinger, B., Martienssen, W.: Adaptive orbit correction in chaos control. Int. J. of Bifurcation and Chaos 5, 1175–1179 (1995)
36. Al-shameri, W.F.H.: Dynamical Properties of the Hénon Mapping. Int. Journal of Math. Analysis 6(49), 2419–2430 (2012)
37. Sprott, J.C.: High-Dimensional Dynamics in the Delayed Hénon Map. EJTP 3(12), 19–35 (2006)
38. Starrett, J.: Time-optimal chaos control by center manifold targeting. Physical Review E 66, 046206 (2002)

# Possibility of Dynamical Process Linearization Using Piecewise-Linear Neural Network

Petr Dolezel and Martin Mariška

University of Pardubice
Studenska 95,
Pardubice, Czech Republic
petr.dolezel@upce.cz,
mariska.martin@gmail.com

**Abstract.** This paper presents a new technique for process identification. Since once nonlinear problem is described by piecewise-linear structure, it may be solved by many well-known techniques, the result of introduced technique provides a set of linear equations, where each of these equations is valid in some region of state space and together, they approximate whole nonlinear process. In the first five paragraphs, the technique is theoretically described and the paper is finished with demonstrative example.

**Keywords:** artificial neural network, linearization, piecewise-linear neural network.

## 1 Introduction

Piecewise-linear functions provide useful tool to deal with nonlinear problems. Once nonlinear problem is modeled by some piecewise-linear function, it is possible to divide it into a set of linear subproblems where each of them can be solved by some efficient algorithm. This idea was originally proposed in [1] and was expanded in [2] or [3]. Among others, this approach can be used in nonlinear process control design. Although majority of control loops is still based on PID control [4], some special control systems are a challenge to design using PID-like controllers. There are several reasons for this difficulty - processes to be controlled may be highly nonlinear, very complex or time-varying. Therefore, more sophisticated control strategies (i.e. adaptive control, robust control or predictive control) were presented in the second half of the twentieth century. Thus, large collection of control techniques, which can handle even highly nonlinear and complex processes, is available these days. However, most of these techniques require a precise mathematical model of the process to be controlled. In next paragraphs, there is introduced a technique, which can efficiently provide currently valid linear model of the process even if the process is highly nonlinear.

I. Zelinka et al. (eds.), *Nostradamus 2014: Prediction, Modeling and Analysis of Complex Systems*, Advances in Intelligent Systems and Computing 289,
DOI: 10.1007/978-3-319-07401-6_22, © Springer International Publishing Switzerland 2014

## 2 Problem Formulation

As mentioned in section one, a majority of recent control techniques require mathematical model of the process to be controlled. It means that a set of equations, which allows us to predict the future behavior of the process, is needed. In linear dynamic system identification, there are available several kinds of models, some of them fully stochastic (time series), others stochastic with exogenous input. Full widely accepted standard is established in [5]. However, in this paper, ARX model (Auto-Regresive model with eXogenous input) is considered - see eq. (1).

$$y(k) = \frac{B(z^{-1})}{A(z^{-1})}u(k) + \frac{1}{A(z^{-1})}v(k). \tag{1}$$

In equation above, $u$ is the input to the process, $y$ is the output, $v$ is the stochastic variable, $A$, $B$ are polynomials of complex variable $z^{-1}$ and $k$ is the discrete time. the crucial task is to determine the polynomials $B(z^{-1})$, $A(z^{-1})$, which are then used by a controller to define suitable control action see e.g. [6] for some possibilities of performing it. In addition, if the process is significantly nonlinear, the coefficients of both polynomials will shift depending on operating point. Therefore, the aim of the article is to introduce a methodology of determining the polynomials $B(z^{-1})$, $A(z^{-1})$ which are currently valid. The methodology is supposed to be efficient enough to be used online. Consider nonlinear SISO process that is to be controlled (Fig. 1). Then, the methodology should work as seen in Fig. 2.

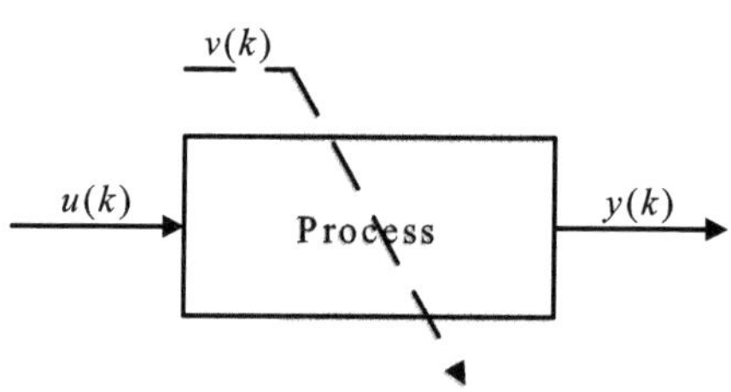


**Fig. 1.** SISO Process

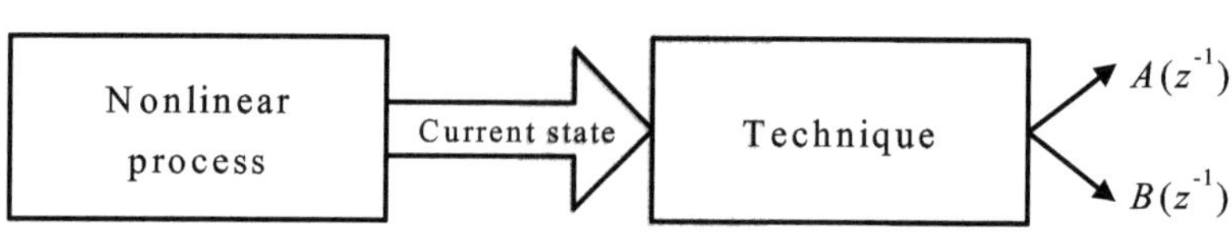


**Fig. 2.** The behavior of the methodology

The idea depicted above is not new, the way of determination of the polynomials $B(z^{-1})$, $A(z^{-1})$ is, anyway. It uses a special topology of artificial neural network and contrary to related techniques, it is computationally simple.

## 3 Artificial Neural Network for Universal Approximation

In 1983, Hornik proved [7] that standard multilayer feedforward network (MFN) with one hidden layer is capable of approximating any real measurable function to any desired degree of accuracy. Topology of MFN with one hidden layer is depicted in Fig. 3, where input layer brings external inputs $x_1$, $x_2$, , $x_P$, hidden layer contains $S$ neurons which process sums of weighted inputs and output neuron processes sum of weighted outputs from hidden neurons. Dataflow between input $i$ and hidden neuron $j$ is gained by weight ${w^1}_{j,i}$. Dataflow between hidden neuron $j$ and output neuron is gained by weight ${w^2}_{1,j}$. Neurons in hidden layer contain squashing activation function, while output neuron contains non-squashing activation function - see [7] for formal definition. For practical applications, continuous, bounded and monotonic activation function is used for neurons in hidden layer and continuous and monotonic activation function is used in output neuron - for some examples, see [8], [9].

Output of the network in Fig. 3 can be expressed by following equations.

$$ {y_a}^1_j = \sum_{i=1}^{P} {w^1}_{j,i}\ x_i + {w^1}_j \tag{2} $$

$$ {y^1}_j = \phi^1\left({y_a}^1_j\right) \tag{3} $$

$$ {y_a}^2_1 = \sum_{j=1}^{S} {w^1}_{1,j}\ {y^1}_j + {w^2}_1 \tag{4} $$

$$ y = \phi^2\left({y_a}^2_1\right) \tag{5} $$

In equations above, $\phi^1(.)$ means activation functions of hidden neurons and $\phi^2(.)$ means output neuron activation function. Apparently, the network has to be well trained to achieve sufficient approximation qualities. In other words, the network is to learn associations between a specified set of input-output pairs (training set). As there were presented many training techniques from simple back-propagation algorithm [10] to some specialized hybrid techniques using evolutionary algorithms [11], they are not defined here. However, the important note is that analytical derivatives of activation functions are required for training by any gradient-based technique.

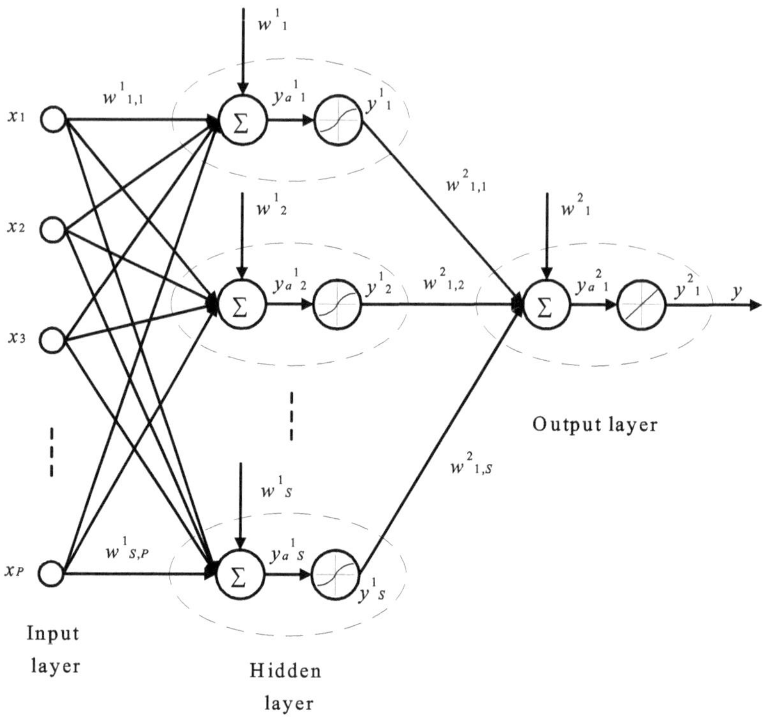


**Fig. 3.** Multilayer feedforward neural network with one hidden layer

# 4 Process Identification by MFN

Process identification is statistical procedure which leads to mathematical model of dynamical process from measured data. Let us narrow the problem down to identify the coefficients of the polynomials $B(z^{-1})$, $A(z^{-1})$, where

$$B(z^{-1}) = [0 + b_1 z^{-1} + b_2 z^{-2} + ... + b_m z^{-m}] \tag{6}$$

$$A(z^{-1}) = [1 + a_1 z^{-1} + a_2 z^{-2} + ... + a_n z^{-n}] \tag{7}$$

The deterministic part of the linear models with exogenous input can be illustrated as seen in Fig. 4.

Contrary to Fig. 4, process identification using ANN (whole procedure is defined in [8]) provides differently-shaped models. Those models are rarely usable for process control, but they are able to model even highly nonlinear processes for MFN are universal approximators. See Fig. 5 where the deterministic part of the NNARX (Neural Network Auto-Regresive model with eXogenous input) model is shown. NNARX model is widely used representative of neural models.

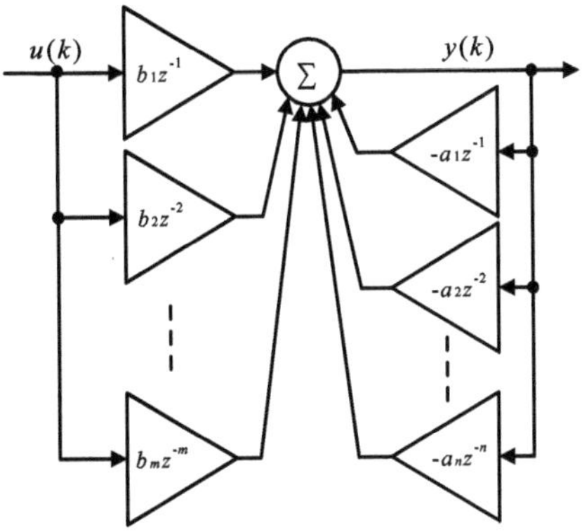


**Fig. 4.** Deterministic part of the models with exogenous input

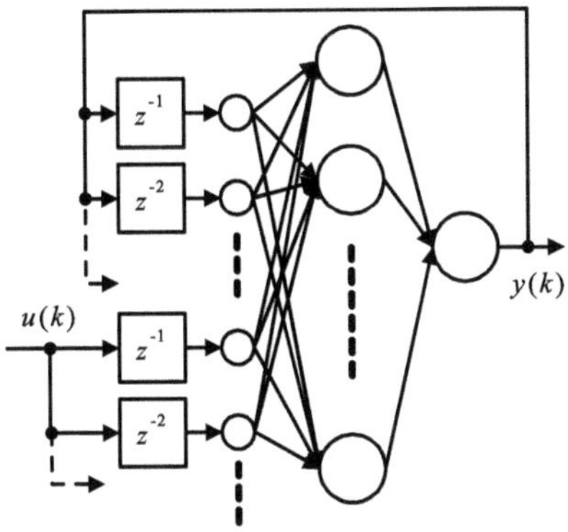


**Fig. 5.** NNARX model

Output of the MFN in Fig. 5 is determined by equations 2 - 5. The thing is that NNARX model is black-box-like structure and it cannot be directly used for control action evaluation. Thus, the procedure of transforming NNARX model into model described in Fig. 4 is proposed in next paragraph.

## 5 Piecewise-Linear Neural Model

As mentioned in section 3, MFN has to contain squashing activation function in hidden layer and non-squashing activation function in output neuron. In addition, these activation functions are expected to be differentiable through some gradient-based training technique can be used. In point of fact, only several types of activation function are really being applied. Actually, in most cases, hyperbolic-tangent or sigmoid activation function are used in hidden layer and linear activation function is used in output neuron - see Fig. 6.

This article offers another approach. It suggests replacing hyperbolic tangent activation function in hidden layer with linear saturated activation function (8).

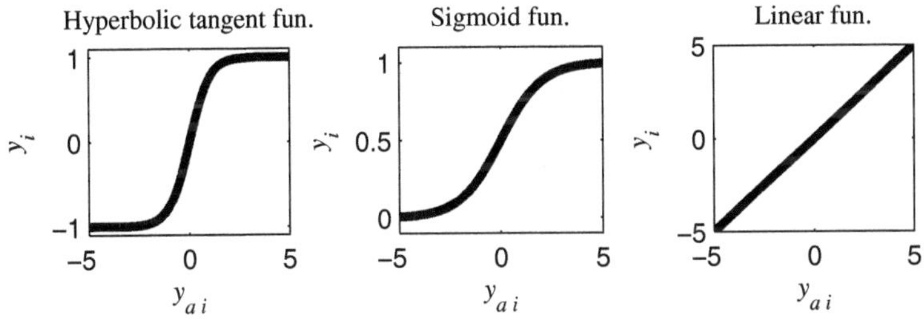


**Fig. 6.** Activation functions

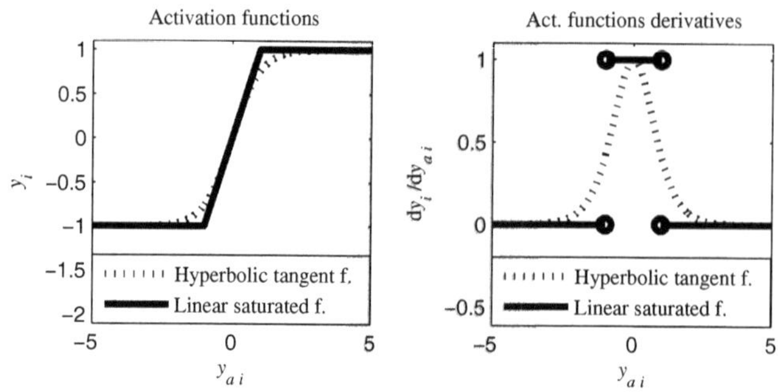


**Fig. 7.** Activation functions comparison

$$y_{\ i} = \begin{cases} 1 & \text{for} \quad y_{a\ i} > 1 \\ y_{a\ i} & \text{for} \ -1 \leq y_{a\ i} \leq 1 \\ -1 & \text{for} \quad y_{a\ i} < 1 \end{cases} \tag{8}$$

Although linear saturated activation function is not fully differentiable, MFN then becomes a piecewise-linear structure. Furthermore, MFN approximation qualities are expected to stay similar through both function resembling courses - see Fig. 7.

Let us presume an existence of NNARX model which uses MFN with linear saturated activation functions in hidden neurons and linear (identic) activation function in output neuron. Apparently, this model acts as piecewise-linear model and one linear submodel turns to another when any hidden neuron becomes saturated or becomes not saturated. Although the output of MFN used in this NNARX model can be evaluated by eqs. (2) - (5), another way for MFN output computing is useful. Let us define saturation vector $\mathbf{v}$ of $S$ elements. This vector indicates saturation states of hidden neurons - see (9).

$$v_i = \begin{cases} 1 & \text{for} \quad {y^1}_i = 1 \\ 0 & \text{for} \ -1 < {y^1}_i < 1 \\ -1 & \text{for} \quad {y^1}_i = -1 \end{cases} \quad i = 1, 2, ..., S. \tag{9}$$

Now, ANN output can be expressed by

$$y(k) = -\sum_{j=1}^{n} a_j y(k-j) + \sum_{j=1}^{m} b_j u(k-j) + c, \tag{10}$$

where

$$a_j = -\sum_{i=1}^{S} {w^2}_{1,i}(1 - |v_i|){w^1}_{i,j} \tag{11}$$

$$b_j = \sum_{i=1}^{S} {w^2}_{1,i}(1 - |v_i|){w^1}_{i,j+n} \tag{12}$$

$$c = w^2{}_1 + \sum_{i=1}^{S} \left( w^2{}_{1,i} v_i + (1 - |v_i|) w^2{}_{1,i} w^1{}_i \right) \tag{13}$$

Thus, difference equation (10) defines MFN output and it is linear in some neighbourhood of current state (in that neighbourhood, where saturation vector **v** stays constant). In other words, if the neural model of any nonlinear process in form of Fig. 5 is designed, then it is simple to determine parameters of linear difference equation which approximates process behaviour in some neighbourhood of current state. Last step is determining of the coefficients of the polynomials $B(z^{-1})$, $A(z^{-1})$ using parameters of equation (10). Let us define

$$\tilde{u}(k) = u(k) - u_0, \tag{14}$$

where $u_0$ is constant. Then, equation (10) turns into

$$y(k) = -\sum_{j=1}^{n} a_j y(k-j) + \sum_{j=1}^{m} b_j \tilde{u}(k-j) + c + \sum_{j=1}^{m} b_j u_0. \tag{15}$$

Equation (15) will become constant term free, if equation eq:u0 is satisfied.

$$u_0 = -\frac{c}{\sum_{j=1}^{m} b_j}. \tag{16}$$

Now, equation (15) can be written in following way.

$$y(k) = \frac{B(z^{-1})}{A(z^{-1})} \tilde{u}(k), \tag{17}$$

where coefficients of the polynomials $B(z^{-1})$, $A(z^{-1})$ are determined by equations (11) and (12), respectively. Equation (17) corresponds to the model in Fig. 4 with respect to equations (14) and (16). A comprehensive diagram of described technique and how it can be used for nonlinear process control is shown in Fig. 8 ($r$ is a set point).

## 6 Demonstrative Example

Described approach has been applied to several simulated models as well as to some laboratory devices. In this paper, let's consider a helicopter model. Helicopter model is twin rotor aerodynamic system (see Fig. 9) which is designed to simulate real copter dynamics. As a plant, it is significantly nonlinear system with two inputs (power of main rotor u and power of tail rotor) and two outputs (vertical elevation and yaw motion). All quantities are normalized to interval [-1; 1]. The point of this section is to design a piecewise-linear model of the vertical elevation part of the helicopter model.

Step-like excitation of the system to be modeled is shown in Fig. 10.

To divide this system into linear subsystems according the algorithm described above, it is necessary to design a neural model of the system in the shape of Fig.

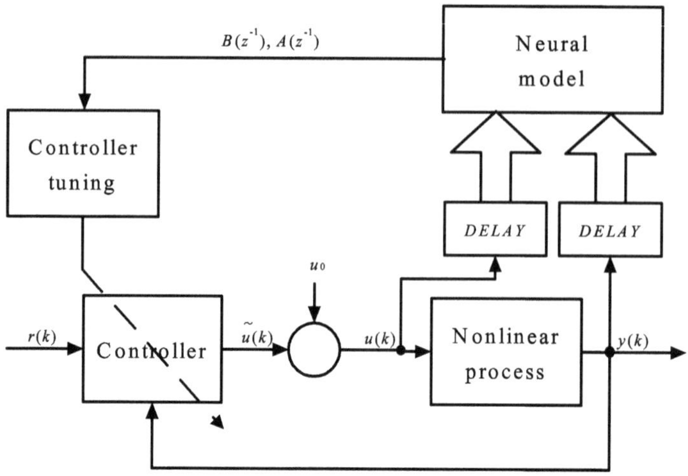


**Fig. 8.** Formal diagram of described technique used for process control

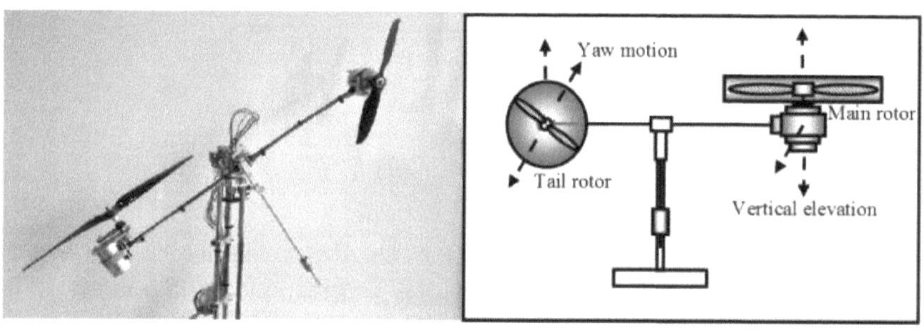


**Fig. 9.** Helicopter model

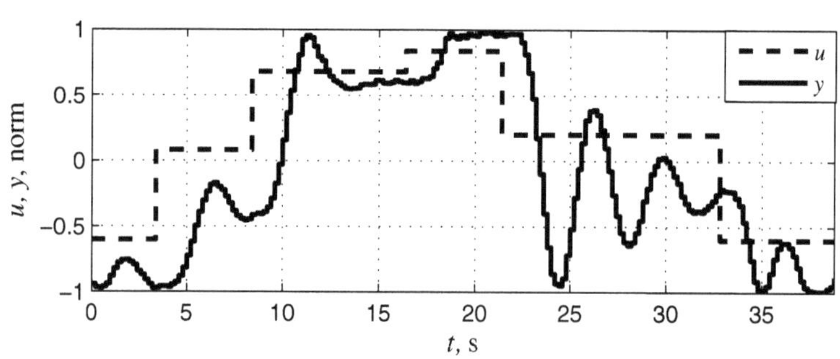


**Fig. 10.** Response to a sum of step functions

5, where neurons in hidden layer of MFN contain linear saturated activation function and output neuron contains linear (identic) activation function. For this particular example, the orders of the polynomials $B(z^{-1})$, $A(z^{-1})$ were chosen according to the eqs. (18), (19).

$$B(z^{-1}) = \left[0 + b_1 z^{-1}\right], \tag{18}$$

$$A(z^{-1}) = \left[1 + a_1 z^{-1} + a_2 z^{-2}\right], \tag{19}$$

Then, the procedure of neural model design involves training and testing set acquisition, neural network training and pruning and neural model validating. As this sequence of processes is illustrated closely in many other publications [8], [9], it is not referred here. Eventually, as the result of the procedure, neural model with four neurons (three in hidden layer, one in output layer) is designed. Except the number of neurons, the important results are the values of the parameters ${w^1}_{i,j}$, ${w^1}_i$, ${w^2}_{1,i}$, ${w^2}_1$, which are necessary for an evaluation of the coefficients $a_j$, $b_j$, and $c$ - see (11), (12) and (13). The resulting MFN is figured in Fig. 11.

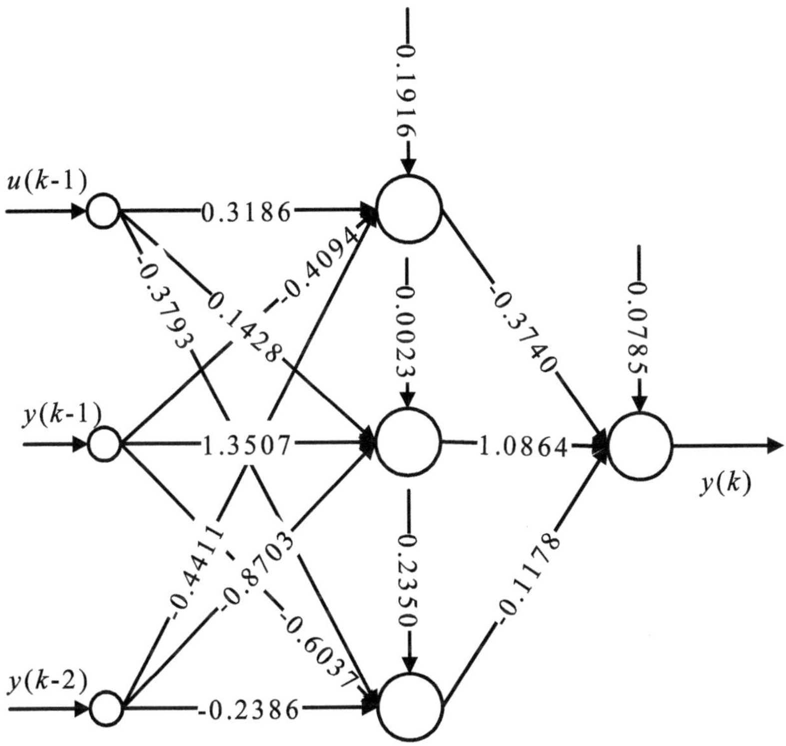


**Fig. 11.** The weights of resulting multilayer feedforward network

As described above, the structure in Fig. 10 can be easily transformed into equation (17). For this particular case, (17) becomes (20).

$$y(k) = \frac{\left[0 + b_1 z^{-1}\right]}{\left[1 + a_1 z^{-1} + a_2 z^{-2}\right]} \tilde{u}(k), \tag{20}$$

where

$$a_1 = -\sum_{i=1}^{3} w^2{}_{1,i}(1 - |v_i|) w^1{}_{i,1}, \tag{21}$$

$$a_2 = -\sum_{i=1}^{3} w^2{}_{1,i}(1 - |v_i|) w^1{}_{i,2}, \tag{22}$$

$$b_1 = \sum_{i=1}^{3} w^2{}_{1,i}(1 - |v_i|) w^1{}_{i,3}, \tag{23}$$

$$\tilde{u}(k) = u(k) + \frac{c}{b_1}, \tag{24}$$

$$c = w^2{}_1 + \sum_{i=1}^{3} \left( w^2{}_{1,i} v_i + (1 - |v_i|) w^2{}_{1,i} w^1{}_i \right) \tag{25}$$

Since there are three neurons in hidden layer of used MFN, vector **V** can potentially gather up to 27 states - see (9). Thus, there are 27 linear models stored in the structure showed in Fig. 11. Transitions between these linear models can be determined by solving the following set of equations.

$$\begin{matrix} y_a{}^1{}_i = -1 \\ y_a{}^1{}_i = 1 \end{matrix} \quad i = 1, 2, 3. \tag{26}$$

Using (2) and considering Fig. 11, equations (26) turn to

$$\begin{matrix} w^1{}_{i,1} u(k-1) + w^1{}_{i,2} y(k-1) + w^1{}_{i,3} y(k-2) + w^1{}_i = -1 \\ w^1{}_{i,1} u(k-1) + w^1{}_{i,2} y(k-1) + w^1{}_{i,3} y(k-2) + w^1{}_i = 1 \end{matrix} \quad i = 1, 2, 3. \tag{27}$$

Solving (27), there is gathered a map of regions, where each linear submodel is valid. The resulting map is shown in Fig. 12. Since the state variables are normalized, the reasonable state space is limited and does not contain all possible regions.

For better illustration, there is figured step-like response of the demonstrative system and its piecewise-linear model - see Fig.13.

In Fig. 13, there are marked and numbered the regions, in which particular linear models are used. The numbering corresponds with the numbers in the Tab. 1.

Hence, the technique is able to determine a set of linear models of the nonlinear process and it is even possible to compute the regions of validity of this linear models.

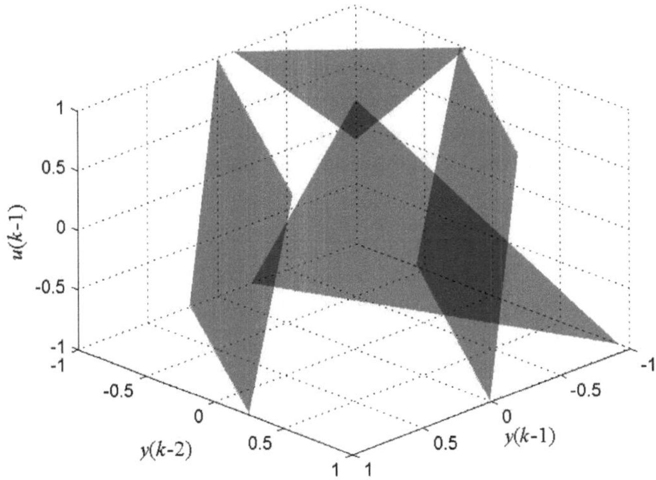


**Fig. 12.** The map of linear regions

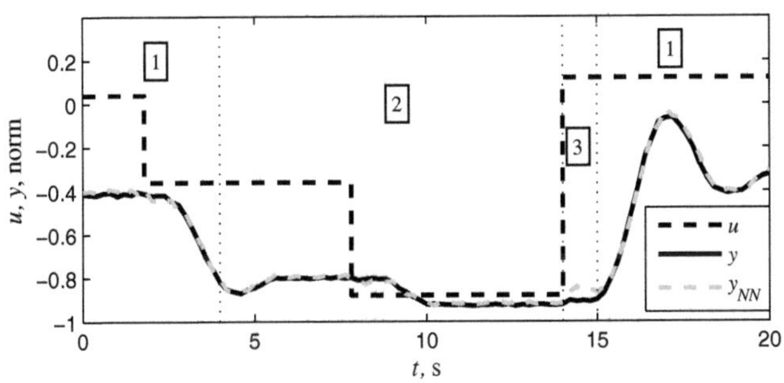


**Fig. 13.** The step-like response of the system and its piecewise-linear neural model

**Table 1.** Used linear submodels - see (20)

| Number | $b_1$ | $a_1$ | $a_2$ | $c$ |
|---|---|---|---|---|
| 1 | 0.0807 | -1.6916 | 0.7524 | -0.0969 |
| 2 | 0.0360 | -1.6204 | 0.7805 | -0.1870 |
| 3 | 0.1999 | -1.5385 | 0.9174 | -0.3993 |

## 7 Conclusions

The aim of this article is to introduce a new technique for process identification using piecewise-linear neural network. The most interesting feature of this approach is that the resulting model effectively stores finite number of linear models (each valid in some region of state space) and these linear models can be locally used e.g. for controller tuning. The technique can be used in two ways. It is possible to determine piecewise-linear model in a shape of Fig. 5 and this structure can be used online for continuous linearization of the process (which is very effective procedure), or the structure of the piecewise-linear model can be divided offline into set of linear models and comprehensively analyzed. Nevertheless, both possibilities bring decent advantages especially for nonlinear process control design.

**Acknowledgments.** The work has been supported by the funds of the IGA, University of Pardubice, Czech Republic. This support is very gratefully acknowledged.

## References

1. Chua, L., Kang, S.M.: Section-wise piecewise-linear functions: Canonical representation, properties, and applications. Proceedings of the IEEE 65, 915–929 (1977)
2. Huang, X., Xu, J., Wang, S.: Nonlinear system identification with continuous piecewise linear neural network. Neurocomputing 77, 167–177 (2012)
3. Breiman, L.: Hinging hyperplanes for regression, classification, and function approximation. IEEE Transactions on Information Theory 39, 999–1013 (1993)
4. Astrom, K., Hagglund, T.: PID controllers: theory, design and tuning. International Society for Measurement and Control, Durham (1995)
5. Ljung, L.: System identification - Theory for the User. Prentice-Hall, New Jersey (1999)
6. Bobal, V., Bohm, J., Fessl, J., Machacek, J.: Digital Self-tuning Controllers. Springer, London (2005)
7. Hornik, K., Stinchcombe, M., White, H.: Multilayer feedforward networks are universal approximators. Neural Networks 2, 359–366 (1989)
8. Haykin, S.: Neural Networks: A Comprehensive Foundation. Prentice-Hall, New Jersey (1999)
9. Nguyen, H., Prasad, N., Walker, C.: A First Course in Fuzzy and Neural Control. Chapman and Hall/CRC, Boca Raton (2003)
10. Rumelhart, D., Hinton, G., Williams, R.: Learning representations by back-propagating errors. Nature 323, 533–536 (1986)
11. Blanco, A., Delgado, M., Pegalajar, M.: A real-coded genetic algorithm for training recurrent neural networks. Neural Networks 14, 93–105 (2001)
12. Nelles, O.: Nonlinear System Identification. Springer, London (2001)

# Bilinear Time Series Model as an Alternative Way of Speaker Modeling*

Oskar Kochana, Patrycja Ksiazek, Michal Olszak, and Ewa Bielinska

The Silesian Technical University, Akademicka 14, 44-100 Gliwice, Poland

**Abstract.** In the paper a class of non-linear time series models is considered, with respect to possible application for speaker recognition. Registered speech signal is is a non-stationary time series. This non-stationarity is usually modeled as autoregressive time series with time varying parameters. In the paper a bilinear approximation of non-stationary autoregressive model is proposed. This way, a model with time varying parameters is approximated by a constant parameters model. Parameters of the bilinear model are assumed to be the speaker features,and are applied for speaker recognition. Effectiveness of the proposed method is compared with classic methods of speaker recognition.

**Keywords:** non-stationary AR models, bilinear time series model, speech analysis, speaker recognition recognition.

## 1 Introduction

In recent years technology development has been substantially adopting biological phenomena and applying them to different technical solutions. Modern engineering is trying to understand these phenomena and possibly model those, using more or less sophisticated mathematical formulas. To use effectively this knowledge appropriate algorithms and high computational power are needed, the possibilities of which increase greatly each year. One of these natural phenomena is human speech, which for several decades has been studied and, what is more, has been a focus of growing attention. Speech recognition and speaker recognition are branches of science that are associated with human speech and are highly related with each other. The speech signal carries important information such as linguistic, which are of interest of speech recognition, or features that may allow unambiguous identification of a speaker. Those features, unique for each person, are usually hard to find.

Registered utterance is a non-stationary time series that, for speech or a speaker recognition, should be represented by a model which mainly contains information either on the content of the utterance or on the specific features of the speaker.

* Publication supported from the Human Capital Operational Programme co-financed by the European Union from the financial resources of the European Social Fund, project no. POKL.04.01.02-00-209/11.

I. Zelinka et al. (eds.), *Nostradamus 2014: Prediction, Modeling and Analysis of Complex Systems*, Advances in Intelligent Systems and Computing 289,
DOI: 10.1007/978-3-319-07401-6_23, 

Non-stationary speech signals used to be modeled by set of locally stationary autoregressive time series models i.e. a threshold model. The reason why the AR models were chosen is that they define a linear regression When the local AR models are of the same structure, the threshold model may be considered as AR model with time varying parameters. Modeling and identification of non-stationary time series have been discussed by, e.g. [9], [7]. In case of speech analysis registered utterance is modeled as a non-stationary auto regressive (AR) series. Usually, to this aim the utterance is segmented into frames in which stationary AR(dA) time series models are identified. Respective models' parameters are then processed in order to obtain the speech or the speaker features. Such process is named feature extraction, a chosen set of features is often considered as a speaker model, and speaker recognition consists in feature matching [4], [5].

There are many different methods of feature extraction and feature matching. Some of them are shortly presented in the following section. In this paper a method of speaker identification proposed by Bielinska [3] is tested and compared with the other methods. The novelty of this method lies in an assumption that the vector of features characterizing the speaker is built of parameters of a bilinear model. To this aim non-stationary AR model is approximated by a constant parameters bilinear model.

## 2 Feature Extraction

Feature extraction is one of the most important steps in speaker recognition systems. It aims at obtaining such model that provides little variability of model's parameters for each particular speaker, and at the same time insure sufficiently high volatility of parameters values among all speakers.

The basic methods of speaker recognizing [1], [2] consist in comparison of the speaker's averaged features to a pattern. The speaker's averaged features are obtained by way of averaging the features over frame boxes, on which the registered signal is divided, during a test session. The pattern consists of equivalent averaged values, gathered for different speakers in process training. The individual techniques of the speaker's identification differ with the choice of the features vector and the way of distance measure between the vectors of features (the classifiers). The most commonly used speaker features are presented below.

### 2.1 Linear Predictive Coding

The idea of Linear Predictive Coding [12] is that the speech signal, $s(n)$, at time $n$, can be approximated as a linear combination of the past $p$ samples

$$s(n) \approx a_1 s(n-1) + a_2 s(n-2) + \cdots + a_p s(n-p) \tag{1}$$

If excitation term, $Gu(n)$, is included, the speech signal is given by:

$$s(n) = \sum_{i=1}^{p} a_i s(n-i) + Gu(n) \tag{2}$$

Respective Z-transform is:

$$S(z) = \sum_{i=1}^{p} a_i z^{-i} S(z) + Gu(z) \tag{3}$$

Hence, discrete transfer function of the vocal tract can be expressed as:

$$H(z) = \frac{S(z)}{GU(z)} = \frac{1}{1 - \sum_{i=1}^{p} a_i z^{-i}} = \frac{1}{A(z)} \tag{4}$$

Stimulation $u(n)$ of the system (2) is either quasi-periodic train of pulses (as excitation of a voiced sound), or a random noise sequence for unvoiced sounds. To determine the set of coefficients, $a_k$, first the value, $\hat{s}(n)$, is estimated from:

$$\hat{s}(n) = \sum_{i=1}^{p} a_i s(n-i) \tag{5}$$

Prediction error e(n) is defined as:

$$e(n) = s(n) - \hat{s}(n) = s(n) - \sum_{i=1}^{p} a_i s(n-i) \tag{6}$$

Owing to the fact that the coefficients: $a_1$, $a_2$, ... $a_p$ are assumed to be constant over short period of time, the speech signal should be first divided into overlapping frames of length 10-20 ms. In every frame vector of coefficients $a_k$ is obtained by minimization of sum of squares of the $e(n)$. Usually, $a_k$ parameters averaged over all the frames form LPC feature vector.

### 2.2 Repetitive Poles

The vocal tract model (4), estimated in each frame, may be represented equivalently as:

$$H(z) = \prod_{k=1}^{p} \frac{1}{1 - p_k z^{-1}} \tag{7}$$

It may be decomposed into two parts

$$H(z) = \prod_{k=1}^{q} \frac{1}{1 - p_k z^{-1}} \prod_{k=q+1}^{p} \frac{G}{1 - p_k z^{-1}} = H_s(z) H_u(z) \tag{8}$$

Poles $p_1$,, $p_q$ characterize the speaker, while poles $p_{q+1}$ ,, $p_p$ characterize the utterance. The feature vector is constructed from the most frequently repetitive poles, with respect to all considered frames.

### 2.3 Cepstrum

In time domain speech signal $s(n)$ is convolution of the stimulation $u(n)$ and an impulse response of the vocal tract $h(n)$. In frequency domain it is a product of corresponding Fourier transforms.

$$S(\omega) = U(\omega)H(\omega) \tag{9}$$

Taking logarithm of (9), the stimulation, which has parameters highly variable in time, can be separated from the vocal tract.

$$\ln S(\omega) = \ln U(\omega) + \ln H(\omega) \tag{10}$$

This is a very useful tool in speaker recognition, as well as in speech recognition.

Cepstrum $C$ of the signal $s(i)$ is defined as the inverse Fourier transform of the logarithm of the signal's spectrum

$$C \equiv \hat{S}(T) = F^{-1}\left(\ln|S(\omega)|\right) \tag{11}$$

Usually, the speaker feature vector C contains several initial cepstral coefficients, averaged over all frames.

### 2.4 LP Cepstrum

Alternatively, cepstral coefficients may be calculated by the following transformations of LP coefficients

$$\begin{aligned} c_1 &= a_1 \\ c_k &= a_k + \sum_{m=1}^{k-1} \tfrac{m}{k} c_m a_{k-m}, \text{for} 2 \leq k \leq dA \\ c_k &= \sum_{m=1}^{k-1} \tfrac{m}{k} c_m a_{k-m}, \text{fordA} + 1 \leq k \leq q \end{aligned} \tag{12}$$

### 2.5 Bilinear Representation of Non-stationary Autoregressive Time Series

In all above mentioned methods non-stationary speech signal was divided into frames. Inside the frame signal was assumed stationary, and was represented by a stationary AR time series model. Set of models' parameters was then processed to obtain vector of features.The new idea presented in [3] lies in representing the whole utterance by a bilinear time series model. Vector of features is built of the bilinear model parameters. Bilinear models are powerful nonlinear models, and a large class of nonlinear systems, including Volterra systems, can be well approximated using this model.

Non stationary time series $s(i)$, for $i = 1 \ldots N$ can be represented as autoregressive $AR(dA)$ time series with time varying parameters.

$$s(i) = \sum_{j=1}^{dA} a_j(i)s(i-j) + \nu(i) \tag{13}$$

where $a_j(i)$ for $j = 1 \ldots dA$ are time varying parameters of the $AR(dA)$ model, and $\nu(i)$ is a series of errors, distributed $N(0, \sigma^2)$.

It was proven in [3] that under the following assumptions:

1. parameters $a_j(i)$ change in the following way

$$a_j(i) = a_j(i-1) + \Delta_j(i) \tag{14}$$

2. $\Delta_j$ is a random innovation signal, with the moving average $MA(dC)$ representation

$$\Delta_j(i) = e_j(i) + \sum_{l=1}^{dC} c_{l,j} e_j(i-l) \tag{15}$$

where $e_j(i)$, $j = 1 \dots dA$ are identically independently distributed normal variables, with mean 0 and variance $\lambda_j^2$, and $c_{l,j}$ are constant, real parameters for $j = 1 \dots dA$ and $l = 1 \dots dC$

3. initial value at time 0 is

$$a_j(0) = \Delta_j(0) \tag{16}$$

signal $s(i)$ described by 13 can be expressed as

$$s(i) = \sum_{j=1}^{dA} \left( s(i-j)e_j^*(i) + c_{1,j}e_j^*(i-1)s(i-j) + \dots + c_{dC,j}e_j^*(i-dC)s(i-j) \right) + \nu(i) \tag{17}$$

where

$$e_j^*(i) = \sum_{k=0}^{i} e_j(k) \tag{18}$$

$$e_j^*(i-l) = \sum_{k=0}^{i} e_j(k-l) \tag{19}$$

Hence, the non-stationary signal $s(i)$ represented previously by $dA$ time varying coefficients $a_j(i)$, under assumptions 14–16, can be represented by $(dA \times dC)$ time constant coefficients $c_{j,k}$ of the bilinear model (17). Method of estimation of the bilinear model parameters proposed in [3] uses technique of MA(dC) identification presented in [10].

## 3 Application of Bilinear Time Series Model for Speaker Recognition

The reported research primarily focused on testing effectiveness of the speaker recognition on the base of the bilinear model of speaker.

Procedure of speaker recognition took place in two stages. In a learning stage a data base for speakers was created. After loading the tested utterance, unvoiced parts of the signal were removed. Particular speaker was represented in a data base by the set of models identified for a number of utterances of the same speaker. In this research there were 5 utterances used as a learning set for each of 10 speakers. In a testing stage, bilinear time series model of a current speaker was identified from a single utterance devoid of unvoiced parts. After that, the speech signal was classified to one of the speakers using either K-Nearest Neighbor algorithm or Maximum Likelihood This procedure was repeated for 5 utterances for each of 10 speakers. Therefore, there was total of 50 trials. Finally the ratio of correct assignments to the number of all tests gives effectiveness.

### 3.1 Project Establishment

All utterances were recorded with the use of built-in microphone (Conexant SmartAudio HD) in a laptop using MATLAB software. The microphone enables two channels recording with up to 24 bits and 96000 Hz sampling frequency. Most of energy of speech signal is contained in 5 Hz up to 4 kHz range so the recording frequency was chosen to 8 kHz. Besides, one channel recording, the data range from -1 to 1, and resolution 16 bits per sample were applied.

The speakers base was collected from volunteer speakers and all utterances were recorded in polish language. The database was divided into:

- various fragments of texts read,
- unique sentence for each speaker,
- common word, "Brzczyszczykiewicz",
- common sentence, "Ewentualnie reagujemy na zagroenie",

In each experiment ten persons took part producing five male and five female utterances, and ten recordings per person, per experiment. Therefore, there was total of four hundred utterances.

### 3.2 Experiment Setup

Resultant bilinear model depends on several initial conditions that should be assumed, therefore influence of the following aspects has been checked:

- values of dA and dC in 17,
- method of estimation MA model 15,
- specific utterances,
- classification method,

In the Fig.1 effectiveness of speaker recognition is shown as a function of dA and dC. Presented results were obtained fot the utterance: "Ewentualnie reagujemy na zagrozenie". The plots shows that the best results were obtained for dA = 4 and dC =3, therefore, in next experiments such values were used.

In the Fig.2 to the left, influence of different methods of estimation of moving average parameters on effectiveness of speaker recognition was checked [13]. The following methods were considered:

- Durbin's method [6],[8]
- Iterative method [3], [11]
- predefined MATLAB function - armax.

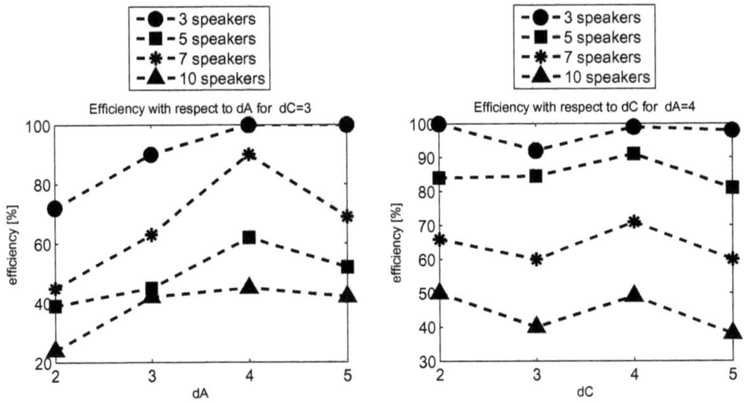


**Fig. 1.** Efficiency with respect to dA (left) and dC (right)

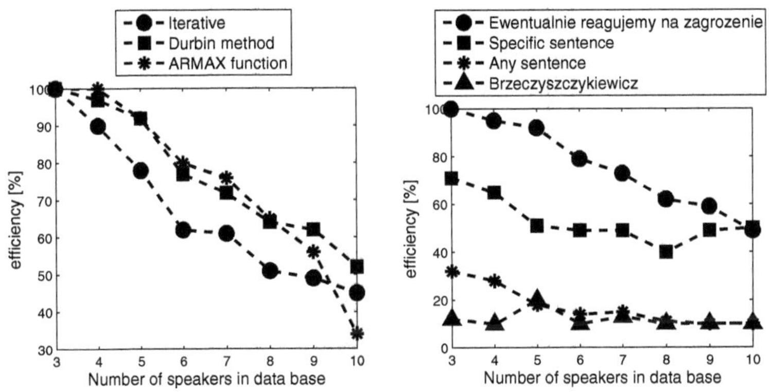


**Fig. 2.** Efficiency for different methods of MA parametrization (left) and for different input speech (right)

It can be seen that all of the methods give similar results but it is worth mentioning that iterative procedure of finding parameters of MA(dC) is much more time consuming then the Durbin's method or the armax function.

The next step of research was checking the effectiveness of the identification with respect to an input utterance, which each time was identified in a closed set - this means every utterance was assigned to one speaker. There were experiments performed as text-independent, the recordings used read sentences, each time different for every person. There were also experiments performed as text-dependent, where speech signal consisted of a defined word, a defined sentence or a sentence unique for each speaker. In the Fig.2 to the right, can be clearly seen that the input speech has significant impact on effectiveness of the method. The utterance: "Ewentualnie reagujemy na zagroenie" gives the best performance because it contains most voiced sounds. In further experiments this speech signal was used to test the method.

Effectiveness of speaker recognition with the use of bilinear speaker model has been tested for different classifications methods. In this project K-nearest Neighbors algorithm was examined with use of two different distance measures: Euclidean distance

$$d(x, y) = \sqrt{\sum_{i=1}^{N} (x_i - y_i)^2} \tag{20}$$

or Mahalonobis distance

$$d(x, y) = \sqrt{\sum_{i=1}^{N} \frac{(x_i - y_i)^2}{s_i^2}} \tag{21}$$

and compared with Maximum Likelihood method. Speaker recognition with use of Maximum Likelihood method proved to be about ten percent better than with the use of K-Nearest Neighbors algorithm in the set of 10 speakers. The application of a more sophisticated pattern comparison technique may enhance the recognition even more.

### 3.3 Filtering

Since bilinear speaker model uses parameters received using recursive method, which may be affected by high frequencies components, the system was examined to check how an application of a low-pass filter influences the results. The

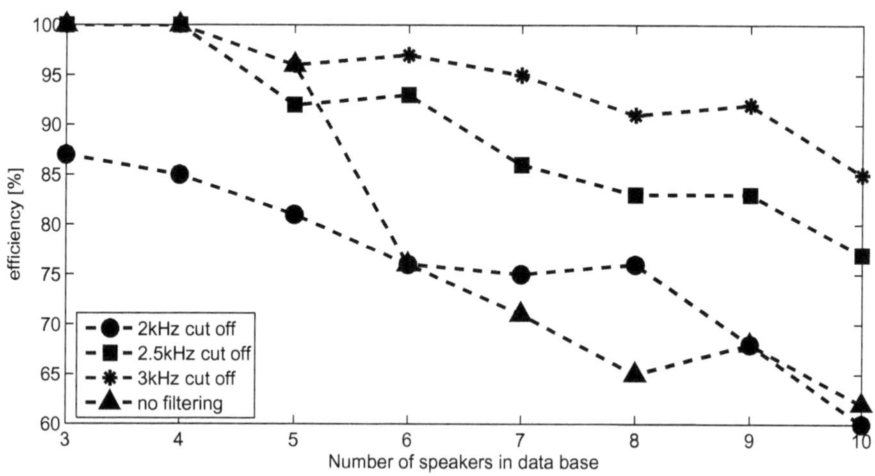


**Fig. 3.** Effectiveness with low-pass filter

filter of order 10 was designed with a use of a predefined MATLAB function $fdesign.lowpass$.

In Fig.3 one can observe that a difference between effectiveness of speaker recognition, performed on the filtered and unfiltered speech signal reaches 20%, which is a meaningful improvement.

### 3.4 Comparison with Other Methods

**Cepstral Coefficients.** The tested speech signal was divided into 32 ms frames overlapped every 8 ms. Each frame was windowed with Hamming window and first 20 Cepstral Coeffcients were calculated according to the formula 12. Ceptral Coeffcients averaged over all frames form a feature vector of a speaker. The pattern recognition technique used was Maximum Likelihood method. In the experiment the sentence: "*Ewentualnie reagujemy na zagrozenie*" was tested. Effectiveness with the use of the Cepstral Coeffcients was about 15 % better than effectiveness with the use of the Bilinear Model.

**Linear Predictive Coding.** The tested speech signal was divided into 64 ms frames overlapped every 16 ms. In every frame autoregressive parameters of A(z) were calculated using MATLAB function *lpc*. The parameters averaged over all frames formed a feature vector of a speaker, corresponding to particular utterance. Recognition efficiency with the use of the LPC coeffcients found to be comparable with the method based on the bilinear model coefficients.

## 4 Conclusions

Previous studies allow the following conclusion:

Bilinear time series models proved to be useful for speaker recognition. Estimation of fixed parameters of bilinear model requires less computational than estimation of the time-varying AR model parameters. However, effective use of the recognition based on bilinear models involves the determination of several parameters, which usually are arbitrary chosen, and there is no simple way of choosing right ones. Among these parameters are: orders of polynomials A(z) and C(z), parameters of the recursive identification method i.e. initial conditions, forgetting factor which has a significant impact on tracking performance, different way of signal filtering, or various techniques of pattern recognition.

Speaker recognition efficiency strongly depends on the phonetic contents of utterance. It was observed that voiced signals have given much better performance of the system. Inclusion of algorithms that detects voiced speech into the speaker recognition system may increase its efficiency.

## References

1. Atal, B.S.: Effectiveness of linear prediction characteristics of the speech wave for automatic speaker identification and verification. J. Acoust. Soc. Am. 55, 1304–1312 (1974)

2. Furui, S.: Recent advances in speaker recognition. Pattern Recognition Letters 18(9), 859–872 (1997)
3. Bielinska, E.M.: Bilinear representation of non-stationary autoregressive time series. In: Proceedings of the International Conference on System Science 2013. Advances in Systems Science, pp. 737–746. Springer, Heidelberg (2013)
4. Bensty, J., Sondhi, M., Huang, Y. (eds.): Springer Handbook of Speech Processing. Springer, Heidelberg (2007)
5. Campbell, J.: Speaker recognition: A Tutorial. Proceeding of the IEEE 85(9) (1977)
6. Durbin, J.: Effcient estimation of parameters in moving-average models. Biometrika 46, 306–316 (1959)
7. Kohlmorgen, J., Lemm, S.: An On-lLine method for segmentation and identification of non-stationary time series. In: Proceeding of Neural Networks for Signal Processing XI, pp. 113–122 (2001)
8. Ludwig, M.: Building on Durbin's method to estimate MA processe. Improving Durbin's method to estimate MA processes, arXiv:1304.7956, `http://mludwig.org/research.html`
9. Ni, H., Yin, H.: Self-organising mixture autoregressive model for non-stationary time series modelling. Int. Journal of Neural Systems. Dec. 18(6), 469–480 (2008)
10. Ozaki, T., Tong, H.: On moving average parameter estimation. In: Proceedings of the 8th Hawaii International Conference on System Science, pp. 224–226 (1995)
11. Pollock, D.S.G.: A Handbook of time series analysis, signal processing and dynamics. Academic Press (1999)
12. Rabiner, L.R.: Fundamentals of Speech Recognitions. Prentice Hall, New Jersey (1993)
13. Sandgren, N., Stoica, P., Babu, P.: On moving average parameter estimation. In: Proceedings of the 20th European Signal Processing Conference (EUSIPCO), pp. 2348–2351 (2012)

# On the Growth of Large Independent Sets in Scale-Free Networks

David Chalupa and Jiří Pospíchal

Institute of Applied Informatics
Faculty of Informatics and Information Technologies
Slovak University of Technology
Ilkovičova 2, 842 16 Bratislava, Slovakia
{chalupa,pospichal}@fiit.stuba.sk

**Abstract.** Independent sets are sets of mutually non-adjacent vertices of a network. For example, independent sets in social networks represent groups of people, who do not know anybody else within the group. In this paper, we investigate the growth of large independent sets in the famous Barabási-Albert (BA) model of scale-free complex networks. We formulate recurrent relations describing the cardinality of typical large independent sets and show that this cardinality seems to scale linearly with network size. This holds not only for the original BA model, where a new vertex brings a constant number of edges to the network, but also when the number of incoming edges is just bounded from above by a constant. Our finding is of a fundamental importance for community detection problems, since vertices of an independent set are naturally unlikely to belong to the same community. In other words, the number of communities in scale-free networks seems to be bounded from below by a linear function of network size.

**Keywords:** independent sets, complex networks, scale-free networks.

## 1 Introduction and Related Work

Social and biological networks, the World Wide Web, research citation networks or language networks represent only several examples of *complex networks*, which are known to exhibit very specific properties [4] and represent a modern research topic with interesting applications such as evolutionary process visualization [6] or time series prediction [5]. It is widely believed that many real-world networks are scale-free. Such networks have a *power-law degree distribution* in the form $P(k) \sim ck^{-\gamma}$, where $P(k)$ is the degree distribution function, which denotes the fraction of vertices with $k$ neighbors, $\gamma$ is a constant exponent describing a particular network [1] and $c$ is a suitable constant. Simply put, many vertices in a scale-free network have only a few neighbors but there are also several vertices with many neighbors. The number of neighbors $k$ is often referred to as the *degree of a vertex.*

Scale-free networks have now been a very popular research topic for more than a decade [3]. However, even before the discovery of scale-free property, many

I. Zelinka et al. (eds.), *Nostradamus 2014: Prediction, Modeling and Analysis of Complex Systems*, Advances in Intelligent Systems and Computing 289,
DOI: 10.1007/978-3-319-07401-6_24, 

complex networks have been known to exhibit the property of *small world*. In a small world, even though the network might be very large, the shortest path between two arbitrary vertices will be much smaller, i.e. the shortest path length will be a sublinear function of the number of vertices [13]. The notoriously well-known example is the "six degrees of separation" theory, which states that two arbitrary people on Earth might reach each other by traversing six edges of the social network of all people.

The aforementioned specificities are related to the *community structure* of complex networks. These networks naturally consist of communities, i.e. dense areas of the networks, which are connected only sparsely. Presumably, connections in the sparse areas are used heavily in shortest paths between vertices of different communities. However, detecting the communities is a non-trivial problem, especially due to difficulties in suitable problem formulation and determining the number of communities to actually search for in the network [10,12].

In this work, we focus on a rigorous study of independent sets in complex networks. Let $G = [V, E]$ be an undirected graph formalizing the complex network. Then, *independent set* $IS$ is a subgraph induced by subset $V_{IS} \subseteq V$ of the network's vertices, which have no edges between each other. In other words, it is an induced subgraph with no pair of mutually adjacent vertices. An independent set, which is not a subset of any other independent set is called *maximal*. Independent set with the largest cardinality is called *maximum independent set*.

Independent sets are naturally related to communities in complex networks. Even though estimating the number of communities to search for is a non-trivial task in community detection, it is reasonable to assume that vertices of independent sets are unlikely to lie within the same community. The computational complexity of searching for independent sets strongly depends on whether maximum or only maximal independent set is searched for. While the maximum independent set problem is NP-hard [9], the maximal independent set problem can be solved in polynomial time [11]. For the NP-hard maximum independent set problem, efficient approximation and heuristic algorithms exist, such as the greedy approximation algorithm [8], GRASP algorithm [7] or local search algorithms [2].

In this paper, we investigate the growth of maximal independent sets in the well-known Barabási-Albert (BA) model of scale-free networks. This model simulates growth of an artificial network. It is described in pseudocode of Algorithm 1. BA model begins with a connected graph and at each time step, one vertex comes to the network and brings $w$ edges with it. These edges are attached to the existing vertices using the preferential attachment rule, which means that the probability of choosing a vertex for attachment is proportional to its current degree. BA model leads to networks with degree distributions, for which it holds that $P(k) \sim k^{-3}$. In the following, we show that even maximal independent sets seem to scale linearly with network size. This implies that also maximum independent set scales linearly and the number of communities in the network is also bounded from below by a linear function of network size.

*Algorithm 1: Barabási-Albert model of scale-free networks*

begin with a connected graph $G_0 = [V_0, E_0]$, $n_0 = |V_0|$, $m_0 = |E_0|$
for $t = 1...(n - n_0)$
  let $V_t = V_{t-1} \cup \{v_t\}$, $n_t = n_{t-1} + 1$
  attach $v_t$ to $w$ vertices from $V_{t-1}$ in the way that the probability of attachment to vertex $v \in V_{t-1}$ is $(\deg(v))_{t-1}/(2m_{t-1})$,
  let $m_t = m_{t-1} + w$

The paper is organized as follows. In Section 2, we formulate the recurrent relations describing the growth of typical maximal independent sets in BA model. In Section 3, we show the empirically determined solutions for the recurrences and propose techniques to solve them analytically. Finally, in Section 4, we give conclusions and a brief discussion of the results.

## 2 Recurrent Relations Describing the Growth of Independent Sets in BA Model

In our investigations, we begin with studying the behavior of a variable $\alpha_L$, which denotes the expected size of a typical maximal independent set in BA model. In the following, we derive recurrent relations, which describe how $\alpha_L$ behaves when the network grows.

### 2.1 Growth of a Typical Maximal Independent Set and Affinity of a New Vertex towards the Independent Set

Let us have an initial connected graph on $n_0$ vertices and $m_0$ edges. At each time step $t$, a vertex comes to the graph and brings a constant number of $w$ edges, which are attached based on the preferential attachment rule. Thus, apparently, in time step $t$, the number of vertices is $n_t = n_0 + t$ and the number of edges is $m_t = m_0 + wt$.

We will now study the properties of a set $IS_t$, which denotes some typical maximal independent set that is prone to be generated in BA model. We will have a sequence $\{\alpha_{L,t}\}_{t=0}^{\infty}$, which denotes the cardinality of such an independent set as the network grows, i.e. $\alpha_{L,t} = |IS_t|$. For the initial value, we naturally have that $\alpha_{L,0} \geq 1$, since one vertex always forms an independent set. A more general bound depends on the initial graph, which is not fixed in the model.

Now, we have to consider the affinity of the new vertex towards the set $IS_t$. For each of the $w$ new edges, the probability of attachment to an existing vertex in the independent set is proportional to the sum of the degrees of vertices in $IS_t$. This measure is called *volume* and will be denoted by $Vol(IS_t) = \sum_{v \in IS_t} (\deg(v))_t$.

The probability of choice of a vertex from $IS_t$ for attachment is $Vol(IS_t)/(2m_t)$. With this in mind, we can see that the recurrent relation describing $\alpha_{L,t+1}$ is:

$$\alpha_{L,t+1} = \alpha_{L,t} + \left(1 - \frac{Vol(IS_t)}{2m_t}\right)^w, \tag{1}$$

since for all $w$ new edges, it must hold that they were attached to vertices outside of the set $IS_t$. Then, the new vertex can be a part of $IS_{t+1}$.

### 2.2 Volume of the Independent Set and Its Growth

In the recurrence for $\alpha_{L,t}$, the crucial question is how $Vol(IS_t)$ behaves. This can be described by recurrent relations, too. The initial value will be $Vol(IS_0)$ and is also dependent on the initial graph. The most general lower bound for this value clearly is $Vol(IS_0) \geq 1$, since the graph is guaranteed to be connected.

We suppose that a vertex comes in time step $t$. Generally, there are two ways how the volume of a fixed independent set can increase.

- We consider the general scenario. The expected number of edges generated from the new vertex to the vertices of $IS_{t+1}$ is $w\frac{Vol(IS_t)}{2m_t}$. This is because the fraction in the value is the probability that the edge will be attached to a vertex in the independent set and there are $w$ possibilities.
- If no edges were attached to the vertices of $IS_t$, then the new vertex will join the independent set. This event occurs with probability $\left(1 - \frac{Vol(IS_t)}{2m_t}\right)^w$. With this probability, the volume of the independent set will be incremented by $w$, since all $w$ new edges will go "outside" of the independent set.

We note that these two sources of incrementation for $Vol(IS_t)$ can be treated as independent. In each of these two cases, there is a different event that causes $Vol(IS_{t+1})$ to be higher than $Vol(IS_t)$. The first case describes the expected contribution to volume by incoming edges at each time step. The second case describes the event that occurs when the actual contribution of the first case is zero. Hence, the recurrent relation describing the behavior of $Vol(IS_{t+1})$ is:

$$Vol(IS_{t+1}) = Vol(IS_t) + w\frac{Vol(IS_t)}{2m_t} + w\left(1 - \frac{Vol(IS_t)}{2m_t}\right)^w. \tag{2}$$

### 2.3 Relative Connectivity of the Independent Set to the Rest of the Graph

Interesting properties of the previous equation can be observed, when we perform substitution $y_t := \frac{Vol(IS_t)}{2m_t}$. Thus, we will work with the fraction of edge endpoints that are in the independent set, rather than the volume (i.e. a relative measure rather than absolute). We will call $y_t$ the *relative connectivity of the independent set* to rest of the graph. In fact, $y_t$ describes the probability that an edge connects to the independent set. As a consequence, $(1 - y_t)^w$ is in fact the *affinity of the new vertex* towards the independent set. This affinity will be the expected contribution to the size of the independent set per one time step.

After the substitution and after we use the fact that $m_t = m_0 + wt$, we obtain the following:

$$2(m_0 + wt + w)y_{t+1} = 2(m_0 + wt)y_t + w[y_t + (1 - y_t)^w]. \tag{3}$$

By tidying the previous equation up, we have:

$$y_{t+1} = \left(1 - \frac{w}{m_0 + wt + w}\right) y_t + \frac{w}{2(m_0 + wt + w)}[y_t + (1 - y_t)^w]. \tag{4}$$

In the following, we will first empirically illustrate how $y_t$ behaves. Consequently, we will prove rigorous results on asymptotical behavior of $y_t$ and show how this recurrence can be solved.

## 3 Solutions to the Recurrent Relations and Techniques to Find Them

We now focus on empirical and analytical study of $y_t$. As a consequence, the affinity of a vertex towards the independent set, i.e. $(1 - y_t)^w$, will also be determined. The following results will establish that both the size of a typical maximal independent set and maximum independent set grow as linear functions of $t$, if $t$ is large enough or a suitable initial graph is chosen.

### 3.1 Empirical Study of the Recurrent Relations

In Fig. 1, we provide the plots for the relative connectivity $y_t$ and the affinity $(1 - y_t)^w$ of the new vertex towards the maximal independent set for values of $w$ between 1 and 6. In these experiments, we set $y_0 = 0.5$. In fact, this value does not seem to influence the value, where $y_t$ tends to converge (for $w = 1$ and 2, we will explain this also analytically). However, $y_0$ might influence, whether $y_t$ reaches this final value in finite time.

From Fig. 1, it is interesting to see that $y_t$ approaches an almost constant value already after ca. 20 iterations. For $w = 1$, the affinity $(1 - y_t)^w$ is a constant function at value 1/2. This naturally makes sense, since for $w = 1$, BA model generates trees and trees are bipartite graphs. On average, one can expect such graphs to have these partitions of roughly equal size. For $w > 1$, the situation is a bit more interesting, but difficult as well. Already for $w = 2$, the value, where $y_t$ would remain constant, will be an irrational number. Thus, for larger values of $w$, $y_t$ seems to just converge to the stable value. In the next section, we show how to approach this recurrence analytically.

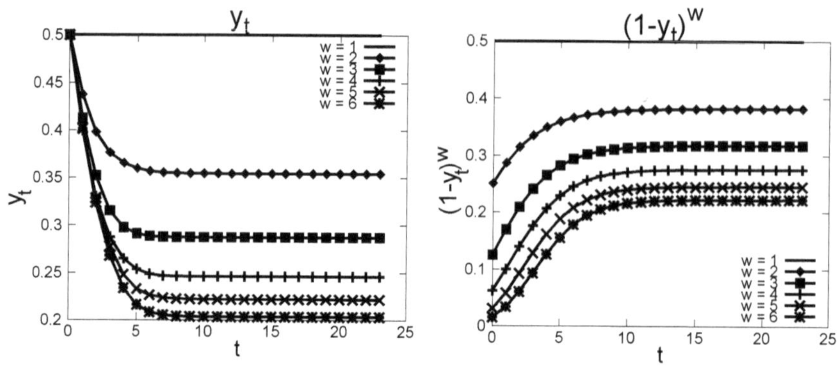


**Fig. 1.** The plots of behavior of the relative connectivity of the independent set $y_t$ (on the left-hand size) and the affinity of the new vertex towards the independent set (on the right-hand side) as functions of the time $t$ (horizontal axes contain $t$, vertical axes contain the respective values).

### 3.2 Analytical Study of the Recurrent Relations

From the basic recurrence on $\alpha_{L,t}$ in (1), we clearly have that:

$$\alpha_{L,t} = \alpha_{L,0} + \sum_{s=0}^{t-1} (1 - y_t)^w. \tag{5}$$

We will now show that for $y_t$, we will have a borderline value $y$, where $y_t$ would stabilize, if it reached $y$ in finite time. For values higher than $y$, $y_t$ will tend to decrease and for values lower than $y$, $y_t$ will tend to increase. Additionally, if $y_t$ reaches the stable value, then the typical maximal independent set will surely scale linearly in the future, since the affinity of new vertices towards the independent set will remain constant.

**Theorem 1.** Let us have a graph generated by BA model with fixed $w$ and a time step $t_0$. For this time step, let us have the value $A$ such that $A = -1/2y_{t_0} + 1/2(1 - y_{t_0})^w$.

If $A > 0$, then $y_{t+1} > y_t$. If $A < 0$, then $y_{t+1} < y_t$. If $A = 0$, then for all values $t \geq t_0$, it will hold that $y_t = y_{t_0}$.

**Proof.** We prove this by considering how formula (4) influences $y_{t+1}$. Apparently, $y_{t+1}$ is determined by first multiplying $y_t$ by $\left(1 - \frac{w}{m_0+wt+w}\right)$, which is in the interval $[0, 1]$ and approaches 1 as $t$ grows. Thus, this is the factor, which lowers the value by $\frac{w}{m_0+wt+w} y_t$. Then, we have the additive factor $\frac{w}{2(m_0+wt+w)} [y_t + (1 - y_t)^w]$.

Hence, the difference $y_{t+1} - y_t$ is:

$$y_{t+1} - y_t = \frac{w}{m_0 + wt + w}[-y_t + 1/2[y_t + (1 - y_t)^w]]. \tag{6}$$

Since $\frac{w}{m_0+wt+w} > 0$, the crucial factor, which determines whether $y_{t+1}$ will be higher or lower than $y_t$, is the one in the parentheses. This will be the factor $A$ we establish. Clearly, if $A > 0$, then $y_{t+1} - y_t > 0$ (for $<$ and $=$, we have analogous outcomes). Therefore:

$$A = -1/2y_t + 1/2(1 - y_t)^w. \blacksquare \tag{7}$$

Theorem 1 shows that once $y_t$ gets stabilized, the typical independent set will scale linearly in expectation. We can also analytically find the values, where $y_t$ and $(1 - y_t)$ are stable. We will denote these values simply by $y$ and $(1 - y)^w$ (since they do not depend on $t$). These values are found by just putting $w$ equal to the respective value and solving the equations from Theorem 1, i.e. we search for $y = y_t$ such that the difference $y_{t+1} - y_t = 0$. This leads to polynomial equations of degree $w$.

Additionally, the stable value $y$ of our recurrence has the interesting property that $y = (1-y)^w$. In other words, once the system gets stabilized, the affinity of a new vertex towards the independent set will actually be equal to the relative connectivity of the independent set to the rest of the graph.

**Theorem 2.** In the stable value $y$ for $y_t$, it holds that $y = (1 - y)^w$.

**Proof.** From Theorem 1, we have that $y_t$ is stable for $t \geq t_0$ if and only if $A = -1/2y_{t_0} + 1/2(1 - y_{t_0})^w = 0$. This clearly implies that $y_{t_0} = (1 - y_{t_0})^w$. $\blacksquare$

For $w = 1$, 2 and 3, the stable values for our recurrence are given by Corollary 1.

**Corollary 1.** For BA model with $w = 1$, it holds that $y = 1/2$. For $w = 2$, it holds that $y = (3 - \sqrt{5})/2$. For $w = 3$, we have that $y = z + 1 - \frac{1}{3z}$, where $z = \sqrt[3]{\sqrt{93}/18 - 1/2}$, i.e. $y \approx 0.31767$. For any $w$, typical maximal independent size is $yt + c$, where $c > 0$ is a suitable constant.

For $w \geq 2$, a problem seems to be that $y_t$ would stabilize in an irrational value. Hence, we can neither assume that the initial graph contains the independent set of the typical size, nor expect that $y_t$ can reach $y$ for finite $t$. All operations in the recurrence are rational. Thus, $y_t$ are rational values for any finite $t$.

To overcome this, we need to show that $\lim_{t\to\infty} y_t = y$. For simplicity, we demonstrate the idea of such a proof for the case when $w = 2$ and $y_0 > y$. We will also assume that the initial graph is simple enough, we will simply take two vertices joined by a single edge, i.e. $m_0 = 1$ and $y_0 = 1/2$. The chosen initial graph represents no practical limitation of the model and simplifies the required proof. The ideas of the proof are quite straightforward to apply to other scenarios, anyway.

Let us have a graph generated by BA model with $w = 2$. Furthermore, let $y_0 = 1/2$ and $m_0 = 1$. We will now show that for this case, it holds that $\lim_{t\to\infty} y_t = y$.

The idea is as follows. From Theorem 1, we have that if $y_t > y$, then $y_{t+1} < y_t$. We know that $y_0 > y$, i.e. $y_t$ "comes" from above. To guarantee that $y_{t+1}$ cannot get lower than $y$, we have to show that $y_t > y \Rightarrow y_{t+1} > y$. This way, we can guarantee that for every $t$, it holds that $y_t > y_{t+1} > y$. This already implies that for any $\epsilon > 0$, the sequence will approach value $y + \epsilon$ at some time, which establishes that the limit is $y$.

For $w = 2$, we consider the difference $y_{t+1} - y_t$ and assume that $y_t \leq 1/2$. We have to find $t$ and $y_t$ such that:

$$y_{t+1} = y_t + \frac{1}{m_0 + 2t + 2}[1 - 3y_t + y_t^2] > y. \tag{8}$$

We tidy up the inequality and separate $t$ from the rest of the factors:

$$\frac{-y_t^2 + (1 - m_0)y_t + (m_0 y + 2y - 1)}{2(y_t - y)} < t. \tag{9}$$

Hence, if $t$ exceeds the value on the left-hand side, we have that our statement is true. We will now simplify the situation by using that $m_0 = 1$ and we will also directly apply that $y = (3 - \sqrt{5})/2$. We obtain:

$$\frac{-y_t^2 + 7/2 - 3/2\sqrt{5}}{2(y_t - (3 - \sqrt{5})/2)} < t. \tag{10}$$

One can easily show that $7/2 - 3/2\sqrt{5} = ((3 - \sqrt{5})/2)^2$. In other words, the numerator will be equal to zero exactly for $y_t = (3 - \sqrt{5})/2$. According to our premises, $y_t > (3 - \sqrt{5})/2$. This implies that the numerator $-y_t^2 + 7/2 - 3/2\sqrt{5}$ is negative. The denominator $2(y_t - (3 - \sqrt{5})/2)$ is positive. Thus, we simply obtain that our statement holds if $t > 0$.

The case when $y_t > 1/2$ does not have to be handled here, since $y_0 = 1/2$. For all values of $t$, it holds that $y_t > y_{t+1} > y$, which proves that the limit is $y$.

We note that the arguments seem to be extensible also to the cases, when $y_0 < y$ or values $w > 2$. However, we have taken the case of the simple initial graph with $y_0 \geq y$, since our aim is to investigate what emerges in the network over time, anyway.

The previous results are important because of their consequences. We now consider $\alpha$ as functions of the number of vertices $n$, rather than time $t$, and bound the size of maximum independent set $\alpha_n$. For $w = 1$, Theorem 1 implies that if we take a graph with $y_0 = 1/2$ as the initial graph, then it holds that $\alpha_n \geq \alpha_{L,n} = 1/2n$.

For $w = 2$, we choose the simplest initial graph with $m_0 = 1$. Since $y_t$ does not reach $y$ in finite time, we fix two constants $\delta, \epsilon > 0$ and upper bound $y_t$ by $y + \epsilon$ for $t \geq t_0$. The second constant $\delta$ will be used to deal with the few initial iterations, where $y_t$ still stabilizes and it is not yet reasonable to bound it below by a constant. Additionally, it also includes the constant factor caused by the fact that in (5), the summation ends for $t-1$. By using this, we have that for $n \geq n_0 + t_0$, it holds that $\alpha_n \geq \alpha_{L,n} = (((3-\sqrt{5})/-2-\epsilon)^2)n - \delta$. This is a simple linear function of the network size.

# 4 Conclusions and Discussion

We investigated the growth of large independent sets in scale-free complex networks. Independent sets are sets of mutually non-adjacent vertices. In the context of social networks, independent sets represent groups of people, in which nobody knows each other. Practically, the size of large independent sets is an interesting estimator for the number of communities within the social network, since the vertices of the independent set are naturally unlikely to belong to the same community.

For the well-known Barabási-Albert (BA) model of scale-free networks, we showed that the size of large independent sets seems to scale linearly with network size, if the network is large enough. This is because the affinity of a new vertex towards a large independent set seems to quickly stabilize closely to a certain value. This value is denoted by $(1-y)^w$, where $w$ is the number of incoming edges per vertex, which is fixed to a constant in BA model, and $y$ is a stable value for a recurrent relation we derived. Interestingly, the values obtained in the recurrence seem to stabilize very quickly in the beginning of the growing process, as soon as approximately 20 vertices join the network.

We also developed techniques to solve the recurrent equations analytically. We showed how to derive polynomial equations of degree $w$ for the stable value $y$ of the recurrence. Interestingly, in the stable value, it holds that $y = (1-y)^w$, i.e. relative connectivity of the independent set to the rest of the graph and the affinity of new vertices towards the independent set stabilize at the same value. For $w \geq 2$, the affinity of a new vertex towards the independent set never reaches the stable value, since the stable value is an irrational number and the affinity should be rational by definition. However, we showed on an example of a growing network with $w = 2$ and a simple initial state that the affinity converges to the stable value $y$ over time. These ideas seem to be generalizable.

We believe that our results are beneficial in several ways. Firstly, using the knowledge about the process of growth of the network might be a way to better understand community detection problems. For example, we can see that in BA model, a reasonable number of communities to search for is a linear function of the network size. Secondly, our results imply that the average size of such a reasonable community in BA model must be a constant, if we assume that there are no growing independent sets inside the communities. This is an outcome definitely worth confronting with other models of growing networks, as well as with practice. It seems natural that communities should be small, even in very large networks. We note that networks generated by BA model have both a power-law degree distribution and the small-world property. Therefore, we can treat them as quite good "estimators" of how the real-world networks are created. On the other hand, an open question remains, whether there is also a mechanism other than growth and preferential attachment, which influences the number of communities and their sizes in the real world.

**Acknowledgement.** This contribution was supported by Grant Agency VEGA SR under the grants 1/0553/12 and 1/0458/13.

## References

1. Albert, R., Barabási, A.L.: Statistical mechanics of complex networks. Rev. Mod. Phys. 74(1), 47–97 (2002)
2. Andrade, D.V., Resende, M.G.C., Werneck, R.F.: Fast local search for the maximum independent set problem. J. Heuristics 18(4), 525–547 (2012)
3. Barabási, A.L.: Scale-free networks: A decade and beyond. Science 325(5939), 412–413 (2009)
4. Barabási, A.L., Albert, R.: Emergence of scaling in random networks. Science 286(5439), 509–512 (1999)
5. Burguillo, J.C., Dorronsoro, B.: Using complex network topologies and self-organizing maps for time series prediction. In: Zelinka, I., Chen, G., Rössler, O.E., Snasel, V., Abraham, A. (eds.) Nostradamus 2013: Prediction, Model. & Analysis. AISC, vol. 210, pp. 323–332. Springer, Heidelberg (2013)
6. Dubec, P., Plucar, J., Rapant, L.: Case study of evolutionary process visualization using complex networks. In: Zelinka, I., Chen, G., Rössler, O.E., Snasel, V., Abraham, A. (eds.) Nostradamus 2013: Prediction, Model. & Analysis. AISC, vol. 210, pp. 125–135. Springer, Heidelberg (2013)
7. Feo, T.A., Resende, M.G.C., Smith, S.H.: A greedy randomized adaptive search procedure for maximum independent set. Oper. Res. 42(5), 860–878 (1994)
8. Halldórsson, M.M., Radhakrishnan, J.: Greed is good: Approximating independent sets in sparse and bounded-degree graphs. Algorithmica 18(1), 145–163 (1997)
9. Karp, R.M.: Reducibility among combinatorial problems. In: Miller, R., Thatcher, J. (eds.) Proceedings of a Symposium on the Complexity of Computer Computations, pp. 85–103. Plenum Press, New York (1972)
10. Leskovec, J., Lang, K.J., Mahoney, M.W.: Empirical comparison of algorithms for network community detection. In: Rappa, M., Jones, P., Freire, J., Chakrabarti, S. (eds.) Proceedings of the 19th International Conference on World Wide Web, WWW 2010, pp. 631–640. ACM, New York (2010)
11. Luby, M.: A simple parallel algorithm for the maximal independent set problem. In: Proceedings of the Seventeenth Annual ACM Symposium on Theory of Computing, STOC 1985, pp. 1–10. ACM, New York (1985)
12. Schaeffer, S.E.: Graph clustering. Comput. Sci. Rev. 1(1), 27–64 (2007)
13. Watts, D.J.: Small Worlds. Princeton University Press (1999)

# Time Series Prediction Based on Averaging Values via Neural Networks

Eva Volna and Martin Kotyrba

University of Ostrava, 30.dubna 22, 70103 Ostrava, Czech Republic

**Abstract.** This paper introduces a development method for time series prediction based on averaging values. The experimental part will focus on teaching more neural networks with the same topology and settings that will solve the same problem (time series). The resulting values for training and test set are averaged depending on how many neural networks are involved in the calculation. The experimental part is focused on testing of periodic time series with different topologies and neural network settings. The results of prediction are compared with ARIMA models.

## 1 Introduction

The advantage of using neural networks for prediction is that they are able to learn from examples only and once their learning is finished, they are able to catch hidden and strongly non-linear dependencies, even when there is a significant noise in the training set. We introduce the topic of prediction using artificial neural networks. In particular, prediction of time series using multi-layer feed-forward neural networks will be described. In this paper, we will try to discuss one particular type of neural networks called backpropagation networks. It is the most popular network for practical applications and a very powerful tool. You can become familiar with our results with the backpropagation algorithm and its modification and various application uses.

## 2 Backpropagation Neural Networks

A neural network is a parallel, distributed information processing structure consisting of processing elements (which can possess a local memory and can carry out localized information processing operations) interconnected together with unidirectional signal channels called connections. Each processing element has a single output connection which branches into as many collateral connections as desired (each carrying the same signal - the processing element output signal). The processing element output signal can be of any mathematical type desired. All of the processing that goes on within each processing element must be completely local: i.e., it must depend only upon the current values of the input signals arriving at the processing element via impinging connections and upon values stored in the processing element's a local memory [1].

I. Zelinka et al. (eds.), *Nostradamus 2014: Prediction, Modeling and Analysis of Complex Systems*, Advances in Intelligent Systems and Computing 289,
DOI: 10.1007/978-3-319-07401-6_25, 

The backpropagation neural network architecture is a hierarchical design consisting of fully interconnected layers or rows of processing units (with each unit itself comprised of several individual processing elements. Backpropagation belongs to the class of mapping neural network architectures and therefore the information processing function that it carries out is the approximation of a bounded mapping or function $f : A \subset R^n \rightarrow R^m$, from a compact subset $\boldsymbol{A}$ of $n$-dimensional Euclidean space to a bounded subset $f[\boldsymbol{A}]$ of $m$-dimensional Euclidean space, by means of training on examples $(x_1, z_1), (x_2, z_2), \ldots, (x_k, z_k)$. . . . It will always be assumed that such examples of a mapping $f$ are generated by selecting $\boldsymbol{x_k}$ vectors randomly from $\boldsymbol{A}$ in accordance with a fixed probability density function $p(\boldsymbol{x})$. The operational use to which the network is to be put after training also assumed to involve random selections of input vectors $\boldsymbol{x}$ in accordance with $p(\boldsymbol{x})$. The backpropagation architecture described in this paper is the basic, classical version (Fig. 3). The backpropagation learning algorithm is composed of two procedures: (a) feed-forward and (b) back-propagation weight training [1].

*Feed-forward.* Assume that each input factor in the input layer is denoted by $x_i$, the $y_j$ and $z_k$ represent the output in the hidden layer and the output layer, respectively. And, the $y_j$ and $z_k$ can be expressed as follows (1):

$$y_j = f(X_j) = f\left(w_{oj} + \sum_{i=1}^{I} w_{ij} x_i\right) \text{ and } z_k = f(Y_k) = f\left(w_{ok} + \sum_{j=1}^{J} w_{ik} y_j\right) \quad (1)$$

where the $w_{0j}$ and $w_{0k}$ are the bias weights for setting threshold values, $f$ is the activation function used in both hidden and output layers, and $X_j$ and $Y_k$ are the temporarily computing results before applying activation function $f$. In this study, a sigmoid function is selected as the activation function. Therefore, the actual outputs $y_j$ and $z_k$ in hidden and output layers, respectively, can be also written as:

$$y_j = f(X_j) = \frac{1}{1+e^{-X_j}} \text{ and } z_k = f(Y_k) = \frac{1}{1+e^{-Y_k}} \quad (2)$$

The activation function $f$ introduces the non-linear effect to the network and maps the result of computation to a domain (0, 1). This sigmoid function is differentiable. The derivative of the sigmoid function in eq. (2) can be easily derived as: $f' = f(1 + -f)$ .*Back-propagation weight training.* The error function is defined as [3], (3):

$$E = \frac{1}{2}\sum_{k=1}^{K} e_k^2 = \sum_{k=1}^{K} (t_k - z_k)^2 \quad (3)$$

Where $t_k$ is a predefined network output (or desired output or target value) and $e_k$ is the error in each output node. The goal is to minimize $E$ so that the weight in each link is accordingly adjusted and the final output can match the desired output. To get the weight adjustment, the gradient descent strategy is employed. In the link between hidden and output layers, computing the partial derivative of $E$ with respect to the weight $w_{jk}$ produces, as (4)

$$\frac{\partial E}{\partial w_{jk}} = -e_k f'^{(Y_k)y_j} = -\delta_k y_j \text{ where } \delta_k = (t_k - z_k) f'(Y_k) \tag{4}$$

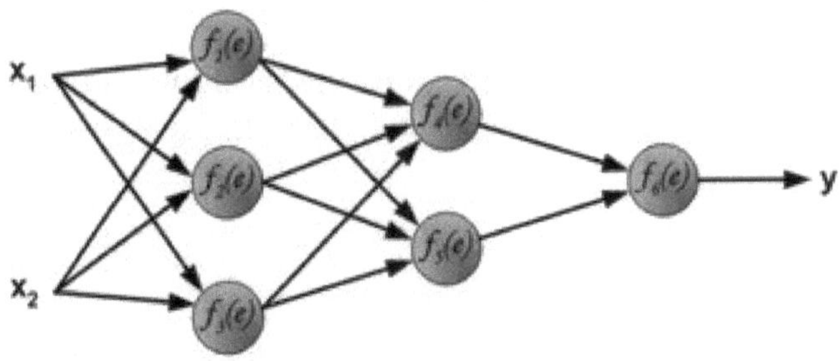


**Fig. 1.** A backpropagation network architecture

The weight adjustment in the link between hidden and output layers is computed by $\Delta w_{jk} = \alpha \times y_j \times \delta_k$, where $\alpha$is the learning rate, a positive constant between0 and 1. The new weight herein can be updated by the following $w_{jk}(n+1) = w_{jk}(n) + \Delta w_{jk}(n)$, where $n$ is the number of iteration. Similarly, the error gradient in links between input and hidden layers can be obtained by taking the partial derivative with respect to $w_{ij}$, as (5):

$$\frac{\partial E}{\partial w_{ij}} = -\Delta_j x_j = f'(X_j) \sum_{k=1}^{K} \partial_k w_{jk} \tag{5}$$

The new weight in the hidden-input links can be now corrected as: $\Delta w_{ij} = \alpha \times x_i \times \Delta_j$ and $w_{ij}(n+1) = w_{ij}(n) + \Delta_j$. Training the BP-networks with many samples is sometimes a time-consuming task. The learning speed can be improved by introducing the momentum term $\eta$ [4]. Usually, $\eta$ falls in the range $\langle 0,1 \rangle$. For the iteration n, the weight change Δ w can be expressed. The back-propagation learning algorithm used inartificial neural networks is shown in many text books [1, 2, 4, 8, 9].

## 3 Our New Approach Based on Averaging Values

First, the results of experiments on the prediction of time series using neural networks with hyperbolic tangent transfer function are described. When evaluating the predicted values of this function, the predicted value oscillates around the true value. Even during completion of the calculation, the value sometimes remains above or under the real value. It is caused by the fact that the weights of neural network are set pseudo-randomly at the beginning of the calculation. It means that "randomly" set weights are sometimes set so well that the neural network may not learn much at all because it already knows the problem or it already has an in-built nonlinear dependence, which the given problem contains(time series ). On the other hand, it may sometimes happen that the weights of neural network are set in a bad way so that the neural network is not able to learn the problem at the same number of epochs or it learns it with a larger error. It means that the same topology and the same initial setup of a neural network (learning coefficient, number of epochs) always leads to different learning on the training set and the prediction of the test set . Sometimes the predicted value is above the real value and sometimes the predicted value after completion of the calculation is under the real value (see Fig. 2), [5, 6, 7].

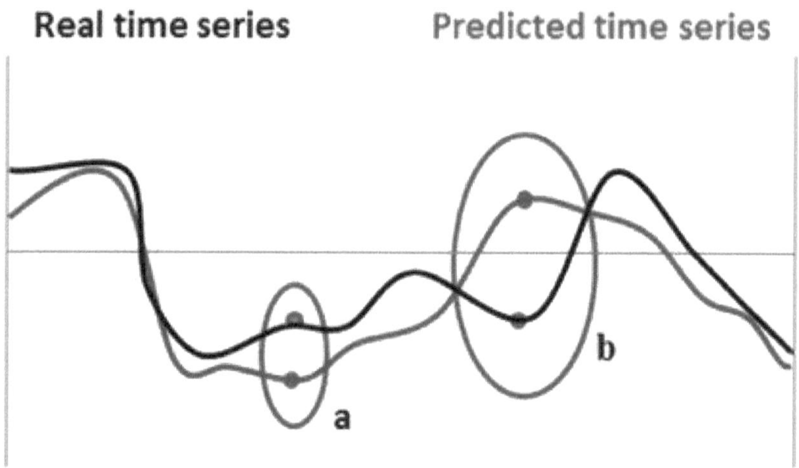


**Fig. 2.** Predicted values above (b) and under (a) the real value

Therefore, we solve the problem of teaching more neural networks. Networks do not learn the problem in parallel, but in serial, i.e.: when the the first neural network learns, next network starts learning, etc. These neural networks have the same topology and configuration (number of epochs, learning coefficient), but each of them has randomly initialized weights. When all the networks learn and test, averaging the values of the training and test set of all such networks occurs. Let us have "n" neural networks which solve the problem (prediction of the time series). To obtain the final field, averages apply this formula (6):

$$means(p) = \frac{\sum_{i=1}^{n}(n(i).a(p).Y)}{k}, p \in (1; number\ of\ patterns) \tag{6}$$

where ***Means*** () - is the field that stores the averaged results, ***a*** () - is the predicted field patterns of the network, ***n*** () - is the array of networks that solves the problem (time series), ***p*** - represents the integer from one to the number of patterns (a training and test set together), ***k*** - is the constant representing the number of sites (eg: in my experiments it is five).

Or take the first pattern (from the training set) of neural networks (if you have five neural networks, so we have five models) and all values are added up and divided by the number of the neural networks that are involved in the calculation (adds up to five values of five neural networks corresponding to the first pattern and will share five). Next, take the second pattern and perform the same operation, and so on until we reach the end of the training set. Do the same on the test set and obtain the resulting field diameters. And finally, this field is used to calculate the errors relative to the actual values. The error of the neural network is a sum of differences in the predicted values and actual values according to the formula (7):

$$E_p = \frac{1}{2} \sum_{i=1}^{number\ of\ patterns} (means(i).Y - real(i).Y)^2 \qquad (7)$$

$E_p$ = error which arises after averaging the corresponding values of all the neural networks that are involved in the calculation.

## 4 Experimental Study

The experimental part included 883 tests on a periodic time series of time consuming approximately 32 hours. It used the time series of the periodic character downloaded from [1].

### 4.1 The Adaptation Phase

The learning coefficient is set from 0.1 to 0.5 in increments of 0.05. The number of the input neurons is set from four to ten for periodic time series. The number of the output neurons is always one. The number of neurons in the hidden layer is set from four to ten. The number of the hidden layers is set from one to two. The number of neural networks solving the given problem is five. The maximum number of epochs devoted to the problem is set to two thousand. The second limiting condition is that if the error on the training set falls below 0.01, the calculation ends. All neural networks solving the problem have the same topology and configuration (coefficient learning, the size of the input size of hidden layers, number of hidden layers, and size of the output). Data containing 101 pieces of information and is divided into two sets. The training set contains 91 values. The test set contains ten values. The training set serves each of the five networks to learn the time series and the test set is used to estimate values from which it then calculates the error Ep, [5, 6, 7].

### 4.2 Size of the Window of the Time Series

Results were sorted by the error on the training set Ep and the top ten results were selected. At periodic time series, the neural network achieves very small errors Ep below 0.01.

**Table 1.** Top ten results for 4 input neurons

| Number of input neurons | Coefficient of learning | Number of hidden neurons | $E_p$ - training | $E_p$ - test |
|---|---|---|---|---|
| 4 | 0.2 | 10 | 0.003519023 | 0.000651 |
| 4 | 0.2 | 8 | 0.003672255 | 0.000401 |
| 4 | 0.15 | 9 | 0.005061269 | 0.001903 |
| 4 | 0.2 | 9 | 0.005250529 | 0.000796 |
| 4 | 0.1 | 7 | 0.005390971 | 0.002024 |
| 4 | 0.25 | 7 | 0.006265312 | 0.001151 |
| 4 | 0.2 | 6 | 0.006643008 | 0.000875 |
| 4 | 0.15 | 10 | 0.006816136 | 0.002397 |
| 4 | 0.15 | 6 | 0.007003977 | 0.000801 |
| 4 | 0.15 | 8 | 0.007248934 | 0.000723 |

For best results with four input neurons (see Table 1) the learning coefficient ranges between 0.1 and 0.25. The number of neurons in the hidden layer is higher than six and the number of hidden layers is always one. Ep error on the training set is less than 0.01.

**Table 2.** Top ten results for7 input neurons

| Number of input neurons | Coefficient of learning | Number of hidden neurons | $E_p$ - training | $E_p$ - test |
|---|---|---|---|---|
| 7 | 0.15 | 10 | 0.005131046 | 0.000648 |
| 7 | 0.1 | 7 | 0.006259021 | 0.000437 |
| 7 | 0.15 | 9 | 0.006313337 | 0.001768 |
| 7 | 0.15 | 7 | 0.007306655 | 0.000666 |
| 7 | 0.1 | 10 | 0.007980901 | 0.001015 |
| 7 | 0.1 | 9 | 0.008092436 | 0.000365 |
| 7 | 0.2 | 9 | 0.011187745 | 0.000556 |
| 7 | 0.15 | 8 | 0.014573762 | 0.001803 |
| 7 | 0.2 | 8 | 0.015802574 | 0.003287 |
| 7 | 0.25 | 8 | 0.034600117 | 0.006913 |

The seven input neurons (see Table 2) have the coefficient of learning ranging between 0.1 0.25. The number of neurons in the hidden layer is higher than seven. The number of the hidden layers and one error on the training set Ep in the case of 6 results is less than 0.01.

**Table 3.** Top ten resultsfor10 input neurons

| Number of input neurons | Coefficient of learning | Number of hidden neurons | $E_p$ - training | $E_p$ - test |
|---|---|---|---|---|
| 10 | 0.2 | 8 | 0.003113092 | 0.001083 |
| 10 | 0.15 | 10 | 0.004574624 | 0.003054 |
| 10 | 0.15 | 8 | 0.004822277 | 0.000335 |
| 10 | 0.15 | 9 | 0.005008596 | 0.001954 |
| 10 | 0.15 | 7 | 0.006241835 | 0.004688 |
| 10 | 0.1 | 9 | 0.008004524 | 0.001274 |
| 10 | 0.1 | 10 | 0.008335235 | 0.002329 |
| 10 | 0.2 | 10 | 0.021063054 | 0.002463 |
| 10 | 0.1 | 8 | 0.026847242 | 0.003149 |
| 10 | 0.2 | 9 | 0.039218375 | 0.007717 |

At ten neurons (see Table 3) the learning coefficient is from 0.1 to 0.2. The number of the hidden neurons is higher than seven. The number of the hidden layers is one Ep and the average error on the training set in the case of seven results is less than 0.01.

**Table 4.** Result error table

| Number of input neu-rons | Coefficient of learning | Number of hidden neurons | Number of results that have error $E_p$ smaller than 0.01 |
|---|---|---|---|
| 4 | 0.1 - 0.25 | 6 - 10 | 10 |
| 7 | 0.1 - 0.25 | 7 - 10 | 6 |
| 10 | 0.1 - 0.2 | 7 - 10 | 7 |

In summary, Table 4 shows the results for periodic time series. Coefficient of learning is mostly around 0.1 to 0.25. The number of the hidden neurons is mostly higher than seven and the number of results which have an error Ep on the training set smaller than 0.01 is higher than five.

### 4.3 Ep - Errors on the Test Set

In the next section, we focused on comparing the errors on the test set for all networks at certain intervals. Errors E1-E5 are individual errors of each neural network on the test set. Each row in Table 5 represents one topology, which means that the network 1-5 in the row have the same topology. Error Ep was formed after averaging the corresponding values of all the neural networks that were involved in the calculation. At first, the top ten results were selected, their error was in the range 0 to 0.0006.

**Table 5.** Errors on the test sets of different neural networks

| $E_p$ | $E_1$ - test | $E_2$ - test | $E_3$ - test | $E_4$ - test | $E_5$ - test |
|---|---|---|---|---|---|
| 0.000208 | 0.012401 | 0.004372 | 0.000772 | 0.001091 | 0.004339 |
| 0.000335 | 0.001099 | 0.001661 | 0.004487 | 0.003134 | 0.006045 |
| 0.000365 | 0.00264 | 0.00238 | 0.001456 | 0.007301 | 0.002701 |
| 0.000375 | 0.001327 | 0.003921 | 0.005708 | 0.001215 | 0.001697 |
| 0.000379 | 0.000782 | 0.004423 | 0.002449 | 0.002169 | 0.000955 |
| 0.000401 | 0.001338 | 0.003676 | 0.002813 | 0.003037 | 0.001221 |
| 0.000437 | 0.002784 | 0.001227 | 0.001888 | 0.001753 | 0.001044 |
| 0.000472 | 0.047804 | 0.024766 | 0.003859 | 0.002925 | 0.000516 |
| 0.000503 | 0.004425 | 0.003076 | 0.002142 | 0.006081 | 0.002057 |
| 0.000544 | 0.007855 | 0.001935 | 0.000625 | 0.002648 | 0.002508 |

Table 5 shows that the best results are in errors after averaging (Ep) below 0.001. Furthermore, we see that none of various neural networks has an error smaller than Ep. The next table contains values for which the error is in the range 0.005 to 0.008.

**Table 6.** Errors on the test sets of different neural networks

| $E_p$ | $E_1$ - test | $E_2$ - test | $E_3$ - test | $E_4$ - test | $E_5$ - test |
|---|---|---|---|---|---|
| **0.005021** | 0.178416 | 0.076424 | 0.003785 | 0.001887 | 0.001005 |
| **0.005817** | 0.001108 | 0.020713 | 0.002823 | 0.101839 | 0.004743 |
| **0.006163** | 0.006726 | 0.008091 | 0.061983 | 0.001667 | 0.025376 |
| **0.006178** | 0.108114 | 0.003969 | 0.002504 | 0.002105 | 0.006089 |
| **0.006523** | 0.009666 | 0.090545 | 0.004462 | 0.011411 | 0.002763 |
| **0.006598** | 0.006697 | 0.000687 | 0.136909 | 0.003945 | 0.045954 |
| **0.006914** | 0.003835 | 0.177705 | 0.007987 | 0.005592 | 0.007578 |
| **0.007117** | 0.001759 | 0.002979 | 0.004167 | 0.005154 | 0.097156 |
| **0.007425** | 0.004142 | 0.152833 | 0.111691 | 0.079456 | 0.023734 |
| **0.007717** | 0.001541 | 0.149562 | 0.004605 | 0.002604 | 0.005401 |

Table 6 shows that the network (or networks) did not learn the problem as well as the others. The red marked ones learned the problem better than Ep, the blacked mark ones worse than Ep. This suggests that even if some of the networks learn the problem worse than the others, the error calculated by averaging the values is acceptable. Therefore it is better to teach more networks and then make their values average [5, 6, 7].

### 4.4 Comparation of Prediction

The number of the predicted values is ten. The maximum number of iterations is set to two thousand and convergence is set to 0.01. The graph of the used periodic time series is shown in Figure 3 downloaded from [3].

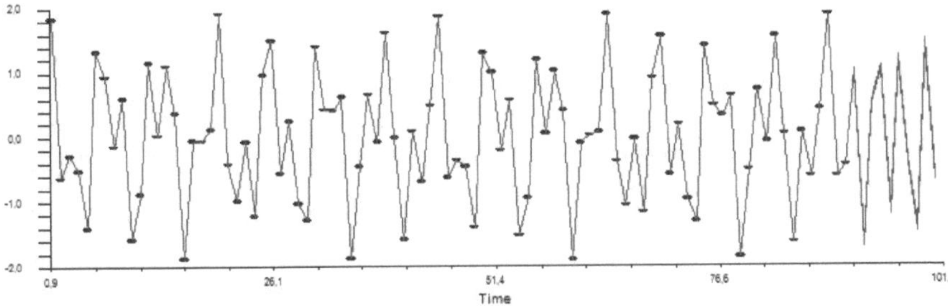


**Fig. 3.** Periodic time series

The error on the test set is calculated from the output file using the following equation (8):

$$E = \frac{1}{2}\sum_{i=1}^{n}(\hat{y}_i - y_i)^2 \quad (8)$$

By using the previous equation, the error on the test set was calculated, which equalled to 2.944186. The method of averaging has an error on the test set 0.013112803. The network topologies were: the number of the input neurons is five, the number of the hidden neurons is ten, the number of the hidden layers is one, the learning coefficient is 0.1 and the number of epochs is 2000. The following figure shows the difference between the predicted and real values using ARIMA and our approach (Fig. 4), [5, 6, 7]:

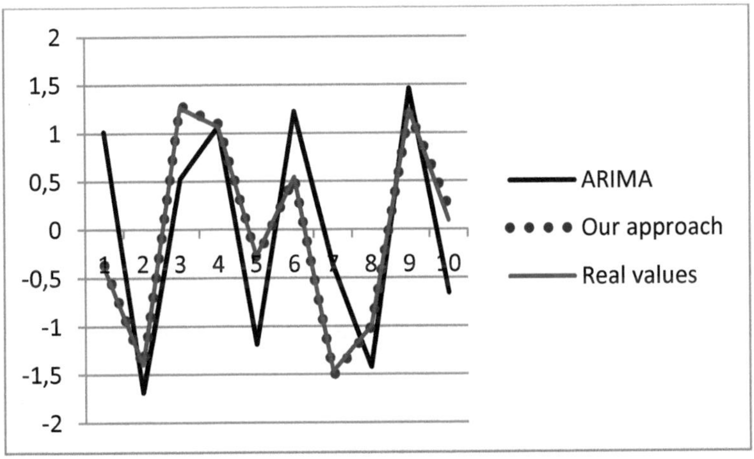


**Fig. 4.** Comparison of prediction results

The output file reveals that the ARIMA models predicted 10 values out of 91. It took 12 iterations. MSE (Mean Square Error) error is 6.884332 E-05.

## 5 Conclusion

The method of averaging achieved very small errors at periodic time series on the training (under 0.01) and test (under 0,001) sets. Finally, the averaging method for periodic time series compared with ARIMA (Box-Jenkins) method. ARIMA models with similar settings as the neural network tried to predict 10 values out of 91 previous. The error calculated on the test set for neural networks did not exceed 0.545674. Although ARIMA models needed twelve cycles to reduce the error on the training set under 0.01, the error on the test set was 2.944186.

**Acknowledgements.** The research described here has been financially supported by University of Ostrava grant SGS16/PRF/2014. Any opinions, findings and conclusions or recommendations expressed in this material are those of the authors and do not necessarily reflect the views of the sponsors.

## References

[1] Fausett, L.: Fundamentals of Neural Network, 1st edn. Prentice-Hall (1994) ISBN: 0-13-334186-0
[2] Kondratenko, V., Kuperin, Y.: Using recurrent neural networks to forecasting of Forex. St. Petersburg State University (2003)
[3] A database from the area of financial forecasting, `http://www.bcpp.cz` (accessed January 5, 2014)
[4] Nugent, C.D., Lopez, J.A., Smith, A.E., Black, N.D.: Reverse engineering of neural network classifiers in medical applications. In: Proceedings of the VII EFOMP Congress, Physi-caMedica, vol. XVII, p. 184 (2001)
[5] Hrnecek, J.: Neural networks for time series prediction, diploma thesis, University of Ostrava (2013)
[6] Volna, E., Janosek, M., Kotyrba, M., Kocian, V.: Pattern recognition algorithm optimization. In: Zelinka, I., Snasel, V., Rössler, O.E., Abraham, A., Corchado, E.S. (eds.) Nostradamus: Mod. Meth. of Prediction, Modeling. AISC, vol. 192, pp. 251–260. Springer, Heidelberg (2013)
[7] Volná, E., Janošek, M., Kocian, V., Kotyrba, M.: "Nonlinear Time Series Analysis via Neural Networks. In: Stavrinides, G., Banerjee, S., Caglar, H., Ozert, M. (eds.) International Interdisciplinary Chaos Symposium. Complexity series. Springer, Heildelberg (2013)
[8] Hao-Tien, L.: An integrated fuzzy time series forecasting system, NY, USA, pp. 10045–10053 (2009)
[9] Lee, K., Booth, D., Alam, P.: A comparison of supervised and unsupervised neural networks in predicting bankruptcy of Korean rms. Expert Systems with Applications 29, 1–16 (2005) ISSN 0957-4174

# ECG Prediction Based on Classification via Neural Networks

Eva Volna and Martin Kotyrba

University of Ostrava, 30. dubna 22, 70103 Ostrava, Czech Republic

**Abstract.** The article deals with ECG prediction based on neural networks classification of different types of time courses of ECG signals. The main objective is to recognise normal cycles and arrhythmias and perform further diagnosis. We proposed two detection systems that have been created with usage of neural networks. The experimental part makes it possible to load ECG signals, preprocess them, and classify them into given classes. Outputs from the classifiers carry a predictive character. All experimental results from both of the proposed classifiers are mutually compared in the conclusion.

## 1 Background

Biometrical data is typically represented as an image or a quantification of measured physiological or behavioural characteristics. As this data should refer to very complex human behaviour or describe very precisely physiological characteristic (typically iris scan, fingerprint, palm vein image, hand scan, voice, walk pattern etc.), this data can easily become very large and hard to process. For this reason, a modern ways of data processing and classification are applied for biometrical data. The leading method is the usage of neural networks [6].

For more than four decades, computers have been used in the classification of the electrocardiogram (ECG) resulting in a huge variety of techniques [1] all designed to enhance the classification accuracy to levels comparable to that of a 'gold standard' of expert cardiology opinion. Included in these techniques are multivariate statistics, decision trees, fuzzy logic, expert systems and hybrid approaches [5]. The recent interest in Neural Networks coupled with their high levels of performance has resulted in many instances of their application in this field [2].

The electrocardiogram is a technique of recording bioelectric currents generated by the heart. Clinicians can evaluate the conditions of a patient's heart from the ECG and perform further diagnosis. ECG records are obtained by sampling the bioelectric currents sensed by several electrodes, known as leads. A typical one-cycle ECG tracing is shown in Fig. 2.

## 2 Basic Principles of ECG Evaluation

ECG scanning has own rules, which are in accordance with the laws of physics. The heart irritation spreads in all directions. In the case that the depolarisation spreads

I. Zelinka et al. (eds.), *Nostradamus 2014: Prediction, Modeling and Analysis of Complex Systems*, Advances in Intelligent Systems and Computing 289,
DOI: 10.1007/978-3-319-07401-6_26, © Springer International Publishing Switzerland 2014

towards the electrode, which is placed on the body surface, a positive deflection is recorded on an ECG monitor. A negative deflection is recorded at the opposite end of the body. The ECG waveform is written with a chart speed of 25 $mm \cdot s^{-1}$. An algorithm describing the curve goes in the following steps. First, we evaluate the shape and rhythm of ventricular complexes or atrial, which can be either regular or irregular. Then we evaluate the frequency of ventricular complexes and atrial fibrillations. Contraction of each muscle of the human body (and thus the heart as well) is associated with electrical changes called depolarization, which can be detected by electrodes. The heart contains two basic types of cells. Myocardial cells, which are responsible for generating the pressure necessary to pump blood throughout the body, and conduction cells, which are responsible for rapidly spreading electrical signals to the myocardial cells in order to coordinate pumping. A graph of an action potential of a muscle of cardiac cells is shown in Fig. 1

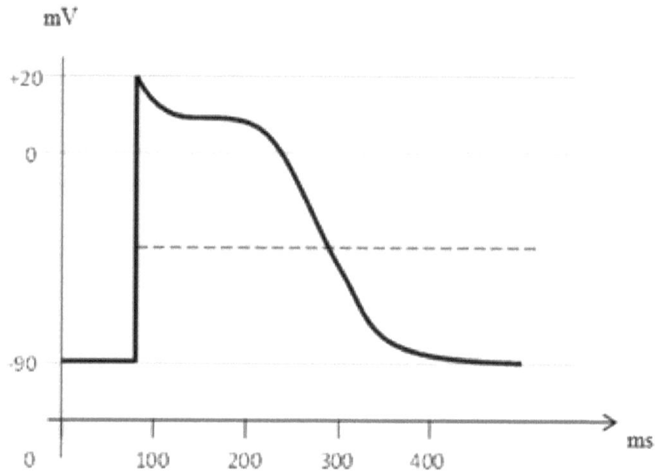


**Fig. 1.** The cardiac action potentials

A normal electrocardiogram is illustrated in Fig. 2. The figure also includes definitions for various segments and intervals in the ECG. The deflections in this signal are denoted in alphabetic order starting with the letter *P*, which represents atrial depolarization. The ventricular depolarization causes the *QRS* complex, and repolarization is responsible for the *T*-wave. Atrial repolarization occurs during the *QRS* complex and produces such a low signal amplitude that it cannot be seen apart from the normal ECG.

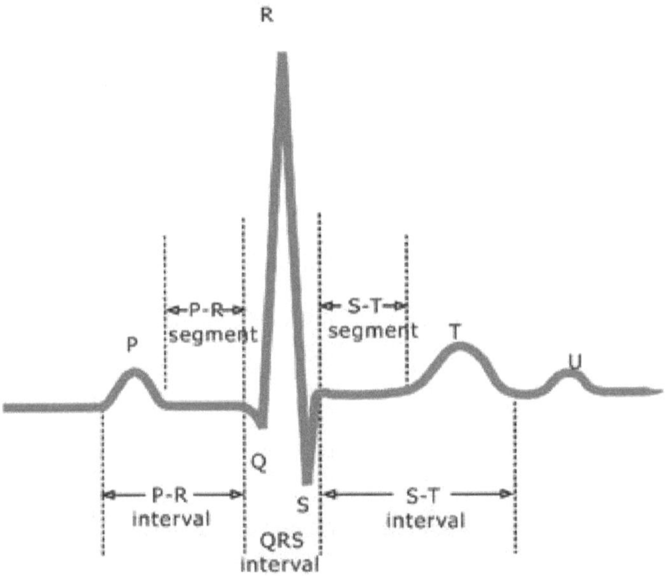


**Fig. 2.** A typical one-cycle ECG tracing (adapted from http://www.ni.com/white-paper/app/largeimage?lang=cs&imageurl=%2Fcms%2Fimages%2Fdevzone%2Ftut%2F2007-07-09_141618.jpg)

## 3 Signal Processing Using Neural Networks

In practice, a relatively reliable diagnostic program stored in ECG monitors has been used, which is a guideline for determining the final diagnosis of heart disorders. This program works according to the principle of IF-THEN rules. The values of the electrical signal are discretized and uploaded into expert systems in the form of thousand rules. The aim of this article is to use a different approach based on the principle of neural networks. The proposed methodology of solution could be summarized into the following steps:

1. A conversion of analog signal from the ECG monitor to a computer.
2. We have used multi-layer networks that are fully connected.
3. We have obtained ECG waveforms in collaboration with the University Hospital in Poruba, specifically at the Department Cardiac Surgery from sick patients and at the Department Traumatology from healthy patients (i.e. 'healthy' with regard to heart diseases).
4. ECG waveforms built training/test sets.
5. Neural network adaptation.
6. Testing phases.

### 3.1 Technical Equipment

ECG measurements were performed using ADDA Junior with converter ADDA Junior, which was connected to a computer via bidirectional parallel cable

(CETRONICS). Technical parameters of the A/D converter (8-bit conversion) were the following:

- 3 measuring ranges
- Measuring of a frequency of AC voltage at any channel
- Autoranging for measuring the frequency of 100Hz, 1kHz, 10kHz
- Input resistance of 300kΩ
- Measurement accuracy 1%

Technical parameters of the /D/A converter (a programmable voltage source ±10V) were the following:

- Maximum current consumption of 15 mA (after optimization 4A at the output)
- Power of the converter ±15 V (stabilized)

## 3.2 Experimental Results

**Time Series Prediction**

The training set consisted of modified ECG waveforms. We used a backpropagation neural network with topology 101-10-1. The output unit represents a diagnose 0/1 – a healthy/sick person. A smaller number of inputs would not be appropriate due to the nature of the ECG waveform. We use 34 ECG time series associated with sick persons and 36 ECG time series associated to healthy persons. 25 time series of each group were used as a training set and the rest as a test set. Fig. 3 shows a comparison of mean values of ECG waveforms for healthy/sick persons. We used the backpropagation method [3, 4] for the adaptation with the following parameters: the learning rate value is 0.1 and momentum is 0. The conducted experimental studies also showed that in each cycle of adaptation is to present an adequate network of training patterns mixed randomly to ensure their greater diversity, but also acts as a measure of system stability. Uniform system in a crisis usually collapses entirely, while in the diversion system through a crisis of its individual parts, but the whole remains functional. The condition of end of the adaptation algorithm specified the limit value of the overall network error, $E < 0.1$.

The test set consisted of 20 samples (11 health and 9 sick persons) that were not included in the training set. The summary results for this type of experiment are shown in a graph in Fig 4. For clarity, the results of testing are given in percentage. The average test error was 0.194. A healthy population was detected with an average error of 0.263 and sick population with an average error of 0.109.

**Pattern Recognition Classifier Leading to Prediction**

For the purpose of adaptation of the pattern recognition classifier, it is necessary to remark that determination of training patterns is one of the key tasks. Improperly chosen patterns can lead to confusion of neural networks. During our experimental work, we made some study included ECG pattern recognition. When creating appropriate patterns of the training set, we used characteristic curves shown as mean values from ECG waveforms for healthy and sick persons (Fig.3). We use two different groups of patterns. Patterns H1-H4 (Fig.5) represent healthy persons and patterns S1-S4 (Fig.6) represent sick persons. The whole training set is shown in Table 1.

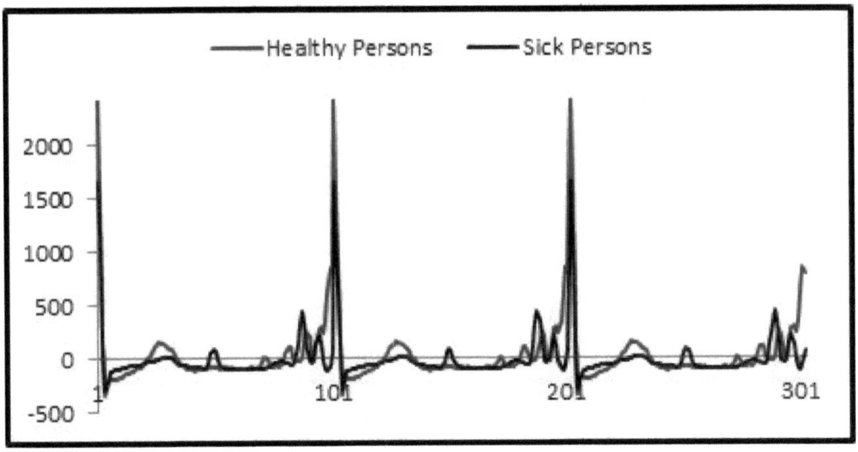


**Fig. 3.** Comparison of mean values of ECG waveforms for healthy/sick persons

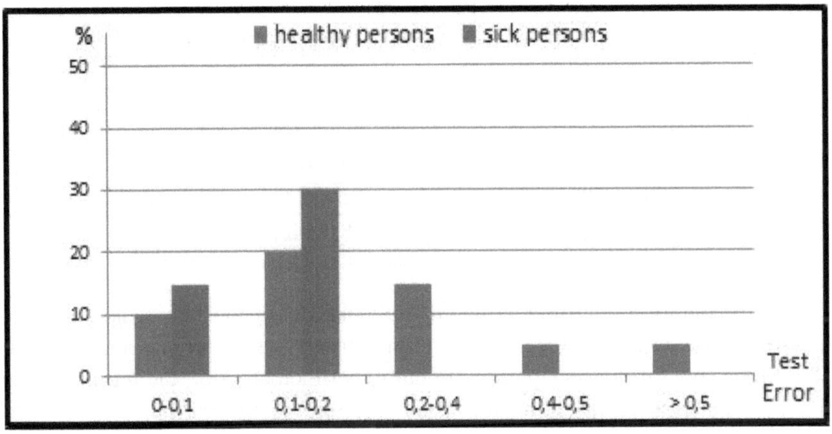


**Fig. 4.** Experimental results – test error for healthy/sick persons

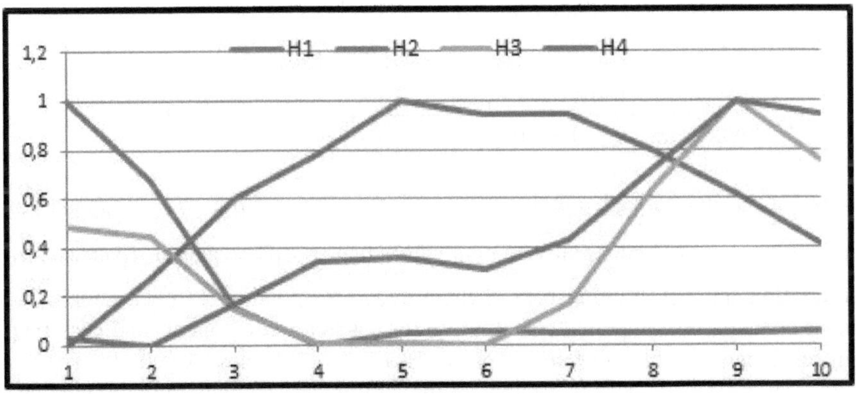


**Fig. 5.** Patterns representing healthy persons

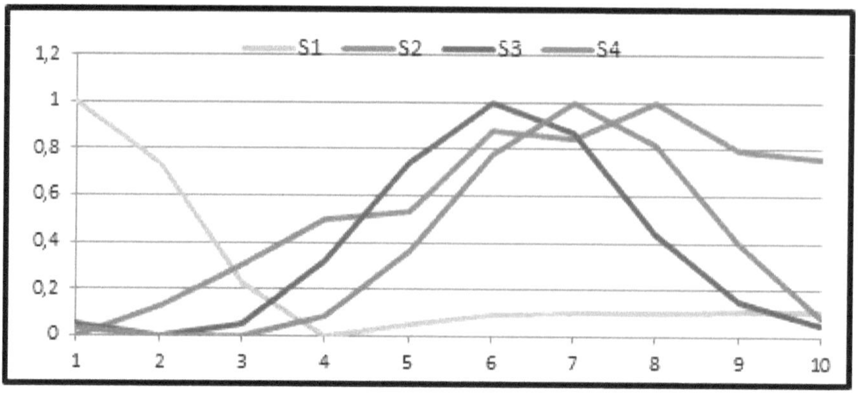


**Fig. 6.** Patterns representing sick persons

**Table 1.** The training set

| Patterns | INPUTS | | | | | | | | | | OUTPUTS | | | | | | | |
|---|---|---|---|---|---|---|---|---|---|---|---|---|---|---|---|---|---|---|
| H1 | 1.000 | 0.672 | 0.155 | 0.000 | 0.045 | 0.057 | 0.049 | 0.049 | 0.053 | 0.055 | 1 | 0 | 0 | 0 | 0 | 0 | 0 | 0 |
| H2 | 0.000 | 0.273 | 0.600 | 0.782 | 1.000 | 0.945 | 0.945 | 0.799 | 0.618 | 0.418 | 0 | 1 | 0 | 0 | 0 | 0 | 0 | 0 |
| H3 | 0.485 | 0.449 | 0.147 | 0.007 | 0.007 | 0.000 | 0.169 | 0.632 | 1.000 | 0.757 | 0 | 0 | 1 | 0 | 0 | 0 | 0 | 0 |
| H4 | 0.035 | 0.000 | 0.170 | 0.338 | 0.356 | 0.309 | 0.430 | 0.719 | 1.000 | 0.946 | 0 | 0 | 0 | 1 | 0 | 0 | 0 | 0 |
| S1 | 1.000 | 0.740 | 0.228 | 0.000 | 0.045 | 0.091 | 0.098 | 0.101 | 0.104 | 0.107 | 0 | 0 | 0 | 0 | 1 | 0 | 0 | 0 |
| S2 | 0.000 | 0.123 | 0.304 | 0.495 | 0.536 | 0.883 | 0.851 | 1.000 | 0.796 | 0.761 | 0 | 0 | 0 | 0 | 0 | 1 | 0 | 0 |
| S3 | 0.044 | 0.000 | 0.045 | 0.319 | 0.748 | 1.000 | 0.868 | 0.440 | 0.154 | 0.050 | 0 | 0 | 0 | 0 | 0 | 0 | 1 | 0 |
| S4 | 0.033 | 0.000 | 0.000 | 0.085 | 0.360 | 0.779 | 1.000 | 0.820 | 0.399 | 0.079 | 0 | 0 | 0 | 0 | 0 | 0 | 0 | 1 |

Pattern recognition classifier is based on backpropagation neural network and is able to recognise wave structures in given time series [7, 8]. Artificial neural networks need training sets for their adaptation. In our experimental work, the training set consisted of 8 patterns representing the basic structure of the various waves in ECG graphs, see Fig. 5 and 6. Input data is sequences always including $n$ consecutive numbers, which are transformed into interval <0, 1> by the formula (1). Samples are adjusted for the needs of backpropagation networks with sigmoid activation function in this way [3, 4].

$$x'_j = \frac{x_j - \min(x_i,\dots,x_{i+n-1})}{\max(x_i,\dots,x_{i+n-1}) - \min(x_i,\dots,x_{i+n-1})},\ (j = i,\dots,i+n-1) \qquad (1)$$

where $x'_j$ is normalized output value of the $j$-th neuron ($j = i,\dots,i+n$-1) and $(x_i,\dots x_{i+n-1})$ are $n$-$1$ consecutive output values that specify sequences (patterns) from the training set (e.g. training pars of input and corresponding output vectors). Input vector contains 10 components. Output vector has got 8 components and each output unit represents one of 8 different types of ECG wave samples. A neural network architecture is 10 - 10 - 8 (e.g. 10 units in the input layer, 10 units in the hidden layer, and 8 units in the output layer). The net is fully connected. Adaptation of the neural network starts with randomly generated weight values.

We used the backpropagation method for the adaptation with the following parameters: the learning rate value is 0.1 and momentum is 0. The conducted experimental studies also showed that in each cycle of adaptation is to present an adequate network of training patterns mixed randomly. The condition of end of the adaptation algorithm specified the limit value of the overall network error, E<0.1.

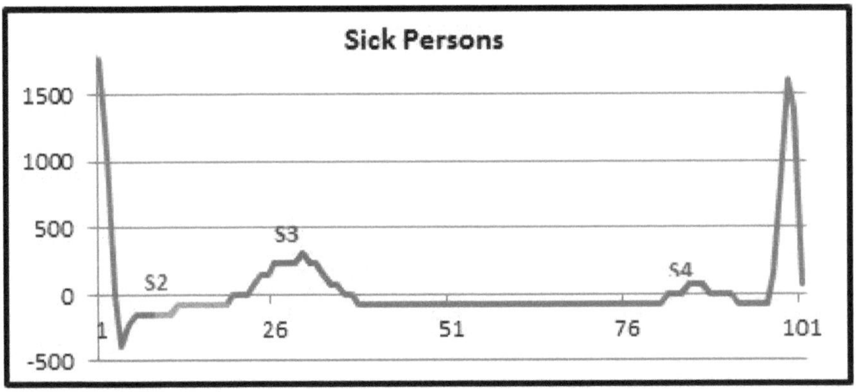


**Fig. 7.** Some recognized patterns that occur in ECG time series

In order to test the efficiency of the method, we applied the same set of data that we used in the previous experimental part. Outputs from the classifier produce sets of values that are assigned to each recognized training pattern in the given test time series. It is important to appreciate what can be considered as an effective criterion related to consensus of similarity. The proposed threshold resulting from our experimental study was determined at least $p$ = 70%. Comparison of the patterns look, how are learned via neural network versus their presentations in test set is represented in Fig. 8. The neural network is able to discover some connections, which are almost imperceptible. Illustration of some recognized patterns that occur in ECG time series is shown in Fig. 7. Outputs from the classifier carry a predictive character. The neural network determines if the time series belongs to a healthy or sick person on the basis of the recognised ECG patterns which appear in the time series history.

The methodology of testing is shown in Fig. 9. This means, if the test pattern S1, S2, S3 or S4 appeared in ECG waveform with probability $p_S \geq p$ ($p$ = 70%), thus it was predicted 'a sick person'. Then we work only with the remaining time series. If the test pattern H1, H2, H3 or H4 appeared in ECG waveform with probability $p_H \geq p$ ($p$ = 70%), thus it was predicted 'a healthy person'. In all other cases, the ECG time series was unspecified. We examined a total of 20 data sets. Each of them contains 101 values that assign 92 possible patterns. The whole number of examined patterns is 1840. The graph in Fig. 10 demonstrates a summary of results, where 'sick persons' represent patterns S1-S4 and 'healthy persons' represent patterns H1-H4. The resulting prediction is based on the methodology, see Fig. 9.

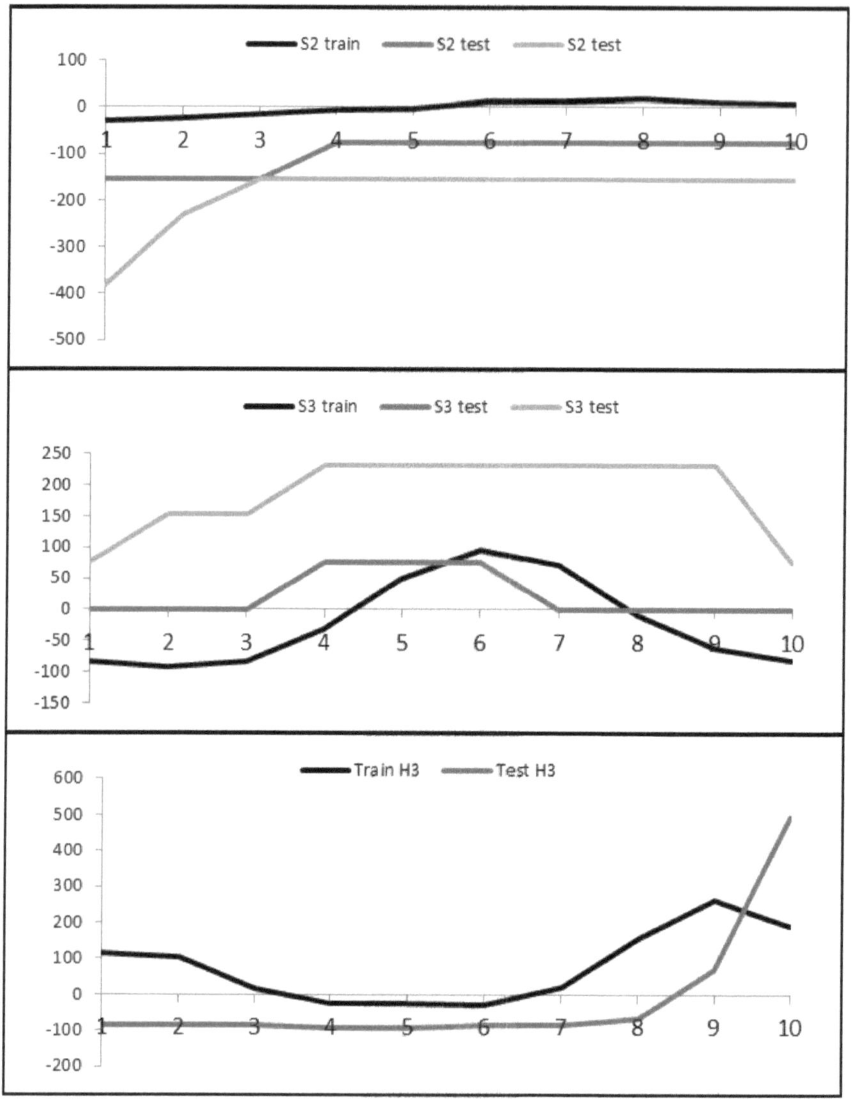


**Fig. 8.** Training patterns their representation in used test sets

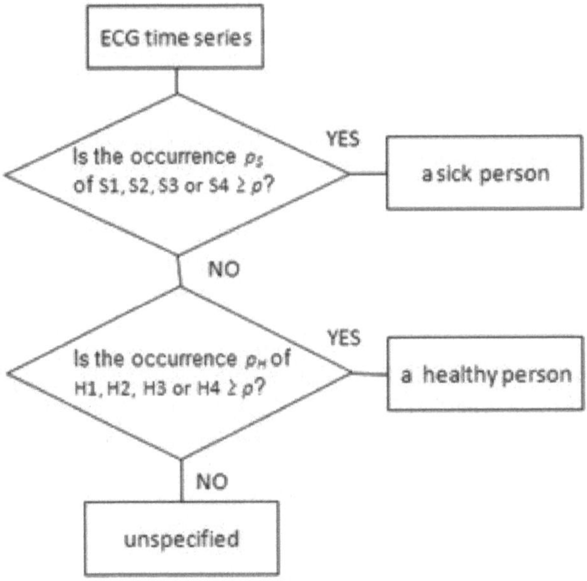


**Fig. 9.** The methodology of testing

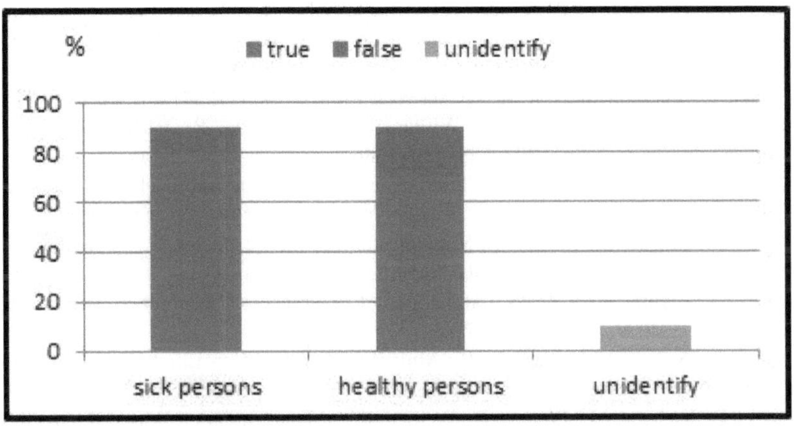


**Fig. 10.** Experimental results – test error

# 4 Conclusion

In this paper, a short introduction into the field of ECG waves recognition using backpropagation neural network has been given. Main objective was to recognise the normal cycles and arrhythmias and perform further diagnosis. We proposed two detection systems that have been created with usage of neural networks. One of them is adapted according to the training set. Here, each pattern represents the whole one

ECG cycle. Then, an output unit represents a diagnose 0/1 – a healthy/sick person. The second one approach uses neural network, which training set contains two different groups of patterns for healthy/sick persons. According to the results of experimental studies, it can be stated that ECG waves patterns were successfully extract in given time series and recognise using suggested method, how as can be seen from figures in result section. It might result in better mapping of the time series behaviour for better prediction.

Both approaches were able to predict with high probability, if the ECG time series as assigns to sick or healthy persons. It is interesting that a sick diagnose was recognised with higher accuracy in both experimental works.

**Acknowledgments.** The research described here has been financially supported by University of Ostrava grant SGS16/PřF/2014. Any opinions, findings and conclusions or recommendations expressed in this material are those of the authors and do not necessarily reflect the views of the sponsors.

## References

1. Nugent, C.D., Webb, J.A.C., Black, N.D., Wright, G.T.H.: Electrocardiogram 2: Classification. Automedica 17, 281–306 (1999)
2. Nugent, C.D., Lopez, J.A., Smith, A.E., Black, N.D.: Reverse engineering of neural network classifiers in medical applications. In: Proc. of the VII EFOMP Congress, Physica Medica, vol. XVII, p. 184 (2001)
3. Fausett, L.: Fundamentals of Neural Network. Prentice Hall (1994) ISBN:0-13-334186-0
4. Rumelhart, D.E., Hinton, G.E., William, R.J.: Learning Representations by Back-Propagation Errors. Nature 323, 533–536 (1986)
5. Russell, S., Norvig, P.: Artificial Intelligence - A Modern Approach, 2nd edn. Prentice Hall (2003)
6. Tangkraingkij, P., Lursinsap, C., Sanguansintukul, S., Desudchit, T.: Selecting Relevant EEG Signal Locations for Personal Identification Problem Using ICA and Neural Network. In: Eighth IEEE/ACIS International Conference on Computer and Information Science, ICIS 2009, pp. 616–621 (2009)
7. Volna, E., Kotyrba, M., Jarusek, R.: Multiclassifier based on Elliott wave's recognition. Computers and Mathematics with Applications 66 (2013) ISSN:0898-1221
8. Volna, E., Kotyrba, M., Jarušek, R.: Prediction by Means of Elliott Waves Recognition. In: Zelinka, I., Snasel, V., Rössler, O.E., Abraham, A., Corchado, E.S. (eds.) Nostradamus: Mod. Meth. of Prediction, Modeling. AISC, vol. 192, pp. 241–250. Springer, Heidelberg (2013)

# Measurable Error Compensation with GPC in a Heat-Exchanger with a Traffic Delay

Stanislav Talaš[1] and Vladimír Bobál[1,2]

[1] Tomas Bata University in Zlín, Department of Process Control, Faculty of Applied Informatics, Nam T.G. Masaryka 5555, 760 01 Zlin, Czech Republic
[2] Tomas Bata University in Zlín, Centre of Polymer Systems, University Institute, Nam T.G. Masaryka 5555, 760 01 Zlin, Czech Republic
{talas,bobal}@fai.utb.cz

**Abstract.** This paper demonstrates use of Model Predictive Control (MPC) to system control with delay. Generalized Predictive Control (GPC) method was selected to demonstrate the ability to both control the system and compensate the measurable disturbance while bot the system and the disturbance are delayed. For the realization the MATLAB/SIMULINK program environment was used with system parameters based on the laboratory system. GPC method was chosen and its ability to compensate the outer disturbance with delay was verified by simulation of system control based on real laboratory model. Control algorithm and simulation were realized in MATLAB/SIMULINK program environment. Results have proven capabilities of GPC method to control and compensate error in stable, oscillatory and non-minimum phase systems with traffic delay. Additionally, real model parameters were selected to test a possibility of realization.

## 1 Introduction

The current efforts in the scientific area of process control tend to focus on satisfaction of demands of maximal productivity of the highest quality products at the lowest cost possible. This combination of requirements directs us to area of optimization. With the computing power of the modern technology a solutions for complex problems can be found in reasonable time.

In the industrial sector of slow and large dimensional systems an advanced optimization approach is with use of model predictive control methods [1]. This term refers to strategy of using internal model for system behavior predictions and computing a sequence of control inputs for optimal performance according to given conditions. Optimality is specified by value of function containing sum of squares between the desired and predicted trajectories and sum of changes in control input. Therefore the result of minimization of this function provides a performance with maximal precision with minimal change in control value.

Model predictive control methods begin their development in 1980s with publication of Dynamic matrix control (DMC) method [2]. Benefits of this control approach caused its widespread use in world's major industrial companies. The GPC method appeared in 1987 [3], nowadays being one of the most popular MPC algorithms [4].

I. Zelinka et al. (eds.), *Nostradamus 2014: Prediction, Modeling and Analysis of Complex Systems*, Advances in Intelligent Systems and Computing 289,
DOI: 10.1007/978-3-319-07401-6_27, 

Handling the measurable disturbance is one of major advantages of MPC methods, since it can be included in the internal.

The purpose of this paper is to give an insight on capabilities of MPC to compensate the effects of the measurable disturbance. A GPC was selected as a representing method.

All the presented control processes were simulated in MATLAB/SIMULINK program environment and multiple systems with various dynamics were tested.

## 2 Model Based Predictive Control

Predictive control is an approach to control a process trough optimization. The main principle is in prediction of future process outputs based on inner model of process. The goal of the control algorithm is to find such a vector of input values that output of model is optimal along the defined time area called horizon. To ensure robustness and stability an approach using feedback called receding horizon strategy is often applied. From vector of input values only the first value is used as an increment $\Delta u(k)$ added to previous input giving current input value $u(k)$. In the next step the entire procedure is repeated with new process output values; this is called the *receding horizon strategy* illustrated in Figure 1.

The area of optimization is defined by values of horizons representing amount of sampling periods from the current time into the future. Values of horizons $N_1$ and $N_2$ limit the area, where the divergence between the desired and the output value is minimized. Horizon $N_u$ limits the distance of steps where the action value is minimized.

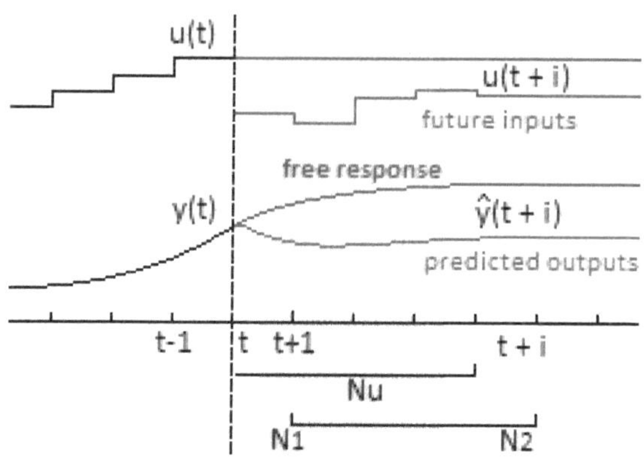


**Fig. 1.** Receding horizon strategy

Calculation of optimal output consists of free response prediction describing system behavior in case of constant input and the forced response with a reaction on a suggested series of inputs. Based on the superposition principle, the sum of these responses results in the future output prediction.

Several methods of model predictive control are used in practice; the main differences are in description of controlled process and in objective function.

The Figure 2 shows a layout of predictive control and a data transfer.

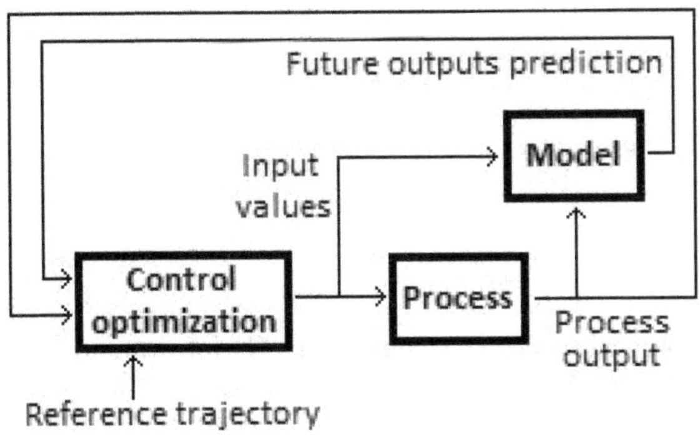


**Fig. 2.** Basic structure of model predictive control

The optimization process is based on the minimization of values involved in control. Their mutual relations are formed by an objective function. The general expression of an objective function is

$$J = \sum_{i=N_1}^{N_2} \left[ \hat{y}(k+i) - w(k+i) \right]^2 + \sum_{i=1}^{N_u} \left[ \lambda(i) \Delta u(k+i-1) \right]^2 \tag{1}$$

where $\delta(i)$ and $\lambda(i)$ are weighting values, usually constants representing a ratio of the minimization between a divergence of output from the desired value and a change of the action value [5].

## 2.1 Generalized Predictive Control

GPC method is based on internal model CARIMA

$$A(z^{-1})y(k) = B(z^{-1})u(k-1) + \frac{e(k)}{\Delta}, \tag{2}$$

where $\Delta$ represents difference form last step $1 - z^{-1}$.

Estimations of future system outputs are therefore calculated by equation

$$\hat{y}(k+d+i) = G_i(z^{-1})\Delta u(k+i-1) + F_i(z^{-1})y(k+1), \tag{3}$$

which can be rewritten as

$$\mathbf{y} = \mathbf{Gu} + \mathbf{F}(z^{-1})y(k) + \mathbf{G}'(z^{-1})\Delta u(k-1), \tag{4}$$

where

$$\mathbf{y} = \begin{bmatrix} \hat{y}(k+d+1) \\ \hat{y}(k+d+2) \\ \vdots \\ \hat{y}(k+d+N) \end{bmatrix}, \mathbf{u} = \begin{bmatrix} \Delta u(k) \\ \Delta u(k+1) \\ \vdots \\ \Delta u(k+N-1) \end{bmatrix}, \mathbf{G} = \begin{bmatrix} g_0 & 0 & \cdots & 0 \\ g_1 & g_0 & \cdots & 0 \\ \vdots & \vdots & \ddots & \vdots \\ g_{N-1} & g_{N-2} & \cdots & g_0 \end{bmatrix}. \quad (5)$$

Elements depending only on past values **F** and **G'** form the system free response **f** leading to the final formulation of the equation

$$\mathbf{y} = \mathbf{G}\mathbf{u} + \mathbf{f} \quad (6)$$

By rewriting the objective function (2.1) into the following form

$$J = \mathbf{u}^T 2(\mathbf{G}^T\mathbf{G} + \lambda\mathbf{I})\mathbf{u} + 2(\mathbf{f} - \mathbf{w})^T \mathbf{G}\mathbf{u} + (\mathbf{f} - \mathbf{w})^T \quad (7)$$

To find the minimal value a derivation of $J$ by u is set equal to zero, creating an equation for an optimal control procedure

$$\mathbf{u} = (\mathbf{G}^T\mathbf{G} + \lambda\mathbf{I})^{-1}\mathbf{G}^T(\mathbf{w} - \mathbf{f}) \quad (8)$$

From the calculated sequence only the first value $\Delta u(k)$ is used for actual step. In each of following steps the calculation is repeated [6], [7].

### 2.2 Generalized Predictive Control with Measurable Disturbances

Description of measurable disturbances are added into internal model

$$A(z^{-1})y(k) = z^{-d}B(z^{-1})u(k-1) + z^{-d}D(z^{-1})v(k) + \frac{e(k)}{\Delta} \quad (9)$$

with $v(k)$ expressing a value of the measurable disturbance and $D(z^{-1})$ being polynomial numerator of its input-output model describing its behavior.

If the future values of the measurable disturbance are unknown, it can be predicted using trends or simply estimated to be constant $\Delta v(k+i) = 0$.

Predictions of future output are then estimated by following equation

$$\hat{y}(k+d+i) = G_i(z^{-1})\Delta u(k+i-1) + H_i(z^{-1})\Delta v(k+i) + f_i \quad (10)$$

where $f_i$ represents the free response calculated using past values

$$f = F_i(z^{-1})y(k) + G_i'(z^{-1})\Delta u(k-1) + H_i'(z^{-1})\Delta v(k) \tag{11}$$

Equations (3.12) and (3.13) can be recursively expressed as matrices [8]

$$\begin{bmatrix} y(k+d+1) \\ y(k+d+2) \\ \vdots \\ y(k+d+N) \end{bmatrix} = \begin{bmatrix} g_0 & 0 & \cdots & 0 \\ g_1 & g_0 & \cdots & 0 \\ \vdots & \vdots & \ddots & \vdots \\ g_{N-1} & g_{N-2} & \cdots & g_0 \end{bmatrix} \begin{bmatrix} \Delta u(k) \\ \Delta u(k+1) \\ \vdots \\ \Delta u(k+N-1) \end{bmatrix} + \begin{bmatrix} h_0 & 0 & \cdots & 0 \\ h_1 & h_0 & \cdots & 0 \\ \vdots & \vdots & \ddots & \vdots \\ h_{N-1} & h_{N-2} & \cdots & h_0 \end{bmatrix} \begin{bmatrix} \Delta u(k) \\ \Delta u(k+1) \\ \vdots \\ \Delta u(k+N-1) \end{bmatrix} + \begin{bmatrix} f_1 \\ f_2 \\ \vdots \\ f_N \end{bmatrix} \tag{12}$$

By including the measurable disturbance into free response $\mathbf{f}' = \mathbf{Hv} + \mathbf{f}$ an equation of prediction in the same form like the basic prediction can be used

$$\mathbf{y} = \mathbf{Gu} + \mathbf{f}' \tag{13}$$

## 3 Experimental Laboratory Heat Equipment

A scheme of the laboratory heat equipment [9] is described in Figure 3. The heat transferring fluid (e. g. water) is transported using a continuously controllable DC pump (6) into a flow heater (1) with max. power of 750 W. The temperature of a fluid at the heater output $T_1$ is measured by a platinum thermometer. Warmed liquid then goes through a 15 meters long insulated coiled pipeline (2) which causes the significant delay (20 – 200 s) in the system. The air-water heat exchanger (3) with two cooling fans (4) and (5) represents a heat-consuming appliance. The speed of the first fan can be continuously adjusted, whereas the second one is of on/off type. Input and output temperatures of the cooler are measured again by platinum thermometers as $T_2$, respective $T_3$. The platinum thermometer $T_4$ is dedicated for measurement of the outdoor-air temperature. The laboratory heat equipment is connected to a standard PC via technological multifunction I/O card MF 624. This card is designed for the need of connecting PC compatible computers to real world signals. The card is designed for standard data acquisition, control applications and optimized for use with Real Time Toolbox for SIMULINK. The MATLAB/SIMULINK environment was used for all monitoring and control functions.

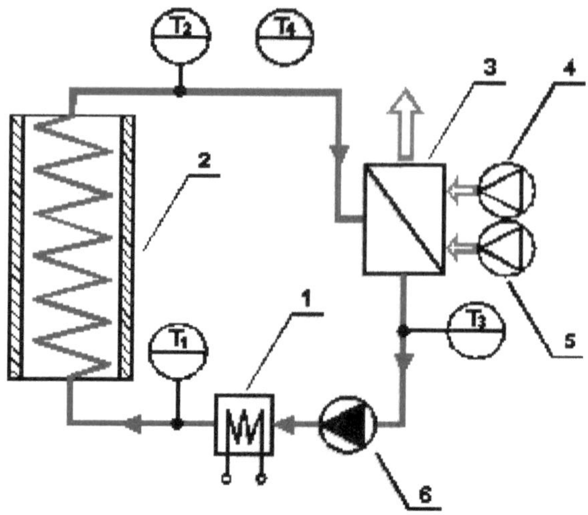


**Fig. 3.** Scheme of laboratory heat equipment

## 4 Simulation Control of Basic Dynamics with Delay

The capability to negate the effects of measurable disturbance was studied on three simple systems representing basic dynamics expanded with the traffic-delay. In order to verify the control algorithm a simulation scheme was created in SIMULINK environment.

To test the general control capabilities of GPC, as well as the rejection of the measurable disturbance, three systems representing different dynamics were chosen. These systems are described with following continuous transfer functions:

Stable non-oscillatory system:

$$G_1(s) = \frac{2}{4s^2 + 5s + 1} e^{-2s}$$

Stable oscillatory system:

$$G_2(s) = \frac{2}{4s^2 + 2s + 1} e^{-2s}.$$

Non-minimum phase system:

$$G_3(s) = \frac{-10s + 2}{4s^2 + 5s + 1} e^{-2s}.$$

Discrete versions with sampling time $T_0 = 2$s are

$$G_1(z^{-1}) = \frac{-0{,}7419z^{-1} + 0{,}0821z^{-2}}{1 + 0{,}4728z^{-1} + 0{,}2076z^{-2}} z^{-2}$$

$$G_2(z^{-1}) = \frac{-0{,}7859z^{-1} + 0{,}3679z^{-2}}{1 + 0{,}6806z^{-1} + 0{,}4834z^{-2}} z^{-2}$$

$$G_3(z^{-1}) = \frac{-0{,}7419z^{-1} + 0{,}0821z^{-2}}{1 - 1{,}0980z^{-1} + 1{,}7780z^{-2}} z^{-2}$$

Considering systems dynamics, the sizes of control horizon $N_u$ and maximum horizon $N_2$ were set to 30 steps, while minimum horizon $N_1$ remained 1. Control performances for corresponding systems are demonstrated with desired trajectory with the shape of step change. Measurable disturbance was realized as a step function with the same dynamic as controlled system and polynomial $D(z^{-1}) = 1$.

Due to presence of the traffic delay of two sampling steps the values of the minimum and the maximum horizons were increased by the value of the delay steps. Also, the control algorithm has to be provided with information of the measurable disturbance in order to compensate upcoming change. Results can be seen in following figures.

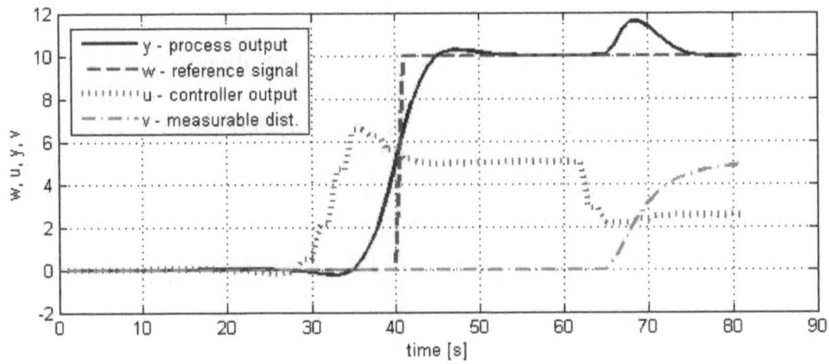


**Fig. 4.** Control of delayed non-oscillatory system $G_1$

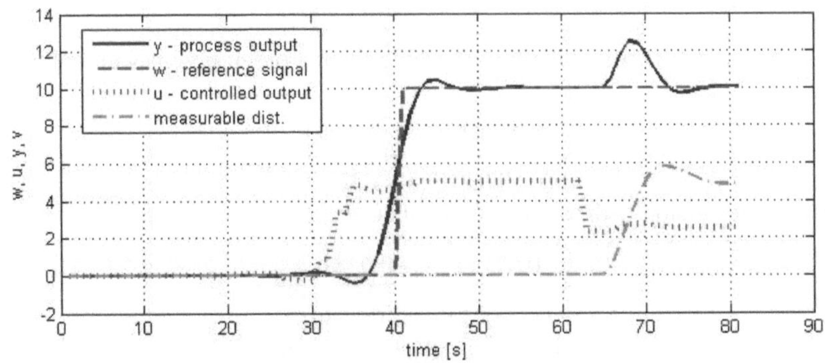


**Fig. 5.** Control of delayed oscillatory system $G_2$

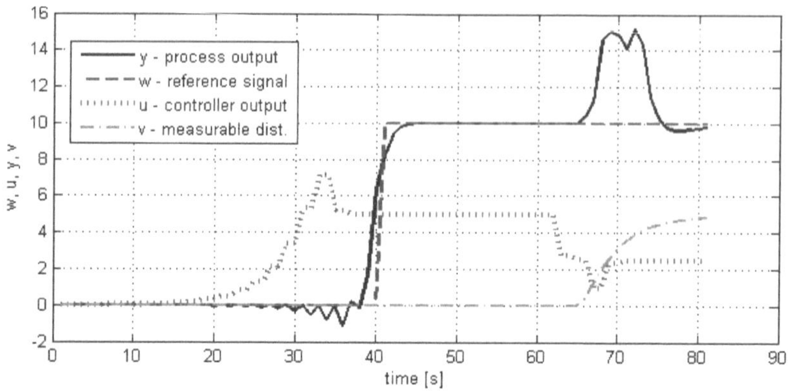


**Fig. 6.** Control of delayed non-minimum phase system $G_3$

Results have proven that GPC algorithm is able to control delayed stable, oscillatory and even non-minimum phase systems.

The ability to compensate measurable disturbance varies depending on each system dynamic. Figure 4 shows the control of the delayed stable non-oscillatory system managing to limit the maximal value caused by the disturbance to 10%. In the case of delayed oscillatory system in Figure 5 is the maximal value increased to 48% due to system overshoot. Significantly higher value occurs with non-minimum phase system in Figure 6 reaching full 100% of disturbance maximum, caused by the nature of the system increasing the error for a limited time.

## 4.1 Simulation Control of the Heat System

Following simulation model is based on identification results of a delayed laboratory heat-system. The real system is constructed as a circulation of water warmed by heater and cooled by fans creating stable circuit with large time constants.
This laboratory model was identified as stable second order system with continuous expression

$$G_4(s) = \frac{0.001614s + 2.664 \cdot 10^{-5}}{s^2 + 0.03017s + 4.075 \cdot 10^{-5}} e^{-120s}$$

and its discrete version with sampling time of 60 seconds

$$G_4(z) = \frac{0.0719z^{-1} - 0.0281z^{-2}}{1 - 1.097z^{-1} + 0.1636z^{-2}} z^{-2}$$

Values of control horizon $N_u$ was set to 30 steps as well as range between maximum horizon $N_2$ and minimum horizon $N_1$, but both values were increased by value of delay $d$. As a tuning mechanism to gain a suitable precision with an appropriate action value, several different settings of weighting values $\delta(i)$ and $\lambda(i)$ ratio was examined.

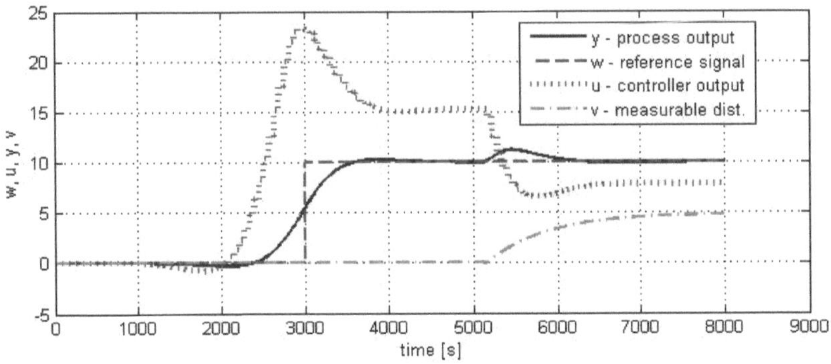


**Fig. 7.** Measurable disturbance compensation in control of the heat system

Figure 7 demonstrates the control performance regulation the error caused by the measurable disturbance to 24%.

## 5 Conclusion

The paper presents results of the simulated control of the laboratory heat system by the GPC. Outcomes of simulations present the functionality of handling measurable disturbances in presence of traffic delay. A significant impact of the ahead known information was proven, as well as the ability of the algorithm to reduce the upcoming error value for the most of basic types of dynamics.

Results have confirmed the ability of GPC to provide a high quality control and its suitability for use in case of real delayed system.

**Acknowledgment.** This article was created with support of Operational Programme Research and Development for Innovations co-funded by European Regional Development Fund (ERDF), national budget of Czech Republic within the framework of the Centre of Polymer Systems project (reg. number: CZ.1.05/2.1.00/03.0111) and in part with support of the Ministry of Education of the Czech Republic under grant No. IGA/FAI/2014/029.

## References

1. Qin, S.J., Badgwell, T.A.: A survey of industrial model predictive control technology. Control Engineering Practice 11, 733–764 (2003)
2. Cutler, C.R., Ramaker, B.L.: Dynamic Matrix Control. In: Proc. Joint Automatic Control Conference, San Francisco, CA, paper WP5-B (1980)
3. Clarke, D.W., Mohtadi, C., Tuffs, P.S.: Generalized Predictive Control – Part I. The Basic Algorithm. Automatica 23, 137–148 (1987)
4. Morari, M., Lee, J.H.: Model predictive control: past, present and future. Computers & Chemical Engineering 23, 667–682 (1999)

5. Normey-Rico, J.E., Camacho, E.F.: Control of Dead-time Processes. Springer, London (2007)
6. Camacho, E.F., Bordons, C.: Model Predictive Control. Springer, London (2004)
7. Haber, R., Bars, R., Schmitz, U.: Predictive Control in Process Engineering: From basics to the applications. Wiley-VCH, Weinheim (2011)
8. Pawlowski, A., Guzmán, J.L., Normey-Rico, J.E., Berenguel, M.: Improving feedforward disturbance compensation capabilities in Generalized Predictive Control. Journal of Process Control 22, 527–539 (2012)
9. Pekař, L., Prokop, R., Dostálek, P.: Circuit heating plant model with internal delays. WSEAS Transactions on Systems and Control 8, 1093–1104 (2009)

# Global and Local Environment State Information as Neural Network Input by Solving the Battleship Game

Ladislav Clementis

Institute of Applied Informatics, Faculty of Informatics and Information Technologies, Slovak University of Technology, Ilkovičova 2, 842 16 Bratislava, Slovakia
clementis@fiit.stuba.sk

**Abstract.** Artificial neural networks are capable of solving various tasks like pattern matching and pattern recognition if well adapted. While solving the Battleship game, we can use additional information about a current state of the environment, like the problem probability model information to enhance the decision process. We show that neural network can be adapted to show the same behavior as the model-based decision making. Neural network response gives us the information about next position in the environment to be shot next. We provide two different neural network input approaches. The first is the global environment state information and the second is bounded local environment state information. Neural networks adapted by data based on local environment states and global environment states show comparable results. We provide simulation results to compare speed of neural network adaptation processes. Results show that adaptation process is faster while using local state information. We used supervised learning based on gradient descent method as a neural network learning technique.

## 1 Introduction

In the area of computer-based optimization tasks [27] we distinguish subsymbolic approaches like neural networks [14,22,24] and symbolic approaches like rule based systems (Learning Classifier Systems [6,12,13]). In both subsymbolic and symbolic approaches, adaptation process, its design and efficiency is concerned [9,10,11].

Reasonable design of intelligent systems is key to their successful application. If some relevant additional information is known, it can be used by system design to increase efficiency, performance and accuracy of these systems [15,18,19].

We provide the example how neural network can be designed to act like reasonable probability based agent to solve probabilistic decision making while solving the simplified Battleship game. We use the simplified Battleship game as a problem of pattern matching. We compare two approaches by their performance, one using global state information, second using local state information.

The simplified version we use in this contribution is a member of *Partially Observable Markov Decision Processes* (*POMDPs*) [4,5,16,24,28]. *POMDP* is *Markov Decision Process* (*MDP*) where problem space is not entirely visible. In the Battleship game, we do not know if hit attempt will result into successful hit so result is not 100% predictable.

I. Zelinka et al. (eds.), *Nostradamus 2014: Prediction, Modeling and Analysis of Complex Systems*, Advances in Intelligent Systems and Computing 289,
DOI: 10.1007/978-3-319-07401-6_28, 

## 2 The Battleship Game

*The Battleship game* is a guessing game for two players. It is known worldwide as a pencil and paper game which dates from World War I. It was published by various companies as a pad-and-pencil game since 1931 [21]. We should mention some known commercial origins of this game:

- The Starex Novelty Co. of NY published game as *Salvo* (1931)
- The Strathmore Co. published game as *Combat, The Battleship Game* (1933)
- The Milton Bradley Company published the pad-and-pencil game *Broadsides, The Game of Naval Strategy* (1943)
- *The Battleship game* was released as a plastic board game by the Milton Bradley Company (1967)

The game has been popularized under the name "*The Battleship game*" by the Milton Bradley Company as it was published as a plastic board game in 1967. The Milton Bradley version is the most popular one version worldwide over decades.

Originally (the Milton Bradley Company version), ships are placed in a battlefield of size of $10 \times 10$. This battlefield includes set of linear (oblong) shaped ships. A player wins, if he reveals al opponent's ships completely before opponent reveals his ships.

### 2.1 The Simplified Battleship Game

For our purpose we provide the simplified version of the Battleship game. Unlike the original Battleship game, our simplified version includes these modifications:

- only single player optimization perspective, minimizing the number of hit attempts (originally two competing players)
- environment size of $n \times n = 7 \times 7 = 49$ "cells", i.e. max. total number of hit attempts (in the worst case) is 49 (originally $n \times n = 10 \times 10 = 100$ cells)
- "L"-shaped patterns with size of 4 cells arranged in $2 \times 3$ shapes, with all 8 permutations allowed (originally 5 linear shaped ships, single *patrol boat* with size of $2 \times 1$, two (*destroyer* and *submarine*) with size of $3 \times 1$, single *battleship* with size of $4 \times 1$ and single *aircraft carrier* with size of $5 \times 1$)

For explanation we provide analogue of the simplified Battleship game with optimization problem terminology:

- player's overview of battlefield - a current state of the environment
- ship hidden in battlefield - a pattern which complete position is our goal to find effectively
- hit attempt - an action performed in environment, revelation of unrevealed cell
- hit (miss) - an environment response

The example state of the environment, which we will use for our purpose is shown in figure 1.

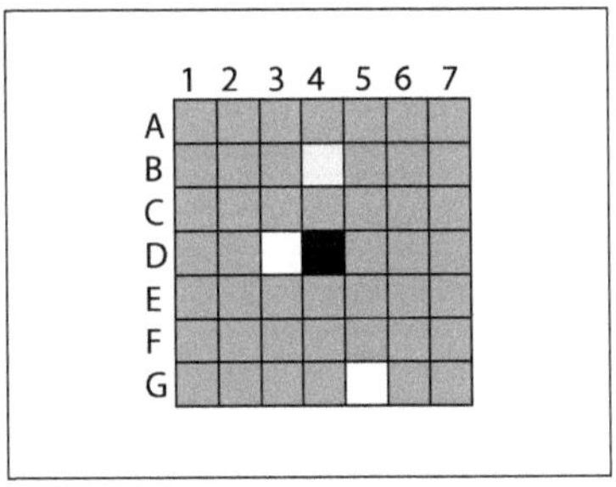


**Fig. 1.** Figure showing the current state of the environment. Gray cells are not revealed. Black and white cells have been already revealed. Each white cell is empty, each black cell is not emty, i.e. contains a part of the pattern.

## 2.2 Probability-Based Heuristics to the Simplified Battleship Game

According to "L"-shaped pattern placed in the environment, there are 8 possible pattern configurations, i.e. pattern permutations $P = \{P_1, P_2, ..., P_8\}$ shown in figure 2. Each of this permutations is possible to be the one present in our current state of the environment shown in figure1 [1,3].

In the current state of the environment shown in figure 1, for each cell we can calculate the probability of being covered by pattern. This calculation is based on the number of possible pattern permutations covering each cell. There are multiple possibilities how pattern can be placed in the environment, according to the current state of the environment. Table 1 is showing that there are totally 17 possible ways how pattern can be present in the environment.

Therefore, each cell is covered with some non-negative number of pattern permutations. If a pattern permutation overlaps the cell, it's covering number is increased by 1. Non-negative covering number ($Coverings_i$) of each cell in the current state of the environment are shown in figure 3.

Normalised $CoveringProbability_i$ of each cell $c_i$ is computed as the number of cell coverings divided by sum of covering numbers in the current state of the environment

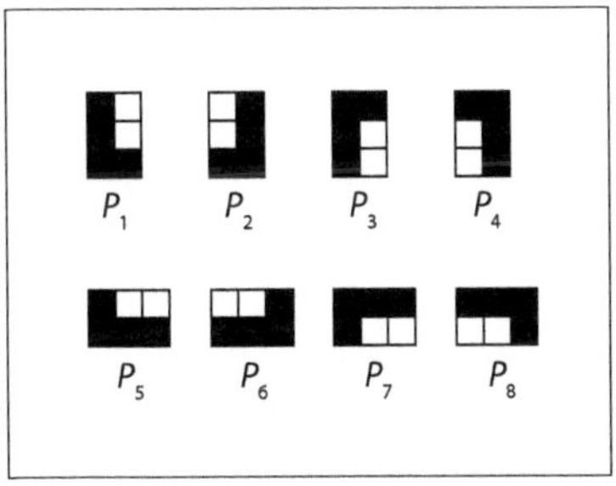


**Fig. 2.** All 8 possible pattern permutations, 4 vertical and 4 horizontal

**Table 1.** Numbers of possible pattern positions with correspondence to the current state of the environment. Numbers correspond to all possible pattern permutations.

| Pattern Permutation Option: | $P_1$ | $P_2$ | $P_3$ | $P_4$ | $P_5$ | $P_6$ | $P_7$ | $P_8$ | $\Sigma$ |
|---|---|---|---|---|---|---|---|---|---|
| Numbers of Possible Placements: | 2 | 3 | 2 | 2 | 2 | 2 | 2 | 2 | 17 |

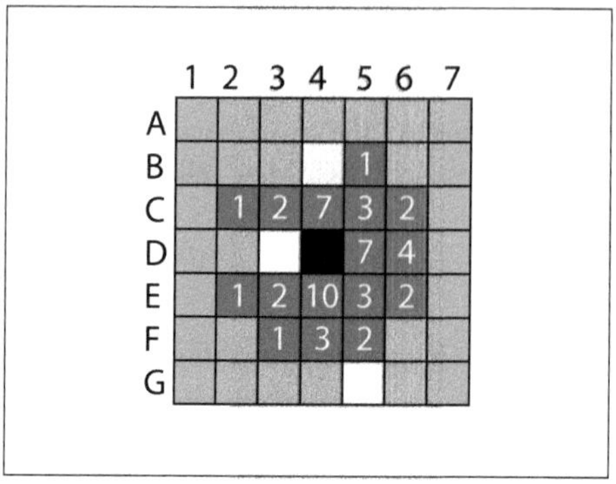


**Fig. 3.** The current state of the environment, enriched by non-zero values of pattern placement covering of each cell

(51). $HitProbability_i$ of each cell $c_i$ is computed as the number of cell coverings divided by sum of possible pattern permutations (17).

Relevant non-zero values of $CoveringProbability_i$ and $HitProbability_i$ are shown in table 2. Note that $HitProbability_i$ values are 3× higher than $CoveringProbability_i$ values, because we are searching for 3 unrevealed parts of the pattern in the current state of the environment.

Choosing cell $E4$, is therefore the highest probability candidate for next hit attempt. Cells $C4$ and $D5$ are considerable too with relatively high $HitProbability_i$ values.

Outputs of neural network adapted by the supervised learning technique were in correspondence with this probability-based heuristics with the precision up to 96.12%. This correlation has proven that this probability-based heuristics is effective, and 3-layer feedforward neural network described in section 3 is adaptable to solve the simplified Battleship game effectively.

## 3 Neural Network by Solving the Battleship Game

We use the 3-layer feedforward neural network shown in figure 4.

We use transformation of the current state of the environment as the two-dimensional neural network input vector $X$ defined by equation 1 [1,2]. If we use the whole environment information, input of the neural network is the global environment state.

$$X = (x_{11}, x_{12}, ..., x_{ij}, ..., x_{nn}) \in \{-1, 0, 1\}^{n \times n} \quad (1)$$

**Table 2.** Relevant non-zero $CoveringProbability_i$ and $HitProbability_i$ values

| Cell $c_i$ | $Coverings_i$ | $CoveringProbability_i$ | $HitProbability_i$ |
|---|---|---|---|
| B1 | 1 | $1/51 = 0.0196$ | $1/17 = 0.0588$ |
| C2 | 1 | $1/51 = 0.0196$ | $1/17 = 0.0588$ |
| C3 | 2 | $2/51 = 0.0392$ | $2/17 = 0.1176$ |
| **C4** | **7** | $7/51 = \mathbf{0.1373}$ | $7/17 = \mathbf{0.4118}$ |
| C5 | 3 | $3/51 = 0.0588$ | $3/17 = 0.1765$ |
| C6 | 2 | $2/51 = 0.0392$ | $2/17 = 0.1176$ |
| **D5** | **7** | $7/51 = \mathbf{0.1373}$ | $7/17 = \mathbf{0.4118}$ |
| D6 | 4 | $4/51 = 0.0784$ | $4/17 = 0.2353$ |
| E2 | 1 | $1/51 = 0.0196$ | $1/17 = 0.0588$ |
| E3 | 2 | $2/51 = 0.0392$ | $2/17 = 0.1176$ |
| **E4** | **10** | $10/51 = \mathbf{0.1961}$ | $10/17 = \mathbf{0.5882}$ |
| E5 | 3 | $3/51 = 0.0588$ | $3/17 = 0.1765$ |
| E6 | 2 | $2/51 = 0.0392$ | $2/17 = 0.1176$ |
| F3 | 1 | $1/51 = 0.0196$ | $1/17 = 0.0588$ |
| F4 | 3 | $3/51 = 0.0588$ | $3/17 = 0.1765$ |
| F5 | 2 | $2/51 = 0.0392$ | $2/17 = 0.1176$ |
| Σ | **51** | $51/51 = \mathbf{1}$ | $51/17 = \mathbf{3}$ |

Vector $X$ member values are based on information about cells in a current state of the environment, thus which cell is revealed (with outcome) and which is unrevealed. Vector $X$ member $x_{ij} = 0$ if position $c_{ij}$ is unrevealed. If position is revealed, $x_{ij} = -1$ value indicates that cell $c_{ij}$ is empty, $x_{ij} = 1$ value indicates that cell $c_{ij}$ is not empty. Formalized:

$$x_{ij} = \begin{cases} -1 & \text{if cell } c_{i,j} \text{ is revealed, empty} \\ 0 & \text{if cell } c_{i,j} \text{ is unrevealed} \\ 1 & \text{if cell } c_{i,j} \text{ is revealed, not empty} \end{cases} \quad (2)$$

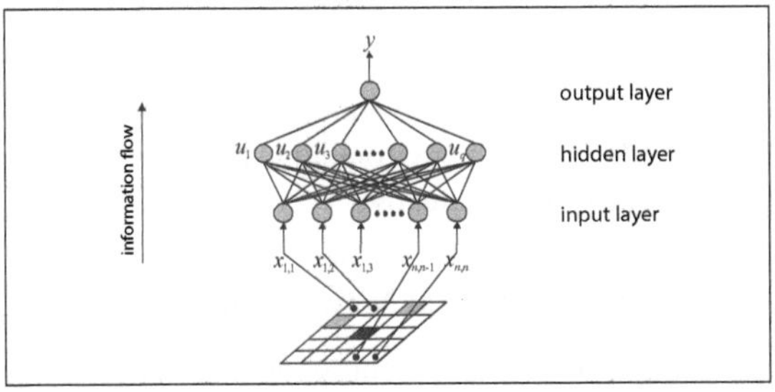


**Fig. 4.** Diagram showing an abstract overview of used 3-layer feedforward neural network

The neural network shown in figure 4 contains the hidden layer defined by equation 3.

$$U = (u_1, u_2, ..., u_\alpha, ..., u_q) \in [0,1]^q \tag{3}$$

The neural network output $y$ defined by equation 4 returns the real number variable from the range $[0,1]$.

$$y = (y_1) \in [0,1]^1 \tag{4}$$

In the current state of the environment, we create a set of all possible next states. Each new state is created as the current state of the environment with single possible hit attempts performed. Each of these possible next states is evaluated by neural network. Result of this evaluation is neural network output activity $y$ from the range $[0,1]$. Thus, the state with the highest neural network output activity $y$ is considered as the one chosen by neural network.

Neural network input to neural network output mapping is modified by changing neuron weights and threshold values by some learning technique. Thus a neural network is adapted.

This simple feed-forward neural network can be adapted to solve the Battleship game effectively. Formally, artificial neuron weights $w_{\alpha;i,j}$ and threshold $\vartheta_\alpha$ values are approximated. There are common approaches used for neural network adaptation process:

1. Supervised learning [25]
2. Unsupervised learning [8,20]
3. Reinforcement learning [7,17,23,24] like Q-Learning [26,28]

For our purpose, we use the supervised learning technique. We use the gradient method of the steepest descend to make the adaptation process simple and straightforward.

Our training set consists of 10000 randomly created environment states, with randomly placed patterns. We have used approach by using global state information in comparison to approach by using local state information.

### 3.1 Global State Information as Neural Network Input

If we use whole vector $X$ shown in equation 1 as an input of neural network, we use global state information. This vector consists of $n \times n = 7 \times 7 = 49$ vector components: $(x_{11}, x_{12}, ..., x_{ij}, ..., x_{nn})$.

Size of input for neural network influences adaptation process. The bigger the input is, more adaptation iterations are needed to adapt neural network sufficiently.

In section 3.2 we provide bounded environment state as an input of neural network to make the adaptation process more effective.

### 3.2 Local State Information as Neural Network Input

While using global state of the environment as neural network input, there is more information that needed. Figure 5 is showing bounded state of the environment used as our example.

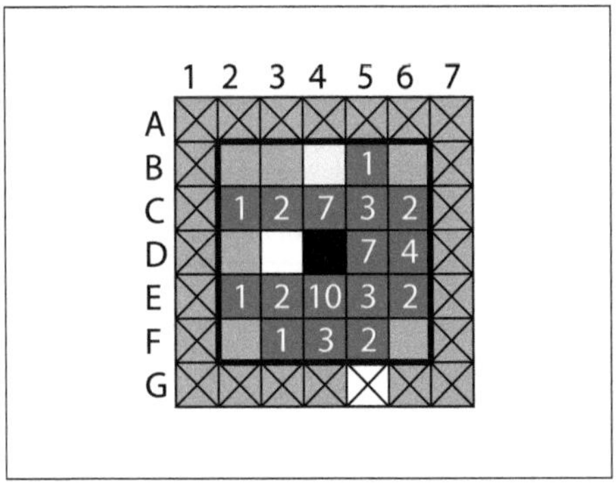


**Fig. 5.** Figure showing bounded state of the environment. Concerned area where the pattern is located is within sub-area highlighted by bold square.

Cells within highlighted sub-area shown in figure 5 is proper local environment sate information. This information is sufficient enough for reasonable decision making.

In our example state of the environment, this sub-area ranges from $B2$ cell to $F6$ cell. While using this sub-area as neural network input, we use reduced vector $X_k$ with size of $k \times k = 5 \times 5 = 25$.

This approach reduces neural network input size from $X = 49$ to $X_k = 25$. Input size is then reduced to $25/49 = 51.02\%$ of its original size.

Simulation results shown in section 4 prove that neural network with reduced input is adapted more effectively. Important fact is that while using reduced input, outputs of adapted neural network do correlate with probabilistic heuristic approach explained in section 2.2.

## 4 Simulation Results

Our results have shown that both approaches, neural network adaptation by using global and/or local environment state information are capable of learning probabilistic model of the simplified Battleship game. Results showing adaptation quality are shown in table 3.

Average simulation results in table 3 have shown that neural network adapted by global state information needed almost 8000 adaptation runs to give output activity correlated to the probabilistic heuristic approach with ratio above 90%. It took less than 1000 adaptation runs ($1000/8000 = 12.5\%$) for neural network adapted by local information state to correlate with probabilistic approach with ratio of 90%.

Neural network weights and thresholds were almost stabilized after 1400 adaptation runs while using local state information. While using global state information, weights and thresholds begin to stabilize after 12000 simulation runs ($1400/12000 = 11,67\%$).

These results are showing that while using local information state information, we can acquire comparable results of neural network adaptation quality (approximately 95%). But while using local state information as neural network input, adaptation process is more effective.

**Table 3.** Table showing the efficiency and of neural network adaptation process while using global and local environment state information. Quality ratio of adaptation process is measured by correlation of neural network output activity with the probabilistic approach. We have performed 100 simulations of adaptation processes. Table includes average values.

| Number of Iterations | Global State Information $X$ | Local State Information $X_k$ |
|---|---|---|
| 0 | 4.156% | 7.367% |
| 100 | 6.845% | 9.431% |
| 200 | 7.702% | 14.057% |
| 300 | 8.212% | 20.620% |
| 400 | 10.940% | 27.391% |
| 500 | 14.763% | 38.609% |
| 600 | 17.318% | 49.160% |
| 700 | 20.046% | 63.514% |
| 800 | 26.736% | 76.603% |
| 900 | 37.820% | 88.481% |
| 1000 | 48.014% | 92.705% |
| 2000 | 61.512% | 94.381% |
| 4000 | 72.694% | 95.835% |
| 6000 | 85.125% | 95.704% |
| 8000 | 92.304% | 95.324% |
| 10000 | 93.548% | 95.513% |
| 15000 | 95.614% | 95.166% |
| 20000 | 96.047% | 95.407% |

## 5 Conclusion

We have compared two alternate approaches of neural network design while solving the Battleship game. We have proven that while using relevant local state information we can acquire comparable results as while using global information.

While using local state information, the neural network input is shorter what results into much faster adaptation speed as while using global state information. Adaptation process was approximately $1/8 = 12.5\%$ more effective while using local state information.

Reasonable complex systems design can rapidly increase their efficiency and performance. If applied to specific problems, additional information can be used.

**Acknowledgement.** This contribution was supported by the VEGA (Slovak Scientific Grant Agency) of the Ministry of Education of the Slovak Republic (ME SR) and of the Slovak Academy of Sciences (SAS) under the contract No. VEGA 1/0458/13 and VEGA 1/0553/12.

## References


1. Clementis, L.: Model driven classifier evaluation in rule-based system. In: Snasel, V., Abraham, A., Corchado, E.S. (eds.) SOCO Models in Industrial & Environmental Appl. AISC, vol. 188, pp. 267–276. Springer, Heidelberg (2013)
2. Clementis, L.: Supervised and reinforcement learning in neural network based approach to the battleship game strategy. In: Zelinka, I., Chen, G., Rössler, O.E., Snasel, V., Abraham, A. (eds.) Nostradamus 2013: Prediction, Model. & Analysis. AISC, vol. 210, pp. 191–200. Springer, Heidelberg (2013)
3. Drugowitsch, J.: Design and Analysis of Learning Classifier Systems: A Probabilistic Approach. SCI, vol. 139. Springer, Heidelberg (2008)
4. Dung, L.T., Komeda, T., Takagi, M.: Mixed reinforcement learning for partially observable markov decision process. In: International Symposium on Computational Intelligence in Robotics and Automation Proceedings, pp. 7–12. IEEE (2007)
5. Dung, L.T., Komeda, T., Takagi, M.: Reinforcement learning for partially observable markov decision process using state clasification. Applied Artificial Intelligence 22(7-8), 761–779 (2008)
6. Halavati, R., Shouraki, S.B., Lotfi, S., Esfandiar, P.: Symbiotic evolution of rule based classifier systems. International Journal on Artificial Intelligence Tools 18(1), 1–16 (2009)
7. Harmon, M., Harmon, S.: Reinforcement learning: A tutorial (1996)
8. Hinton, G., Sejnowski, T.: Unsupervised Learning: Foundations on Neural Computation. Computational Neuroscience Series. MIT Press (1999)
9. Holland, J.H.: Adaptation in Natural and Artificial Systems. The University of Michigan Press (1975)
10. Holland, J.H.: Adaptation in Natural and Artificial Systems. MIT Press (1992)
11. Holland, J.H.: Adaptation in Natural and Artificial Systems: An Introductory Analysis with Applications to Biology, Control and Artificial Intelligence. MIT Press, Cambridge (1992)
12. Holland, J.H., et al.: What is a learning classifier system? In: Lanzi, P.L., Stolzmann, W., Wilson, S.W. (eds.) IWLCS 1999. LNCS (LNAI), vol. 1813, pp. 3–32. Springer, Heidelberg (2000)
13. Kovacs, T., Llorà, X., Takadama, K., Lanzi, P.L., Stolzmann, W., Wilson, S.W. (eds.): IWLCS 2003-2005. LNCS (LNAI), vol. 4399. Springer, Heidelberg (2007)
14. Kriesel, D.: A Brief Introduction to Neural Networks, Zeta version (2007), `http://www.dkriesel.com` (January 2014)
15. Lacko, P., Kvasnicka, V., Pospichal, J.: An emergence of game strategy in multiagent systems. International Journal of Computational Intelligence and Applications 4(3), 283–298 (2004)
16. Li, H., Liao, X., Carin, L.: Region-based value iteration for partially observable markov decision processes. In: Proceedings of the 23rd International Conference on Machine Learning, ICML 2006, pp. 561–568. ACM (2006)
17. Mohri, M., Rostamizadeh, A., Talwalkar, A.: Foundations of Machine Learning. Adaptive Computation and Machine Learning Series. MIT Press (2012)
18. Qudrat-Ullah, H., Spector, J.M., Davidsen, P.I.: Complex Decision Making: Theory and Practice. Understanding Complex Systems. Springer (2008)
19. Qudrat-Ullah, H., Spector, J.M., Davidsen, P.I.: Complex Decision Making: Theory and Practice. New England Complex Systems Institute Book Series. Springer (2010)
20. Sanger, T.D.: Optimal unsupervised learning in a single-layer linear feedforward neural network. Neural Networks 2(6), 459–473 (1989)
21. Sevenster, M.: Battleships as a decision problem. ICGA Journal 27(3), 142–149 (2004)
22. Smith, M.: Neural Networks for Statistical Modeling. Thomson Learning (1993)

23. Sutton, R.S., Barto, A.G.: Reinforcement Learning: An Introduction. Adaptive computation and machine learning. MIT Press (1998)
24. Takita, K., Hagiwara, M.: A pulse neural network reinforcement learning algorithm for partially observable markov decision processes. Systems and Computers in Japan 36(3), 42–52 (2005)
25. Vapnik, V.: The Nature of Statistical Learning Theory. Statistics for Engineering and Information Science Series. Springer (2000)
26. Watkins, C.J.C.H., Dayan, P.: Q-learning. Machine Learning 8(3-4), 279–292 (1992), `http://jmvidal.cse.sc.edu/library/watkins92a.pdf` (January 2014)
27. Weise, T.: Global Optimization Algorithms - Theory and Application, 2nd edn. Self-Published (2009), `http://www.it-weise.de/` (January 2014)
28. Wiering, M.A., Kooi, T.: Region enhanced neural q-learning for solving model-based partially observable markov decision processes. In: The 2010 International Joint Conference on Neural Networks (IJCNN), pp. 1–8. IEEE (2010)

# Multi Agent Environment for Modelling and Testing of Cooperative Behaviour of Agents

Martin Mariška and Petr Dolezel

Faculty of Electrical Engineering and Informatics, University of Pardubice, Studenska 95
mariska.martin@gmail.com, petr.dolezel@upce.cz

**Abstract.** The article introduces a new universal software environment for multi agent simulation. The environment presents the model of reality and it is suitable for testing of the existing or new methods, especially in the field of agent coordination, cooperation or collaboration. The software is aimed to be used in any Java based simulation tool and to provide a consistent approach for gathering comparable results among these tools. Moreover, it can be applied in problems such as coordinated decision making, attitude alignment problem, robot position synchronization, dynamic obstacle avoidance and more others. The concept of this tool helps to create programs which are easily transferable and applicable to real robotic platforms.

## 1 Introduction

For centuries, mankind has been seeking the aid of machines to help with tasks that are too difficult or impossible to solve. Therefore, the humans aim their re-search at areas of robotics. Robot technology has grown significantly in the last 30 years with more and more examples and applications seen every day. Although there is still a considerable amount of research continuing in humanoid robots today, a major focus is devoted to mobile robotics or autonomous vehicles. These robots take the form of all types of vehicles including cars, planes, boats, submarines and even spacecrafts that can operate without the need for human control. Autonomous vehicles are useful especially in areas where a human presence is particularly expensive or dangerous. Some of the most famous autonomous vehicles in recent history have been the exploration rovers that were sent to explore the surface of the planet Mars. Those two vehicles, Spirit and Opportunity, captured the attention of the world with their ability to go where we as humans could not.

As robotics technology evolves, there is a growing need for multiple vehicles to work together to solve more complex tasks, which leads to a new problem. The problem is how to combine and reuse specific capabilities of each robot in group. Collective intelligence (CI) is one of possible solutions. The CI is scientific branch focused at intelligent group behaviour like coordination and cooperation of individual entities [6]. The approaches and algorithms for achieving coordination or cooperation are still evolving and are typically modelled and used in multi agent systems [1, 2]. Multi agent paradigm is one of methods suitable for modelling complex, distributed and

I. Zelinka et al. (eds.), *Nostradamus 2014: Prediction, Modeling and Analysis of Complex Systems*, Advances in Intelligent Systems and Computing 289,
DOI: 10.1007/978-3-319-07401-6_29, © Springer International Publishing Switzerland 2014

dynamic systems and it preserves the simple structure of individual components in the model [3, 4, 8]. The multi agent system is typically composed of the environment, the population of agents, their mutual communication and the interactions between agents and environment [7]. Each environment defines its own interactions and creates specific constraints for entities in it as well as humans are restricted by the physical laws of real world. Therefore, we developed the environment for agents as the model of real world. That allows developers and researchers to easily simulate and to consistently compare the existing or new methods of collaboration and cooperation, collective planning and decision making solutions.

## 2 Intent of Environment

The main purpose of the presented environment is to provide unified and consistent tool for testing and result comparison of different techniques that achieve required agent's behaviour. Today, there is still no tool which would allow us to implement various techniques and their support testing in one environment in order to compare the results of experiments easily.

The next objective is to create flexible and open version of the environment and provide it for many researchers. The developed environment is written in the Java programming language to accomplish both objections. Java fulfils the portability across platforms, and is also one of languages suitable for implementation of multi agent systems [9]. For even better usability we are planning to add a socket interface. This could also make the environment absolutely independent of the subsequent programming language.

The level of abstraction for real environment modelling is aimed on two dimensions only. The reason for choosing two dimensions is the possibility of clear visualization and graphical reconfiguration of the environment in a short time. However, it still provides many possibilities for finding suitable problems to test new methods. Moreover, two dimensions are easily and clearly viewable on all currently available display devices and researchers do not need to learn how to operate complicated modelling tools. The solutions created in this environment are easily transformable to the real robotic platform. The important assumption is that condition in reality are supposed to be the same as the basic conditions of the environment such as orientation in two dimensions or moving in grid-based environment.

## 3 Capabilities and Properties of Environment

The main advantages of the environment are straightforward and fast reconfiguration. They are achieved thanks to using the 2D abstraction level and graphical user interface, as it is presented in Fig. 1. The actual environment configuration can be persisted and it can be automatically reloaded at the start of new simulation. The next advantage is the environment visualization and graphical representation of all elements in it, including agents. The actual state of the environment can be visualized during the simulation.

The environment has the set of prepared blocks for creating specific problems. Each block simulates some real world components. For example, if the wall block occupies a field, any other entity cannot enter it. A box block is an object that can be moved by agents or serves only as a mark in indirect communication between agents. The use of blocks is not limited by number and all types of blocks can be combined variously.

Other important features of the environment are external events and signals. The developers can observe the external events and they are allowed to react in specific situations occurred in the environment. The signals are used for gathering statistical information about individual agents or the whole environment and can be viewed using a graphical interface.

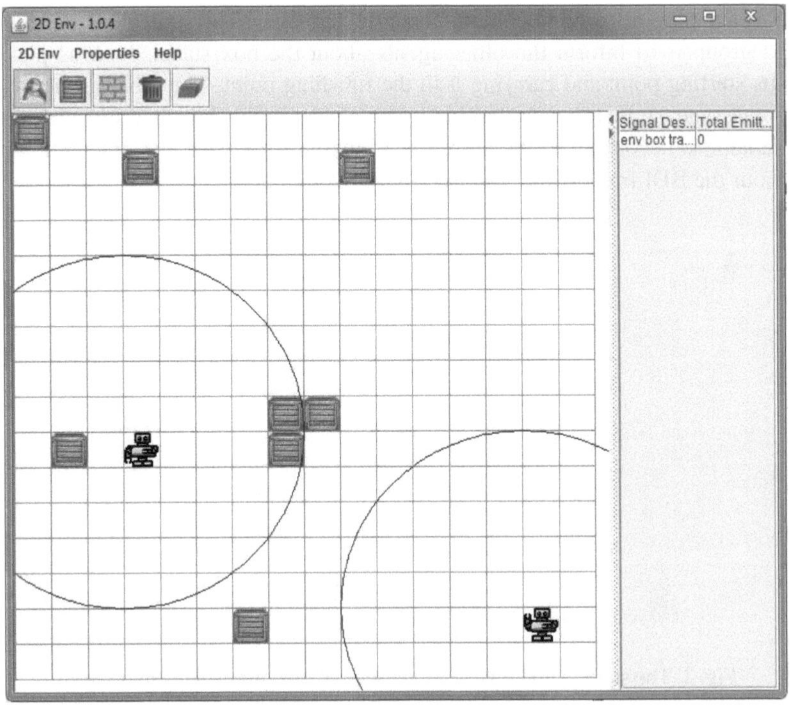


**Fig. 1.** Visualization and control window of presented multi agent environment with 2 robots

The modelling problems associated with space orientation and object transportation are not the only use of the environment. The application on typical multi agents problems such as coordinated decision making, attitude alignment, flocking, rendezvous problem, robot position synchronization or obstacle avoiding is also possible.

# 4 Validation Experiment

The aim of the validation experiment is to show that the designed environment properly approximates the behaviour of real environments. The example verifies the correct functionality of the environment and also presents the benefits of cooperative behaviour, even in simple cases. More information about verification and validation of agent based models is in [5].

The selected problem is designed for a group of agents. Group has 4 members in experiment. All members of the group have the same behaviour, and all of them start with the same problem description. The basic task for the agents is to move all boxes from the defined starting point A to the destination point B as it is presented in Fig. 2 and Fig. 4. The unique description of each box is the part of starting information for agents. The quality of the problem solution is measured in time units needed to move all boxes to their destinations. The environmental restrictions are that each agent can carry only one box and can be only on one field at the same time.

We compare two groups of agents with almost the same behaviour. The second group of agents is allowed to communicate by messages. The strategy of agents in the second group is to inform the other agents about the box status, such as picking it from its starting point and carrying it to the finishing point. The other agents respond to this message, so that they are no longer interested in this box and they can reschedule the actions according to the currently available issue. For implementation of the behaviour the BDI framework was used in this example.

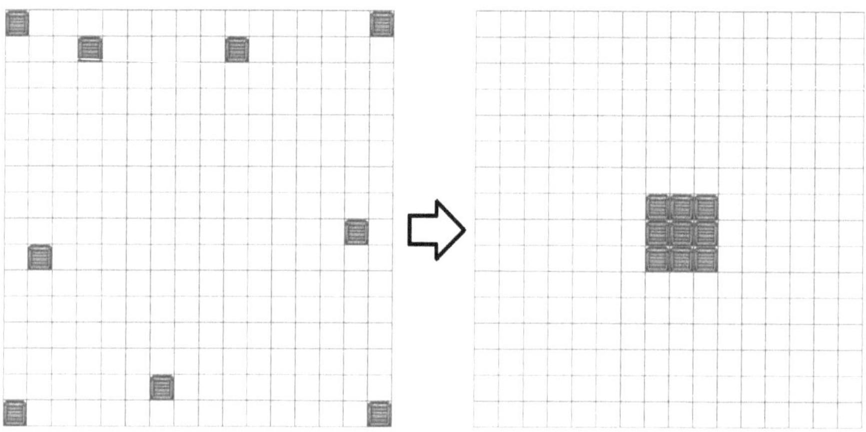

**Fig. 2.** The starting and final configuration of box problem in environment

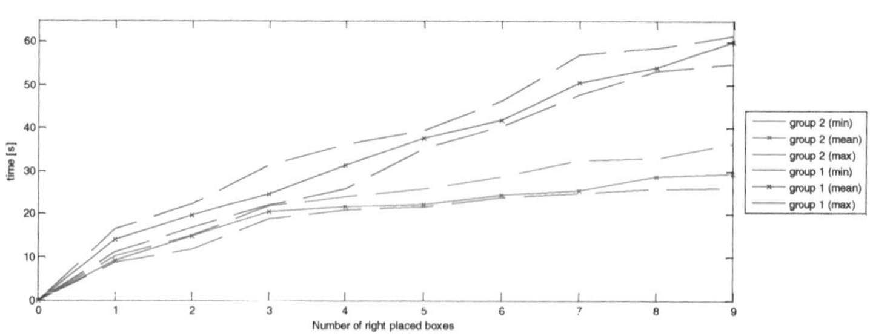


**Fig. 3.** Number of boxes at correct positions in time. *Group 2* is using cooperative behaviour.

The experiment confirms the original assumption, that the second group of agents will be faster. This is due to the fact that if agent passes the information about the box by using direct communication with other agents, it will reduce the travel time of other agents. The agents of the first group receive the information about the transported box only if they occur near to its defined starting point. This piece of information is only based on data from the environment if the agents of the first group do not find a box on its defined starting point, they assume that the box has been carried off by any other agent and they continue replacing other boxes. Results of experiment are in Fig. 3.

The travel time approves the conformity of the model with reality. All agents in the environment can move with the same speed and if they have to travel further, it will take them longer time.

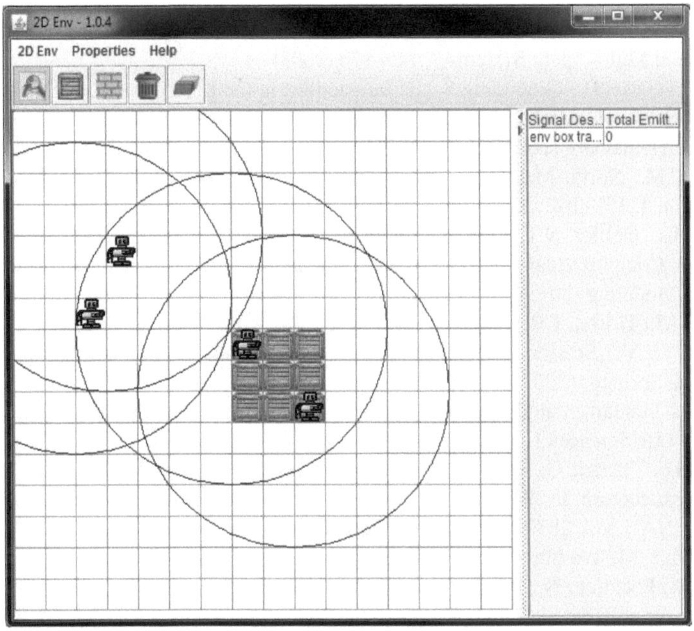


**Fig. 4.** Group of robots with solved problem in visualization. (blue circle is sensing range)

## 5 Conclusion

The main aim of this article is to introduce a new universal software environment for multi agent simulations. The environment is a model of real environment and has a predefined behaviour approximating the real one. For supporting the validity and usage of the environment, there is presented an example in this paper. The environment is suitable for modelling of the problems associated with space orientation such as coordinated decision making, attitude alignment problem, rendezvous problem, robot position synchronization, dynamic obstacle avoidance and others. Researchers can use the software for testing of different techniques and regularly comparing the results between methods or even different simulation tools. The presented environment can be used in any simulation tool or modelling system based on Java language and in any operation system. Its benefits are the possibility to visualize the current state of the environment, easy control and use, and large number of options for creating the suitable environmental configuration for modelled problem.

**Acknowledgments.** The work has been supported by the funds of the IGA, University of Pardubice, Czech Republic. This support is very gratefully acknowledged.

## References

1. Brooks, C.H., Durfee, E.H.: Congregation Formation in Multiagent Systems. Autonomous Agents and Multi-Agent Systems 7, 145–170 (2003)
2. Klein, F., Giese, H.: Grounding Social Interactions in the Environment. In: Weyns, D., Van Dyke Parunak, H., Michel, F. (eds.) E4MAS 2005. LNCS (LNAI), vol. 3830, pp. 139–162. Springer, Heidelberg (2006)
3. Macal, C.M., North, M.J.: Tutorial on Agent-based Modeling and Simulation. Journal of Simulation 4, 151–162 (2010)
4. North, M.J., Collier, N.T., Ozik, J., Tatara, E., Altaweel, M., Macal, C.M., Bragen, M., Sydelko, P.: Complex Adaptive Systems Modeling with Repast Simphony. Complex Adaptive Systems Modeling (2013)
5. North, M.J., Howe, T.R., Collier, N.T., Vos, J.R.: A Declarative Model Assembly Infrastructure for Verification and Validation. In: Advancing Social Simulation: The First World Congress. Springer (2007)
6. Sulis, W.: Fundamental Concepts of Collective Intelligence. Nonlinear Dynamics, Psychology, and Life Sciences 1, 35–53 (1997)
7. Weyns, D., Vizzari, G., Holvoet, T.: Environments for Situated Multi-agent Systems: Beyond Infrastructure. In: Weyns, D., Van Dyke Parunak, H., Michel, F. (eds.) E4MAS 2005. LNCS (LNAI), vol. 3830, pp. 1–17. Springer, Heidelberg (2006)
8. Wooldridge, M.: An Introduction to MultiAgent Systems (2009)
9. Wong, D., Paciorek, N., Walsh, T., DiCelie, J., Young, M., Peet, B.: Concordia: An infrastructure for collaborating mobile agents. In: Rothermel, K., Popescu-Zeletin, R. (eds.) MA 1997. LNCS, vol. 1219, pp. 86–97. Springer, Heidelberg (1997)

# Traffic Simulation Study: Traffic Intersection in the Village of Dobrá

Jan Vrobel and Michal Janošek

University of Ostrava, 30. dubna 22, 702 00 Ostrava, Czech Republic

**Abstract.** This article aims at presenting the results of a simulation study of the traffic intersection in the village of Dobrá near Frýdek-Místek. The article deals with creation of a model of the particular intersection, from object and system definition to model implementation and its validation with a real system. After the model is validated, several experiments are presented on the model with the aim to examine the influence of a number of interventions on the model, such as average speed of vehicles, average number of stops, and average travel time. A simulation tool for the experimental study was the CityTrafficSimulator, which is based on a so-called car-following model.

## 1 Traffic Simulation

Traffic simulation is one of the often used techniques for investigating traffic systems. It is a subdiscipline of informatics which started to develop after the First World War, but more rapid development occurred after the Second World War with modern information technologies and due to increasing traffic as well (Pursula 1999). Traffic situation is quite a complex system which is designed purely by men. A simulation study does not provide exact information on how the particular traffic situation will look like in future but it shows us some prediction how this system should look like or behave. Precision of this study relies on many factors; chosen point of view (bottom-up, top-down), level of abstraction, selected elements of the system and their properties, model validation and verification, measurements accuracy, discretization error, etc. Concerning the basic point of view, the bottom-up approach can grasp the modelled reality best way because we are able to model each particular element of the modelled reality. Microscopic simulation tools best suit to this purpose (O'Donoghue 2001, Li and O'Donoghue 2013).

A simulation study which is done in advance can answer a lot of questions, such as how a closure can influence the traffic, how to control traffic flow through a crossroad to be as smooth as possible, where it is best to place a parking lot, where to establish mass transport lines, or even other questions such as an impact on the environment, parking fee revenues, etc (Pelánek, 2011). For most of today's transport systems of large cities, crossroads or big intersections, it is impossible to imagine the traffic to operate without any previous simulation study. Generally, the term traffic simulation does not only encompass road traffic simulation, but it also includes railway, naval or air traffic as well. We can count with pedestrians or other important elements participating in traffic. There may be situations where a first look logical solution will not be

I. Zelinka et al. (eds.), *Nostradamus 2014: Prediction, Modeling and Analysis of Complex Systems*, Advances in Intelligent Systems and Computing 289,
DOI: 10.1007/978-3-319-07401-6_30, © Springer International Publishing Switzerland 2014

ideal. For example, an increase of vehicles speed could lead to a decrease of driving fluency in cities. Simulation models are typically based on mathematical equations, which control the entire operation. Most of the available tools are built on invented principles and most tools make use of the car-following model (Newell 2002).

The car-following model is the most used mathematical model. Most companies have altered it to suit their needs but all modifications are built on same basis (Pelánek 2011). This model controls agent's behaviour (representing a particular vehicle) towards already released agent. Agents do not know the situation of the wide surroundings but only the closest area in the vicinity of the agent. In the case that the first agent stops or slows down, the following agent has to alter its speed to avoid a collision.

## 2 Simulation Tool – CityTrafficSimulator

A simulation tool for the experimental study was the CityTrafficSimulator (Schulte 2014). For our case, the possibilities of the tool are sufficient, there is no need to do any programming because it is possible to create a model using GUI or XML file. Although the program is relatively simple compared to competitors it is easy to handle and it can still work with hundreds of vehicles. It is free to download from the manufacturer's site under a GPL license.

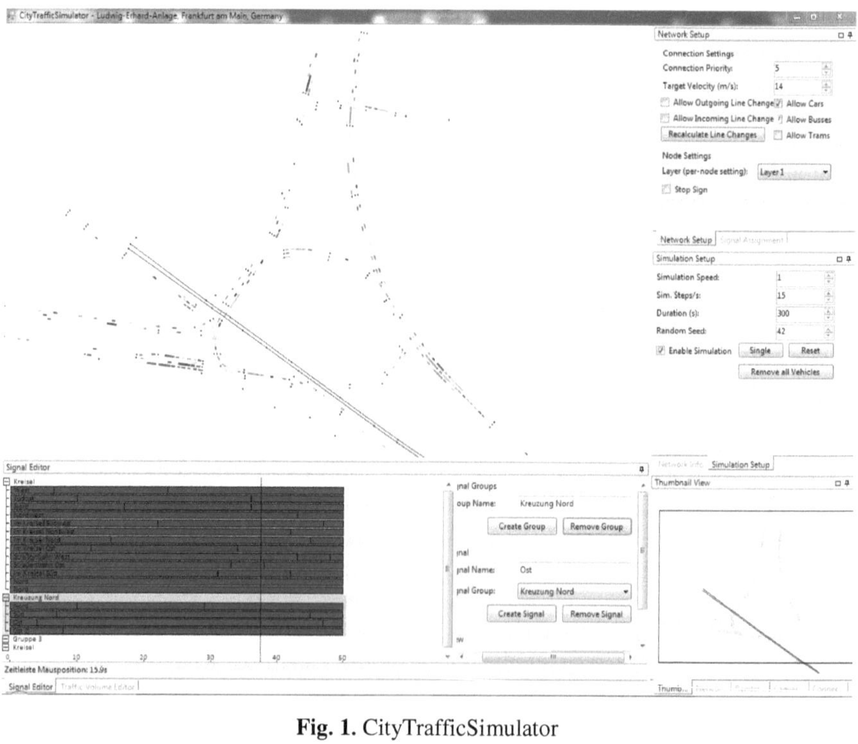


**Fig. 1.** CityTrafficSimulator

Four types of vehicles are supported – cars, trucks, buses and trams. To compute vehicles' moves, the program uses general car-following model. To approach the real situation, it is offered to load bitmap as a background for a model. The tool evaluates simple statistics such as average speed, number of stops of vehicles, etc.

# 3 Creation of a Model

For our model the traffic intersection at the edge of the village of Dobrá was chosen (Figure 2). Most of the time there is no problem with traffic. But it changes with peak hours when employees from the recently-built automotive factory come to work or depart the factory. This peak occurs with regard to the three-shift operation 6 times a day, because the employees coming to work have to arrive before those leaving their shift. Another disadvantage is that the intersection was not originally built as a highway junction, which it also overloads it.

**Fig. 2.** Traffic intersection in the village of Dobrá (maps.google.com)

**Examined Subject Definition**
Intersection description (Figure 2):

- Arrival/departure from village of Nošovice/factory (red).
- Arrival/departure from village of Dobrá (yellow).
- Arrival/departure from village of Vojkovice (blue).
- Arrival/departure from village of Pazderná (green).
- The main road is between red and green arrow.

Intersection problems:

- Lot of accidents – due to the increased level of drivers' nervousness.
- Drivers wait more than 5 minutes on the side road, inexperienced drivers or heavy vehicles cannot cross the road.
- Buses from side roads are delayed.
- Too high speed of drivers going on the main road (speeding).
- Bad view angle due to acoustic barriers.

Goals:

- To reduce accident rate.
- To try to make traffic flow as smooth and fast as possible.

Our task is to create a model of this intersection and to subject the model to impacts of several interventions to the traffic situation. There are several possibilities: to alter the speed limit, to build up traffic lights, roundabout, or others. Our study will focus on peak hours caused by arriving and departing employees from and to the factory. We are not going to solve any continuity to other intersections nearby, for these cases a solution can be seen in using self-organising traffic lights (Gershenson 2007).

**Definition of the Examined System**

Our point of view to this intersection is the transport system. With this definition we will work on model's construction. The level of abstraction will be individual vehicles; all roads are considered as a connection between two points.

**Creation of the Current Image about the System**

To create an image of the real system, it was necessary to conduct several terrain measurements. Several measurements were done during peak hours and off-peak hours, each lasting 30 minutes (length of the peak time). The measurements were done for all directions and the result is presented in (Table 1). Trucks and buses are presented in a short form as heavy vehicles. At first glance, it is possible to notice that there is a rapid increase of vehicles during peak times (from Nošovice direction).

**Model Creation**

The model was created on the basis of the measurement in the preceding step (Table1). The model reflects approximately the same size as the measured surface in reality. The intersection is located in the village, so the speed is set to 50km/h. Since it is a simple intersection, there is only one main and one side road.

**Model Validation**

Data in the rough draft model is based on the real measured data, however the model does not reflect the real situation. Each of the traffic simulation software uses a different computational algorithm. These algorithms can distort the situation because they simplify the reality as they do not consider a lot of factors, i.e. violation of traffic rules, aggressiveness or non-standard behaviour of drivers, etc. This model is a vital example. Real drivers on the main road often exceed the maximum allowed speed so that vehicles would not branch out at the maximum speed. Another discrepancy is rapid acceleration of vehicles in the model; most drivers in the real situation are not able to accelerate so quickly. Thus, it is evident that the model is not right and it is necessary do to corrections.

**Table 1.** 30-minute traffic density measurement

**Direction from Nošovice (red)**

| destination | Vojkovice (blue) | Dobrá (yellow) | Highway (green) | sum | heavy vehicles |
|---|---|---|---|---|---|
| off-peak | 15 | 35 | 87 | 137 | 30 |
| peak | 41 | 221 | 334 | 596 | 36 |

**Direction from highway (green)**

| destination | Nošovice (red) | Vojkovice (blue) | Dobrá (yellow) | sum | heavy vehicles |
|---|---|---|---|---|---|
| off-peak | 114 | 43 | 48 | 205 | 33 |
| peak | 137 | 53 | 54 | 244 | 18 |

**Direction from Dobrá (yellow)**

| destination | Nošovice (red) | Vojkovice (blue) | Highway (green) | sum | heavy vehicles |
|---|---|---|---|---|---|
| off-peak | 21 | 56 | 24 | 101 | 4 |
| peak | 46 | 51 | 32 | 129 | 6 |

**Direction from Vojkovice (blue)**

| destination | Nošovice (red) | Dobrá (yellow) | Highway (green) | sum | heavy vehicles |
|---|---|---|---|---|---|
| off-peak | 2 | 37 | 40 | 79 | 5 |
| peak | 5 | 39 | 61 | 105 | 6 |

Correction 1 – branching out speed
For vehicles turning off the main road, the speed was reduced from 14m/s to 9m/s, because the driver has to slow down. The same situation is when a vehicles branches out to the main road. Here the driver has to be cautious because he has to look around.

Correction 2 – traffic density
This correction addresses the problem of traffic density, which, after entering the measured values, does not correspond to reality. Using the same values of vehicles per hour, the real system exhibits congestions, which is in contradiction with the model, because the traffic flow is quite smooth and there are no congestions. To cope with this issue, global traffic density rate coefficient was experimentally set to 1.2. Other corrections were made for each individual branch (Table 2) as well.

After both corrections the behaviour of the model reflects the current real situation much better. We can state that the model is valid for our needs. When calculating the number of vehicles per hour from the model and transfer the number into a real situation, it is necessary to take into account all specified coefficients.

**Table 2.** Traffic density adjustments

| From | Original (vehicles/h) | New (vehicles/h) | Traffic density coeff. |
|---|---|---|---|
| Nošovice (red) | 1192 | 1506 | 1.26 |
| Dobrá (yellow) | 258 | 254 | 0.98 |
| Vojkovice (blue) | 210 | 343 | 1.63 |
| Highway (green) | 448 | 635 | 1.42 |

# 4 Experiment

In this chapter, a simulation of the above-mentioned intersection is done in order to find a better solution how to control traffic flow during peak hours. Two proposals have been made. The first one simply proposes to reduce the speed limit in the area of the intersection. The second one suggests to use traffic lights with a specified time interval based on traffic density. A final comparison is made as well.

**Reduction of Speed**

The basic idea was to reduce the speed limit in order to give a chance to branch out to vehicles on the side road. Several speed reductions have been examined. One interesting option was to set the speed limit to 9m/s. Most of the time this option is an advantage for vehicles coming from the side road (Figure 3). On the main road, the traffic density was slightly reduced. Overall, the results look fine but when we repeat the experiment several times, it is not such advantage, as presented in the final result (Figure 6).

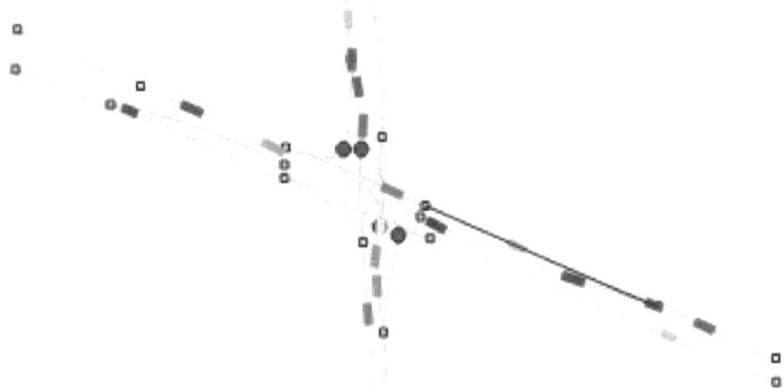

**Fig. 3.** Modelled area during peak hours with speed reduction after 400s

**Traffic Lights**

The basic idea for this option is to assign appropriate time interval for vehicles coming on the main road and side road based on the measured traffic density. The speed limit is set to 14m/s to allow all vehicles to pass the intersection as quickly as permitted. For this kind of intersection, it is sufficient to use simple traffic lights signalling without direction arrows. Nevertheless, using this simple signalling provides very promising results and no congestions are formed on either side. Vehicles on the side road do not have to wait long time to pass the intersections and the average speed on the main road is nearly the same. Using this method of control, the intersection is able to withstand significantly more traffic than without traffic lights. Based on the ratio of vehicles counted per hour on the main road vs. side road, the turn-taking for traffic lights was set. Most of the time (75%) vehicles on the main road are allowed to pass, and 25% of time is reserved for vehicles on the side road (Figure 4).

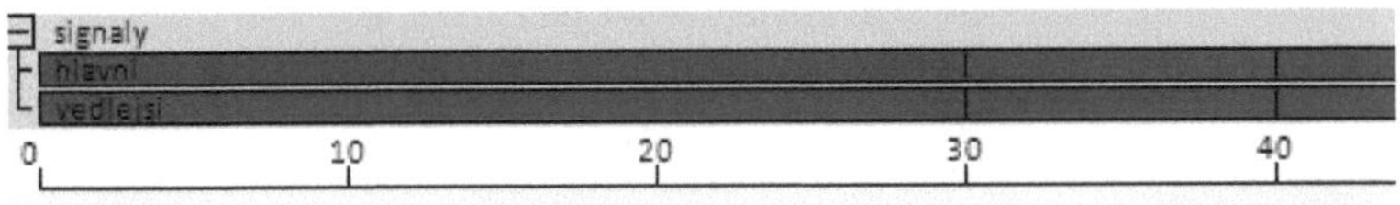


**Fig. 4.** Traffic lights settings

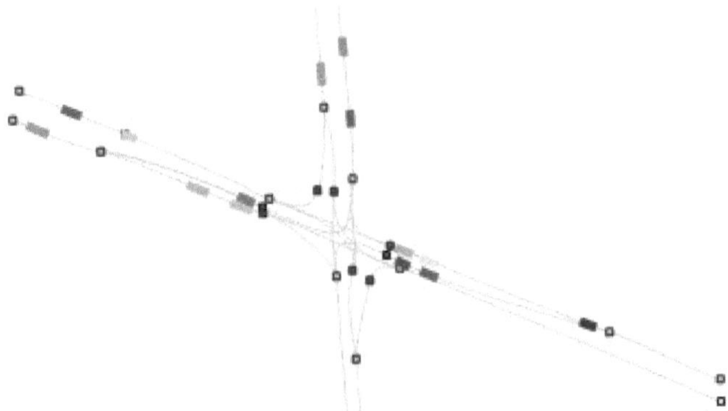

**Fig. 5.** Modelled area during peak hours with traffic lights after 400s

(Figure 5) presents the final state with the above-stated setting after 400s. At first sight it is obvious that this is a more effective method than in the previous option (Figure 3).

## 5 Conclusion and Future Work

(Figure 6) shows that in the original real system, vehicles on the main road reach an average speed of 7m/s while vehicles on the side roads reach an average speed of about 3m/s. The average number of stops on the side road is multiple time greater compared to the main road. The travel time for vehicles coming from the side road is unacceptably long.

The first proposal consisting in speed limit reduction leads to a worse result on every road compared to the real system. The average speed is lower and the average number of stops is increased. The solution also exhibits large fluctuations of values and does not provide good stability of the situation. Therefore, this solution is not ideal.

In contrast, the second presented option with traffic lights leads to much better results. There are no congestions and most of the time all vehicles waiting at red lights pass the intersection during the green light time interval. The average speed of vehicles on the main road is slightly reduced, so as the average number of stops is very slightly increased. On the side road, the average speed is slightly increased but what is more important, the average number of stops is dramatically reduced. As the overall result considering all braches shows, the average speed is reduced from 7m/s to 6m/s but the average stops per experiment (400 seconds) is reduced from 3 to 1. Therefore, the solution with traffic lights seems to bring better results.

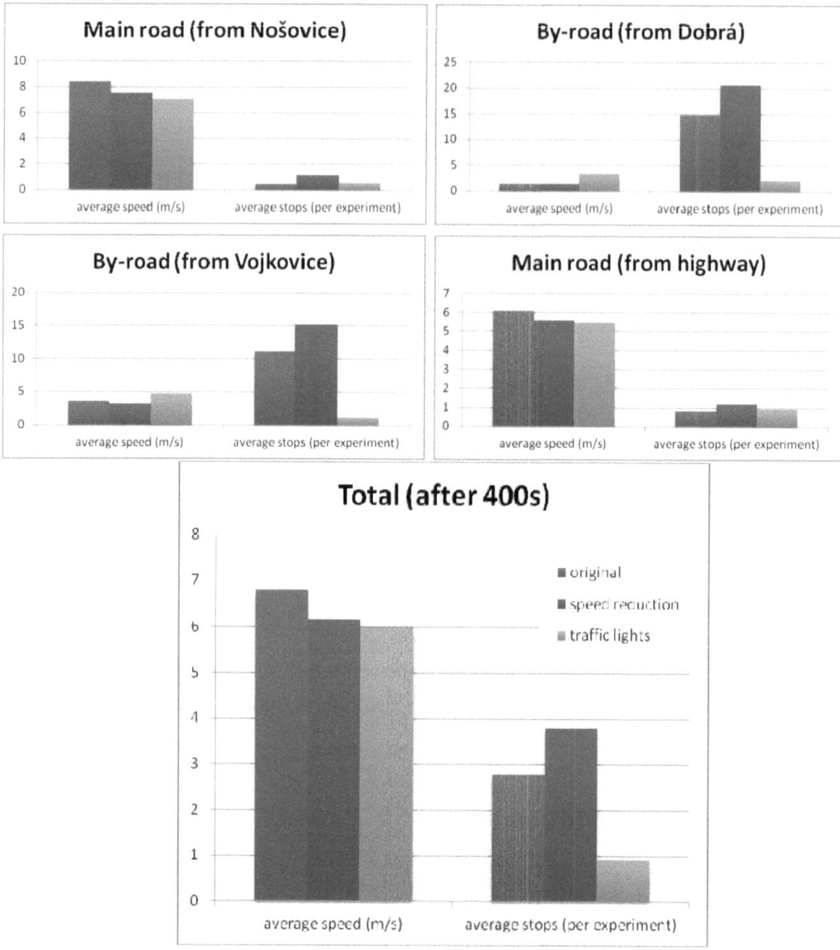


**Fig. 6.** Experiment results after 400s, averaged

It is important to note again that the solution has been examined only for peak hours. During off-peak time, the traffic lights could be switched off. Our next work will focus on investigating various time intervals of traffic lights, i.e. a possibility of adaptive traffic lights for different conditions (traffic density in different parts of day).

**Acknowledgments.** The research described here has been financially supported by University of Ostrava grant SGS16/PřF/2014. Any opinions, findings and conclusions or recommendations expressed in this material are those of the authors and do not necessarily reflect the views of the sponsors.

## References

Gershenson, C.: Design and Control of Self-organizing Systems. CopIt ArXives, Mexico (2007) ISBN: 978-0-9831172-3-0.
Li, J., O'Donoghue, C.: A survey of dynamic microsimulation models: uses, model structure and methodology. International Journal of Microsimulation 6(2), 3–55 (2013)
Newell, G.F.: A simplified car-following theory: a lower order model. Institute of Transportation Studies, University of California, Berkeley (2002)
O'Donoghue, C.: Dynamic microsimulation: A methodological survey. Brazilian Electronic Journal of Economics 4(2), 77 (2001)
Pelánek, R.: Modelování a simulace komplexních systémů, Masaryk University, 236 p. (2011) ISBN: 978-80-210-5318-2
Matti, P.: Simulation of Traffic Systems – An Overview. Journal of Geographic Information and Decision Analysis 3(1) (1999)
Schulte, C.: CityTraficSimulator. Christian Schulte zu Berge, `http://www.cszb.net/` (cited: February 12, 2014)

# Mathematical Models of Multivariable Systems

Vladimír Jehlička

University of Pardubice, Jan Perner Transport Faculty, Department of Informatics in Transport, Studentská 573, 530 10 Pardubice, Czech Republic

**Abstract.** The paper is focused on the build of a mathematical models of multivariable systems by the method of experimental identification. The created model is used for predicting the static and dynamic behavior of the controlled system in the closed loop. Dynamic properties of systems are described by the differential equations. In the experimental part are identified the parameters of the mathematical model of rectifying column. As an example, the multivariable controlled system, in this case is described the dependence of concentration distilled mixture on change the flow of reflux and flow of vapor.

**Keywords:** Mathematical models, multivariable systems, experimental identification, rectification column.

## 1 Introduction

During the design of control algorithms of technological processes and in many other cases [1-8], it is necessary to know the static and dynamic characteristics of controlled systems.

In describing the static and dynamic properties of the controlled system can be proceed by mathematical-physical analysis or by experimental identification. In the first case it is necessary to know how to mathematically describe all physical phenomena taking place in the system.

In the second case, we evaluate the measured experimental values of input and output variables. These methods are theoretically described for example in the work [9, 10]. Experimental identification of the values of the parameters of the model is based on the evaluation of the response of the system to precisely-defined waveforms of the values of the input variables of the system. The internal structure of the model should be chosen. If it has been previously made at least some mathematical-physical analysis of the controlled system, then it is possible to select the internal structure of the model based on the analysis.

This article builds on the work [11], in which is described the process of creating the model of one-dimensional deterministic system. All the theoretical results are extended to multivariable systems.

I. Zelinka et al. (eds.), *Nostradamus 2014: Prediction, Modeling and Analysis of Complex Systems*, Advances in Intelligent Systems and Computing 289,
DOI: 10.1007/978-3-319-07401-6_31, © Springer International Publishing Switzerland 2014

## 2 Identification of Multivariable Systems

In paper [11] is showen one-dimensional linear or in restricted around the operating point linearized deterministic system with lumped constant parameters, which can be using the z-transform equation to describe

$$A(z^{-1})\,y(k) = z^{-q}B(z^{-1})\,u(k) \tag{1}$$

where

$A(z^{-1}) = 1 + a_1 z^{-1} + a_2 z^{-2} + \ldots + a_n z^{-n}$

$B(z^{-1}) = b_1 z^{-1} + b_2 z^{-2} + \ldots + b_m z^{-m}$

$n$, $m$ are the degrees of polynomials $A(z^{-1})$, $B(z^{-1})$,

$y(k)$ is the value of the output variable $y(t)$ in the $k$-th step ($t$ is time),

$u(k)$ is the value of the input variable $u(t)$ in the $k$-th step,

$k$ represents the time value as a multiple of the sampling interval $\Delta t$,

$q$ represents the time delay $q$ sampling interval.

Writing $y(k)$ means the value of the output variables $y(t)$ at time $t = k\,\Delta t$, i.e. in $k$-th step. Symbolic notation $z^{-1}y(k)$ is to be understood as the value of thc output variable $y(t)$ in the previous step, i.e. shifted back in time by one sampling interval, which can be expressed by the equation.

$$z^{-1}y(k) = y(k-1) \tag{2}$$

Likewise

$$z^{-n}y(k) = y(k-n) \tag{3}$$

represents a shift in time on $n$ sample intervals.

Another possible way to edit the model (1) lies in the introduction of the coefficient $s$, in which is included a non-zero mean value of the output and the input variables in a given working point.

$$A(z^{-1})\,y(k) = z^{-q}B(z^{-1})\,u(k) + s \tag{4}$$

The real systems but are not usually one-dimensional systems. Assume a system which contains $r$ entry and $p$ output variables. Each input variable can affect on each output variable. The internal structure of the system can be very diverse. If we do not have information about the internal structure of the modeled multivariable system, then the mathematical model can be written in matrix form

$$\boldsymbol{y}(k) = \boldsymbol{G}(z^{-1})\boldsymbol{u}(k) \tag{5}$$

where

$$\boldsymbol{y}(k)=\left(y_1(k) \quad y_2(k) \quad \dots \quad y_r(k)\right)^{\mathrm{T}}$$

$$\boldsymbol{u}(k)=\left(u_1(k) \quad u_2(k) \quad \dots \quad u_p(k)\right)^{\mathrm{T}}$$

$$\boldsymbol{G}\left(z^{-1}\right)=\begin{pmatrix} G_{11}\left(z^{-1}\right) & G_{12}\left(z^{-1}\right) & \dots & G_{1p}\left(z^{-1}\right) \\ G_{21}\left(z^{-1}\right) & G_{22}\left(z^{-1}\right) & \dots & G_{2p}\left(z^{-1}\right) \\ \dots & \dots & \dots & \dots \\ G_{r1}\left(z^{-1}\right) & G_{r2}\left(z^{-1}\right) & \dots & G_{rp}\left(z^{-1}\right) \end{pmatrix}.$$

For systems where the number of input quantities, is the same as the number of output variables, from rectangular matrix $\boldsymbol{G}\left(z^{-1}\right)$ becomes a square matrix.

Each element $G_{ij}\left(z^{-1}\right)$ of matrix $\boldsymbol{G}\left(z^{-1}\right)$ represents the transfer function of the one-dimensional system, where the input variable is $u_j(k)$ and output variable is $y_i(k)$.

$$y_j(k)=G_{ij}\left(z^{-1}\right)u_i(k) \tag{6}$$

If we write an appropriate transfer function using polynomials, then according to equation (1) can be multivariable deterministic systems to describe by the equation

$$\boldsymbol{A}\left(z^{-1}\right)\boldsymbol{y}(k)=z^{-q}\boldsymbol{B}\left(z^{-1}\right)\boldsymbol{u}(k) \tag{7}$$

where

$$\boldsymbol{A}\left(z^{-1}\right)=\begin{pmatrix} A_{11}\left(z^{-1}\right) & A_{12}\left(z^{-1}\right) & \dots & A_{1r}\left(z^{-1}\right) \\ A_{21}\left(z^{-1}\right) & A_{22}\left(z^{-1}\right) & \dots & A_{2r}\left(z^{-1}\right) \\ \dots & \dots & \dots & \dots \\ A_{r1}\left(z^{-1}\right) & A_{r2}\left(z^{-1}\right) & \dots & A_{rr}\left(z^{-1}\right) \end{pmatrix}$$

$$\boldsymbol{B}\left(z^{-1}\right)=\begin{pmatrix} B_{11}\left(z^{-1}\right) & B_{12}\left(z^{-1}\right) & \dots & B_{1p}\left(z^{-1}\right) \\ B_{21}\left(z^{-1}\right) & B_{22}\left(z^{-1}\right) & \dots & B_{2p}\left(z^{-1}\right) \\ \dots & \dots & \dots & \dots \\ B_{r1}\left(z^{-1}\right) & B_{r2}\left(z^{-1}\right) & \dots & B_{rp}\left(z^{-1}\right) \end{pmatrix}$$

Let us consider the operating point of the system, which in steady state is characterized by a vector of input variables $\boldsymbol{u}_w$ and by a vector of output variables $\boldsymbol{y}_w$. Model of multidimensional deterministic system around the working point can be written with uses of deviations of input and output variables.

$$\boldsymbol{A}\left(z^{-1}\right)\Delta\boldsymbol{y}(k)=z^{-q}\boldsymbol{B}\left(z^{-1}\right)\Delta\boldsymbol{u}(k) \tag{8}$$

where

$$\Delta \boldsymbol{y}(k) = \boldsymbol{y}(k) - \boldsymbol{y}_w$$

$$\boldsymbol{y}_w = \begin{pmatrix} y_{w1} & y_{w2} & \dots & y_{wr} \end{pmatrix}^{\mathrm{T}}$$

$$\Delta \boldsymbol{u}(k) = \boldsymbol{u}(k) - \boldsymbol{u}_w$$

$$\boldsymbol{u}_w = \begin{pmatrix} u_{w1} & u_{w2} & \dots & u_{ww} \end{pmatrix}^{\mathrm{T}}$$

Another possible way to edit the model (7) lies in the introduction of the vector s, in which is included a non-zero mean value of the output and the input variables in a given working point

$$\boldsymbol{A}\left(z^{-1}\right)\boldsymbol{y}(k) = z^{-q}\boldsymbol{B}\left(z^{-1}\right)\boldsymbol{u}(k) + \boldsymbol{s} \tag{9}$$

where

$$\boldsymbol{s} = \begin{pmatrix} s_1 & s_2 & \dots & s_r \end{pmatrix}^{\mathrm{T}}$$

Similarly we can edit model (8).

$$\boldsymbol{A}\left(z^{-1}\right)\Delta \boldsymbol{y}(k) = z^{-q}\boldsymbol{B}\left(z^{-1}\right)\Delta \boldsymbol{u}(k) + \boldsymbol{s} \tag{10}$$

This model approximates the identified multivariable system with a certain error $\boldsymbol{e}_r(k)$, which we call the error of equation

$$\boldsymbol{A}\left(z^{-1}\right)\Delta \boldsymbol{y}(k) = z^{-q}\boldsymbol{B}\left(z^{-1}\right)\Delta \boldsymbol{u}(k) + \boldsymbol{s} + \boldsymbol{e}_r(k) \tag{11}$$

For the estimation of the parameters of the model may be used the non-recursive or recursive methods. In the following experimental section was used recursive least squares method, which minimizes the sum of squared errors of equation

$$\sum_{k=n+q+1}^{N} \boldsymbol{e}_r^2(k) \rightarrow \min. \tag{12}$$

New estimates $\hat{\boldsymbol{\gamma}}_{N+1}$ of the parameters in step $N$+1 are calculated from previous estimates $\hat{\boldsymbol{\gamma}}_N$ calculated in step $N$ by adding corrections $\boldsymbol{\Gamma}_{N+1}$

$$\hat{\boldsymbol{\gamma}}_{N+1} = \hat{\boldsymbol{\gamma}}_N + \boldsymbol{\Gamma}_{N+1} \tag{13}$$

In paper [11] are given formulas for calculating the estimated parameters of the model one-dimensional deterministic system, where

$$\hat{\boldsymbol{\gamma}}_N = \begin{pmatrix} a_{1N} & a_{2N} & \dots & a_{nN} & b_{1N} & b_{2N} & \dots & b_{nN} \end{pmatrix}^{\mathrm{T}} \tag{14}$$

Assume a multi-variable deterministic system which contains $p$ input variables and $r$ output variables. Then the model parameters can be written in matrix

$$\hat{\gamma}\left(z^{-1}\right)=\begin{pmatrix} A_{11}\left(z^{-1}\right) & A_{12}\left(z^{-1}\right) & \dots & A_{1r}\left(z^{-1}\right) & B_{11}\left(z^{-1}\right) & B_{12}\left(z^{-1}\right) & \dots & B_{1p}\left(z^{-1}\right) \\ A_{21}\left(z^{-1}\right) & A_{22}\left(z^{-1}\right) & \dots & A_{2r}\left(z^{-1}\right) & B_{21}\left(z^{-1}\right) & B_{22}\left(z^{-1}\right) & \dots & B_{2p}\left(z^{-1}\right) \\ \dots & \dots & \dots & \dots & \dots & \dots & \dots & \dots \\ A_{r1}\left(z^{-1}\right) & A_{r2}\left(z^{-1}\right) & \dots & A_{rr}\left(z^{-1}\right) & B_{r1}\left(z^{-1}\right) & B_{r2}\left(z^{-1}\right) & \dots & B_{rp}\left(z^{-1}\right) \end{pmatrix} \tag{15}$$

In step $N+1$ will be to measure the value of output $y(N+1)$ and input $u(N+1)$ variables, which will stored to the matrix

$$\boldsymbol{f}_{N+1}=(-\boldsymbol{y}(N) \quad \dots \quad -\boldsymbol{y}(N-1) \quad \boldsymbol{u}(N-q) \quad \dots \quad \boldsymbol{u}(N+1-q-n)) \tag{16}$$

where

$$\boldsymbol{m}_N=\boldsymbol{P}_N\boldsymbol{f}_{N+1}\left(1+\boldsymbol{f}_{N+1}^{\mathrm{T}}\boldsymbol{P}_N\boldsymbol{f}_{N+1}\right)^{-1} \tag{17}$$

To calculate the matrix $\boldsymbol{P}=\left(\boldsymbol{F}^{\mathrm{T}}\boldsymbol{F}\right)^{-1}$ pays recurrent relationship

$$\boldsymbol{P}_{N+1}=\boldsymbol{P}_N-\boldsymbol{m}_N\boldsymbol{f}_{N+1}^{\mathrm{T}}\boldsymbol{P}_N \tag{18}$$

## 3 Experimental Part

The experimental part of the work was carried out on a model of 7 floor rectification column in the distillation of a binary mixture of methanol-water. The rectification column is a two dimensional system, which has two input values ($R$ = flow of reflux, $V$ = flow of vapor) and the two output variables ($x_7$ = concentration of the liquid at the seventh floor column, $x_1$ = concentration of the liquid at the first floor column).

Assume that in the area of the selected working point can be rectification column considered as linear deterministic system which has two input and two output variables. Then it can be described by the equation

$$\begin{pmatrix} A_{11}\left(z^{-1}\right) & A_{12}\left(z^{-1}\right) \\ A_{21}\left(z^{-1}\right) & A_{22}\left(z^{-1}\right) \end{pmatrix}\begin{pmatrix} \Delta x_7(k) \\ \Delta x_1(k) \end{pmatrix}=z^{-q}\begin{pmatrix} B_{11}\left(z^{-1}\right) & B_{12}\left(z^{-1}\right) \\ B_{21}\left(z^{-1}\right) & B_{22}\left(z^{-1}\right) \end{pmatrix}\begin{pmatrix} \Delta R(k) \\ \Delta V(k) \end{pmatrix}+\begin{pmatrix} e_{r1}(k) \\ e_{r2}(k) \end{pmatrix} \tag{19}$$

If we assume the same denominator in the transfer functions $G_{11}\left(z^{-1}\right)$ and $G_{21}\left(z^{-1}\right)$, respectively $G_{22}\left(z^{-1}\right)$ and $G_{12}\left(z^{-1}\right)$ in the matrix $\boldsymbol{G}\left(z^{-1}\right)$, then the matrix $\boldsymbol{A}\left(z^{-1}\right)$ will be diagonal and the equation (19) can be rewritten as

$$\begin{pmatrix} A_{11}\left(z^{-1}\right) & 0 \\ 0 & A_{22}\left(z^{-1}\right) \end{pmatrix}\begin{pmatrix} \Delta x_7(k) \\ \Delta x_1(k) \end{pmatrix}=z^{-q}\begin{pmatrix} B_{11}\left(z^{-1}\right) & B_{12}\left(z^{-1}\right) \\ B_{21}\left(z^{-1}\right) & B_{22}\left(z^{-1}\right) \end{pmatrix}\begin{pmatrix} \Delta R(k) \\ \Delta V(k) \end{pmatrix}+\begin{pmatrix} e_{r1}(k) \\ e_{r2}(k) \end{pmatrix} \tag{20}$$

In the context of the verification of properties of identification algorithms have been studied all four transmissions with one input and one output variable ($R \to x_7, R \to x_1, V \to x_7, V \to x_1$), and transmission with two input and one output variable ($R,V \to x_7, R,V \to x_1$).

In the experimental identification of the two dimensional system, the values of both input variables have been changed by pseudorandom binary signal (PRBS), with the same parameters as in the work [11] , ie the length of the period $P_T = 63$ and the sampling interval $\Delta t = 8\,\text{min}$ . To avoid adding, respectively subtracting of the effects of the same PRBS on both inputs system, the PRBS were mutually shifted of half a period.

The identification was carried out in the vicinity of the working point with parameters:

$x_F = 0.25$ mol. div. (concentration of feed flow),
$F = 19{,}3$ mol $\text{min}^{-1}$ (flow of feed),
$R = 8{,}0$ mol $\text{min}^{-1}$ (flow of reflux),
$V = 11{,}5$ mol $\text{min}^{-1}$ (flow of vapor).

The values of the flow reflux $R$ and flow vapor $V$ were changed according to the parameters of the PRBS with amplitude of 0.5 mol $\text{min}^{-1}$. This means that the flow of the reflux acquired values of $R$ = 7.5 mol $\text{min}^{-1}$, respectively $R$ = 8.5 mol $\text{min}^{-1}$. The flow of the vapor acquired values of $V$ = 11.0 mol $\text{min}^{-1}$, respectively $V$ = 12.0 mol $\text{min}^{-1}$.

In figure 1 is a recording during measurement of input variables *R*, *V* and output variable $x_7$ for one period PRBS.

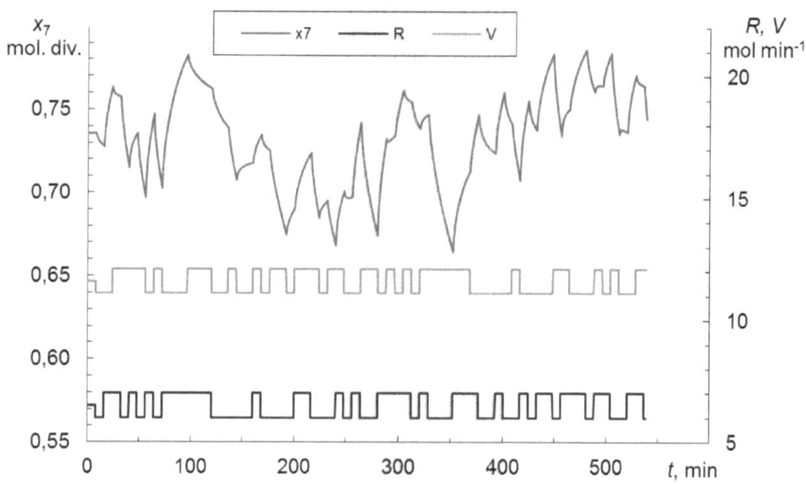


**Fig. 1.** Identification of rectification column

In the context of experiments were identified transmissions of the $R,V \rightarrow x_7$ and $R,V \rightarrow x_1$. Models were in the form of differential equations of first, second and third order. As the optimal models showed 2nd order. Effect of initial values of matrix elements $\boldsymbol{P}_0$ is the same as in the identification of one-dimensional system.

System can be identified in a limited area of the working point as model of second order.

$$\begin{pmatrix} x_7(k) \\ x_1(k) \end{pmatrix} = \begin{pmatrix} \dfrac{0{,}02z^{-1}-0{,}01z^{-2}}{1-0{,}78z^{-1}+0{,}08z^{-2}} & \dfrac{-0{,}01z^{-1}+0{,}01z^{-2}}{1-0{,}78z^{-1}+0{,}08z^{-2}} & \dfrac{0{,}26}{1-0{,}78z^{-1}+0{,}08z^{-2}} \\ \dfrac{0{,}02z^{-1}-0{,}01z^{-2}}{1-0{,}84z^{-1}+0{,}04z^{-2}} & \dfrac{-0{,}02z^{-1}+0{,}01z^{-2}}{1-0{,}84z^{-1}+0{,}04z^{-2}} & \dfrac{0{,}06}{1-0{,}84z^{-1}+0{,}04z^{-2}} \end{pmatrix} \begin{pmatrix} R(k) \\ V(k) \\ 1 \end{pmatrix}$$

Static characteristic of this model can be represented in a three dimensional view as the plane. Static characteristic of nonlinear system, which we identifies, represents the surface that is not a plane.

Table 1 shows the values of the concentration of the liquid on the 7th floor of the column in the steady state, for both static characteristics. In each cell of the table is first given value calculated from the nonlinear model of the rectification column, and then is given value of linear model. The values of all concentrations are in *mol. div.*, flow rates are in *mol. min*$^{-1}$.

**Table 1.** Static characteristics for the 7th floor column

| | R = 7,5 | R = 8,0 | R = 8,5 |
|---|---|---|---|
| V = 10,5 | 0,860 0,859 | 0,876 0,874 | 0,889 0,888 |
| V = 11,0 | 0,845 0,849 | 0,864 0,863 | 0,879 0,878 |
| V = 11,5 | 0,827 0,839 | 0,850 0,853 | 0,868 0,868 |

In the table 2 are shown the same way the concentration values of the liquid to 1 floor of column.

**Table 2.** Static characteristics for the 1th floor column

| | R = 7,5 | R = 8,0 | R = 8,5 |
|---|---|---|---|
| V = 10,5 | 0,263 0,261 | 0,289 0,287 | 0,312 0,313 |
| V = 11,0 | 0,237 0,242 | 0,267 0,267 | 0,292 0,292 |
| V = 11,5 | 0,205 0,220 | 0,240 0,246 | 0,270 0,272 |

As an example of dynamic properties of the identified system, and the properties of obtained model is shown in figure 2 transient response to the change flow of reflux. We see that the response of a linear model is among the characteristics of an identified nonlinear system.

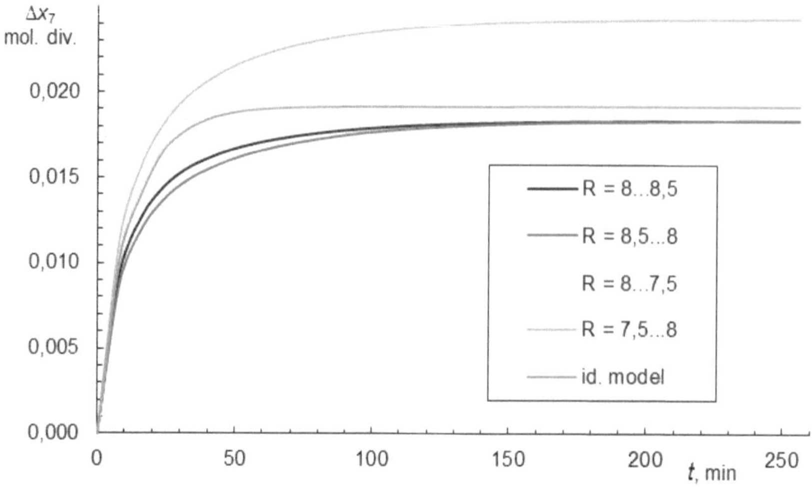


**Fig. 2.** Transient characteristics of identified system and of model

# 4 Conclusion

After comparing the corresponding data in table 1 and 2, it can be stated that in a limited area working point can be nonlinear static characteristic with sufficient accuracy to approximate the plane corresponding to the static characteristics of the linear model.

If we compare the dynamical properties of nonlinear identified system and dynamical properties obtained linear model, then we can say, that in a limited area working point is very good agreement of the model with an identified system.

The created model multidimensional deterministic system can also be used for the design of algorithms for digital control of the system in the closed loop.

# References

[1] Wang, J., Wang, H.-W., Gu, H.: A novel recursive subspace identification approach of closed-loop systems. Mathematical and Computer Modelling of Dynamical Systems 19(6), 526–539 (2013), `http://www.tandfonline.com/doi/abs/10.1080/13873954.2013.801355` (cited), doi:10.1080/13873954.2013.801355

[2] Chen, L., Tulsyan, A., Huang, B., Liu, F.: Multiple model approach to nonlinear system identification with an uncertain scheduling variable using EM algorithm. Journal of Process Control. 23(10), 1480–1496 (2013), `http://linkinghub.elsevier.com/retrieve/pii/S0959152413001820` (cited), doi:10.1016/j.jprocont.2013.09.013

[3] Cadavid, J.G., Godoy-Silva, R.D., Narvaez, P.C., Camargo, M., Fonteix, C.: Biodiesel production in a counter-current reactive extraction column: Modelling, parametric identification and optimisation. Chemical Engineering Journal 228, 717–723 (2013), http://linkinghub.elsevier.com/retrieve/pii/S1385894713006591 (cited), doi:10.1016/j.cej.2013.05.040

[4] Torkamani, S., Butcher, E.A., Khasawneh, F.A.: Parameter identification in periodic delay differential equations with distributed delay. Communications in Nonlinear Science and Numerical Simulation 18(4), 1016–1026 (2013), http://linkinghub.elsevier.com/retrieve/pii/S1007570412004182 (cited), doi:10.1016/j.cnsns.2012.09.001

[5] Abbassi, F., Belhadj, T., Mistou, S., Zghal, A.: Parameter identification of a mechanical ductile damage using Artificial Neural Networks in sheet metal forming. Materials 45, 605–615 (2013), http://linkinghub.elsevier.com/retrieve/pii/S0261306912006632 (cited), doi:10.1016/j.matdes.2012.09.032

[6] Victor, S., Malti, R., Garnier, H., Oustaloup, A.: Parameter and differentiation order estimation in fractional models. Automatica 49(4), 926–935 (2013), http://linkinghub.elsevier.com/retrieve/pii/S0005109813000277 (cited), doi:10.1016/j.automatica.2013.01.026

[7] Valderrama, J.O., Faúndez, C.A., Toselli, L.A.: Advances on modeling and simulation of alcoholic distillation. Part 1: Thermodynamic modeling. Food and Bioproducts Processing 90(4), 819–831 (2000), http://linkinghub.elsevier.com/retrieve/pii/S0960308512000302 (dostupné), doi:10.1016/j.fbp.2012.04.004

[8] Valderrama, J.O., Toselli, L.A., Faúndez, C.A.: Advances on modeling and simulation of alcoholic distillation. Part 2: Process simulation. Food and Bioproducts Processing 90(4), 832–840 (2000), http://linkinghub.elsevier.com/retrieve/pii/S0960308512000296 (dostupné), doi:10.1016/j.fbp.2012.04.003

[9] Isermann, R.: Digitale Regelsysteme: Band II, Stochastische Regelungen, Mehrgrössenregelungen, Adaptive Regelunungen, Anwendungen. 2. überarb. u. erw. Aufl. Springer, Berlin (1987) ISBN 35-401-6597-5

[10] Isermann, R.: Digitale Regelsysteme: Band I, Grundlagen Deterministische Regelungen. 2. überarb. und erw. Aufl. Springer, Berlin (1987) ISBN 35-401-6596-7

[11] Jehlička, V.: Mathematical Models of Controlled Systems. In: Zelinka, I., Chen, G., Rössler, O.E., Snasel, V., Abraham, A. (eds.) Nostradamus 2013: Prediction, Model. & Analysis. AISC, vol. 210, pp. 257–264. Springer, Heidelberg (2013)

# Risk Modeling in Process Industries by Stochastic Petri Nets

Radim Briš and Ondřej Grunt

Department of Applied Mathematics, Faculty of Electrical Engineering and Computer Science, VŠB-Technical University of Ostrava
radim.bris@vsb.cz, ondrej.grunt@vsb.cz

**Abstract.** Modeling the risk to safety of personnel in offshore industry is often realized by the application of Event Trees. The risk is then defined as a product of event frequency and its consequences. However, steady-state methods like Event Trees are not suitable for modeling time-dependent probability of fatality based on time-dependent events. This article presents a new possible approach to modeling the risk to safety of personnel by Stochastic Petri nets (SPN). Stochastic Petri nets models for small leak occurrence on an offshore platform is shown, based on a realistic example from the offshore industry. The probabilities of fatalities were obtained from the simulation by using the Moca-RP software and compared to probabilities obtained by Event Trees and direct Monte Carlo simulation methods.

## 1 Introduction

Personnel in process industries is exposed to the risk of fatality. The risk of fatality is normally estimated by an application of Event Tree method. An event tree displays a sequence of events with their corresponding probabilities, resulting probabilities of fatalities are represented by the probability values assigned to the Event Tree leaves, each leaf representing a particular event scenario.

However, tracing contributions of fatality probabilities in an Event Tree is difficult. While it is possible to construct an Event Tree large enough to enable us to find the most important contributions, such an Event Tree would be very difficult to manage. Another limitation is that the Event Tree method is a steady-state method, therefore not suitable for modeling time-dependent processes.

In the offshore industry, fatality probabilities often depend on personnel reactions and their consequences. Due to the lack of aforementioned limitations bound with the Event Tree method, Stochastic Petri nets were considered to be used to construct a suitable model of the risk to safety of personnel. First, the steady state representation of a time-dependent event was modeled using SPN as an initial step of the substitution of the Event Tree method in the risk modeling. Then, the dynamic representation of the same event was modeled, demonstrating the potential of SPN-based risk modeling. In this article, a small leak occurrence scenario described in [1] was chosen as an example.

I. Zelinka et al. (eds.), *Nostradamus 2014: Prediction, Modeling and Analysis of Complex Systems*, Advances in Intelligent Systems and Computing 289,
DOI: 10.1007/978-3-319-07401-6_32, © Springer International Publishing Switzerland 2014

Probabilities of fatalities due to small leak occurrence obtained by the Moca-RP simulation of Stochastic Petri nets were compared to results obtained by the application of Event Tree and direct Monte Carlo methods, presented in [1].

## 2 Petri Nets

Petri nets is a modeling language for the description of parallel and distributed systems, designed by Carl Adam Petri. Petri net is a directed bipartite graph, its nodes are divided into two groups : *places* and *transitions*. Places are nodes representing current condition of the system (e.g. 'power on', 'power off') and are indicated by circles, transitions are nodes representing events occurring in the system (e.g. 'powering on', 'powering off') and are signified by bars. Places and transitions are connected by directed arcs; however a place can only be connected via arc to a transition and not to another place (i.e. each event leads to a new condition of the system). Similarly, transition cannot be connected via arc to another transition.

Places in a Petri net may contain a number of *tokens*. Distribution of tokens over the places in Petri net is called a *marking*, which represents a current configuration of the net. Token proceeds to another place by firing an enabled transition. Transition is enabled only if the input place contains a sufficient number of tokens, these tokens are then transferred from the input place to an output place. Basic Petri nets model is nondeterministic, therefore any enabled transition may fire immediately after a sufficient number of tokens is present in the input place.

A Petri net can be defined as follows:

**Definition 1.** *A* ***Petri net*** *is a four-tuple* $PN = (P, T, F, M0)$ *where:*

1. ***P*** *is a set of places.*
2. ***T*** *is a set of transitions.*
3. ***F***, $F \subset (P \times T) \cup (T \times P)$ *is a set of arcs.*
4. ***M0*** *is the initial marking.*

However, for modeling the risk to safety of personnel, the nondeterminism of basic Petri nets is a significant issue. Every reaction of personnel to an occurring event during a scenario contributes to the probability of fatality. For this reason, a decision concerning which of the multiple transitions may fire must be confined to a set of carefully chosen laws.

### 2.1 Stochastic Petri Nets

Stochastic Petri nets (further referred to as SPN) is a form of Petri nets in which a probabilistic delay is assigned to each transition. After the delay is over, transition is allowed to fire.

A Stochastic Petri net can be defined as follows:

**Definition 2.** *A* ***Stochastic Petri net*** *is a five-tuple* $SPN = (P, T, F, M0, \Lambda)$ *where:*

1. ***P*** *is a set of places.*
2. ***T*** *is a set of transitions.*
3. ***F***, $F \subset (P \times T) \cup (T \times P)$ *is a set of arcs.*
4. ***M0*** *is the initial marking.*
5. $\Lambda$ *is the array of transition firing rates* $\lambda$.

An extension for SPN exists, the Generalized Stochastic Petri nets, which adds two new modeling elements : *immediate transitions* and *inhibition arcs.* Immediate transition is a transition firing immediately after being enabled, its firing then takes precedence over transitions with assigned delay. Inhibitor arcs disable a transition as long, as the input place contains at least one token (i.e. serves for imposing an additional constraint to firing that transition).

## 3 Application of SPN

The application in this article is based on a typical offshore hydrocarbon installation (well-described example of an offshore production installation can be seen in [1]).

Risks from potential small leaks of the produced hydrocarbons were used as an example for the application. Small leaks were given precedence over the larger leaks due to substantially higher estimations of the frequencies of their release.

### 3.1 Small Leak Scenario

After a small leak occurrence, personnel is alarmed and starts escaping. On the installation, there are primary, secondary and tertiary evacuation routes, each leading to lifeboats or life-rafts. In case of damage or blockade of all evacuation routes, personnel evacuates to the sea. In any case, escaped personnel is then gathered by a standby vessel. Fatality in this scenario is a direct result of the loss of probability of evacuation.

However, there is always a possibility that the leak may ignite. Consequences of the ignition depend on time it occurs. Immediate ignition results in a jet fire, which may cause fatality to the surrounding personnel, while delayed ignition may result in an explosion, possibly damaging or blocking the evacuation routes. In any event, resulting fire may escalate outside the zone, damaging the evacuation routes in the process. Any of these possible events strongly contribute to the probability of fatality.

### 3.2 Steady-State SPN Model of Small Leak Scenario

As aforementioned in the beginning of this article, MOCA-RP simulation software was used for creating the model of the Small Leak scenario. In addition to the standard SPN elements, variables declared by the user may be utilized, setting values is done by firing a transition with corresponding calling function, or may be used in place of a random value for determining delay of a transition. Another useful feature is a possibility to create *repeating places*, a set of pointers to a given place, improving lucidity of the model.

All used transitions (with the exception of the last pair of transitions) utilize exponential firing law with either fixed or variable-based rate.

Presented model of the Small Leak scenario is divided into several blocks: Main block, 3 Escalation blocks and 4 Evacuation blocks. The purpose of the Main block (Figure 1) is to determine whether a small leak occurrence ends with a fatality or not. Each token[1] in place *SmallLeak* represents one small leak occurrence on the installation. As aforementioned, the most important possible event is an ignition of the leak, this is resolved by the first pair of transitions. If the ignition occurred, its time is then determined by another set of transitions.

Purpose of the Evacuation blocks is to alter the probability of fatality, alteration is based on availability of the lifeboats. In case of ignition, escalation blocks determine the nature of fire escalation and damage to the offshore platform caused by ignited leak. Token is then added to place *ZoneProgressed*.

When a token arrives in place *ZoneProgressed*, transitions determining the success of the evacuation are enabled. Firing any of these transitions results in a transfer of token to a place according to the outcome of the evacuation (e.g. if the scenario ended with a fatality, token arrives in place *Fatality*).

The last pair of transitions provides cumulative frequencies of the fatalities and resets values of used variables. Another token then proceeds through the net, until all tokens are contained by place *End*.

### 3.3 Dynamic SPN Model of Small Leak Scenario

For the construction of time-dependent SPN model of Small Leak scenario, Personnel and Installation Actions table from [1] was used as a moot point. The table describes escape of personnel to lifeboat station at the Accommodation Platform. Personnel has 15 minutes to arrive at the lifeboat station. If any personnel fails to arrive, search party is formed and given 15 minutes to search for and rescue missing personnel. 30 minutes after leak occurrence, lifeboats are launched and personnel evacuates.

All possible events during escape of the personnel are divided in six time intervals, every possible event is described and linked with corresponding ignition case and occurrence time. However, for creating SPN model of the Small Leak scenario, quantification of contributions of fatality probabilities is required.

[1] In MOCA software, tokens are called Jets. Number assigned to a place is identifier for the software algorithms.

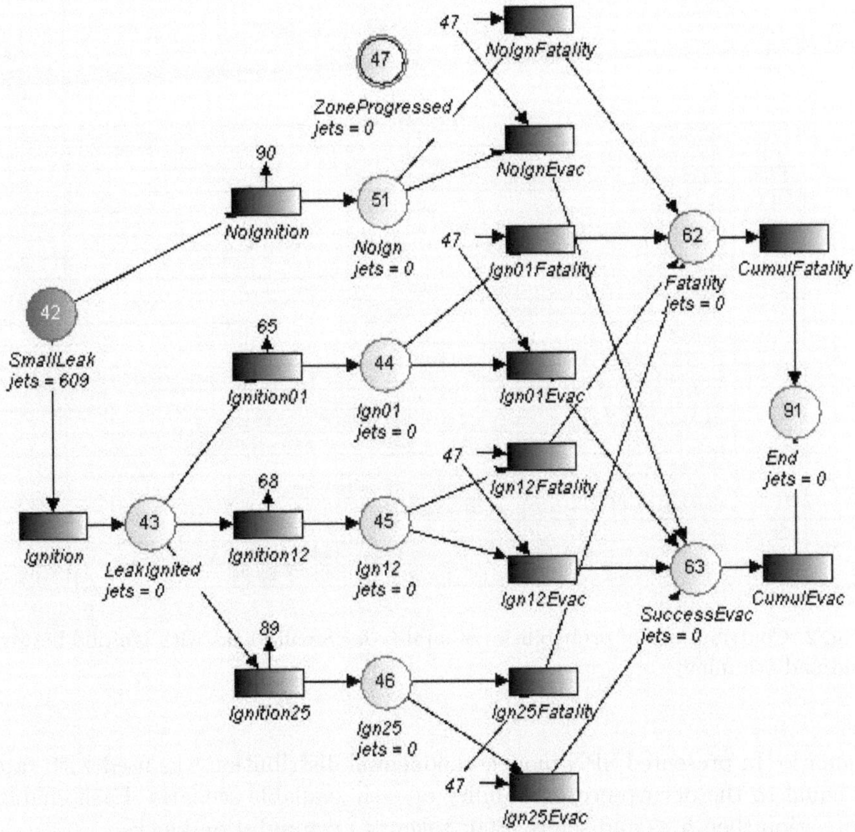


**Fig. 1.** SPN model : Main block

With realistic data from offshore industry available, tables containing different scenarios with corresponding probability of fatality contributions in Small Leak scenario were made, one for each ignition case. As an example, table for 2 to 5 minutes ignition case is shown in Fig. 2.

In the dynamic SPN model, three blocks of scenarios were made, one block for each ignition case. No ignitions cases were not modeled.

When token leaves starting place in each block, immediate transition fires and token is send to place distributing tokens amongst all available scenarios for given ignition case. It should be noted that mentioned immediate transition is substituted in model by timed transition with delay determined by Dirac law with parameter $\delta = 0$.

If suitable delay law is chosen, SPN model allows to 'split' the arriving token amongst multiple transitions. In this way, token is divided between all enabled transition. This division is then based on occurrence probability for each available

| No. | Time T (mins) | Person Location and Activity at Time T | Prob of Fatal at Time T |
|---|---|---|---|
| 17 | 0-1 | In Zone 8. | 0 |
| | 1-25 | Escaping from Zone 8 to Muster Area on AP. | 0 |
| | 25-30 | Embarkation to and evacuation by lifeboats from AP. | 0 |
| 18 | 0-1 | In Zone 8. | 0 |
| | 1-25 | Escaping from Zone 8 to Muster Area on AP. | 0 |
| | 25-30 | Lifeboats on AP not available. Route to secondary evac not available. Evacuation to the sea. | 0.5 |
| 19 | 0-1 | In Zone 8. | 0 |
| | 1-25 | Escaping from Zone 8 to Muster Area on AP. | 0 |
| | 25-30 | Embarkation to and evacuation by lifeboats from AP. | 0 |
| 20 | 0-1 | In Zone 8. | 0 |
| | 1-25 | Escaping from Zone 8 to Muster Area on AP. | 0 |
| | 25-30 | Lifeboats on AP not available. Route to secondary evac not available. Evacuation to the sea. | 0.5 |
| 21 | 0-1 | In Zone 8. | 0 |
| | 1-25 | Escaping from Zone 8 to primary and secondary means of evacuation. | 0.1 |
| | 25-30 | Embarkation to and evacuation by primary and secondary means of evacuation. | 0.1 |
| 22 | 0-1 | In Zone 8. | 0 |
| | 1-15 | Escaping from Zone 8 to primary and secondary means of evacuation. | 0.2 |
| | 15-30 | Lifeboats on AP not available. Route to secondary evac not available. Evacuation to the sea. | 0.5 |
| 23 | 0-1 | In Zone 8. | 0 |
| | 1-15 | Trapped during escape from Zone 8 to primary and secondary means of evacuation. | 0.2 |
| | 15-30 | No escape to any lifeboats or life rafts available. Evacuation to the sea. | 0.7 |

**Fig. 2.** Contributions of probabilities of fatality for Small Leaks with ignition between 2nd and 5th minute

scenario. In presented SPN model, exponential distribution was used with rates $\lambda$ equal to the occurrence probability of each available scenario. Each enabled transition then fires and sends $1800^2$ tokens to computation blocks.

In computation blocks (Figure 3 shows computation block for scenario 23[3]), time-dependent probability of fatality of given scenario is calculated. Each computation block has to simulate its own time for its own probability calculation. In this SPN example, time flow is represented by variable *Time*, incremented by a set of transitions with Dirac delay law with $\delta = 1$. When a transition is enabled by a token, transition fires, variable *time* is raised by 1 and distribution function values $F(time)$ and $F(time - 1)$ are computed. The probability of fatality value is then adjusted by difference $F(time) - F(time - 1)$. This approach was chosen due to internal logic of the MOCA-RP software, which evaluates all variables once a transition fires.

In this paper, it is assumed that the probability of fatality follows exponential distribution. Therefore:

$$F(t) = 1 - e^{-\lambda t} , \tag{1}$$

where $t$ is a value of variable *time* and $\lambda$ is a rate parameter for current phase of evacuation. Correct phase of evacuation for $\lambda$ computation is determined by

[2] As mentioned in part 3.3, personnel has 30 minutes to evacuate the platform, therefore each token represents one second in simulation.

[3] Description of scenario 23 can be viewed in Fig. 2

a set of conditions utilizing time thresholds given by ignition case tables (shown in 2 for 2 to 5 minute ignition case).

Let $x$ be length of current phase in seconds and $F(x)$ be probability of fatality at the end of the current phase. Then rate parameter $\lambda$ is given by following equation:

$$\ln(1 - F(x)) = -\lambda x \; . \tag{2}$$

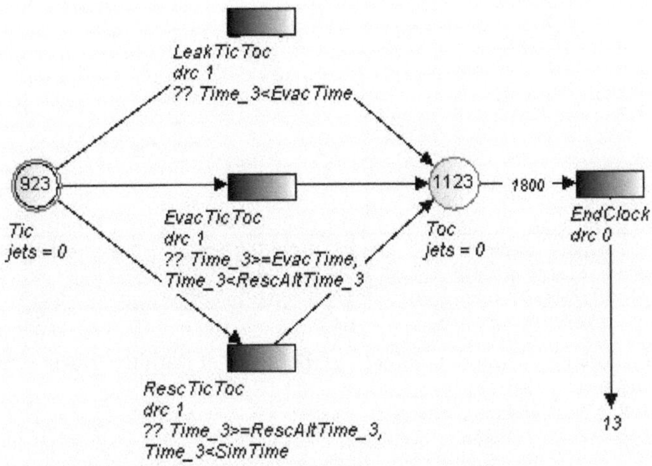


**Fig. 3.** SPN model : Computation block for scenario 23

In this way, probability of fatality is computed for each available scenario (e.g., resulting probability of fatality for single scenario 23 is shown in Fig. 4). However, this does not reflect occurrence probability of those scenarios, thus the previously mentioned token 'splitting' was used. For each scenario, computed probability of fatality was multiplied by scenario occurence rate (i.e. portion of token received by corresponding computation block). By adding these modified probabilities of fatality, the overall probability of fatality was constructed, resulting time-dependent probability of fatality graph is shown in Fig. 5.

## 4 Results

For computation of the probabilities of fatalities, MOCA-RP computing module was used. For comparison with the results presented in [1] by application of the Event Tree and Monte Carlo methods, same sample was chosen, consisting of 609 occurrences of hydrocarbon small leaks.

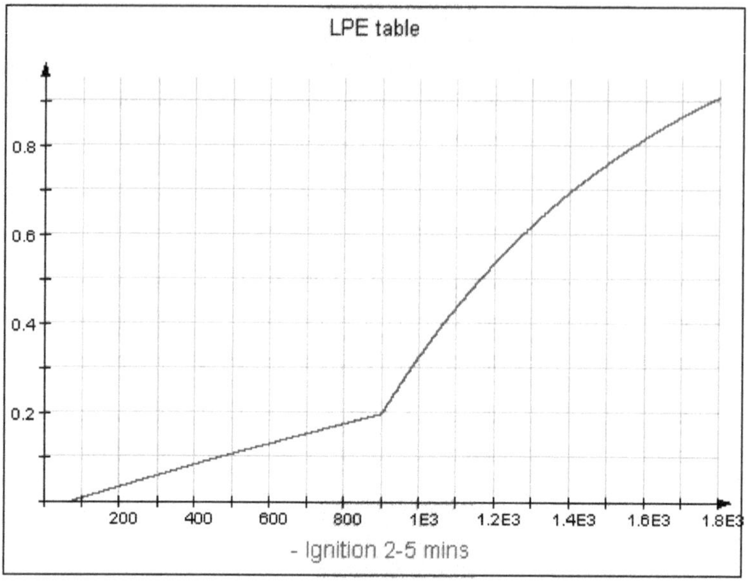


**Fig. 4.** SPN model : Time-dependent probability of fatality during Scenario 23 (Figure 2)

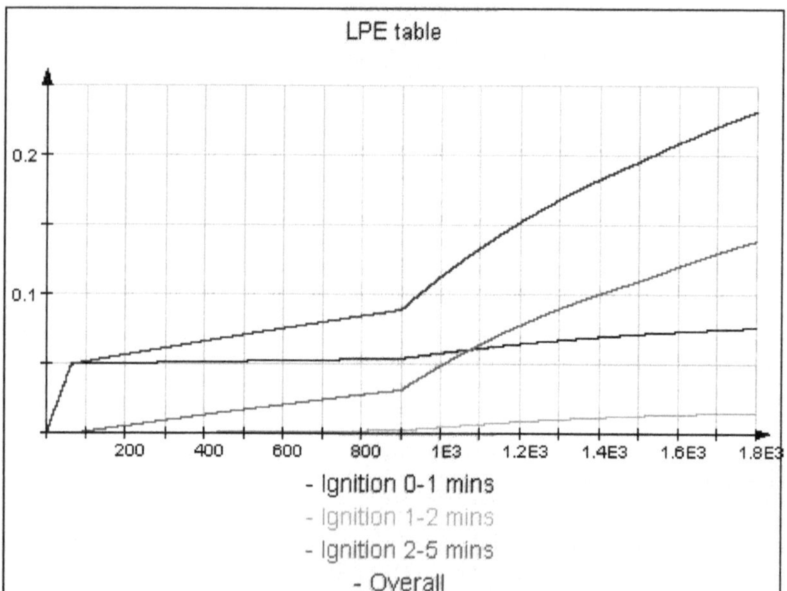


**Fig. 5.** SPN model : Overall time-dependent probability of fatality of Small Leak Scenario

### 4.1 Steady-State SPN Model

Four cases were distinguished by nature of the ignition.

**Table 1.** Resulting probabilities of fatalities obtained by the SPN and the ET model

| Ignition Case | SPN model | ET model |
|---|---|---|
| No ignition | 1.24e-5 | 9.8e-6 |
| 0 and 1 minute | 2.65e-4 | 2.6e-4 |
| 1 and 2 minutes | 8.51e-5 | 8.5e-5 |
| 2 and 5 minutes | 1.08e-3 | 1.1e-3 |
| Total | 1.444e-3 | 1.423e-3 |

As could be expected, delayed leak ignition is the most dangerous event, resulting in an explosion and thus strongly contributing to the loss of possibility of evacuation. This expectation is in accordance with presented results in Table 1.

As can be seen in Table 1, the difference between the probabilities of fatalities in every ignition case are minimal, with the most notable (albeit small) difference between the no ignition cases. Thus the results of the SPN method and Event Tree method are very close, with the cumulative probabilities of fatality differing by a margin of $1.1e-5$.

### 4.2 Dynamic SPN Model

Probability values at the time of 1800 seconds were used and compared to values obtained by DMC model presented in [1]. Probability values obtained by SPN model were multiplied by ignition case frequencies, resulting in comparison shown in Table 2.

**Table 2.** Resulting probabilities of fatalities obtained by the SPN and the MC model

| Ignition Case | SPN model | DMC method |
|---|---|---|
| 0 and 1 minute | 3.02e-4 | 2.6e-4 |
| 1 and 2 minutes | 8.82e-5 | 8.5e-5 |
| 2 and 5 minutes | 9.59e-4 | 1.1e-3 |
| Total | 1.349e-3 | 1.423e-3 |

Considering results of SPN and DMC methods in Table 2, the most significant difference measured between ignition cases was $1.71e-4$ (2 to 5 minute ignition cases). Cumulative probabilities of fatality are differing by a margin of $7.4e-5$. Given author's limited insight into the MOCA-RP's internal logic, specifically MOCA-RP's handling of rounding, these differences could be further mitigated. However, despite the differences, resulting probabilities of fatality could be regarded as fairly accurate and thus allowing SPN to be considered as a viable alternative to other time-dependent probability modeling methods.

## 5 Conclusion

In this article, a new approach for modeling risk to safety of personnel in process industries was presented in the application of SPN. Benefits of SPN over normally used Event Trees were discussed and SPN formalism was presented.

One static and one dynamic model representing Small Leak scenario on a typical offshore installation were used to compute the probabilities of fatalities and results were compared with those obtained by Event Tree model and Direct Monte Carlo method. As the differences were low, the SPN can be considered a suitable substitution for Event Trees in process industry. In comparison with DMC method, the SPN method results were less accurate, thought by a small margin. Thus, the SPN method could be regarded as a viable alternative to DMC method in process industry. Proposed further work in this area is construction of SPN models for Medium and Large Leak scenarios, as well as gaining a better understanding of MOCA-RP's possibilities for modeling other dynamic scenarios from process industries.

**Acknowledgements.** This article was supported by the European Regional Development Fund in the IT4Innovations Centre of Excellence project (CZ.1.05/1.1.00/02.0070).

## References

1. Briš, R., Medonos, S., Wilkins, C., Zdráhala, A.: Time-Dependent Risk Modeling of Accidental Events and Responses in Process Industries. Reliability Engineering and System Safety 125, 54–66 (2014)
2. Signoret, J.-P., Dutuit, Y., Cacheux, P.-J., Collas, S., Foleaux, C., Thomas, P.: Reliability block diagrams driven Petri Nets. Reliability Engineering and System Safety 113, 61–75 (2013)
3. Ajmone Marsan, M., Balbo, G., Conte, G., Donatelli, S., Franceschinis, G.: Modeling with Generalised Stochastic Petri Nets. Universita degli Studi di Torino (1994)

# Impact of Hurst Exponent on Indicator Based Trading Strategies

Tomáš Vantuch

VSB-Technical University of Ostrava, 17. listopadu 15 708 33,
Ostrava-Poruba, Czech Republic
tomas.vantuch.st@vsb.cz

**Abstract.** Appearance of chaotic behavior that covers stock market trading, creates a lot of doubts of its analysis and predictions. However the chaos theory is applicable in a lot of studies of this kind. Stochastic and nonlinear systems can be viewed as deterministic through the prism of chaos theory. This article describes the experiment of analysis of stock market titles by Hurst exponent to find out the randomness or long range memory in the generating of prices. The other question is to figure out the impact of application of the Hurst exponent in two simple indicators like RSI and CCI. Since the usage of the evolution algorithm in stock market prediction and for optimization input parameters is very common it is used in this article too.

**Keywords:** chaos, Hurst, market, RSI, CCI, evolution.

## 1 Introduction

The effective market theory [2] describes a market, whose current price always reflects all the information and in no way takes into consideration its past behavior. Investors in this kind of market do not execute risk trades and they are able to evaluate and interpret all the information. Since the market has no memory, the price creation is very hard to predict in the view of statistic methods. On the other hand, the fractal market theory [2], describes a market where the investors represent prices in various ways, they are not always rational and they tend to be risky. The price itself is made by its past evolution and in this case we can say that the market has a long term memory. The real market stands somewhere between these two theoretical markets, which provides an opportunity for analysis.

In the field of analysis of nonlinear systems with chaotic behavior, there is a very powerful theory named the chaos theory [1]. In comparison to other statistical methods, which are only capable of analyzing linear systems, the theory of chaos can work with nonlinear systems, that seems to be stochastic, but most of them are strictly deterministic. The chaos theory has a wide field of use, from the pseudo-random number generators with the use of optimization of evolution algorithms to the field of the evaluation of time series in terms of predictability.

I. Zelinka et al. (eds.), *Nostradamus 2014: Prediction, Modeling and Analysis of Complex Systems*, Advances in Intelligent Systems and Computing 289,
DOI: 10.1007/978-3-319-07401-6_33, © Springer International Publishing Switzerland 2014

There are many kinds of evaluations, that can determine whether the time series has a long range memory and thus if its behavior depends on its past evolution or if it does not have a long range memory and it is just a random walk time set. Hurst exponent is known for making decisions if the time set is randomness or if it contains long or short range memory [3]. The most popular algorithm for the Hurst exponent calculation is the R/S analysis described by the English hydrologist Hurst, who studied conditions of the river Nile [3,4]. The R/S analysis is better known and used than others, nevertheless it has a weaknesses of short trends [3]. The modifications like V/S statistic [6] or modified R/S [5] do not guarantee more precise results [3], so the standard R/S is used in this experiments too.

If it is the case that we are dealing with a market time set that has long range memory, we are able to make predictions for its future prices. In the field of market prediction many indicators could be used for technical analysis [9]. By patterns from this indicators we are able to buy or sell to make profit. One of the most known indicators is the RSI indicator (relative strength index) [14], that is able to determine whether the market is overbought or oversold and this indicates the right events for buy/sell strategy. It is very popular in automated trading because of its simplicity. The other indicator used in this article is CCI (commodity channel index) developed by Donald Lambert [13]. Based on the behavior of CCI, many non official patterns like TLB, HLTB, ZLR were created, so they could be used as alternative to other indicators too. Each of these indicators has its own set of input attributes. To improve the results, the employment of an evolution algorithm is needed.

The most widely known algorithm used for the optimization of input parameters is the genetic algorithm [11,12]. It is based on the iterative creation of generations and the evaluation of its descendants. Each one is made by a parents crossover and an added mutation. Every crossover is made by the best evaluated candidates to find a "good enough"solution. To optimize a trading strategy as it was used in previous articles [10], it is one of the best approach to split whole time set into shorter time periods and make optimization on the most actual historical data, to develop more adaptable strategy.

## 2 Experiment Design

The first task is to analyze historical data of market titles by the Hurst exponent to find out if they are predictable or not. We need to ask the question – what time set do we need to use as the input for the calculation of the exponent? There is an option to calculate the exponent from close prices (typical approach of time set evaluation) or from calculated values of RSI and CCI indicators. In this experiment both options were tested.

The next task is the optimization of trading rules based on indicators as it was described in previous experiments [11]. So if the market has a long range memory, the goal will be to develop two simple market strategies based on RSI and CCI indicators. The next phase is to optimize them by genetic programming to obtain better results. In the second half, these strategies are extended by the inclusion of the Hurst exponent and simply manage the strategy activity by the value of this exponent. The length of time series for the Hurst exponent and a threshold near 0.5 value of the exponent will be new variables for the genetic optimizer. So in the last phase there will be a new optimization of the new strategies and evaluation if this extension was a good improvement.

### 2.1 Data Set

For this experiment historical data of stock markets titles as YHOO, GOLD, CSCO, EBAY and DELL were used. Each of them consists of records known as candles, that contains information of some period of time, in this case one day. The candle contains information about the highest price of the period (HIGH), the lowest price of the period (LOW), the first price of the period (OPEN), the last price (CLOSE), the quantity of traded instruments (VOLUME) and of course the time stamp of the period. Each of these titles has a historical data of fourteen years. All the historical data was taken from google finance [https://www.google.com/finance].

### 2.2 Software

The trading platform Ninja Trader 7 was used for implementation, testing and collecting results of trading strategies and Hurst exponent. It is a powerful tool for stock market prices visualization, evaluation of strategies and it has its own API for development in C# programming language. This software is free to use on site http://www.ninjatrader.com/.

The success rate of trading strategy is judged by many parameters like count of trades, ratio of winning and losing trades, final profit, profit loss ratio etc. In case of evaluating of the Hurst exponent's impact on trading strategies based on indicators, a single parameter which counted the trades and the win or lose ratio was used. The final profit is not important for this experiment since it can be controlled in real conditions by money management strategy that was not applied.

Ninja Trader has two options for evaluating the trading strategy. There is a simple backtesting method, where the user only sets all the input parameters for his strategy and software will run it on whole historical data to simulate trading. Results are shown immediately and user can change the input values anytime to improve his strategy.

The other, more sophisticated way, is to use the evolution algorithm, the genetic optimizer, that is embedded in the software. The genetic optimizer is able to get the range for input parameters from user and find the "good enough" set of inputs in reasonable time. The genetic optimizer has one key feature that is very suitable for trading and it is its running in adjusted lengths of periods. Every time there is an optimization period for example one month of historical data and when the optimized parameters are found, they are used in next month

**Table 1.** The key parameters of adjustment of walk forward testing and its value ranges

| | |
|---|---|
| Hurst exponent threshold | <55,70> (in using divided by 100) |
| Window size for exponent | <20,80> |
| CCI window size | <14,60> |
| RSI window size | <4,40> |
| Parameters for evolution algorithm: | |
| Count of generations | 5 |
| Crossover rate | 80% |
| Generation size | 25 |
| Mutation rate | 2% |
| Optimization period | 28 days |
| Testing period | 28 days |

and this process is repeated through the whole historical data set. Adjustment of main parameters that was applied in tests is shown on Table 1.

### 2.3 Commodity Chanel Index

CCI indicator transforms the prices into one value and this value is fluctuating in time [13]. The values of indicator are not normalized and can be calculated by rule written in Fig. 1.

$$CCI = \frac{1}{0.015} \frac{p_t - SMA(p_t)}{\sigma(p_t)} \tag{1}$$

If p is the typical price counted as an average of high, low and close price, SMA is simple a moving average of typical prices and everything is divided by 0.015 times of standard deviation of typical prices.

There are many patterns based on the CCI behavior. The best know is using the scale of -100 and 100 as events for buying or selling. If CCI crosses the +100 border from downside, it is an event to open the long position and if it crosses +100 from upside, we have to close the long position. The same scenario but reversed applies for short positions on the border of -100.

### 2.4 Relative Strength Index

The next indicator used is the RSI [14]. It has a normalized output value in a range between 1 and 100. The calculation of this indicator can be seen in Fig. 2. If the value of RSI is higher than 70, we can say that the market is overbought or if the value of RSI is lower than 30, we can say that the market is oversold. In these cases we are executing a long or a short position to react to market circumstances.

The calculating of RSI is:

$$RSI = 100 - (\frac{100}{1 + RS}) \tag{2}$$

The RS is an average of x days up closes divided by an average of x days down closes.

### 2.5 Hurst Exponent

For the analysis of the chaotic behavior of the time series, the so called Hurst exponent H [2] is used. It can determined when the collection has a long range memory or if it is random. The calculation of the H by R/S statistic [3,4] can be seen by Fig 3, 4, 5, 6, 7:

$$X(t,\tau) = \sum_{i=1}^{t}(x_i - \overline{x}(\tau)), t = 1, 2..., \tau; \tag{3}$$

$$\overline{x}(\tau) = \frac{1}{\tau}\sum_{i=1}^{\tau} x_i; \tag{4}$$

$$S(\tau) = \sqrt{\frac{1}{\tau}\sum_{i=1}^{\tau}(x_i - \overline{x}(\tau))^2} \tag{5}$$

$$R(\tau) = maxX(t,\tau) - minX(t,\tau); 1 \le t \le \tau \tag{6}$$

$$H(\tau) = \frac{log(\frac{R(\tau)}{S(\tau)})}{log(\tau)} \tag{7}$$

The result values range between 0 and 1 and the values near these borders mean that the selected time set has a long range memory and conversely, the values near 0.5 indicate randomness.

## 3 Analyze of Historical Data

As we can see the resulted Hurst exponent, depends on the window size that we set. In case of small length of set, the resulted exponents were more often indicating randomness but in the case of greater length of set, they were qualifying more often as having a long range memory.

The figures 1, 2 and 3 show charts for the Hurst exponent distribution with window sizes of 20, 50 and 100 with the outputs from RSI indicator were used as inputs for the exponent calculation.

Very similar behavior of exponent's distribution was in the rest of titles too when inputs for exponent was close prices or outputs from CCI indicator.

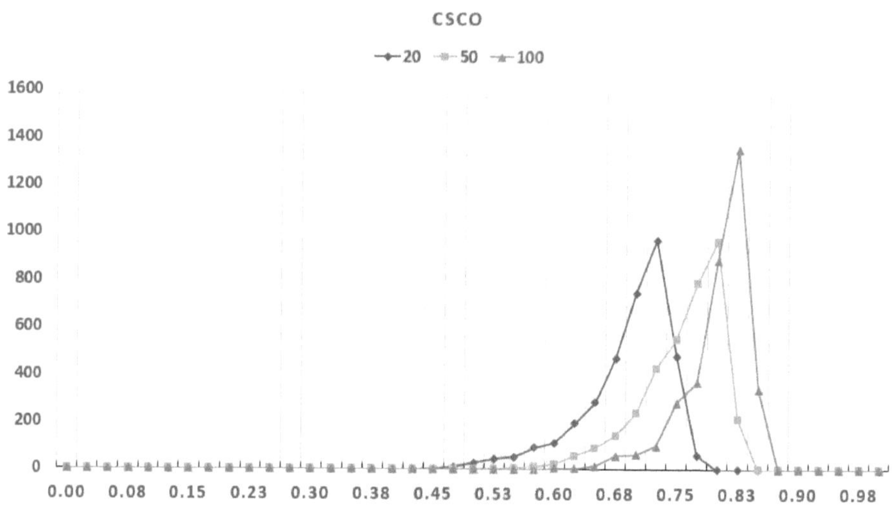


**Fig. 1.** Distribution of Hurst exponent for market title CSCO

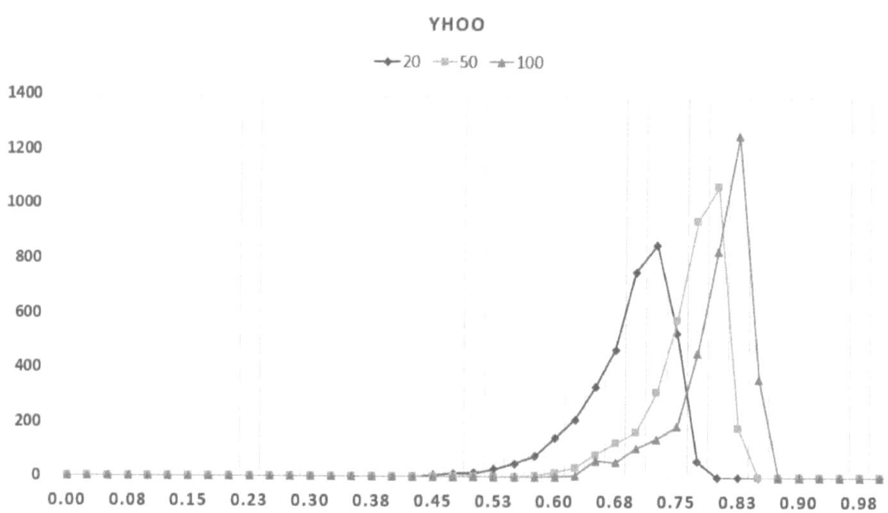


**Fig. 2.** Distribution of Hurst exponent for market title YHOO

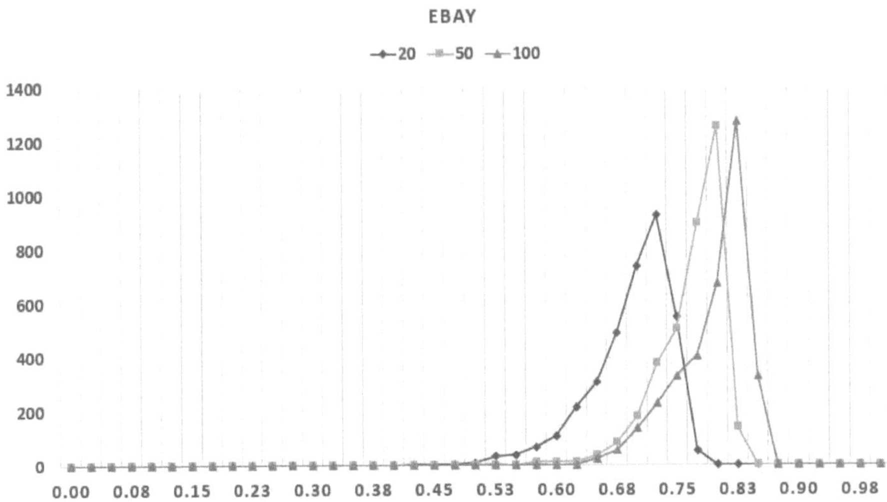


**Fig. 3.** Distribution of Hurst exponent for market title EBAY

# 4 Testing of Trading Strategies

The test was applied in two steps. In the first step, the trading strategy based just on indicators was optimized by a genetic algorithm and applied on the whole historical data set. The main result was to obtain the best rate of winning and losing trades and prevent huge lost of the count of trades.

The second step consists of extending strategies by Hurst exponent and in times qualified as randomness the trading was stopped. We present a table of

**Table 2.** Results of implemented strategies optimized by genetic algorithm when the input data for exponent was outputs from indicator (input from RSI for RSI test and input from CCI for CCI test)

| Indicator | Market Title | Default version | | Extended version | | period length |
|---|---|---|---|---|---|---|
| | | W/L rate | Trades | W/L rate | Trades | |
| RSI | YHOO | 43.85% | 130 | 53.9% | 141 | 10 years |
| | GOLD | 56.5% | 115 | 61.02% | 118 | 14 years |
| | CSCO | 50.35% | 143 | 49.66% | 143 | 14 years |
| | DELL | 53.85% | 117 | 51.18% | 117 | 14 years |
| | EBAY | 50% | 110 | 55.56% | 117 | 14 years |
| CCI | YHOO | 45.10% | 286 | 45.56% | 270 | 10 years |
| | GOLD | 44.12% | 272 | 44.62% | 262 | 14 years |
| | CSCO | 40.21% | 291 | 41.04% | 307 | 14 years |
| | DELL | 40.58% | 276 | 39.35% | 271 | 14 years |
| | EBAY | 40.92% | 303 | 39.8% | 296 | 14 years |

**Table 3.** Results of implemented strategies optimized by genetic algorithm when the input data for exponent was candles close prices

| Indicator | Market Title | Default version | | Extended version | | period length |
|---|---|---|---|---|---|---|
| | | W/L rate | Trades | W/L rate | Trades | |
| RSI | YHOO | 43.85% | 130 | 51.54% | 130 | 10 years |
| | GOLD | 56.5% | 115 | 50.46% | 109 | 14 years |
| | CSCO | 50.35% | 143 | 48.2% | 139 | 14 years |
| | DELL | 53.85% | 117 | 52.63% | 114 | 14 years |
| | EBAY | 50% | 110 | 53.85% | 104 | 14 years |
| CCI | YHOO | 45.10% | 286 | 47.87% | 282 | 10 years |
| | GOLD | 44.12% | 272 | 42.07% | 271 | 14 years |
| | CSCO | 40.21% | 291 | 42.33% | 300 | 14 years |
| | DELL | 40.58% | 276 | 39.21% | 278 | 14 years |
| | EBAY | 40.92% | 303 | 41.42% | 309 | 14 years |

results of all the strategies. In Table 2 the results when the input time series for Hurst exponent was the outputs from indicators and in Table 3 as inputs the close prices was used.

# 5 Conclusions

The main goal of this paper was to use the Hurst exponent for the analysis of stock market titles and to try improve two basic technical analysis indicators.

Results of the Hurst exponent show, that the stock market can have long term memory and all new generated prices belong on the past behavior too. On the other hand, we can say that the size of the selected collection of prices influences the final distribution of the Hurst exponent.

Final comparison of trading strategies shows that there is no rule about that Hurst exponent has to improve strategy based on indicators. In some cases an improvement was seen, but in others the results worsened. It did not matter if the inputs for the calculation of the exponent has to be close prices or outputs from the indicators. In both cases the results remained similar and no improvement was confirmed.

The final question is whether there is a room for improvement. A lot could be done in terms of the improvement of the genetic algorithm, testing on more precise data series or expanding the range of input parameters. All of these factors can show relatively more valuable results.

**Acknowledgment.** The following grants are acknowledged for the financial support provided for this research: Grant Agency of the Czech Republic - GACR P103/13/08195S, is partially supported by Grant of SGS No. SP2014/159, VŠB - Technical University of Ostrava, Czech Republic, by the Development of human resources in research and development of latest soft computing methods and their application in practice project, reg. no. CZ.1.07/2.3.00/20.0072 funded by Operational Programme Education for Competitiveness.

## Reference

1. Zelinka, I., Chadli, M., Davendra, D., Senkerik, R., Pluhacek, M., Lampinen, J.: Hidden Periodicity - Chaos Dependance on Numerical Precision. In: Zelinka, I., Chen, G., Rössler, O.E., Snasel, V., Abraham, A. (eds.) Nostradamus 2013: Prediction, Model. & Analysis. AISC, vol. 210, pp. 47–59. Springer, Heidelberg (2013)
2. Dostál, P.: Pokročilé metódy rozhodování v podnikatelství a veřejné správě. Akademické nakladatelství (2012)
3. He, L.-Y., Qian, W.-B.: A Monte Carlo simulation to the performance of the R/S and V/S methods-Statistical revisit and real world application, Beijing 100083, China (2012)
4. Lo, A.W.: Long term memory in stock market prices. Econometrica 59, 1279–1313 (1991)
5. Lo, A.W., Mackinlay, A.C.: A Non-Random Walk Down Wall Street. Princeton University Press (1996)
6. Giraitis, L., Kokoszka, P., Leipus, R., Teyssiere, G.: Rescaled variance and related tests for long memory in volatility and levels. Journal of Econometrics 112, 265–294 (2003)
7. Pluchino, A., Rapisarda, A., Tsallis, C.: Noise, synchrony, and correlations at the edge of chaos. Physical Review E 87(2) (2013), doi:10.1103/PhysRevE.87.022910
8. He, L.-Y., Fan, Y., Wei, Y.-M.: The empirical analysis for fractal features and long-run memory mechanism in petroleum pricing systems. International Journal of Global Energy Issues 27(4), 492–502 (2007)
9. Ke, J., Chen, Y.: Modeling and Simulation of the Artificial Stock Market Trading System. Natural Sciences Publishing Cor. (2013)
10. Bodas-Sagi, D.J., Fernández-Blanco, P., Hidalgo, J.I., Soltero-Domingo, F.J.: A parallel evolutionary algorithm for technical market indicators optimization. Springer Science+Business Media Dordrecht (2012)
11. Potvina, J.-Y., Sorianoa, P., Vall, M.: Generating trading rules on the stock markets with genetic programming. Computers & Operations Research 31, 1033–1047 (2004)
12. Grefenstette, J.J.: Optimization of Control Parameters for Genetic Algorithms. IEEE Transactions on Systems, Man, and Cybernetics SMC-16(1) (January/February 1986)
13. Davies, D.W.: Defining The Commodity Channel Index
14. Deng, S., Sakurai, A.: Foreign Exchange Trading Rules using a Single Technical Indicator from Multiple Timeframes. In: 2013 27th International Conference on Advanced Information Networking and Applications Workshops (2013)

# Entropy and Market Prediction with Technical Indicators

Marian Bielik

VSB-Technical University of Ostrava, 17. listopadu 15 708 33,
Ostrava-Poruba, Czech Republic
marian.bielik@vsb.cz

**Abstract.** Technical indicators are often used in automated and non-automated trading strategies. In this pager, we will use one of them with the combination of entropy. We will use entropy to find predictable market part and then we will use technical indicator to predict market movements.

**Keywords:** Entropy, MACD, genetics algorithm, chaos, market.

## 1 Introduction

Entropy is one of the most known terms in the theory of information. Everywhere, where we speaks about probability of possible states of one system, there we can see entropy. The father of theory of information is Claude Elwood Shannon, who defined entropy in 1984. Detailed description of entropy can be found in [1]. We will use a normalized entropy from Eq. 1

$$H_{norm} = \frac{H_1}{H_{max}} = -\sum_{i=1}^{m} \log_m p_i \tag{1}$$

as described in [2] and in [5]. Similar article about using entropy in finance is in [10].

We will use the entropy to detect parts of markets, where the markets oscillate. In this parts of markets, we apply technical indicator MACD Moving average convergence/divergence. This is one of the most used indicators in trading systems that uses exponential moving averages. More about this method can be found in [3].

The markets data that we will use for testing are daily data from Yahoo. We will use 10 years historical market daily data of Oracle, Apple, Microsoft and 5 years old minutes data of Collagate-Palmolive (CL).

Strategies in this paper were tested using NinjaTrader trading system. This trading platform supports multicore strategy testing with genetics optimization and itself is capable of providing results that are later used for purposes of this article. Genetic algorithms are one of the most known algorithms used for optimizations of input parameters in market strategies [6,7]. Optimization of trading strategies is discussed in [8].

I. Zelinka et al. (eds.), *Nostradamus 2014: Prediction, Modeling and Analysis of Complex Systems*, Advances in Intelligent Systems and Computing 289,
DOI: 10.1007/978-3-319-07401-6_34, 

## 2 Experiment Design

Strategy counts entropy from X market bars back, where X will be natural number optimized by genetics algorithm implemented in NinjaTrader. The entropy will be number between 0 and 1 according normalized entropy. 0 indicates, that market is maximum predictable. On the other hand, 1 indicates chaotic behavior and then market is not predictable at all. If the entropy will be bellow specific number , then our MACD indicator will be applied.

First the strategy detects predictable markets data and then apply MACD indicator on this data. This indicator checks predictable markets data and opens or closes bearish and bullish market positions. Unlike using only MACD, this strategy reduces amount of market positions. As a result, the strategy should generate less loosing positions.

Genetics algorithm optimized our strategy parameters to get maximum percent profitable strategy. MACD will be set upped to MACD(12,26,9) discussed in [4,9]. All prices from historical data will be closed price.

We will not use any stop loss or profit target settings or any money. We will compare entropy and MACD based strategy with strategy using only MACD indicator.

## 3 Results

As described in previous section, we will use normalized entropy, sliding window and number that indicates using MACD, optimized by genetic algorithm, called entropy index.

Firstly we will start with Apple historical market data retrieved from Yahoo. Figure 1 shows 10 year period of historical Apple (AAPL) daily data.

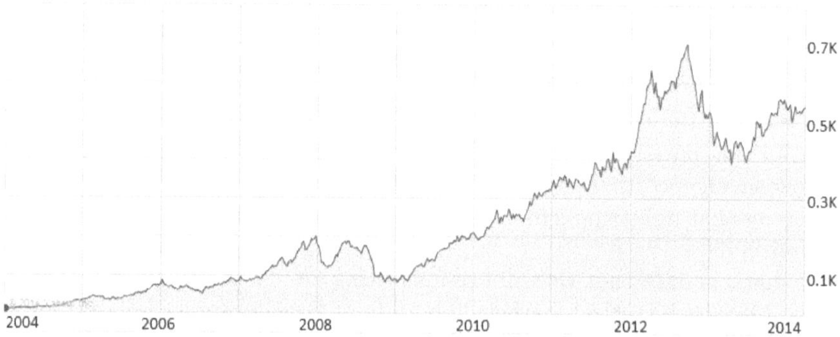


**Fig. 1.** Apple historical daily data

Table 1 shows result from NinjaTrader after applying MACD. As we can see from Table 1, there are more losing trades, than winnings. This strategy can predict market movement with 38.4%.

Table 1 shows also strategy with entropy applied on the same data as we used in strategy without entropy. This strategy reduces total trades from 185 to only 5 trades. This means that from entropy perspective, this market is not predictable. Entropy window size was set to 21 (bars back). We used market data that had entropy below 0.99. From Table 1 we can deduce, that this market has almost whole historical data with 0.99-1 entropy index.

**Table 1.** Results using Apple historical data

| Strategy | MACD | Entropy+MACD |
|---|---|---|
| Percent Profitable | 38.4% | 60% |
| Total of Trades | 185 | 5 |
| Winning Trades | 71 | 3 |
| Losing Trades | 114 | 2 |

Equity curve is important for us too, because it shows, if the strategy is robust during whole time period. Figure 2 shows the results using strategy with entropy. Y-axis shows Profit/Loss ratio. As we mentioned in previous section, this strategy does not use any stop loss, profit target or any other money management. This settings can influence Profit/Loss ratio, but for comparing strategies we can leave this settings.

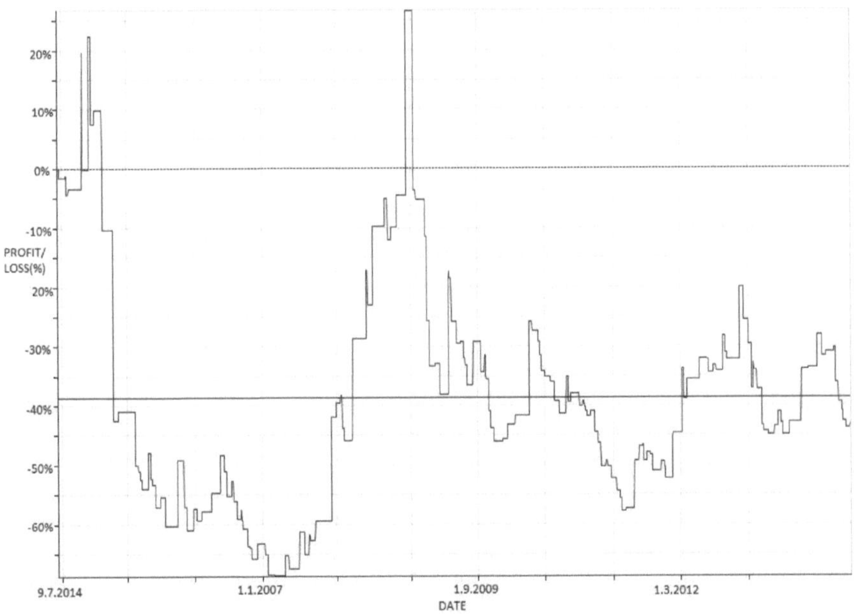


**Fig. 2.** Equity of Apple historical daily data using entropy

The second market will be Cisco historical market data. Figure 3 shows graph of 10 year period of historical Cisco (CSCO) daily data.

**Fig. 3.** Cisco historical daily data

Applying MACD indicator to this market we will get data from Table 2. Results are similar to Apple historical data. Equity curve, without entropy, from Figure 4 is not profitable. Whole time interval in this strategy is in bigger and bigger lose.

Next we will use this data with strategy with entropy, results are in Table 2. Genetics algorithm uses this strategy with window size 15 and entropy index 0.99. This strategy has improvements in all aspects. Also equity curve, using entropy, from Figure 5 is profitable. This is still not ideal equity curve, because as we can see, at the end of the graph, equity has decreasing trend.

**Table 2.** Results using Cisco historical data

| Strategy | MACD | Entropy+MACD |
|---|---|---|
| Percent Profitable | 33.15% | 55.56% |
| Total of Trades | 199 | 27 |
| Winning Trades | 60 | 15 |
| Losing Trades | 139 | 12 |

From Figure 3 we can see, that historical data may not be increasing to have profitable strategy.

Last daily historical data that we will use to compare our strategies are Microsoft data from Figure 6.

The results using MACD are in Table 3. The results are similar to Apple and Cisco data and equity is not profitable.

We will apply entropy based strategy on this data. Results can be seen in table 3. Entropy window size was set to 16 and entropy index to 0.99.

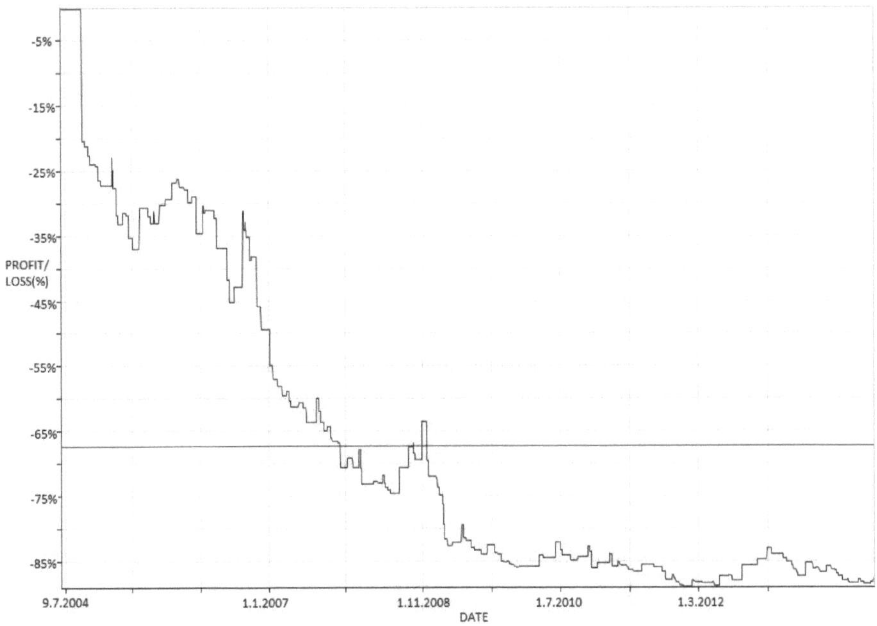


**Fig. 4.** Equity of Cisco historical daily data without entropy

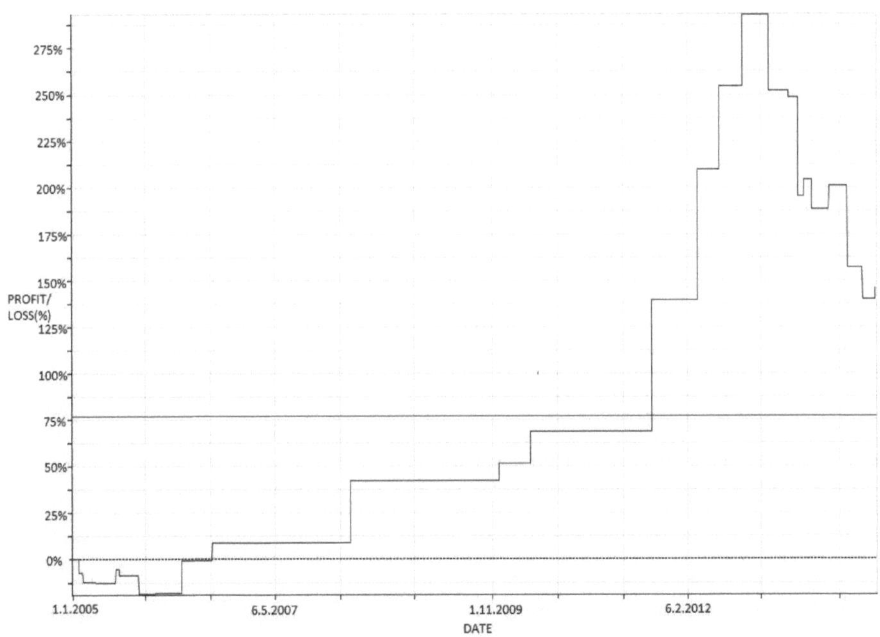


**Fig. 5.** Equity of Cisco historical daily data using entropy

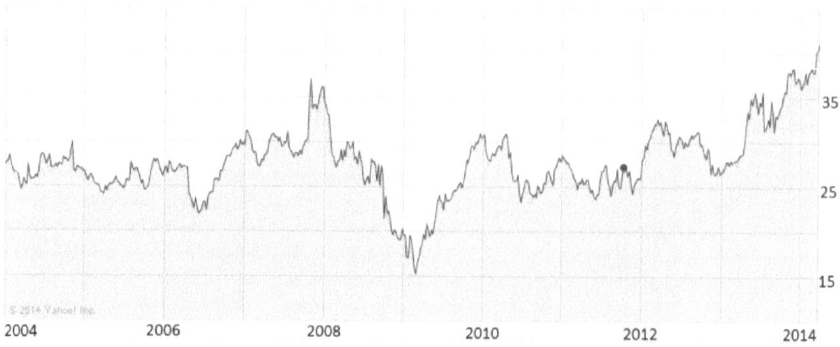


**Fig. 6.** Microsoft historical daily data

**Table 3.** Results using Microsoft historical data

| Strategy | MACD | Entropy+MACD |
|---|---|---|
| Percent Profitable | 36.65% | 65.22% |
| Total of Trades | 191 | 23 |
| Winning Trades | 70 | 15 |
| Losing Trades | 121 | 8 |

Equity curve in Figure 7, using entropy, is profitable, but still not ideal. It was not profitable only at the end of the year 2008 and in the middle of the year 2012.

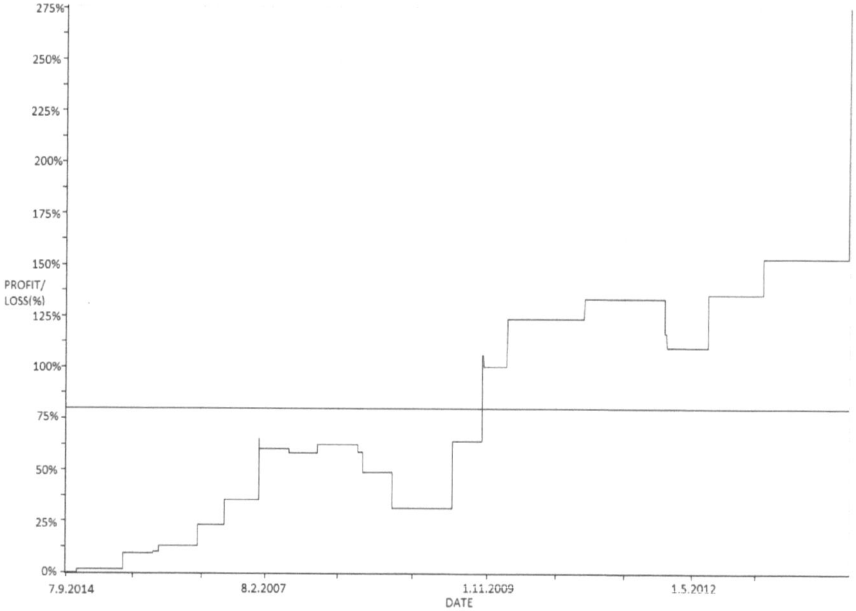


**Fig. 7.** Equity of Microsoft historical daily data without entropy

**Table 4.** Results using CL historical data

| Strategy | MACD | Entropy+MACD |
|---|---|---|
| Percent Profitable | 39.15% | 45.9% |
| Total of Trades | 106264 | 1608 |
| Winning Trades | 41605 | 738 |
| Losing Trades | 64659 | 870 |

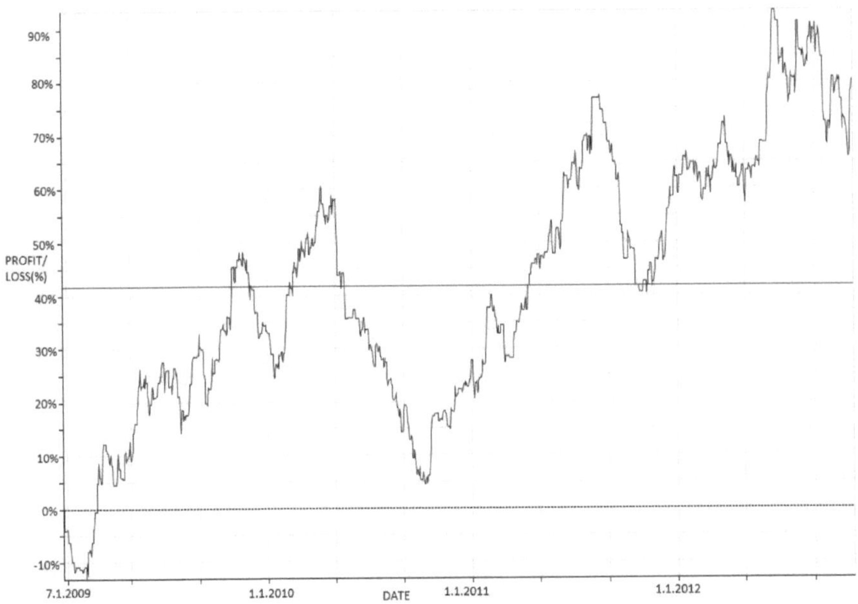


**Fig. 8.** Equity of CL historical daily data using entropy

All historical data were on daily bases. In all cases, strategy utilizing entropy dramatically decreased number of trades. Now we will apply this stratedy on historical minutes data. For this purpose we will use Colgate-Palmolive (CL) historical minutes data from last 5 years. Strategy using only MACD is showing results in Table 4. The number of trades has increased, but percent profitable is similar to previous results.

Now we compare this results with our proposed strategy. The results using entropy are in Table 4. Entropy index was set to 0.85 and entropy window was set to 20 by genetics algorithm. We can see increase in percentage of profitable trades and decreased number of trades as in previous experiments.

Equity curve in Figure 8 is similar to equity curve without entropy. Both are increasing and decreasing that is no good.

## 4 Conclusion

MACD itself is based on exponential averages and because of that is sensitive on chaotic markets. Entropy detects markets that can be predictable and helps MACD to predict market movements. All previous experiments has demonstrated an impact entropy on prediction market movements using MACD, mainly decreased amount of trades.

It was demonstrated that percentage of profitable trades was increased. We can summarize fact that entropy improved MACD prediction, see for example Tables 1-4.

Future research will be focused on implementing money management with stop loss and profit target into MACD strategy that uses entropy.

**Acknowledgment.** The following grants areacknowledged for the financial support provided for this research: Grant Agency of the Czech Republic - GACR P103/13/08195S, is partially supported by Grant of SGS No. SP2014/159, VSB - Technical University of Ostrava, Czech Republic, by the Development of human resources in research and development of latest soft computing methods and their application in practice project, reg. no. CZ.1.07/2.3.00/20.0072 funded by Operational Programme Education for Competitiveness, co-financed by ESF and state budget of the Czech Republic.

## References

1. Shannon, C., Weaver, W.: The Mathematical Study of Communication (1949)
2. Masisi, L., Nelwamondo, V., Marwala, T.: The use of entropy to measure structural diversity, Cambridge, Massachusetts
3. Appel, G.: The Moving Average Convergence Divergence Method, Signalert (1979)
4. Appel, G.: Technical Analysis - Power Tools for Active Investors by Gerald Appel. FT Press (March 2005)
5. Zelinka, I., Chadli, M., Davendra, D., Senkerik, R., Pluhacek, M., Lampinen, J.: Hidden Periodicity - Chaos Dependance on Numerical Precision. In: Zelinka, I., Chen, G., Rössler, O.E., Snasel, V., Abraham, A. (eds.) Nostradamus 2013: Prediction, Model. & Analysis. AISC, vol. 210, pp. 47–59. Springer, Heidelberg (2013)
6. Potvina, J.-Y., Sorianoa, P., Vall, M.: Generating trading rules on the stock markets with genetic programming. Computers and Operations Research (2004)
7. Grefenstette, J.J.: Optimization of Control Parameters for Genetic algorithms. IEEE Transactions on Systems, Man, and Cybernetics (January/February 1986)
8. Bodas-Sagi, D.J., Fernandez-Blanco, P., Hidalgo, J.I., Soltero-Domingo, F.J.: A parallel evolutionary algorithm for technical market indicators optimization. Springer Science+Business Media Dordrecht (2012)
9. Colby, R.W.: The Encyclopedia Of Technical Market Indicators, 2nd edn. McGraw-Hill (October 2002)
10. Zhou, R., Cai, R., Tong, G.: Applications of Entropy in Finance (November 2013)

# Dynamics and Efficiency of an Impact Damper

Marek Lampart and Jaroslav Zapoměl

Department of Applied Mathematics & IT4Innovations,
VŠB - Technical University of Ostrava, Czech Republic,
Department of Mechanics, VŠB - Technical University of Ostrava, Czech Republic
marek.lampart@vsb.cz,
jaroslav.zapomel@vsb.cz
http://www.vsb.cz

**Abstract.** The main aim of this paper is to focus on dynamics of the electromechanical system flexibly coupled with a baseplate and damped by an impact element. The model is constructed with three degrees of freedom in the mechanical oscillating part, two translational and one rotational. The system movement is described by three mutually coupled second-order ordinary differential equations. Here, the most important nonlinearities are: stiffness of the support spring elements and internal impacts. As it is shown in the main results, the impact damping device attenuates vibrations of the rotor frame in a wide range of the excitation frequency and the system is showing periodic and chaotic behavior.

**Keywords:** electromechanical system, inertia damper, impact damper, nonlinear stiffness, impacts, vibration attenuation.

## 1 Introduction

The impacts of solid bodies are an important mechanical phenomena. Their presence can be observed during a large number of natural and technological processes. The impacts are characterized by short duration body collisions, very large impact forces and by near sudden changes of the system state parameters. The experience and theoretical analyses show that the behavior of the impact systems is highly nonlinear, highly sensitive to initial conditions and instantaneous excitation effects, leading frequently to irregular vibrations and nearly unpredictable movements. The behavior of each system where the body collisions take place is different, and therefore, each of them must be investigated individually.

Because of the practical importance, a good deal of attention is focused on analysis of vibro-impact systems, where the vibrations are governed by the momentum transfer and mechanical energy dissipation through the body collisions. This is utilized for impact dampers applied to attenuate high-amplitude oscillations, such as those appearing in subharmonic, self-excited and chaotic vibrations.

Even though the problem of impacts is very old, the new possibilities of its investigation enabled by efficient computational simulations appeared at the end

I. Zelinka et al. (eds.), *Nostradamus 2014: Prediction, Modeling and Analysis of Complex Systems*, Advances in Intelligent Systems and Computing 289,
DOI: 10.1007/978-3-319-07401-6_35, © Springer International Publishing Switzerland 2014 

of the 20th century. [1] studied the dynamic behavior of an impact damper for vibration attenuation of an externally loaded and self-excited cart moving in one direction. The damping was produced by impacts of a point body colliding with the cart walls. A number of authors have dealt with the so-called non-ideal problem, which means that the power of the source exciting the oscillator is limited. The article of [2], who extended Chaterjee's model with the cart by attaching a rotor driven by a motor, belongs in this category. Application of a non-ideal model to the gear rattling dynamics was done by [3]. A new mechanical model with clearances for a gear transmission was reported by [6]. Their model has time varying boundaries and impacts between two gears occur at different locations. The horizontal movement of a cart excited by a rotating particle and damped by an impact damper formed by a point body bouncing on the cart walls was investigated by [7]. The computational simulations confirmed a chaotic character of the oscillations. Consequently, the influence of the impact body to the cart mass ratio on its suppression was examined. The non-ideal impact problem completed by flexible stops was analyzed by [8]. The main objection was to study the character of the vibration of the oscillator excited by an unbalanced rotor driven by a motor of limited power. The mutual interaction of a mechanical and electrical system was analyzed by [9]. The investigated unbalanced rotor of an electric motor was attached to a cantilever beam. The amplitude of its bending vibration was limited by a flexible stop. The performed simulations were aimed at studying the character of the induced vibration and at fluctuations of the current in the electric circuit. Vibrations of a rotor supported by bearings with nonlinear stiffness and damping characteristics considering its impacts against the stationary part were investigated by [10]. The shaft was represented by a beam like-body and rotation of the disc was taken into account. The impacts were described both by collisions of rigid bodies utilizing the Newton theory and by impacts with soft stops. Vibration reduction of an electromechanical system by an impact damper having rigid stops was investigated by [4]. Emphasis was put on observing the influence of the inner impacts on the character and reduction of the system vibration dependent on the geometric parameters.

In this paper, a system formed by a rotor and its casing flexibly coupled with a baseplate and of an impact body, which is separated from the casing by two, lower and upper, gaps is analyzed. The rotor is driven by a motor of limited power and from this point of view the investigated model system can be classified as non-ideal. A new contribution of the presented work consists of investigating the system oscillations as a result of a combined time variable loading caused by two sources, the rotor unbalance and the baseplate vibrations, and in investigating the interaction between the motor and its feeding electric circuit. Emphasis is put on observing the influence of the inner impacts on the character and reduction of the system vibration dependent on the width of the upper and lower clearances between the rotor frame and the impact body. The investigated system is of great practical importance as it represents a simplified model of a rotating machine, which is excited by a ground vibration and unbalance of the rotating parts and damped by an impact damper. Results of the performed simulations contribute to better understanding of the dynamic behavior of such technological devices and of impact systems with complicated loading, in general.

## 2 The Vibrating System

The investigated system consists of a rotor (body 1, Figure 1), of its casing (body 2, Figure 1) and of a baseplate (body 3, Figure 1), with which the rotor casing is coupled by a spring and damping element (body 4, Figure 1). The casing and the baseplate can move in a vertical direction and the rotor can rotate and slide together with its casing. Vibration of the baseplate and unbalance of the rotor are the main sources of the casing excitation. To attenuate its oscillation an impact damper was proposed. It consists of a housing fixed to the rotor casing (body 2, Figure 1) and of an impact element (body 4, Figure 1), which is coupled with the housing by a linear spring. The impact body can move only in a vertical direction and is separated from the housing by the lower and upper clearances that limit its vibration amplitude. The rotor is loaded by an external moment produced by a DC motor. Its behavior is described by a moment characteristic, which enables implementation of the influence of the electric parameters of the motor feeding circuit into the mathematical model of the investigated impact system.

The task was to analyze the influence of the upper and lower clearances and the mass of the impact body respectively on attenuation of the rotor frame oscillation and character of its motion.

In the computational model all bodies are considered as absolutely rigid except the contact areas between the impact element and the rotor frame. The plate springs coupling the rotor casing and the baseplate have nonlinear cubic characteristic

$$F_K = k_1 \Delta + k_3 \Delta^3 \tag{1}$$

where $F_K$ is the spring force, $k_1$, $k_3$ are the stiffness parameters and $\Delta$ is the spring deformation (compression or extension). The damper between the rotor frame and the baseplate and the spring coupling the impact body with the damper housing are linear. The Hertz theory has been accepted to describe the impacts. The nonlinear contact stiffness and damping were linearized in the expected range of the contact deformation.

The investigated system has three mechanical degrees of freedom. Its instantaneous position is defined by three generalized coordinates: Y - vertical displacement of the rotor casing, $\mathrm{Y}_t$ - vertical displacement of the impact body and $\Phi$ - angular rotation of the rotor:

$$\begin{aligned}(m+m_R)\ddot{\mathrm{Y}} + m_R e_T \cos(\Phi)\ddot{\Phi} &= m_R e_T \dot{\Phi}^2 \sin(\Phi) - b(\dot{\mathrm{Y}} - \dot{y}_z) - \\ &\quad -k_3(\mathrm{Y} - y_z)^3 - k_1(\mathrm{Y} - y_z) - \\ &\quad -k_t(\mathrm{Y} - \mathrm{Y}_t) - (m+m_R)g - F_{I1} - F_{I2}, \\ m_t \ddot{\mathrm{Y}}_t &= F_{I1} + F_{I2} - k_t(\mathrm{Y}_t - \mathrm{Y}) - m_t g, \\ (J_{RT} + m_R e_T^2)\ddot{\Phi} + m_R e_T \cos(\Phi)\ddot{\mathrm{Y}} &= -m_R g e_T \cos(\Phi) + M_Z - k_M \dot{\Phi}\end{aligned} \tag{2}$$

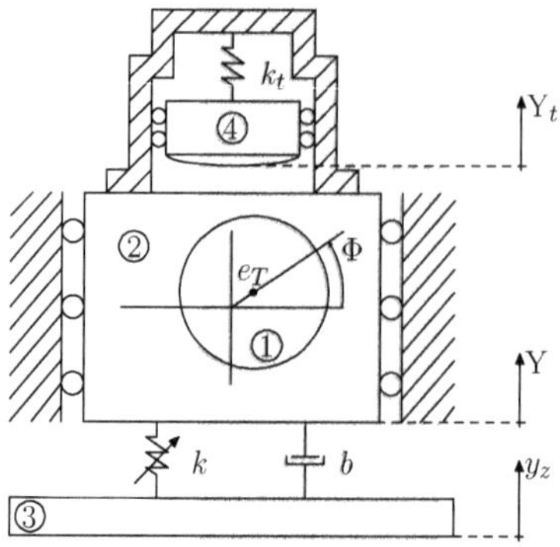


**Fig. 1.** Model of vibrating system

where $\ddot{\mathrm{Y}}$, $\ddot{\mathrm{Y}}_t$ and $\ddot{\Phi}$ denote the second derivative of Y, $\mathrm{Y}_t$ and $\Phi$ respectively, and $\dot{\mathrm{Y}}$ the first derivative of Y with respect to time.

It can be assumed, without loss of generality, that in the beginning, the system is at rest and takes the equilibrium position with no contacts between the impact body and the rotor frame. Then the initial conditions are given as follows

$$\begin{aligned}\dot{\mathrm{Y}}(0) = 0,\ \dot{\mathrm{Y}}_t(0) = 0,\ \dot{\Phi}(0) = 0,\ \Phi(0) = 3/2\,\pi,\\ k_3(\mathrm{Y}(0) - y_z(0))^3 + k_1(\mathrm{Y}(0) - y_z(0)) +\\ +k_t(\mathrm{Y}(0) - \mathrm{Y}_t(0)) + (m + m_R)g = 0,\\ k_t(\mathrm{Y}_t(0) - \mathrm{Y}(0)) + m_t g = 0.\end{aligned} \tag{3}$$

Let us simulate the vibration of the baseplate with the map

$$y_z(t) = A\,(1 - \mathrm{e}^{-\alpha t})\sin(\omega t) \tag{4}$$

where $A$ is the amplitude, $\alpha$ is the constant determining how fast the vibration of the baseplate becomes a steady state and $\omega$ stands for the excitation frequency, so $y_z(0) = 0$ and $\dot{y}_z(0) = 0$.

The last two conditions of (3) could be simplified and the following lemma, proved by [4], could be used for their solution.

**Lemma 1.** *Let $k_1$, $k_3$, $k_t$, $m$, $m_t$, $m_R$ and $g$ be real positive constants. Then there is only one real solution of the system*

$$k_3\mathrm{Y}^3(0) + k_1\mathrm{Y}(0) + k_t(\mathrm{Y}(0) - \mathrm{Y}_t(0)) + (m + m_R)g = 0, \tag{5}$$

$$k_t(\mathrm{Y}_t(0) - \mathrm{Y}(0)) + m_t g = 0. \tag{6}$$

*Moreover, the real solution equals*

$$\begin{aligned}\mathrm{Y}(0) &= \sqrt[3]{-q + \sqrt{q^2 + p^3}} + \sqrt[3]{-q - \sqrt{q^2 + p^3}},\\ \mathrm{Y}_t(0) &= \sqrt[3]{-q + \sqrt{q^2 + p^3}} + \sqrt[3]{-q - \sqrt{q^2 + p^3}} - \frac{m_t g}{k_t}\end{aligned} \tag{7}$$

*where $p = k_1/3k_3$ and $q = ((m + m_R + m_t)g)/2k_3$.*

# 3 Main Results

It was shown by [4] that there is a resonance peak for the baseplate excitation frequency of $\omega = 102$ rad s$^{-1}$ for the system parameters summarized in Table 1. Next, parameters $c_1$ and $c_2$ are assumed to be equal for simplicity and together with the mass of the impact element $m_t$ were taken as variables. In the following, the situations are discussed in detail, dependent on clearances between the rotor casing and the impact element, the mass of the impact element and the baseplate excitation frequency.

**Table 1.** Parameters of the system (2)

| value | quantity | format | description |
|---|---|---|---|
| $m$ | 100 | kg | mass of the damping body |
| $m_R$ | 40 | kg | mass of the rotor |
| $m_t$ | 25 | kg | mass of the impact element |
| $k_1$ | $1.5 \times 10^5$ | N m$^{-1}$ | linear stiffness coefficient |
| $k_3$ | $6 \times 10^{10}$ | N m$^{-3}$ | cubic stiffness coefficient |
| $J_{RT}$ | 5 | kg m$^2$ | moment of inertia of the rotor |
| $b$ | $1.5 \times 10^3$ | N s m$^{-1}$ | damping coefficient of the suspension |
| $k$ | | N m$^{-1}$ | stiffness of the suspension spring |
| $k_t$ | $8 \times 10^4$ | N m$^{-1}$ | coupling stiffness of the impact element |
| $e_T$ | 2 | mm | eccentricity of the rotor center of gravity |
| $\Phi$ | | rad | rotation angle of the rotor |
| $M_Z$ | 100 | N m | starting moment |
| $k_M$ | 8 | N m s rad$^{-1}$ | negative of the motor characteristic slope |
| $\alpha$ | 1 | s$^{-1}$ | parameter of the baseplate excitation |
| $\omega$ | 102 | rad s$^{-1}$ | baseplate excitation frequency |
| $A$ | 1 | mm | amplitude of $y_z$ |
| $k_c$ | $4 \times 10^7$ | N m$^{-1}$ | contact stiffness |
| $b_c$ | $3 \times 10^3$ | N s m$^{-1}$ | coefficient of contact damping |

In the linear system it would always be possible to find the mass of the impact element that would work then as an inertia damper and considerably attenuate the vibration of the rotor frame without any impacts occurring. The amplitude of vibrations of the rotor frame would be considerably attenuated and no impacts would take place. As the investigated system is nonlinear this can be reached, but with some limitations. In both cases, linear and nonlinear, such a damper can be effective only in a small range of excitation frequencies.

To extend the interval, the mass of the damping element would have to be changed and such manipulation is not easy to accomplish from the technological point of view.

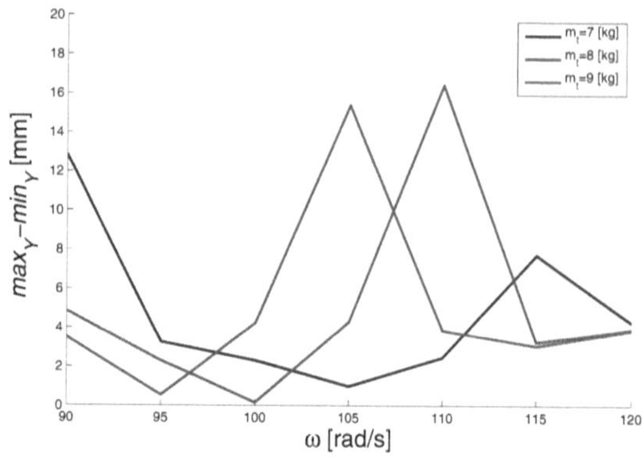


**Fig. 2.** Peak-to-peak vibrations amplitudes of the rotor frame dependent on the excitation frequency for the mass of the impact element $m_t$ from 7 kg to 9 kg. There are no impacts in all these cases.

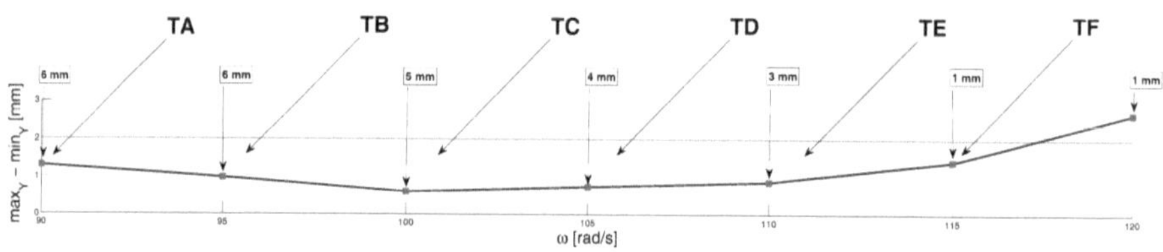


**Fig. 3.** Optimal choice of the clearance $c_1 = c_2$ for the maximal attenuation of vibrations of the rotor body. The attenuation is significant for the range of the excitation frequency $\omega$ from 85 rad s$^{-1}$ to 117 rad s$^{-1}$ and the mass of the impact element $m_t = 25$ kg.

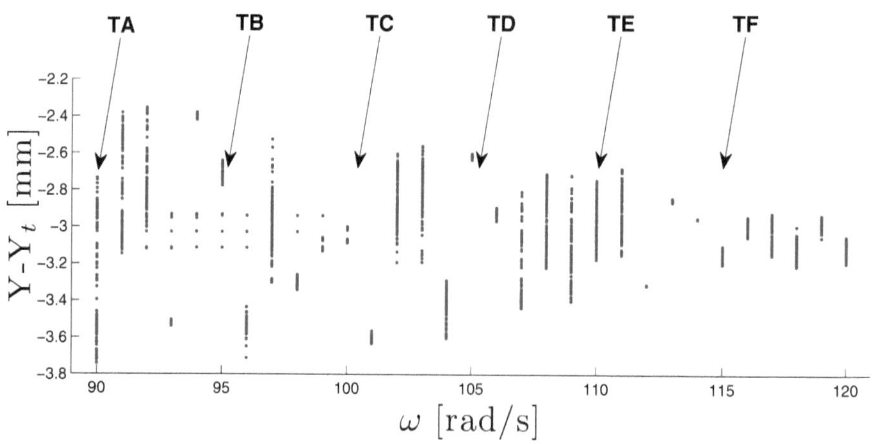


**Fig. 4.** Bifurcation diagram with respect to the angular frequency $\omega$

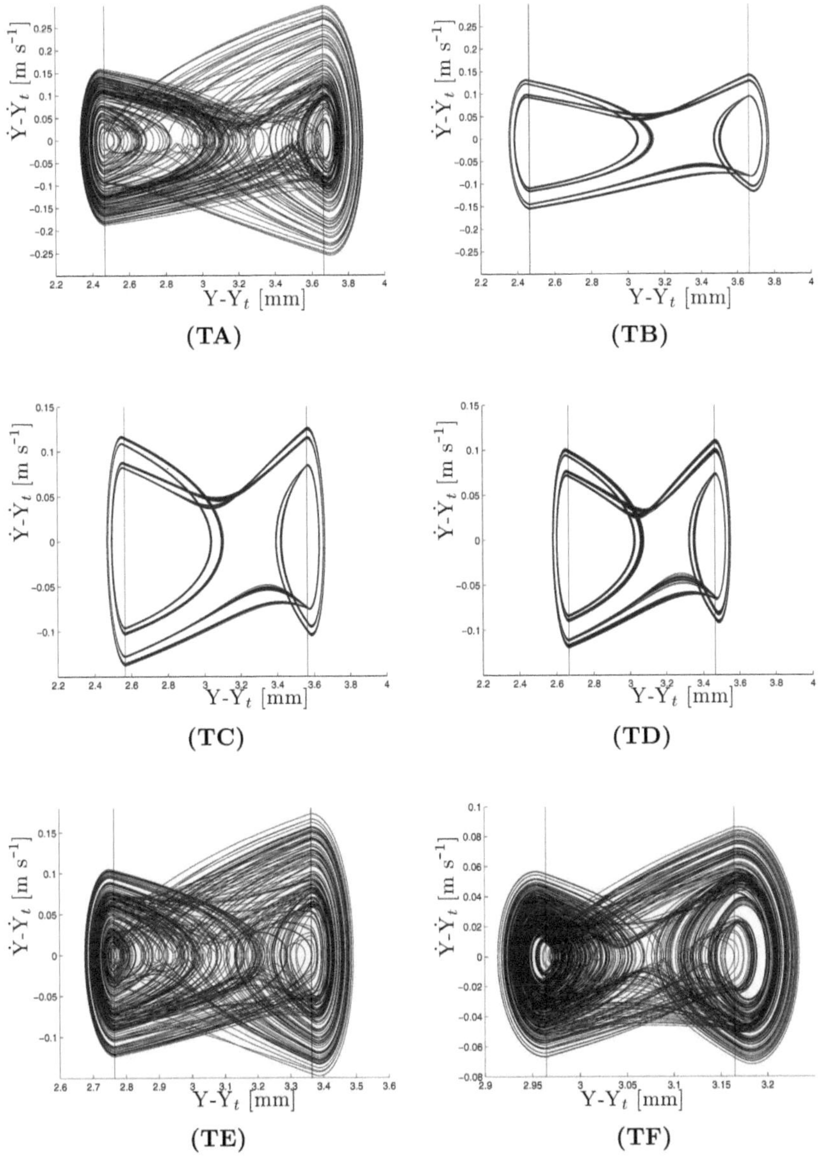


**Fig. 5.** Phase portraits, $\mathrm{Y} - \mathrm{Y}_t$ versus $\dot{\mathrm{Y}} - \dot{\mathrm{Y}}_t$ with respect to the change of the upper and lower clearances between the rotor casing and the impact body $c_1 = c_2$ and the angular frequency:
(TA) $\omega = 90$ rad s$^{-1}$ and $c_1 = c_2 = 6$ mm, (TB) $\omega = 95$ rad s$^{-1}$ and $c_1 = c_2 = 6$ mm, (TC) $\omega = 100$ rad s$^{-1}$ and $c_1 = c_2 = 5$ mm, (TD) $\omega = 105$ rad s$^{-1}$ and $c_1 = c_2 = 4$ mm,
(TE) $\omega = 110$ rad s$^{-1}$ and $c_1 = c_2 = 3$ mm, (TF) $\omega = 115$ rad s$^{-1}$ and $c_1 = c_2 = 1$ mm.

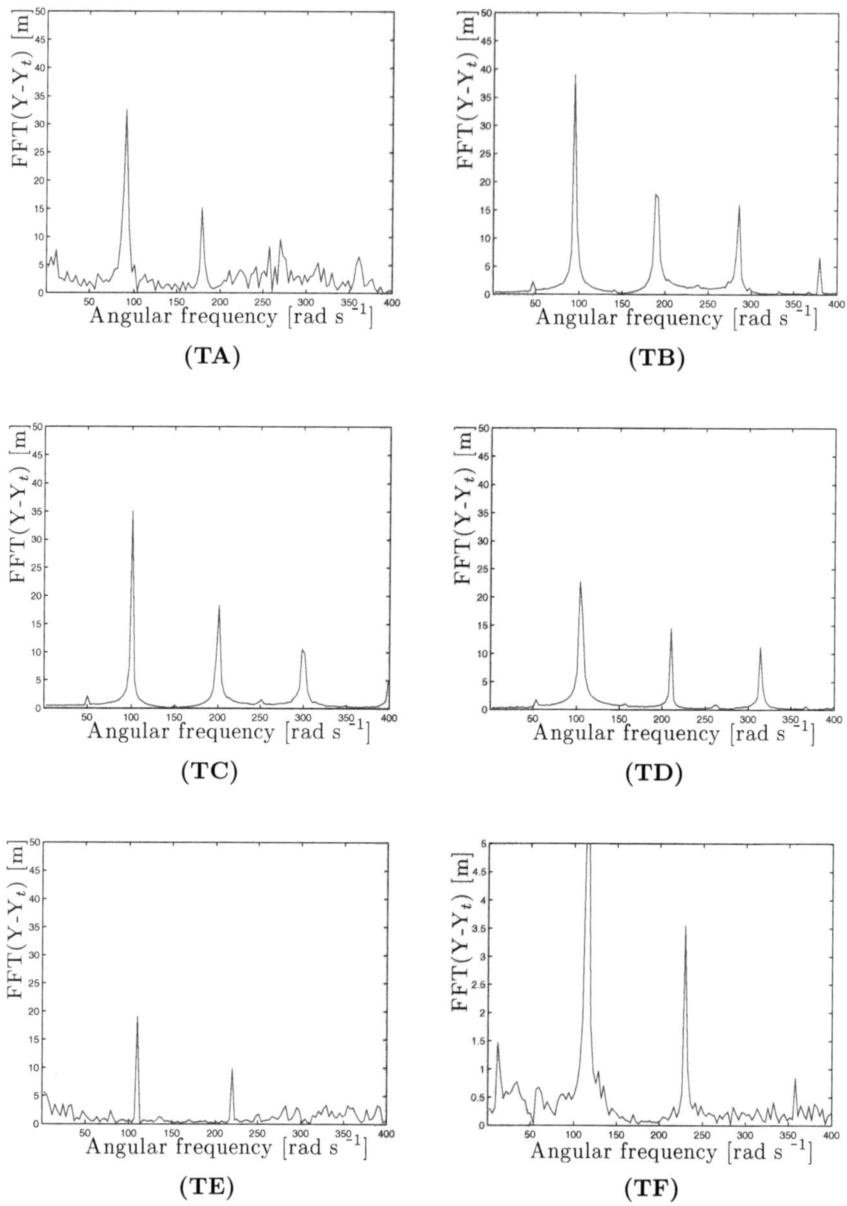


**Fig. 6.** Fourier spectra with respect to the change of the upper and lower clearances between the rotor casing and the impact body $c_1 = c_2$ and the angular frequency: (TA) $\omega = 90$ rad s$^{-1}$ and $c_1 = c_2 = 6$ mm, (TB) $\omega = 95$ rad s$^{-1}$ and $c_1 = c_2 = 6$ mm, (TC) $\omega = 100$ rad s$^{-1}$ and $c_1 = c_2 = 5$ mm, (TD) $\omega = 105$ rad s$^{-1}$ and $c_1 = c_2 = 4$ mm, (TE) $\omega = 110$ rad s$^{-1}$ and $c_1 = c_2 = 3$ mm, (TF) $\omega = 115$ rad s$^{-1}$ and $c_1 = c_2 = 1$ mm.

On the contrary it is easy to find a technological solution of the damping device that would make it possible to actively change the clearance width dependent on the excitation frequency to get maximum attenuation of the rotor vibrations. This is suitable especially in the cases when the excitation frequency is not constant but when it can slightly vary in a wider frequency interval.

Figure 3 shows that it is possible to change the clearance width dependent on the excitation frequency intentionally to get maximal attenuation. More precisely, the simulations show (see Figure 2) that for the mass of the impact element $m_t = 8$ kg (corresponding to the resonance peak) the range of the excitation frequency is only (96.5, 102, 5) where the attenuation is remarkable (i.e. $\max_Y - \min_Y < 2$ mm.) In this situation the inertia damper works in the range of 6 rad $s^{-1}$. On the other hand, in Figure 3 it is confirmed that for the mass of the impact element $m_t = 25$ kg and the controlled clearance width the range of excitation frequency overlaps 30 rad $s^{-1}$. That is if the damping element is in active control, the clearance is changed dependent on the excitation frequency, attenuation is meaningful for the excitation frequency in the interval (87, 117).

The Fourier spectra (Figure 6), bifurcation diagram (Figure 4) and phase trajectories (Figure 5) show that the movement is formed by a number of harmonic components having the basic, super-harmonic, sub-harmonic and combination frequencies on which there are superposed further motions with frequencies forming the sided bands of the dominant frequencies. Their mutual ratio indicates the irregularity of the vibration. For tests (marked in Figure 3) (TB), (TC) and (TD) the vibration is periodic or almost periodic, see Figure 5. For simulations (TA), (TE) and (TF) the regularity of the motion sharply increases. The Fourier spectra and the phase trajectories show that the oscillation is chaotic (Figures 5 and 6) which is also confirmed by the Bifurcation diagram in Figure 4.

## 4 Conclusions

In this paper, it was developed and analyzed a new electromechanical system damped by impact element with soft stops dependent on parameters, namely the weight of the impact element, the clearance between the impact body and rotor casing and finally the excitation frequency. This model was inspired by real frequently occurring technological problems when electromechanical rotating machines are excited by a combined loading produced by the rotor unbalance and ground vibrations. The equations of motions were solved numerically by the explicit Runge-Kutta method. The computational simulations showed that the vibration of the baseplate played a key role here and proved that application of the impact body arrived at a significant decrease of vibration amplitude of the rotor frame.

It was observed that for given parameters of the model, the damping element can work as an inertia (there are no impacts) or as an impact damper (impacts occur). Both situations were investigated and commented on. The ranges of parameters for which the attenuation is meaningful were detected. In the case of the inertia damper, the range of the excitation frequency is very narrow. On the other hand, it is quite wider in the case of the impact damping device. Finally, if the damping element is in active control, the clearance is changed dependent on the excitation frequency, attenuation is meaningful in the interval which is five times wider then those in the case of the inertia damper. It was also proved by simulations that movements corresponding to the inertia damper is showing periodic or chaotic pattern.

**Acknowledgments.** This work was supported by the European Regional Development Fund in the IT4Innovations Centre of Excellence Project (CZ.1.05/1.1.00/02.0070). The work was also supported by the Grant Agency of the Czech Republic, Grant No. P201/10/0887.

## References

1. Chaterjee, A.K., Mallik, A., Ghosh, A.: On impact dampers for non-linear vibration systems. Journal of Sound and Vibration 187, 403–420 (1995)
2. Warminski, J., Balthazar, J.M., Brasil, R.M.L.R.F.: Vibrations of a non-ideal parametrically and self-excited model. Journal of Sound and Vibration 245, 363–374 (2001)
3. de Souza, S.L.T., Caldas, I.L., Balthazar, J.M., Brasil, R.M.L.R.F.: Analysis of regular and irregular dynamics of a non ideal gear rattling problem. Journal of Brayilian Society of Mechanical Sciencies 24, 111–114 (2002)
4. Lampart, M., Zapomĕl, J.: Dynamics of the electromechanical system with impact element. Journal of Sound and Vibration 332, 701–713 (2013)
5. Lampart, M., Zapomĕl, J.: Dynamical properties of the electromechanical system damped by impact element with soft stops. International Journal of Applied Mechanics 6, 1450016 (2014)
6. Luo, A.C.J., O'Connor, D.: Periodic motions with impacting chatter and stick in a gear transmission system. ASME Journal of Vibration and Acoustics 131 (2009)
7. de Souza, S.L.T., Caldas, I.L., Viana, R.L., Balthazar, J.M., Brasil, R.M.L.R.F.: Impact dampers for controlling chaos in systems with limited power supply. Journal of Sound and Vibration 279, 955–967 (2005)
8. Zukovic, M., Cveticanim, L.: Chaos in non-ideal mechanical system with clearance. Journal of Vibration and Control 8, 1229–1246 (2009)
9. Půst, L.: Electro-mechanical impact system excited by a source of limited power. Engineering Mechanics 6, 391–400 (2008)
10. Zapomĕl, J., Fox, C.H.J., Malenovský, E.: Numerical investigation of a rotor system with disc-housing impact. Journal of Sound and Vibration 243, 215–240 (2001)

# Chaos in Nitrogen Dioxide Concentration Time Series and Its Prediction

Radko Kříž

University of Pardubice, Pardubice, Czech Republic
Radko.Kriz@upce.cz

**Abstract.** This paper is aimed at analysis of Nitrogen Dioxide ($NO_2$) concentration time series. At first we will estimate the time delay and the embedding dimension, which is needed for the Lyapunov exponent estimation and for the phase space reconstruction. Subsequently we will compute the largest Lyapunov exponent, which is one of the important indicators of chaos. Then we will calculate the 0-1 test for chaos. Finally we will compute predictions using a radial basis function to fit global nonlinear functions to the data. The results indicated that chaotic behaviors obviously exist in $NO_2$ concentration time series.

**Keywords:** Chaos theory, Nitrogen dioxide, Time series analysis, Phase space reconstruction, Prediction, Gaussian radial basis function.

## 1 Introduction

The term *air quality* describes the level of air pollution, which may affect human health, vegetation, entire ecosystems and materials. Air pollution is influenced by the emission of pollutants from various sources as a consequence of human activity (e.g. transportation, combustion). Pollutants emitted from a source are transported in the atmosphere and can thus affect the air quality in both the immediate vicinity of the pollution source and the broader territory.

$NO_2$ is one of the major air pollutants. There is some evidence that long-term exposure to $NO_2$ at concentrations above 40–100 $\mu g.m^{-3}$ may decrease lung function and increase the risk of respiratory symptoms [1]. The most prominent sources of $NO_2$ are internal combustion engines [2], thermal power stations and, to a lesser extent, pulp mills. Butane gas heaters and stoves are also sources. The excess air required for complete combustion of fuels in these processes introduces nitrogen into the combustion reactions at high temperatures and produces nitrogen oxides ($NO_x$). Limiting $NO_x$ production demands the precise control of the amount of air used in combustion. The burning of biomass creates additional pollutants including $NO_x$ [3]. $NO_2$ is a large scale pollutant, with rural background ground level concentrations around 30 $\mu gm^{-3}$ in some areas, not far below unhealthy levels. $NO_2$ plays a role in atmospheric chemistry, including the formation of tropospheric ozone. Environmental legislation creates continual pressure on finding new solutions that are both economically advantageous and environmentally friendly [4].

I. Zelinka et al. (eds.), *Nostradamus 2014: Prediction, Modeling and Analysis of Complex Systems*, Advances in Intelligent Systems and Computing 289,
DOI: 10.1007/978-3-319-07401-6_36, © Springer International Publishing Switzerland 2014

Recent advances in the modeling of air pollutants have led to the application of concepts including nonlinear modeling techniques. Numerous new methods of time series analysis have been developed for dealing with nonlinear data e.g. Abarnabel [5], Kantz et Schreiber [6], Mikšovský et Raidl [7]. The concepts from chaos theory reveals that, if the knowledge of a system in terms of its present and past states is available, then it is possible to make predictions of system's future behavior. The approaches based on chaos theory are widely acceptable due to the assumption that, it is possible to predict the future state of the system based on the single scalar time series assuming that, all the information regarding the external forcing factors is contained in that single time series [8]. Chaos theory allows for the reconstruction of phase space from time series, which can be used for specifying the system states. Thus, the dynamics of the system can be investigated by studying the dynamics of the movement of the phase space points [9].

The modeling of $NO_2$ concentration has been studied in several works based on different unconventional approaches such as neural networks [8,10,11], Petri nets [12], hierarchical fuzzy inference systems [13], ship plume parameterization [14].

## 2 Methodology

In short, we will describe the basic definitions and the basic methods for examining the input data.

### 2.1 Phase Space Reconstruction

The main goal in nonlinear time series analysis is to determine whether or not a given time series is of a deterministic nature. If it is, then further questions of interest are: What is the dimension of the phase space supporting the data set? Is the data set chaotic? [15]

The key to answering these questions is embodied in the method of phase space reconstruction, which has been rigorously proven by the embedding theorems of Takens [16]. Takens' theorem transforms the prediction problem from time extrapolation to phase space interpolation [17].

Let there be given a time series $x_1, x_2, \ldots, x_N$ which is embedded into the $m$-dimensional phase space by the time delay vectors. A point in the phase space is given as:

$$Y_n = x_n, x_{n-\tau}, \ldots, x_{n-(m-1)\tau} \quad n = 1, 2, \ldots, N-(m-1)\tau, \tag{1}$$

where $\tau$ is the time delay and $m$ is the embedding dimension. Different choices of $\tau$ and $m$ yield different reconstructed trajectories. Kodba et al. [18] discuss How we can determine optimal $\tau$ and $m$. Fraser and Swinney [19] introduced the mutual information between $x_n$ and $x_{n+\tau}$ as a suitable quantity for determining $\tau$. The mutual information between $x_n$ and $x_{n+\tau}$ quantifies the amount of information we have about the state $x_{n+\tau}$ presuming we know the state $x_n$. Now we can define mutual information function:

$$I(\tau) = -\sum_{h=1}^{j}\sum_{k=1}^{j} P_{h,k}(\tau)\ln\frac{P_{h,k}(\tau)}{P_h P_k}, \tag{2}$$

where $P_h$ and $P_k$ denote the probabilities that the variable assumes a value inside the $h^{th}$ and $k^{th}$ bins, respectively, and $P_{h,k}(\tau)$ is the joint probability that $x_n$ is in bin $h$ and $x_{n+\tau}$ is in bin $k$. The first minimum of $I(\tau)$ then marks the optimal choice for the time delay.

The embedding dimension $m$ can be chosen using the "false nearest neighbors'" method. This method measures the percentage of close neighboring points in a given dimension that remain so in the next highest dimension. The minimum embedding dimension capable of containing the reconstructed attractor is that for which the percentage of false nearest neighbors drops to zero for a given tolerance level $\mu$.

In order to calculate the fraction of false nearest neighbors the following algorithm is used according to Kennel et al. [20]. Given a point $p(i)$ in the $m$-dimensional embedding space, one first has to find a neighbor $p(j)$, so that

$$\|p(i) - p(j)\| \leq \mu . \tag{3}$$

We then calculate the normalized distance $R_i$ between the $(m + 1)th$ embedding coordinate of points $p(i)$ and $p(j)$ according to the equation:

$$R_i = \frac{|x_{i+m\tau} - x_{j+m\tau}|}{\|p(i) - p(j)\|}. \tag{4}$$

If $R_i$ is larger than a given threshold $R_{tr}$ , then $p(i)$ is marked as having a false nearest neighbor. Equation (4) has to be applied for the whole time series and for various $m = 1, 2, \ldots$ until the fraction of points for which $R_i > R_{tr}$ is negligible [18].

### 2.2 The Largest Lyapunov Exponent

Lyapunov exponent $\lambda$ of a dynamical system is a quantity that characterizes the rate of separation of infinitesimally close trajectories. Quantitatively, two trajectories in phase space with initial separation $\delta Z_0$ diverge.

$$\delta Z(t) \approx e^{\lambda t}|\delta Z_0| \tag{5}$$

The largest Lyapunov exponent can be defined as follows:

$$\lambda = \lim_{\substack{\delta Z_0 \to 0 \\ t\to\infty}} \frac{1}{t}\ln\frac{|\delta Z(t)|}{|\delta Z_0|}. \tag{6}$$

The limit $\delta Z_0 \to 0$ ensures the validity of the linear approximation at any time. Largest Lyapunov exponent determines a notion of predictability for a dynamical system. A positive largest Lyapunov exponent is usually taken as an indication that the system is chaotic [21].

We have used the Rosenstein algorithm [22], which counts the largest Lyapunov exponent as follows:

$$\lambda_1(i) = \frac{1}{i\Delta t} \cdot \frac{1}{(M-i)} \sum_{j=1}^{M-i} \ln \frac{d_j(i)}{d_j(0)}, \tag{7}$$

where $d_j(i)$ is distance from the $j$ point to its nearest neighbor after $i$ time steps and $M$ is the number of reconstructed points. For more information see [22].

## 2.3 Correlation Dimension

Everybody suspects what is dimension that is Euclidean dimension. Briefly, Euclidian dimension is given by the number of phase variables. However, for a deeper understanding of behavior of dynamical systems we must define fractal dimension. There are many specific definitions of fractal dimension. Generally, the fractal dimension $D$ is a statistical quantity that gives an indication of how completely a fractal appears to fill space, as one zooms down to finer and finer scales [23]. Let $S$ be a set of points in a space of Euclidean dimension $d$. We now consider certain hypercubes of side $r$, and calculate the minimum number of such cells, $N(r)$, necessary to "cover" $S$. Let definite capacity dimension (Kolmogorov dimension), which is typical and common, be an example of fractal dimension:

$$D = \lim_{r \to 0} \frac{\ln(N(r))}{\ln r^{-1}}. \tag{8}$$

Notice that fractal dimension is a real number. A non-integer dimension does not imply chaotic dynamic, but all strange attractors must have non-integer fractal dimensions [23].

In practice, capacity dimension cannot be computed easily. A different approach has been designed by Grassberger et Procaccia [24]. The method is based on the concept of correlation dimension $D_C$ suggested by Grassberger et Procaccia [24]. $D_C$ describes the dimensionality of the underlying process in relation to its geometrical reconstruction in phase space. $D_C$ quantifies the "strangeness" of an attractor [25]. $D_C$ is calculated using the fundamental definition. Define the correlation integral $C(\varepsilon)$ for set of $M$ data:

$$C(\varepsilon) = \frac{1}{M(M-1)} \sum_{\substack{i,j=1 \\ i \neq j}}^{M} \Theta(\varepsilon - \|x_i - x_j\|), \tag{9}$$

where $\Theta$ is the Heaviside step function.

$$\Theta(x) = \begin{cases} 0 & x < 0 \\ \frac{1}{2} & x = 0 \\ 1 & x > 0 \end{cases} \tag{10}$$

Euclidean metric is used for all calculations in this paper. $C(\varepsilon)$ is related to the radius $r$ by the power law as

$$\underset{\substack{\varepsilon \to 0 \\ N \to \infty}}{C(r)} \propto \varepsilon^{D_C}. \tag{11}$$

When a lower limit exists, the correlation dimension is then defined as

$$D_C = \lim_{\substack{\varepsilon \to 0 \\ M \to \infty}} \frac{\partial \ln(C(\varepsilon))}{\partial \ln(\varepsilon)}. \tag{12}$$

In real cases, the limitation imposed in the above equation is not possible due to the limited amount of available data. Hence, $D_C$ can be obtained by plotting $lnC(\varepsilon)$ against $ln(\varepsilon)$. The slopes of the curves for different embedding dimensions $m$ give the values of $D_C$. An important inference that can be made from $D_C$ is that the process is stochastic, if it does not saturate for increasing $m$ [24]. The saturation of $D_C$ at a certain values of $m$ indicates that the process generating time series is not random but rather deterministic [26].

The $C(\varepsilon)$ can be regarded as an average density of points where the local density is obtained by a kernel estimator with a step kernel

$$\Theta(\varepsilon - r). \tag{13}$$

A natural modification for small point sets is to replace the sharp step kernel by a smooth kernel function of bandwidth ε. Very practical is the replacement of the step kernel by the Gaussian kernel

$$e^{-\frac{r^2}{4\varepsilon^2}}. \tag{14}$$

The resulting Gaussian kernel correlation integral $C_G(\varepsilon)$ has the same scaling properties as the usual $C(\varepsilon)$ [26]. Gaussian kernel correlation integral $C_G(\varepsilon)$ can be obtained from usual $C(\varepsilon)$ via

$$C_G(\varepsilon) = \frac{1}{2\varepsilon^2} \int_0^\infty d\tilde{\varepsilon} e^{-\frac{\tilde{\varepsilon}^2}{4\varepsilon^2}} \tilde{\varepsilon} C(\tilde{\varepsilon}). \tag{15}$$

If $C(\varepsilon)$ is given at discrete values of ε, the Gaussian kernel correlation integral $C_G(\varepsilon)$ can be carried out numerically by interpolating $C(\varepsilon)$ with pure power laws [25].

## 2.4 Entropies

Entropies are an information theoretical concept to characterize the amount of information needed to predict the next measurement with a certain precision. The most popular one is the Kolmogorov-Sinai entropy. When analyzing time series we are usually dealing with distributions of delay vectors with delay $\tau$ in an $m$-dimensional reconstructed phase space. The $m$ dependence of correlation integral $C_q$ in the limit of large m can then be expressed as

$$\underset{\substack{\varepsilon \to 0 \\ m \to \infty}}{C_q(m,\varepsilon)} = \alpha m e^{-(q-1)h_q \tau m} \varepsilon^{(q-1)D_q}, \tag{16}$$

which defines the order $q$ entropy $h_q$. [25] Second order entropy is called Kolmogorov entropy. An algorithm for the determination of the Kolmogorov entropy is given in Cohen et Procaccia [27].

## 2.5 The 0-1 Test for Chaos

A new test for the presence of deterministic chaos was developed by Gottwald et Melbourne [28]. Their '0 - 1' test for chaos takes as input a time series of measurements, and returns a single scalar value usually in the range 0 - 1. The '0 - ' test does not depend on phase space reconstruction but rather works directly with the time series given. The input is the time-series data and the output is 0 or 1, depending on whether the dynamics is non-chaotic or chaotic.

Briefly, the 0-1 test takes as input a scalar time series of observations $\varphi_1, \dots, \varphi_N$. According Dawes et Freeland [29], first we must fix a real parameter $c$ and construct the Fourier transformed series:

$$z_n = \sum_{j=1}^{n} \phi_j e^{ijc}, \quad n = 1,\dots,N . \tag{17}$$

Then we compute the smoothed mean square displacement:

$$M_c(n) = \frac{1}{N-p} \sum_{j=1}^{N-p} \left| z_{j+n} - z_j \right|^2 - \left( \sum_{k=1}^{N} \frac{\phi_k}{N} \right)^2 \frac{1-\cos nc}{1-\cos c} . \tag{18}$$

Finally we estimate correlation coefficient to evaluate the strength of the linear growth

$$r_c = \frac{\operatorname{cov}(n, M_c(n))}{\sqrt{\operatorname{cov}(n,n)\operatorname{cov}(M_c(n), M_c(n))}} . \tag{19}$$

## 2.6 Prediction

Most properties of chaotic systems are much more easily determined from the governing equations than from a time series. Unfortunately, the governing equations are usually not known, except for well controlled laboratory experiments. Analyzing an empirical model, and maybe synthetic time series data generated from it, can provide a valuable consistency test for the results of time series analysis. The best we can hope for when fitting a model to data is that the result comes close to the real underlying dynamics. Nevertheless, chaotic dynamical systems generically show the phenomenon of structural instability. This means that models with very similar characteristics may exhibit qualitatively different global dynamics, such as near to an attractor crisis. [6]

According [8], using the reconstructed phase space for m and τ, a functional relationship $f$ between the current state $X(t)$ and future state $X(t + T)$ can be given as

$$X(t+T) = f(X(t)), \tag{20}$$

where $T$ represents the number of time steps ahead that one wishes to perform the prediction. Function $f$ represents the approximation to unknown dynamical system. It is shown that for sufficiently large values of the embedding dimension and if some additional conditions are satisfied, the reconstructed trajectory has the same topological and geometrical properties as the system's phase space trajectory [16]. This means that, if the conditions of Takens embedding theorem are met, this mapping captures some of the properties of the unknown dynamical system. The predictive mapping can be expressed as

$$X(t+T) = f_p(X(t)) . \tag{21}$$

The aim is to find the predictor $f_p$, so that $x(t + T)$ can be predicted based on the reconstructed time series. If the time series is chaotic, then $f_p$ is necessarily nonlinear. Several local and global approaches are available in the literature to find the function $f_p$ [30].

The local linear fits are very flexible, but can go wrong on parts of the phase space where the points do not span the available space dimensions and where the inverse of the matrix involved in the solution of the minimization does not exist. Moreover, very often a large set of different linear maps is unsatisfying. Therefore many authors suggested fitting global nonlinear functions to the data, i.e. to solve

$$\sigma^2 = \sum_n (s_{n+1} - f_p(s_n))^2 , \tag{22}$$

where $f_p$ is now a nonlinear function in closed form with parameters $p$, with respect to which the minimization is done. The results depend on how far the chosen ansatz $f_p$ is suited to model the unknown nonlinear function, and on how well the data are deterministic at all. [26] A radial basis function (RBF) is a real-valued function whose value depends only on the distance from the origin, or alternatively on the distance from some other point $x_i$, called a center, so that

$$\Phi(x, x_i) = \Phi(\|x - x_i\|) . \tag{23}$$

Gaussian kernel is used in this analysis.

$$\Phi(x, x_i) = e^{-c^2 \|x - x_i\|^2} . \tag{24}$$

Hence the prediction made is:

$$x_{t+1} = a_0 + \sum_{i=1}^{n} a_i e^{-(c\|x - x_i\|)^2} . \tag{25}$$

## 3 Analysis of Nitrogen Dioxide Concentration Time Series

### 3.1 Input Data

This study was based on daily averaged nitrogen dioxide concentration (NO2) data from Pardubice-Rosice from 1.1.2005 to 31.12.2011. We used the mean of this data set to fill in several missing values. The data were provided by the Czech Hydrometeorological Institute for a related diploma thesis that is aimed at nonlinear and chaotic behavior of air pollutants time series. Pardubice is the capital of the Pardubice Region and lies on the river Elbe, 100 km east of Prague (Fig. 1).

**Fig. 1.** Map of Czech Republic

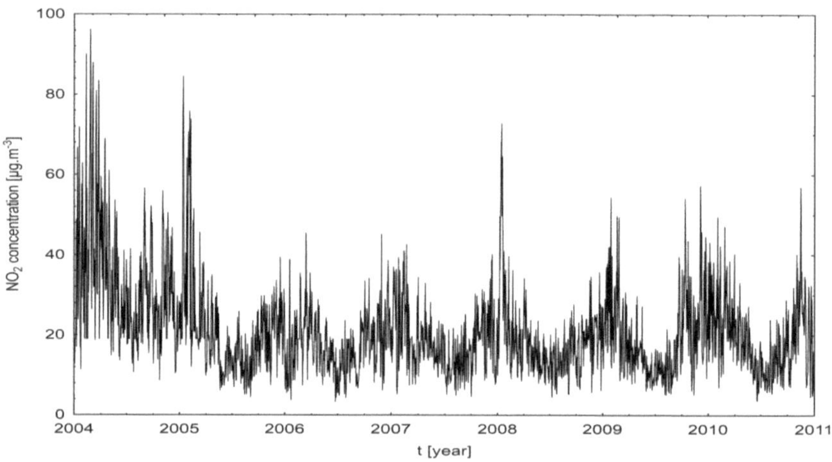


**Fig. 2.** Daily averaged nitrogen dioxide concentration ($NO_2$) data from Pardubice

## 3.2 Calculation of the Time Delay and the Embedding Dimension

In this chapter we will use the mutual information approach to determine the time delay $\tau$ and the false nearest neighbor method to determine the minimal sufficient embedding dimension $m$. This approach is described in chapter 2.1. $\tau$ is estimated from the graph in Fig. 3a. The first minimum of the mutual information function $I(\tau)$ (2) marks the optimal choice for the time delay. Thus, the time delay $\tau$ is 5. The embedding dimension $m$ is chosen using the "false nearest neighbors'" method, estimated from the graph in Fig. 3b. The minimum embedding dimension capable of containing the reconstructed attractor is that for which the percentage of false nearest neighbors drops to zero for a given tolerance level $\mu$. Thus, the embedding dimension $m$ is 7, but the value 6 can be sufficient.

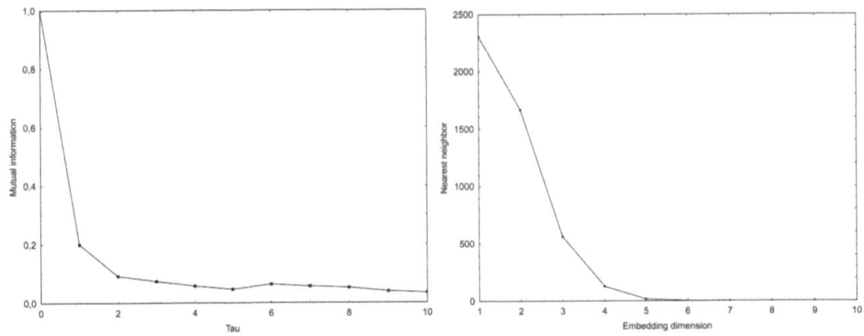


**Fig. 3.** a) Mutual average information for the $NO_2$ series b) Fraction of false nearest neighbors for the $NO_2$ series

### 3.3 Calculation of the Largest Lyapunov Exponents

In this chapter we calculate the largest Lyapunov exponent as was shown above. We used the Rosenstein algorithm. The calculation of the largest Lyapunov exponent depends on the estimation of the embedding dimension. The value of the largest Lyapunov exponent was estimated at 0,0014 for embedding dimension 6 and 0,0013 for embedding dimension 7. A positive largest Lyapunov exponent is one of the necessary conditions for chaotic behavior. Notice that the largest Lyapunov exponent is relatively small. Consequently, the rate of $NO_2$ concentration evolution is rather slow, showing that it is possible to accurately make a short-term forecast.

### 3.4 Estimating the Correlation Dimension and the Kolmogorov Entropy

In this chapter we estimate the correlation dimension. Relationship between the Gaussian kernel correlation integral $C_G(\varepsilon)$ and radius $\varepsilon$ on ln-ln scale with embedding dimensions $m$ from 1 to 8 is shown in Fig 4. The saturation value of the correlation dimension for $NO_2$ concentration time series is 4,77. The value of the Kolmogorov entropy nitrogen dioxide concentration time series was calculated at 7,26 for embedding dimension 7.

### 3.5 Results of the 0-1 Test for Chaos

In this chapter we calculate the correlation coefficient as was shown in chapter 2.4. The value of the correlation coefficient was computed at 0,98. The correlation coefficient is near to 0 for non-chaotic data and near 1 for chaotic data. The value of correlation coefficient 0,98 is closer to 1. Hence we can assume to chaotic behavior in the $NO_2$ concentration time series.

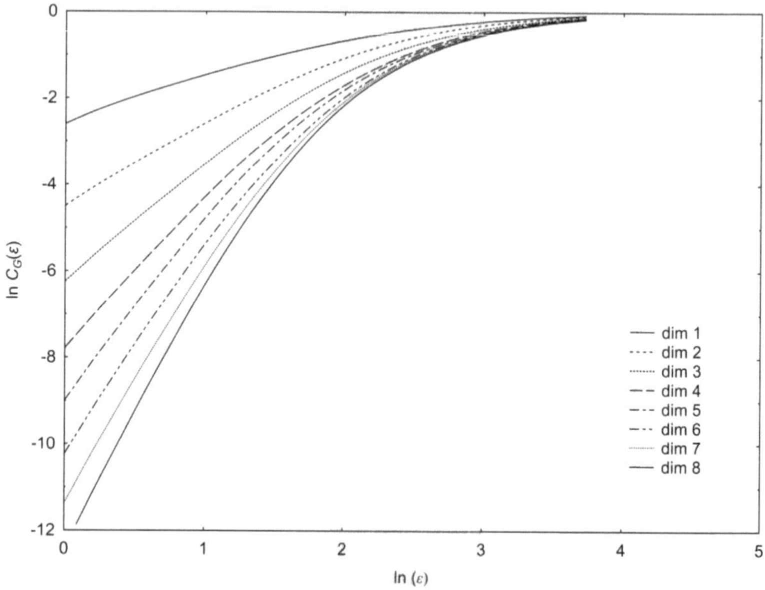


**Fig. 4.** Ln $C_G(\varepsilon)$ versus ln($\varepsilon$) plots for nitrogen dioxide concentration time series

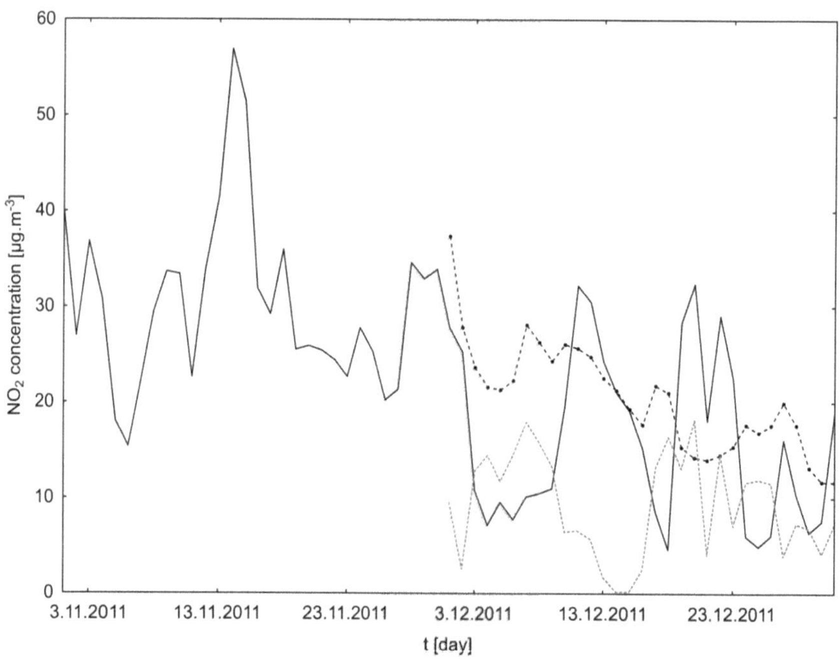


**Fig. 5.** Prediction (dash line) of nitrogen dioxide concentration time series using RBF and RMSE (dot line)

### 3.6 Prediction

Finally we will compute predictions using a Gaussian radial basis function to fit global nonlinear functions to the data. Prediction is displayed in Fig. 5. with root mean square error (RMSE). Prediction is not ideal, but RMSE is always less than mean of all data set.

## 4 Conclusions

We have shown in this paper that the nitrogen dioxide concentration time series is chaotic and contains long memory. First, we computed the values of the time delay $\tau = 5$ and the embedding dimension $m = 7$. The estimated largest Lyapunov exponent is 0,0013. If the correlation dimension is low, the largest Lyapunov exponent is positive and the Kolmogorov entropy has a finite positive value, chaos is probably present. The application of 0-1 test suggests the presence of chaos as well. From these estimations it can be concluded that nitrogen dioxide concentration time series is chaotic. Finally we have computed predictions using a Gaussian radial basis function to fit global nonlinear functions to the data. In the future we would like to focus on the proper statistical significance for nonlinearity and on forecasting using neural networks.

## References

[1] World Health Organization. Health Aspects of Air Pollution with Particulate Matter, Ozone and Nitrogen Dioxide, p. 48 (January 13-15, 2003)

[2] Son, B., Yang, W., Breysse, P., Chung, T., Lee, Y.: Estimation of occupational and non-occupational nitrogen dioxide exposure for Korean taxi drivers using a micro-environmental model. Environmental Research 94, 291–296 (2004)

[3] Baťa, R., Půlkrábková, P.: The importance of Modelling the Environmental Impacts of a Biomass Based Electric Power Generation for public safety. WSEAS Transactions on Environment & Development 9(4) (2013)

[4] Baťa, R., Kadlecová, P.: Modeling of economic and environmental impacts of energy generation using of waste paper. E+ M Ekonomie a Management 2011(2) (2011)

[5] Abarbanel, H.D.I.: Analysis of Observed Chaotic Data, 272 p. Springer, New York (1996)

[6] Kantz, H., Schreiber, T.: Nonlinear time series analysis, 304 p. Cambridge University Press, Cambridge (1997)

[7] Miksovsky, J., Raidl, A.: Testing the performance of three nonlinear methods of time series analysis for prediction and downscaling of European daily temperatures. Nonlinear Processes in Geophysics 12(6), 979–991 (2005)

[8] Chelani, A.B., Singh, R.N., Devotta, S.: Nonlinear dynamical characterization and prediction of ambient nitrogen dioxide concentration. Water, Air, and Soil Pollution 166(1-4), 121–138 (2005)

[9] Abarbanel, H.D., Brown, R., Sidorowich, J.J., Tsimring, L.S.: The analysis of observed chaotic data in physical systems. Reviews of Modern Physics 65(4), 1331 (1993)

[10] Hájek, P., Olej, V.: Air quality modelling by Kohonen's self-organizing feature maps and LVQ neural networks. WSEAS Transactions on Environment and Development 4(1), 45–55 (2008)
[11] Kolehmainen, M., Martikainen, H., Ruuskanen, J.: Neural networks and periodic components used in air quality forecasting. Atmospheric Environment 35(5), 815–825 (2001)
[12] Baťa, R.: Modeling of LCA-chain segment for Biofuels As an instrument for the protection of the population. WSEAS Transactions on Environment & Development 9(2) (2013)
[13] Hájek, P., Olej, V.: Air quality indices and their modelling by hierarchical fuzzy inference systems. WSEAS Transactions on Environment and Development 10, 661–672 (2009)
[14] Huszar, P., et al.: Modeling the regional impact of ship emissions on NO x and ozone levels over the Eastern Atlantic and Western Europe using ship plume parameterization. Atmospheric Chemistry and Physics 10(14), 6645–6660 (2010)
[15] Henry, B., Lovell, N., Camacho, F.: Nonlinear dynamics time series analysis (2001); Akay (ed.)
[16] Takens, F.: Detecting strange attractors in turbulence. In: Dynamical Systems and Turbulence, Warwick, pp. 366–381. Springer, Heidelberg (1980)
[17] Kříž, R.: Chaotic Analysis of the GDP Time Series. In: Zelinka, I., Chen, G., Rössler, O.E., Snasel, V., Abraham, A. (eds.) Nostradamus 2013: Prediction, Model. & Analysis. AISC, vol. 210, pp. 353–362. Springer, Heidelberg (2013)
[18] Kodba, S., Perc, M., Marhl, M.: Detecting chaos from a time series. European Journal of Physics 26(1), 205 (2005)
[19] Fraser, A.M., Swinney, H.L.: Independent coordinates for strange attractors from mutual information. Physical Review A 33(2), 1134 (1986)
[20] Kennel, M.B., Brown, R., Abarbanel, H.D.: Determining embedding dimension for phase-space reconstruction using a geometrical construction. Physical Review A 45(6), 3403 (1992)
[21] Lorenz, H.W.: Nonlinear dynamical economics and chaotic motion. Springer, Berlin (1993)
[22] Rosenstein, M.T., Collins, J.J., De Luca, C.J.: A practical method for calculating largest Lyapunov exponents from small data sets. Physica D: Nonlinear Phenomena 65(1), 117–134 (1993)
[23] Kříž, R., Kratochvíl, Š.: Analyses of the Chaotic Behavior of the Electricity Price Series. In: ISCS 2013: Interdisciplinary Symposium on Complex Systems, pp. 215–226. Springer, Heidelberg (2014)
[24] Grassberger, P., Procaccia, I.: Characterization of strange attractors. Physical Review Letters 50(5), 346–349 (1983)
[25] Schreiber, T.: Interdisciplinary application of nonlinear time series methods. Physics Report 308(1), 1–64 (1999)
[26] Hegger, R., Kantz, H., Schreiber, T.: Practical implementation of nonlinear time series methods: The TISEAN package. Chaos 9(2), 413–435 (1999)
[27] Cohen, A., Procaccia, I.: Computing the Kolmogorov entropy from time signals of dissipative and conservative dynamical systems. Phys. Rev. A 31, 1872 (1985)
[28] Gottwald, G.A., Melbourne, I.: A new test for chaos in deterministic systems. Proceedings of the Royal Society of London. Series A: Mathematical, Physical and Engineering Sciences 460(2042), 603–611 (2004)
[29] Dawes, J.H.P., Freeland, M.C.: The '0–1 test for chaos' and strange nonchaotic attractors (2008) (preprint)
[30] Farmer, D.J., Sidorowich, J.J.: Predicting chaotic time series. Phys. Rev. Lett. 59, 845–848 (1987)

# Fuzzy Clustering and Loan Risk Prediction

Petr Dostál and Stanislav Škapa

Brno University of Technology, Faculty of Business and Management,
Institute of Informatics, Kolejní 2906/4, 612 00 Brno, Czech Republic

**Abstract.** Fuzzy logic has had successful applications in risk management. There are different methods used as a support for risk decision making. The article deals with the fundamentals of fuzzy clustering method. The case study of the use of this method is presented on the loan risk prediction. The risk prediction plays an important role nowadays.

## 1 Introduction

The soft computing plays very important roles especially in risk management, because it helps to reduce risk that can lead to higher profits and to success in the competitive fight. There are many tasks in risk management where a clustering helps to make a correct decision making in business. A cluster analysis or a clustering is the task of grouping a set of objects in such a way that objects in the same group (called cluster) are more similar (in some sense or another) to each other than to those in other groups (clusters). Popular notions of clusters include groups with low distances among the cluster members. The application of the fuzzy logic model is realized on the case of loan risk prediction by means of fuzzy clustering. The fuzzy clustering could be use, not only neural networks or genetic algorithms. The program MATLAB® with Fuzzy Logic Toolbox is used. The fuzzy applications in economy and business are described in [1], [2], [4] and [5].

## 2 Theory

The fuzzy logic theory is described in many books such as [15], [16]. The fuzzy theory of clustering is mentioned in [3]. Cluster analysis or clustering is the task of grouping a set of objects in such a way that objects in the same group (called cluster) are more similar (in some sense or another) to each other than to those in other groups (clusters).

In hard clustering, data is divided into distinct clusters, where each data element belongs to exactly one cluster. In fuzzy clustering (also referred to as soft clustering), data elements can belong to more than one cluster, and associated with each element is a set of membership levels. These indicate the strength of the association between

I. Zelinka et al. (eds.), *Nostradamus 2014: Prediction, Modeling and Analysis of Complex Systems*, Advances in Intelligent Systems and Computing 289,
DOI: 10.1007/978-3-319-07401-6_37, © Springer International Publishing Switzerland 2014

that data element and a particular cluster. Fuzzy clustering is a process of assigning these membership levels, and then using them to assign data elements to one or more clusters.

One of the most widely used fuzzy clustering algorithms is the fuzzy $c$-means algorithm described in [3]. The fuzzy $c$-means algorithm attempts to partition a finite collection of $n$ elements $X=\{x_1, x_2, \ldots, x_n\}$ into a collection of $c$ fuzzy clusters with respect to some given criterion. Given a finite set of data, the algorithm returns a list of $c$ cluster centers where each element and a partition matrix $W = w_{ij}$ $[0,1]$, $i=1,2, \ldots, n$, $j = 1, 2, \ldots, c$, where each element $w_{ij}$ tells the degree to which element $x_i$ belongs to cluster $c_j$. the fuzzy $c$-means aims to minimize an objective function. The standard function is

$$w_k(x) = \frac{1}{\sum_j \left(\frac{d(center_{k,x})}{d(center_{j,x})}\right)^{2/(m-1)}}, \tag{1}$$

this differs from the $k$-means objective function by the addition of the membership values $u_{ij}$ and the fuzzifier $m$. The fuzzifier $m$ determines the level of cluster fuzziness. A large $m$ results in smaller memberships $w_{ij}$ and hence, fuzzier clusters. In the limit $m = 1$, the memberships $w_{ij}$ converge to 0 or 1, which implies a crisp partitioning. In the absence of experimentation or domain knowledge, $m$ is commonly set to 2. The basic fuzzy $c$-means algorithm, given $n$ data points $(x_1, x_2, \ldots, x_n)$, to be clustered, a number of $c$ clusters with $(c_1, c_2, \ldots, c_n)$ and $m$ the level of cluster fuzziness with.

In fuzzy clustering, every point has a degree of belonging to clusters, as in fuzzy logic, rather than belonging completely to just one cluster. Thus, points on the edge of a cluster may be in the cluster to a lesser degree than points in the center of cluster. An overview and comparison of different fuzzy clustering algorithms is available.

Any point $x$ has a set of coefficients giving the degree of being in the $k$-th cluster $w_k(x)$. With fuzzy $c$-means, the centroid of a cluster is the mean of all points, weighted by their degree of belonging to the cluster

$$c_k = \frac{\sum_x w_k(x)x}{\sum_x w_k(x)}. \tag{2}$$

The degree of belonging, $w_k(x)$, is related inversely to the distance from $x$ to the cluster center as calculated on the previous pass. It also depends on a parameter $m$ that controls how much weight is given to the closest center. The fuzzy applications in risk management are described in [13].

## 3 Risk Prediction

The application of the fuzzy clustering is realized on the case of loan risk prediction. The solved risk management problem is based on sorting of customers according their parameters. In other words, we have to find the risk customer according it's salary, loan add age. The data of the case study are following. The data are characterized by customer's parameters such as *Salary*, *Loan* and *Age*. The data are represented by 17 objects. See Table 1.

**Table 1.** Cluster data

| *Order* | *Salary* | *Loan* | *Age* |
|---|---|---|---|
| 1 | 55000 | 0 | 45 |
| 2 | 11000 | 0 | 65 |
| 3 | 9500 | 0 | 63 |
| 4 | 8300 | 0 | 78 |
| 5 | 10800 | 100000 | 30 |
| 6 | 12300 | 0 | 54 |
| 7 | 12600 | 88000 | 26 |
| 8 | 15200 | 0 | 58 |
| 9 | 15500 | 0 | 52 |
| 10 | 8800 | 15000 | 20 |
| 11 | 4900 | 520000 | 19 |
| 12 | 6400 | 0 | 77 |
| 13 | 42000 | 0 | 48 |
| 14 | 10300 | 0 | 25 |
| 15 | 7400 | 0 | 79 |
| 16 | 13400 | 0 | 55 |
| 17 | 5200 | 900000 | 18 |

The output will be the classification of clients according their characteristic to clusters. The software MATLAB and its Fuzzy Logic Toolbox is used for the software applications. The example presents the objects recorded in MS Excel format in *FCr.xlsx* file. This task is solved by the program *FCr.m*. See Table 2.

**Table 2.** M-file *FCr.m*

```
fd=xlsread('FCr.xlsx','data');
plot3(fd(:,1),fd(:,2),fd(:,3),'o','color','k','markersize',7,'LineWidth',2)
title('Data');xlabel('Salary');ylabel('Loan');zlabel('Age')
grid
[center,U,objFcn] = fcm(fd,3);
figure;plot(objFcn)
title('Fitness Function Values')
xlabel('Iteration Count');ylabel('Fitness Function Value')
maxU = max(U);
index1 = find(U(1, :) == maxU);
index2 = find(U(2, :) == maxU);
index3 = find(U(3, :) == maxU);
figure
center; c1='x'
fd(index1,:); c2='d'
fd(index2,:); c3='*'
fd(index3,:)
plot3(fd(:,1),fd(:,2), fd(:,3), 'o','color','k','markersize',7)
hold on;grid
stem3(center(1,1),center(1,2),center(1,3),'marker',
'x','color','g','markersize',10,'LineWidth',2)
stem3(center(2,1),center(2,2),center(2,3),'marker','d','color','r','markersize',10,'LineW
idth',2)
stem3(center(3,1),center(3,2),center(3,3),'marker','*','color','b','markersize',10,'LineW
idth',2)
view(30,30)
line(fd(index1, 1), fd(index1,2), fd(index1,3),'linestyle','none','marker', '+','color','g');
line(fd(index2,1),fd(index2,2),fd(index2,3),'linestyle','none','marker', 'd','color','r');
line(fd(index3,1),fd(index3,2),fd(index3,3),'linestyle','none','marker', '*','color','b');
title('Loan Risk Prediction');
```

The program is started using the command *FCr* in the MATLAB program environment. The number of clusters is set up to 3. During the calculation the iteration count is displayed. When the calculation is finished the output results, the coordinates of centroids and assign of product to centroids are displayed. See Table 3.

**Table 3.** Results of calculation

*Iteration count = 1, obj. fcn = 366546304306.694340*
*Iteration count = 2, obj. fcn = 169967938341.514860*
*Iteration count = 3, obj. fcn = 65075451317.648163*
……
*Iteration count = 33, obj. fcn = 57455801071.301353*
*Iteration count = 34, obj. fcn = 57455801071.301315*
*Iteration count = 35, obj. fcn = 57455801071.301308*

*center = 1.0e+05 **

| | | |
|---|---|---|
| 0.0513 | 8.0654 | 0.0002 |
| 0.1165 | 1.0997 | 0.0003 |
| 0.1580 | 0.0237 | 0.0006 |

*c1 =* ♦

*ans =*

| | | |
|---|---|---|
| 4900 | 520000 | 19 |
| 5200 | 900000 | 18 |

*c2 =*

*ans =*

| | | |
|---|---|---|
| 10800 | 100000 | 30 |
| 12600 | 88000 | 26 |

*c3 =*.

*ans =*

| | | |
|---|---|---|
| 55000 | 0 | 45 |
| 11000 | 0 | 65 |
| 9500 | 0 | 63 |
| 8300 | 0 | 78 |
| 12300 | 0 | 54 |
| 15200 | 0 | 58 |
| 15500 | 0 | 52 |
| 8800 | 15000 | 20 |
| 6400 | 0 | 77 |
| 42000 | 0 | 48 |
| 10300 | 0 | 25 |
| 7400 | 0 | 79 |
| 13400 | 0 | 55 |

The program displays the graph where each customer is represented by circle according its *Salary*, *Loan* and *Age*. See Fig.1.

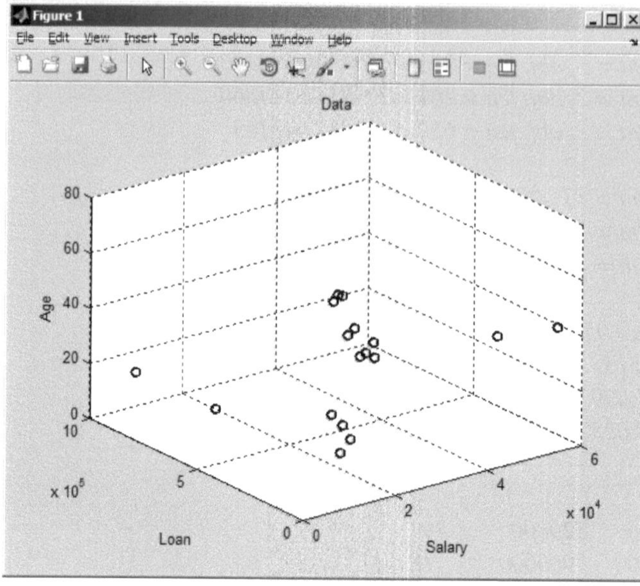


**Fig. 1.** Three-dimensional graph – customers

It is suitable to search the fitness function values dependent on number of iteration. The graph presents good process of iteration. See Figure 2.

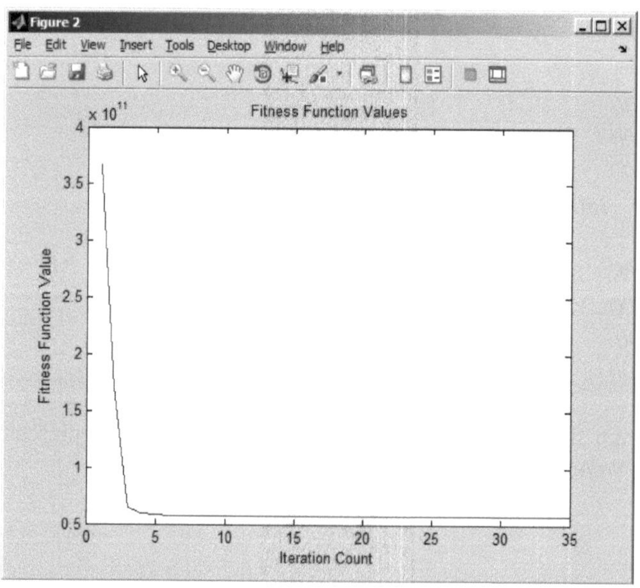


**Fig. 2.** Fitness function values

The results are presented by coordinates of clusters and assignment of customers to thc clusters. A threc-dimensional stem graph is drawn. See Figure 3.

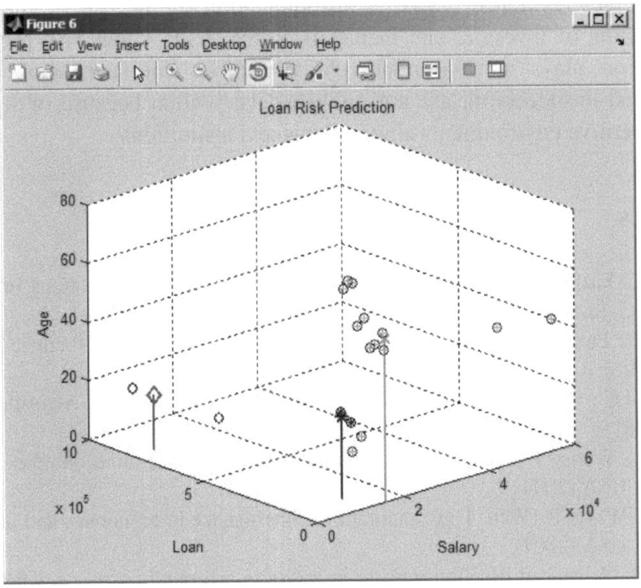


**Fig. 3.** Graph – customer's clustering

# 4 Conclusion

The results are presented by centroids of three clusters marked ♦, X, * and assignment of goods to the clusters. The results presents the case where the cluster • includes the group of customers with low salary, of low age and high loan, the cluster * includes the group of customers with middle salary, of low age and middle loan and the cluster X includes the group of customers with various salary, of middle and high age and no loan. The fuzzy model enables to evaluate risk customers according their salary, loan and age. The tasks from practice lead to multi-dimensional ones, where their graphical presentation is impossible: the image of the solution is in a hyper sphere, for example the variables could be income, property, debt, sex, age, status, children, job, territory, religion etc.

It was used other clustering methods in business field. The use of neural network clustering methods was described in [10] and genetic algorithm in [7]. The neural network method gives better results in comparison with genetic algorithms in the case when number of clusters is higher (more than five). The fuzzy clustering method is comparable with mentioned ones, but the main advantage is in the fact that the data elements belong to more than one cluster, and associated with each element is a set of membership levels.

There are other fuzzy applications in decision making in business [6], optimization in business [8], forecasting in business [9], in management [11], in finance [12] and planning [14].

The example mentioned above is an application of the use of fuzzy logic for decision making of investors and managers in the field of loan risk prediction. The risk prediction plays an important role nowadays, but such applications, both successful and unsuccessful, are not published very often because of secrecy in the highly competitive environments among firms and institutions.

## References

1. Aliev, A., Fazlollahi, B., Aliev, R.: Soft computing and its applications in business and economics. Springer, USA (2004)
2. Altroc, C.: Fuzzy logic & neurofuzzy - applications in business & finance. Prentice Hall, USA (1996)
3. Bezdek, J.C.: Pattern Recognition with Fuzzy Objective Function Algorithms. Springer, USA (1981)
4. Chen, S., Wang, P., Wen, T.: Computational intelligence in economics and finance. Springer, USA (2004)
5. Chen, S., Wang, P., Wen, T.: Computational intelligence in economics and finance, vol. II. Springer, USA (2007)
6. Dostál, P.: Advanced decision making in business and public services. CERM Academic Publishing House, Czech Republic (2011)
7. Dostál, P.: The use of soft computing for optimization in business, economics, and finance. In: Meta-Heuristics Optimization Algorithms in Engineering, Business, Economics, and Finance. IGI Globe, USA (2012a)
8. Dostál, P.: The use of optimization methods in business and public services. In: Handbook of Optimization. Springer, USA (2012b)
9. Dostál, P.: The use of soft computing methods for forecasting in business, their applications in practice. In: Nostradamus: Modern Methods of Prediction, Modeling and Analysis of Nonlinear Systems. Springer, USA (2013a)
10. Dostál, P.: The Use of Soft Computing in Management. In: Handbook of Research on Novel Soft Computing Intelligent Algorithms: Theory and Practical Applications. IGI Globe, USA (2013b)
11. Gil-Lafuente, A.M., Gil-Lafuente, J., Merigó-Lindahl, J.M.: Soft computing in management and Business Economics. Springer, USA (2012)
12. Ribeiro, R., Yager, R.: Soft Computing in Financial Engineering. Springer, USA (1999)
13. Ruan, D., Fedrizzi, M.: Soft Computing for Risk Evaluation and Management, vol. 76. Springer, USA (2001)
14. Vasant, P.: Application of fuzzy linear programming in production planning. Fuzzy Optimization and Decision Making 2(3), 229–241 (2003)
15. Zadeh, L.A.: Fuzzy sets. Information and Control 8, 338–353 (1965)
16. Zadeh, L.A.: A Definition of Soft Computing - adapted from L.A. Zadeh (2012), http://www.soft-computing.de/def.html (retrieved January 10, 2013)

# Sensor Fusion: An Application to Localization and Obstacle Avoidance in Robotics Using Multiple IR Sensors

Rahul Sharma, Honc Daniel, and František Dušek

Department of Process control, Faculty of Electrical Engineering and Informatics,
University of Pardubice, Pardubice, Czech Republic,
rahul.sharma@student.upce.cz,
{daniel.honc,frantisek.dusek}@upce.cz

**Abstract.** Sensor fusion brings the advantage of combining data from various sensors and there by generating a more accurate prediction or estimation of data. Over dependency of sensor and estimation from unreliable data are the most challenging tasks in mobile robotics. In this paper, a framework of sensor fusion technique is presented. The data from the multiple sensors are fused together and the parameters and crash time are estimated. The experiment results show that the sensor fusion technique provides solution to over dependency of sensor and problems with estimation of data from unreliable data. The technique finds application in obstacle avoidance and localization of mobile robots.

## 1 Introduction

For an autonomous mobile robot, the most challenging task is the perception of the environment using sensors. In most of the cases the robots tend to depend heavily on dedicated sensors, which often tend to be unreliable. So, for autonomous robots in fault tolerant applications, sensor fusion techniques help to perceive the environment in a better way.

Sensor fusion is a technique by which the data from multiple sensors are fused together to get an exact information. The fusion of sensor data can be from redundant sensors or complementary sensors. The simplest case of sensor fusion is to combine the data from the sensors, by averaging the sensor reading, if all the sensors have same belief. If the sensors have different belief, a weighted average would help to some extent. However, this simple combining of sensor data would not work, when the system is complex or requires a precise data.

A range of sensor fusion techniques are reviewed in [1,4,6]. Over the years, many techniques of sensor fusion have been emerged. Kalman Filter and Extended Kalman Filter are the most researched technique in sensor fusion for robotic navigation [2,3].

This paper focuses on a simple sensor fusion technique of redundant sensors (IR range finder), in which data from multiple sensors are fused together to determine the three parameters, namely perpendicular distance from center of robot to the wall, distance to the wall and angle between the horizontal axis of robot and obstacle wall. Another important parameter which needs to be considered in applications like ob-

I. Zelinka et al. (eds.), *Nostradamus 2014: Prediction, Modeling and Analysis of Complex Systems*, Advances in Intelligent Systems and Computing 289,
DOI: 10.1007/978-3-319-07401-6_38, © Springer International Publishing Switzerland 2014

stacle avoidance is the crash time, the time left before crashing in to the wall or obstacle. The crash time can be predicted from the current velocity and acceleration of robot and distance to the wall from the robot. Some application of the proposed method is also presented.

The paper is organized as follows: section 2 describes the calibration of the IR range finder, followed by mathematical modelling of the system in section 3. The design of the proposed system is stated in section 4, followed by experimental results in section 5 and conclusion in section 6.

## 2 Characterization of IR Sensor

SHARP IR sensor (GP2Y0A02) is chosen as the range sensor. The output voltage of the sensor is a nonlinear function of the distance between the object and the receiver. The distance value depends on a non-linear way from the sensor analogical value. The best function to fit the sensor curve is given by the expression:

$$y = \frac{Ax}{x+B} + C \tag{1}$$

Where x is the analog voltage output from the sensor, y is the distance to the object from the sensor.

Fig. 1 shows the real acquired values and the result of best fit that makes it possible to compute the distance, measuring the analogical Sharp's value. The best relation was obtained optimizing, on MATLAB, the sum of the quadratic error between the real acquired values and the fitting curve. The parameters A,B and C are optimised by MATLAB *lsqcurvefit* function.

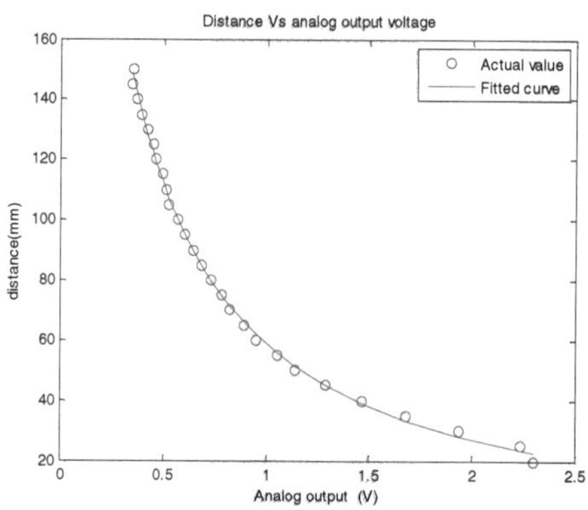


**Fig. 1.** Plot of distance Vs ADC volt after curve fitting by *lsqcurvefit* function in MATLAB (Comparison between actual data and fit data)

## 3 Mathematical Modelling

$x_k$ and $y_k$ are the points on the line (wall), where the IR sensor beam gets reflected. The parameters of interest are $d$, $\emptyset$ and $d_k$, where $d$ is the distance from the robot to the wall, $d_k$, is the perpendicular distance from the origin of the robot to the wall and $\emptyset$ is the angle between horizontal axis of robot and axis parallel to the wall.

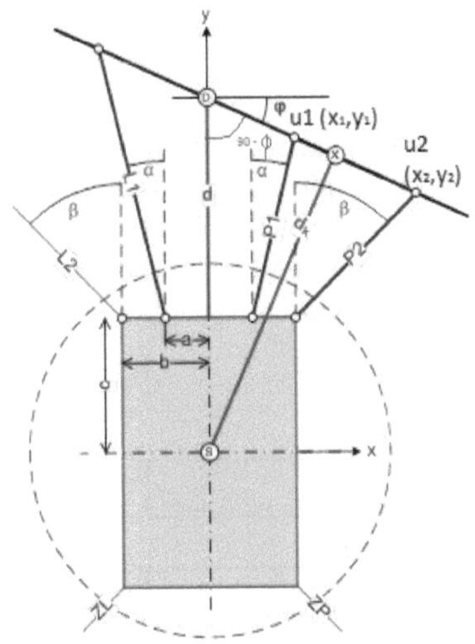


**Fig. 2.** Layout of sensor deployment

$\alpha$ and $\beta$ are the angle at which the IR sensors are mounted on the robot. $p_k$ is the depth to the wall from the robot measured by sensor $k$. $a$,$b$ and $c$ are the distances from the axes of robot to the sensors as shown in Fig. 2.The parameters can be derived from the following expressions.

$$\begin{bmatrix} x_i \\ x_j \\ y_i \\ y_j \end{bmatrix} = \begin{bmatrix} sin\alpha & 0 & 0 & 0 \\ 0 & sin\beta & 0 & 0 \\ 0 & 0 & cos\alpha & 0 \\ 0 & 0 & 0 & cos\beta \end{bmatrix} * \begin{bmatrix} p_i \\ p_j \\ p_i \\ p_j \end{bmatrix} + \begin{bmatrix} a \\ b \\ c \\ c \end{bmatrix} \tag{2}$$

$$\begin{bmatrix} y_i \\ y_j \end{bmatrix} = \begin{bmatrix} x_i & 1 \\ x_j & 1 \end{bmatrix} * \begin{bmatrix} m \\ n \end{bmatrix} \tag{3}$$

This is in the form of,

$$Y = F * X \,, \tag{4}$$

Applying least square estimation (LSE) method to eq. (4)

$$X = (F^T F)^{-1} F^T Y \tag{5}$$

$$d = n - c \tag{6}$$

$$\emptyset = \tan^{-1}(m) \tag{7}$$

$$d_k = (d + c) * \sin(\Pi/2 - \emptyset) \tag{8}$$

From each pair of sensors the corresponding parameters can be calculated and these values can be represented as $p_{ijk}$, where $i$,$j$,$k$ represents the parameter ($d$, $d_k$ or $\emptyset$), first sensor and second sensor respectively. The variables $m$ and $n$ represents slope and intercept respectively.

# 4 Design of Sensor Fusion System

Fig. 3 shows the basic design of sensor fusion system adopted in the paper. The first step is preprocessing of the signal acquired from the sensor. In this stage, a series of sample are averaged to get an initial rough data. Using this data ($p_i$), the parameters ($p_{ijk}$) are calculated using eq. (1) – (8). These set of parameter have to be validated, for knowing the reliability of the sensors. At the data validation stage, the data is validated and a weighing factor $v_{ijk}$ is calculated. The following sections will explain the stages in detail.

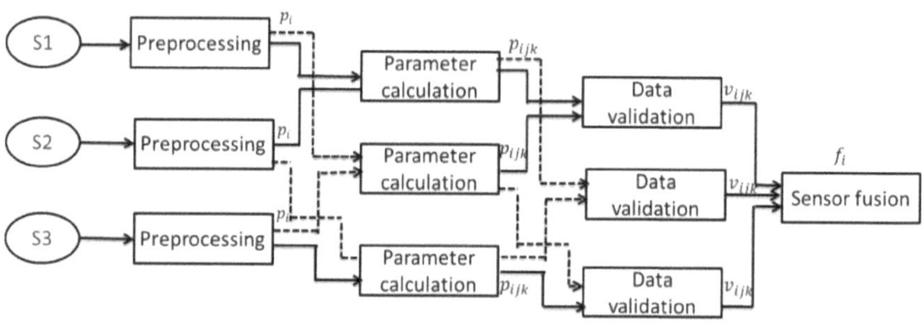


**Fig. 3.** Sensor fusion system

## 4.1 Data Validation

Data from the sensors has to be validated before proceeding to sensor fusion. The data from the sensor is only allowed to fuse, if it confirms certain criteria. The criteria are correlation coefficient and closeness coefficient. Correlation coefficient is given by the expression,

$$Cor(j,\mathrm{k}) = \frac{N\sum_0^N P_{ijk}P_{ikj} - \sum_0^N P_{ijk}\sum_0^N P_{ikj}}{[N\sum_0^N P_{ijk}{}^2 - (\sum_0^N P_{ijk})^2]^{\frac{1}{2}}\ [N\sum_0^N P_{ikj}{}^2 - (\sum_0^N P_{ikj})^2]^{\frac{1}{2}}} \tag{9}$$

$Cor(j,\mathrm{k})$ is the correlation of parameter $P_i$ , calculated from sensor $j$ and sensor $k$ and N is the number of samples. Closeness coefficient [5] is given by the expression,

$$\gamma = 1 - \frac{|P_{ij} - P_{ji}|}{Closeness} \tag{10}$$

Where $Closeness$ is the parameter which allows choosing the data which are closer.

When the new data (parameters) falls in a range, say magnitude of correlation coefficient is above 0.75 and closeness coefficient is less than 0.8, the data is selected. Otherwise the data is rejected and not chosen for data fusion. This allows the system to choose only reliable sensor data by comparing with all the other sources. A weighing factor, $v_{ijk}$ gives the weightage of each parameter, so that the corresponding parameter can either selected or rejected. This factor is calculated by the following expression

$$v_{ijk} = \begin{cases} 0, & if\ any\ of\ the\ criteria\ fails \\ 1, & if\ both\ the\ criteria\ are\ satisfied \end{cases}$$

$v_{ijk}$, will help to choose the most reliable parameters for fusion, calculated from different sensors.

### 4.2 Sensor Fusion

The parameters which confirms the criteria of data validation are selected and fused together [7] by the following expression

$$f_i = \sum_{i=0,j=0,k=0}^{i=3,j=n,k=n} \frac{\sigma_{kj}^2}{\sigma_{jk}^2 + \sigma_{kj}^2} p_{ijk} * v_{ijk} \tag{11}$$

Where is the final parameter, $\sigma_{kj}^2$ is the variance and $p_{ijk}$ is the parameter *i*, calculated from sensor *j* and *k*.

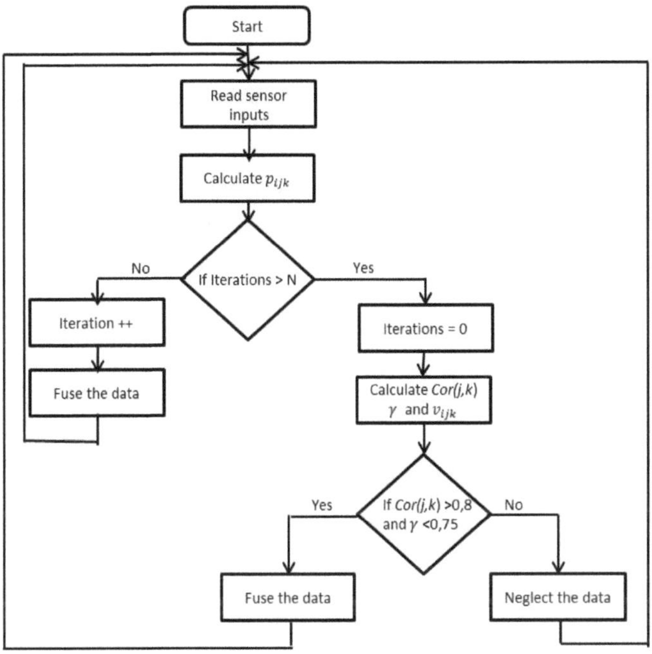


**Fig. 4.** Flow chart of sensor fusion system

Fig. 4 illustrates the process, step by step. For making the system faster, a variable N (period), is introduced to validate data on regular intervals, thereby avoiding the sensor validation process in each iteration.

### 4.3 Prediction of Crash Time

For autonomous robots, the time to crash is an important criterion. If robot can predict the crash time, it could avoid the obstacle by making the appropriate decision. The crash time can be predicted from the parameters estimated by sensor fusion. The crash time T, derived from basic laws of motion, is given by the expression

$$T = \frac{-v \pm \sqrt{v^2 - 2a(\frac{z}{2} * \tan\emptyset + d)}}{a} \quad (12)$$

Where $v$, $a$ and z represents velocity of robot, acceleration of robot and width of the robot respectively.

# 5 Experimental Results

Experiments were conducted using Arduino Due board and Sharp GP2Y0A02, IR range finders. Three IR range sensors were fixed on the front side of the robot with a=0, b=10, c=10, α=0 and β=5. Three sets of parameters were calculated from the sensor readings. As a first step the IR sensor data were preprocessed by averaging a sample set of 10. This data was then validated by using the correlation coefficient threshold as 0.75 and closeness coefficient as 0.8. If the data fall in the range of the criteria, it was selected to fusion. Three sets of data were then fused together as explained in section sensor fusion.

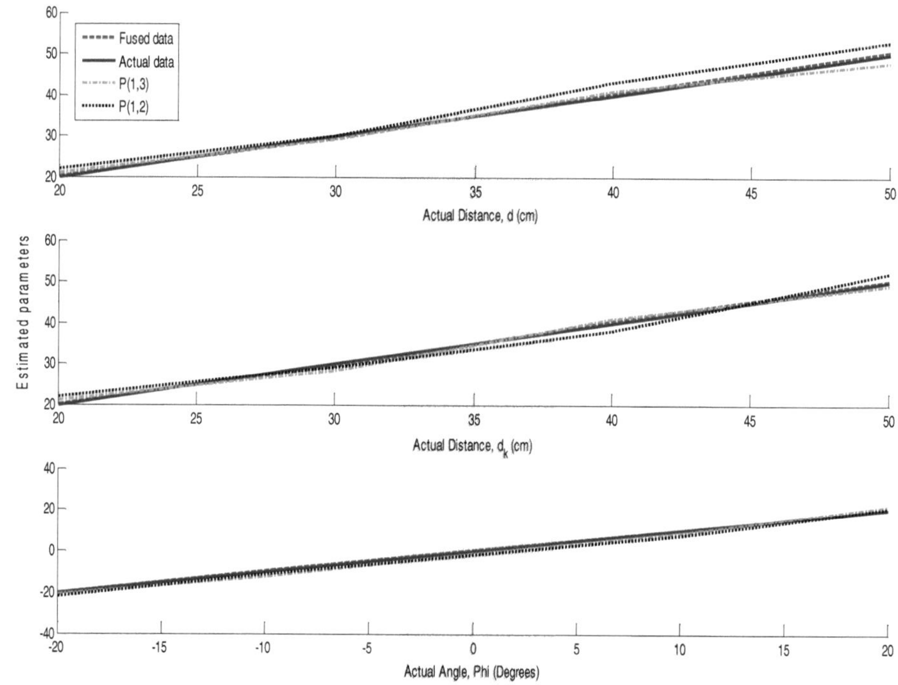


**Fig. 5.** Comparison of fused and non-fused data a) parameter d, b) parameter dk, c) parameter ∅, where P(1,2) is the parameters calculated from sensor 1 and sensor 2, P(1,3) is the parameters calculated from sensor 1 and sensor 3

Several experiments were conducted by varying distances and angles. Fig. 5 shows the variation from actual data to fused and non-fused data from a series of experiments. The fused data tends to follow more closely to actual data, compared to non-fused data P(1,2) and P(1,3), which had larger variations.

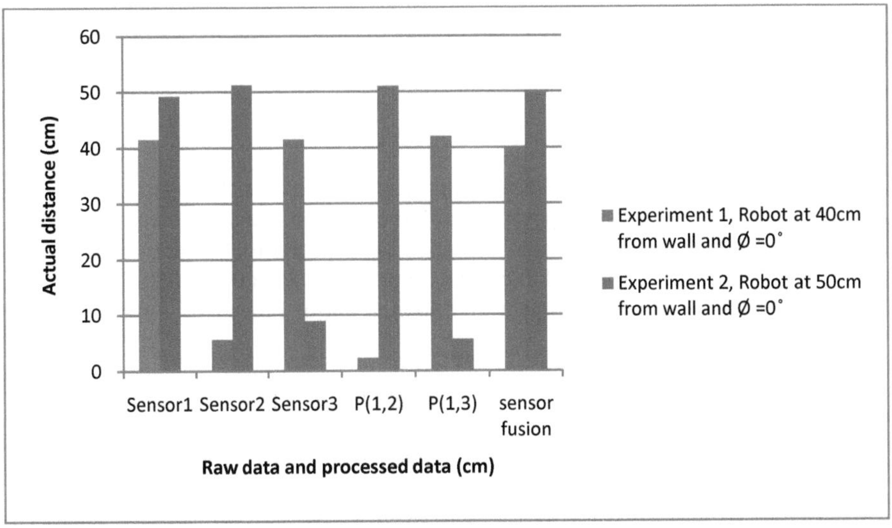


**Fig. 6.** Experiments with faulty sensors (Actual vs Observed and calculated variables)

Experiments were also conducted by making one or more sensors faulty intentionally. Fig. 6 shows that even if sensor data is unreliable, the sensor fusion system is still able to estimate an accurate data. For example, in experiment 1, sensor 2 was made faulty and by fusing the data from all the three sensors, the system could reject the data from the sensor, identifying the reliability of sensor through data validation process, and estimate the data which was closer to the actual data. Crash time was also predicted by measuring current velocity and acceleration of robot.

It was noticed that, the proposed system performed better even in case of unreliable data from faulty sensors. Furthermore, the treat of over dependency of dedicated sensors was also avoided to some extent, by making use of multiple sensors instead of dedicated sensor. This will allow the robots to perform better when working in unknown dangerous environments. In another words the system works better where there is a requirement of fault tolerant system.

The proposed system finds two applications in robotics. Firstly, in case of decision making during obstacle avoidance, where the parameter Ø, can be used to take a right decision. For example, if the Ø is negative, the robot must take a LEFT turn and vice versa. The second application is localization of robot. The Ø will give a better orientation estimate with the help of multiple sensor input. The approximation from multiple sensor input will give more accurate estimate of orientation of the robot.

# 6 Conclusion

The proposed system is modeled, simulated and tested in various environments. The results show that the system performs better by making use of multiple sensor inputs. The fusion of multiple sensor inputs allows the robot to work in fault tolerant applications. The system finds applications in obstacle avoidance decision making and localization (orientation estimate) of robots. When combined with absolute positioning sensors (GPS) or dead-reckoning sensors (IMU), the can be further extended to real world applications in robotics.

**Acknowledgments.** This research was supported by Institutional support of The Ministry of Education, Youth and Sports of the Czech Republic and by the project "Support for internships and professional activities by innovation of tertiary education at DFJP and FEI University of Pardubice, Reg. No: CZ.1.07/2.4.00/17.0107" in Modern control methods team - development and application of predictive control methods using artificial intelligence.

# References

1. Elmenreich, W.: A review on system architectures for sensor fusion applications. In: Obermaisser, R., et al. (eds.) SEUS 2007. LNCS, vol. 4761, pp. 547–559. Springer, Heidelberg (2007)
2. Foxlin, E.: Inertial head-tracker sensor fusion by a complementary separate-bias Kalman filter. In: Proceedings of the IEEE 1996 Virtual Reality Annual International Symposium, pp. 185–194. IEEE (March 1996)
3. Jetto, L., Longhi, S., Venturini, G.: Development and experimental validation of an adaptive extended Kalman filter for the localization of mobile robots. IEEE Transactions on Robotics and Automation 15(2), 219–229 (1999)
4. Kam, M., Zhu, X., Kalata, P.: Sensor fusion for mobile robot navigation. Proceedings of the IEEE 85(1), 108–119 (1997)
5. Naji, H.R., Weir, J., Wells, B.E.: Applying the multi-agent paradigm to reconfigurable hardware: a sensor fusion example. In: Second International Workshop on Intelligent Systems Design and Application, pp. 207–212. Dynamic Publishers, Inc. (January 2002)
6. Nicosevici, T., Garcia, R., Carreras, M., Villanueva, M.: A review of sensor fusion techniques for underwater vehicle navigation. In: OCEANS 2004, MTTS/IEEE TECHNO-OCEAN 2004, vol. 3, pp. 1600–1605. IEEE (2004)
7. Panich, S., Afzulpurkar, N.: Sensor Fusion Techniques in Navigation Application for Mobile Robot. In: Sensor Fusion-Foundation and Applications, pp. 101–120 (2011)

# Estimation of Angles Yaw, Pitch and Roll in a Special Transformation Problem

Martin Svoboda, Jaroslav Marek, and Jana Heckenbergerová

Faculty of Electrical Engineering and Informatics, Department of Mathematics and Physics, University Pardubice, Studentská 95, 532 10 Pardubice, Czech Republic
{jaroslav.marek,martin.svoboda,jana.heckenbergerova}@upce.cz

**Abstract.** The aim of this contribution is to present a possible approach how to estimate the unknown parameters yaw, pitch and roll by regression models in transformation from the G-sensor coordinate system to the vehicle system. The unknown angles can be determine from acceleration vector measurement in coordinate system of sensor when vehicle is parked and when it is accelerating in a forward direction. Estimation of rotation matrix based on a long-time measurement looks like a simple task. But even small change of one acceleration component or one measurement that does not meet the prerequisites (acceleration only and only in a forward direction, plane ride) could mean significant shift in estimators of unknown angles. Nonlinear regression model and its linearization can help to improve uncertainty of estimated parameters. Feasible solution could be provided as well by utilization of some heuristic algorithms.

## 1 Introduction

The passengers safety is currently one of the most important issues that automotive industry is dealing with. Operation of active protection systems against the direct impact (airbags) is controlled pursuant with sensor data. Cars are frequently equipped with a three-axes accelerometer (G-sensor) providing measurements of actual acceleration vector. In the data processing many problems have to be solved. One of them is the correct orientation of the G-sensor. Due to technical difficulties it is sometimes impossible to install the G-sensor consistently with the vehicle coordinate system. Data transformation from the sensor coordinate system, denoted as $S_{\mathbf{Y}}$, to the vehicle system ($S_{\mathbf{X}}$) is therefore required. Resulting acceleration data then correspond to the actual vehicle movement [1].

In this paper, rotation matrix is used for transformation of G-sensor coordinate system. Therefore three independent rotation angles have to be estimated from long-time sensor data. Because of the direct utilization in the automotive industry, it seems to be wiser to use yaw, pitch and roll angles from flight dynamics instead of Euler angles describing general spinning motion. We can determine unknown angles from measurement of acceleration of parked vehicle in coordinate system of sensor and from acceleration of vehicle in a forward direction with plane ride. We suggest to utilize linear regression model for this

I. Zelinka et al. (eds.), *Nostradamus 2014: Prediction, Modeling and Analysis of Complex Systems*, Advances in Intelligent Systems and Computing 289,
DOI: 10.1007/978-3-319-07401-6_39, © Springer International Publishing Switzerland 2014

estimation. Even small change in one acceleration coordinate in sensor system or one measurement which does not meet the prerequisites (acceleration only and only in a forward direction, plane ride) could mean significant shift in estimators of unknown angles. Therefore proposed methodology can provide powerful tool for G-sensor calibration and data validation.

Paper is divided into six main sections, the following one describes transformation of coordinate system using rotation matrix and three major angles. Sections 3 and 4 describe long-time time series and illustrate building of linear regression model for statistical estimation of yaw, pitch and roll angles. Numerical results are summarized in 5. The last section brings main conclusions and outlines possible directions for future work.

## 2 Transformation of Coordinate System and Definition of Yaw, Pitch and Roll Angles

In many problems we need to use different coordinate systems in order to reveal different vector quantities. In this section such a vector transformation is described. We will show how the components of a vector are transformed when the reference frame (vector basis) is changed [2].

Let us consider two different orthogonal, right-hand sided, reference frames $S_{\mathbf{X}} = \{\mathbf{X}_1, \mathbf{X}_2, \mathbf{X}_3\}$ and $S_{\mathbf{Y}} = \{\mathbf{Y}_1, \mathbf{Y}_2, \mathbf{Y}_3\}$. A vector $\mathbf{b}$ in coordinate system $S_{\mathbf{Y}}$ can be transformed to coordinate system $S_{\mathbf{X}}$ by considering the 9 angles that define the relationships between two systems. Transformation of vector $\mathbf{b} = (b_1, b_2, b_3)$ from $S_{\mathbf{Y}}$ to vector $\mathbf{a} = (a_1, a_2, a_3)$ in coordinate system $S_{\mathbf{X}}$ can be written in matrix form as

$$\mathbf{a} = \begin{pmatrix} a_1 \\ a_2 \\ a_3 \end{pmatrix} = \begin{pmatrix} \cos(\Theta_{11}), \cos(\Theta_{12}), \cos(\Theta_{13}) \\ \cos(\Theta_{21}), \cos(\Theta_{22}), \cos(\Theta_{23}) \\ \cos(\Theta_{31}), \cos(\Theta_{32}), \cos(\Theta_{33}) \end{pmatrix} \cdot \begin{pmatrix} b_1 \\ b_2 \\ b_3 \end{pmatrix}, \tag{1}$$

where $a_1, a_2, a_3$ are components of the vector $\mathbf{a}$ in the $S_{\mathbf{X}}$ coordinate system and $\Theta_{ij}$ are angles between the coordinate directions $\mathbf{Y}_i$ and $\mathbf{X}_j$. Transformation matrix is usually denoted by $\mathbf{T}$ or $\mathbf{R}$ like in this case.

It is important to remark that only three of nine transformation angles $\Theta_{ij}$ are independent. Thus, the general orientation of a coordinate system can be described by a sequence of three rotations about coordinate axis. One particular set of such rotations, called Euler angles, leads to description of general spinning motion. In our case, because of the direct utilization in the automotive industry, it seems to be wiser to use yaw, pitch and roll angles. These angles are critical flight dynamics parameters and correspond to rotations in three dimensions about the vehicle's center of mass. Roll represents the rotation about $x$-axis heading forward, pitch shows the incline of rising or descending as rotation about $y$-axis and yaw determines turning angle about $z$-axis directing down [3]. Rotation transformation matrices $\mathbf{R}_z, \mathbf{R}_y$ and $\mathbf{R}_x$ are written as

$$\mathbf{R}_z(\alpha) = \begin{bmatrix} \cos(\alpha) & -\sin(\alpha) & 0 \\ \sin(\alpha) & \cos(\alpha) & 0 \\ 0 & 0 & 1 \end{bmatrix}, \mathbf{R}_y(\beta) = \begin{bmatrix} \cos(\beta) & 0 & \sin(\beta) \\ 0 & 1 & 0 \\ -\sin(\beta) & 0 & \cos(\beta) \end{bmatrix},$$

$$\mathbf{R}_x(\gamma) = \begin{bmatrix} 1 & 0 & 0 \\ 0 & \cos(\gamma) & -\sin(\gamma) \\ 0 & \sin(\gamma) & \cos(\gamma) \end{bmatrix},$$

where $\alpha$ is yaw, $\beta$ is pitch and $\gamma$ is roll. Resulting transformation matrix $\mathbf{R}$ is given by step by step rotation about $x$, $y$ and $z$-axis so it can be written in form:

$$\mathbf{R} = \mathbf{R}_z \cdot \mathbf{R}_y \cdot \mathbf{R}_x, \tag{2}$$

where

$$\begin{aligned}
R_{1,1} &= \cos(\alpha)\cos(\beta), \\
R_{1,2} &= -\sin(\alpha)\cos(\gamma) + \cos(\alpha)\sin(\beta)\sin(\gamma) \\
R_{1,3} &= \sin(\alpha)\sin(\gamma) + \cos(\alpha)\sin(\beta)\cos(\gamma), \\
R_{2,1} &= \sin(\alpha)\cos(\beta), \\
R_{2,2} &= \cos(\alpha)\cos(\gamma) + \sin(\alpha)\sin(\beta)\sin(\gamma), \\
R_{2,3} &= -\cos(\alpha)\sin(\gamma) + \sin(\alpha)\sin(\beta)\cos(\gamma), \\
R_{3',1} &= -\sin(\beta), \\
R_{3,2} &= \cos(\beta)\sin(\gamma), \\
R_{3,3} &= \cos(\beta)\cos(\gamma).
\end{aligned}$$

Transformation matrix $\mathbf{R}$ is therefore a function of unknown rotation parameters $\alpha, \beta, \gamma$. Our milestone is estimation of these angles from G-sensor data. Statistical methodology using regression model will be described in following sections.

# 3 Measurements and Direct Estimation of Rotation Angles

Datasets provided to us by automotive industry facility contain measurements $\mathbf{Y}^0 = [x_0, y_0, z_0]_{S_\mathbf{Y}}$ of parked vehicle accelerations in G-sensor coordinate system and G-sensor measurements $\mathbf{Y}^i = [x_i, y_i, z_i]_{S_\mathbf{Y}} \forall i = 1, \ldots, M$ of accelerating vehicle during regular ride. In vehicle coordinate system these measurements correspond to the vector $\mathbf{U}^0 = [0, 0, -G]_{S_\mathbf{X}}$ for parked vehicle and $\mathbf{U}^j = [A_j, 0, -G]_{S_\mathbf{X}} \forall j = 1, \ldots, N$ for vehicle accelerating in a forward direction with plane ride. Other G-sensor measurements can not be used for rotation angles estimation and have to be filtered from original time series.

From the acceleration data can be constructed the time series $(x_1, x_2, \ldots, x_M)'$, $(y_1, y_2, \ldots, y_M)'$, $(z_1, z_2, \ldots, z_M)'$ as we illustrated in figures 1 and 2. The first figure shows whole time series for vehicle ride. It is clear that a huge part of data do not meet the prerequisites (acceleration only and only in a forward direction, plane ride). Such data have to be filtered before rotation parameters estimation. In second figure, filtered acceleration time series are visualized. These data are used later in regression model.

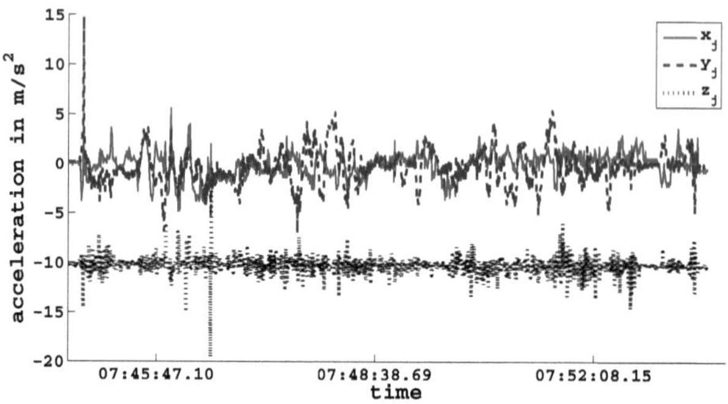


**Fig. 1.** Acceleration data: solid line for $x$-coordinate, dash line for $y$ and dash-dot line for $z$. The regular vehicle ride that takes a few minutes and does not meet the prerequisites about forward direction.

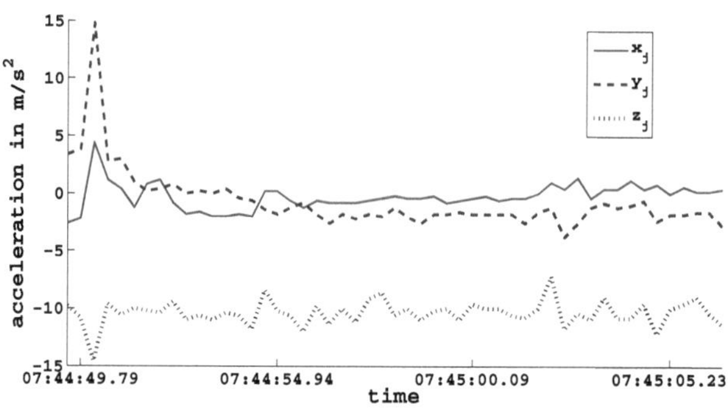


**Fig. 2.** Filtered acceleration data: solid line for $x$-coordinate, dash line for $y$ and dash-dot line for $z$. The vehicle ride that meets the prerequisites about forward direction at a plane ride.

The transformation between the G-sensor and vehicle coordinate systems is defined by the vector parameter $\boldsymbol{\Theta} = (\alpha, \beta, \gamma)'$ of rotation matrix $\mathbf{R}$:

$$\phi(\mathbf{Y}^j, \boldsymbol{\Theta}) = \mathbf{R}(\boldsymbol{\Theta}) \cdot (x_j, y_j, z_j)' = (A_j, 0, -G)', \ j = 0, 1, \ldots, N. \tag{3}$$

The forward acceleration $A_j$ can be evaluated from formula

$$A_j = \sqrt{(x_j - x_0)^2 + (y_j - y_0)^2 + (z_j - z_0)^2}, \tag{4}$$

and value of constant G can be estimated using formula

$$\hat{G} = \sqrt{x_0^2 + y_0^2 + z_0^2}.$$

From filtered datasets, angles yaw, pitch and roll can be directly estimated from transformation equation (3) as:

$$\widetilde{\beta}_j = \text{asin}\frac{z_0 - z_j}{A_j}, \widetilde{\gamma}_j = \text{acos}\frac{-z_0}{G\cos(\beta)}, \widetilde{\alpha}_j = \text{acos}\frac{x_j - x_0}{A_j\cos(\beta)}. \tag{5}$$

The quality of such estimators is very low. Moreover a problem with argument of inverse trigonometric function, that do not lie in its definition domain, can frequently occur.

## 4 Regression Model

Our goal is to find the estimators of the unknown angles yaw, pitch and roll based on the G-sensor measurements using the regression model.

We have to estimate unknown parameters of rotation matrix in floating window. It means that $\mathbf{Y}^0$ and four consecutive measurements $\mathbf{Y}^j$, $\mathbf{Y}^{j+1}$, $\mathbf{Y}^{j+2}$, $\mathbf{Y}^{j+3}$, $j = 1, \ldots, N-3$ are used for estimation evaluation of parameters $\alpha_j$, $\beta_j$, $\gamma_j$. Results are then visualized as angular time series of these estimators for $j = 1, \ldots, N-3$.

We will get the model of indirect measurement of the vector parameter in the nonlinear form (3). After linearization, cf. [4] we can write our model in form of

$$\mathbf{Y} \sim_{15} \left(\mathbf{F}\boldsymbol{\Theta}, \sigma^2\mathbf{V}\right), \tag{6}$$

where $\mathbf{Y} = (\mathbf{Y}^0, \mathbf{Y}^j, \mathbf{Y}^{j+1}, \mathbf{Y}^{j+2}, \mathbf{Y}^{j+3})'$, $\boldsymbol{\Theta}_j = [\alpha, \beta, \gamma]'$ is the vector of the unknown parameters,

$$\{\mathbf{F}\}_{i\cdot} = \left(\frac{\partial\phi(x_i, y_i, z_i, \boldsymbol{\Theta}^0)}{\partial\alpha}, \frac{\partial\phi(x_i, y_i, z_i, \boldsymbol{\Theta}^0)}{\partial\beta}, \frac{\partial\phi(x_i, y_i, z_i, \boldsymbol{\Theta}^0)}{\partial\gamma}\right), \tag{7}$$

is the known design matrix and the value of $\sigma = 0.2$ has been adopted from the documentation protocol of the measurement device.

The observation vector is then in form of

$$(\xi - \boldsymbol{\Theta}^0) = \begin{pmatrix} \mathbf{Y}^0 - \phi(\mathbf{Y}^0, \boldsymbol{\Theta}^0) \\ \mathbf{Y}^j - \phi(\mathbf{Y}^j, \boldsymbol{\Theta}^0) \\ \mathbf{Y}^{j+1} - \phi(\mathbf{Y}^{j+1}, \boldsymbol{\Theta}^0) \\ \mathbf{Y}^{j+2} - \phi(\mathbf{Y}^{j+2}, \boldsymbol{\Theta}^0) \\ \mathbf{Y}^{j+3} - \phi(\mathbf{Y}^{j+3}, \boldsymbol{\Theta}^0) \end{pmatrix} \tag{8}$$

with the variance matrix given by

$$\begin{aligned} \mathrm{Var}\left((\xi - \boldsymbol{\Theta}^0)\right) &= \sigma^2 \mathbf{V}, \\ \mathbf{V} &= \mathrm{Diag}((1,1,1,2,2,2,2,2,2,2,2,2,2,2,2)). \end{aligned}$$

**Theorem 1.** *BLUE*

$$\widehat{\delta\boldsymbol{\Theta}} = \left(\mathbf{F}'\mathbf{V}^{-1}\mathbf{F}\right)^{-1} \mathbf{F}'\mathbf{V}^{-1}(\xi - \mathbf{f}_0) \sim N_3\left[\delta\boldsymbol{\Theta}; \sigma^2\left(\mathbf{F}'\mathbf{V}^{-1}\mathbf{F}\right)^{-1}\right]. \tag{9}$$

*Proof.* See [5].

In the same manner it is possible to study estimator of parameters in a two stage regression model, where we must respect uncertainty in connecting measurement of vector $\mathbf{Y}^0$. Further details about this problem can be found in [6].

## 5 Numerical Results

An important criterion for choosing suitable part of the acceleration measurement is stability of a sequence of measurements.

Since we have time series of parked vehicle acceleration measurements, we can evaluate the observation vector $\mathbf{Y}^0 = (0.1077, 0.2154, -10.1769)'$ and its covariance matrix

$$\mathrm{Var}(\mathbf{Y}^0) = \begin{pmatrix} 0.0199, 0.0047, 0.0006 \\ 0.0047, 0.0062, 0.0028 \\ 0.0006, 0.0028, 0.0138 \end{pmatrix}.$$

After selecting one time step, e.g. $j = 30$ with corresponding time 07:44:48.76, we can compute following rotation vector estimators and variance matrix

$$\widehat{\boldsymbol{\Theta}} = (139.54, 12.29, -2.66)', \ \mathrm{Var}(\widehat{\boldsymbol{\Theta}}) = \begin{pmatrix} 9.6060, & -1.2354, & -4.6169 \\ -1.2354, & 0.3859, & 0.6035 \\ -4.6169, & 0.6035, & 2.4718 \end{pmatrix}.$$

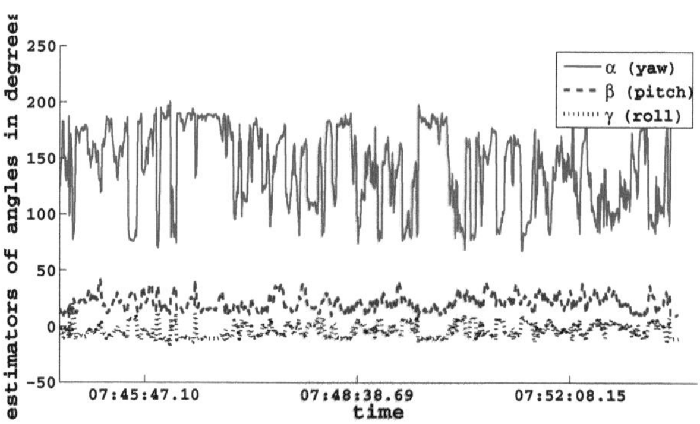


**Fig. 3.** Estimators of angles yaw, pitch and roll. The ride, which takes a few minutes and which does not meet the prerequisites (acceleration only and only in a forward direction, plane ride).

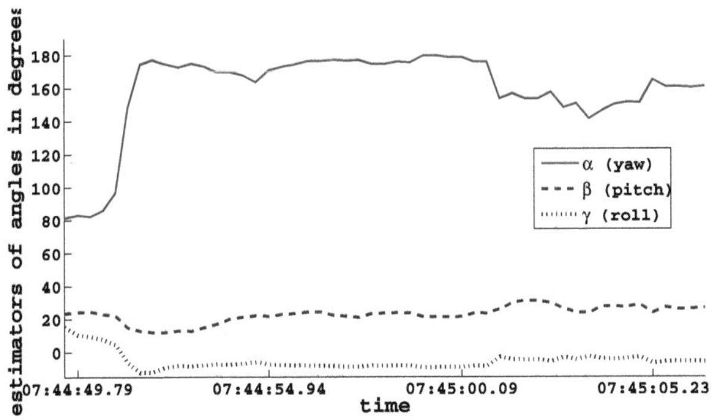


**Fig. 4.** Estimators of angles yaw, pitch and roll from filtered dataset, where the prerequisites are fulfilled.

Angular time series of yaw, pitch and roll estimators, illustrated in Fig. 3 and 4, evidently show that estimators of rotation angles are not identical. Mean value of time series of yaw, pitch and roll estimators through whole time period are $\hat{\alpha} = 142.92$, $\hat{\beta} = 21.63$, $\hat{\gamma} = -3.39$.

Analyzed acceleration time series does not provide consistent value of rotation angles yaw, pitch and roll. Using described model, we can only estimate their mean and variance. Illustrating boxplots are depicted in figure 5.

Feasible solution could provide some heuristic algorithm. In our case, function

$$\sum_{j=1}^{3} \left( (\mathbf{R}(\alpha, \beta, \gamma) \cdot \mathbf{Y}^0) - (0, 0, -G)' \right)_j \tag{10}$$

was minimized on area centered in averages and bounded by standard deviations. Resulting approximation of minimum was $\tilde{\alpha} = 120.21$, $\tilde{\beta} = 0.16$, $\tilde{\gamma} = -1.47$.

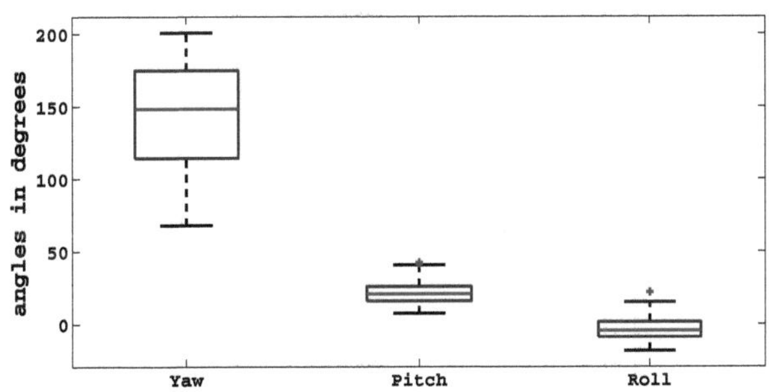


**Fig. 5.** Boxplot for estimators of angles yaw, pitch and roll obtained from regression model.

## 6 Concluding Remarks

In this paper, time series of acceleration data are analyzed. Datasets provided by automotive industry facility contain acceleration vector measurements of parked vehicle and accelerating vehicle during regular ride in G-sensor coordinate system. Main task is to estimate rotation angles yaw, pitch and roll, so the time series can be transformed and represented in vehicle coordinate system.

Rotation transformation matrix can be estimated using regression model, when data fulfill the movement condition (acceleration only and only in a forward direction, plane ride). This condition is barely fulfilled during the whole analyzed time interval as illustrated in figure 1. Moreover the estimators of angles are varying a few grades because of data dispersion.

The fluctuation of the three rotation angles is however evidently higher than these conditions would lead to. In spite of that, our estimators of angles yaw, pitch and roll received form the regression model show a lower dispersion than arithmetic mean of directly counted angles by formulas (5).

Application of heuristic algorithms on minimization problem of a criterion function (10) introduced in the last paragraph shows the new approach to given problem and leads us in our research to different directions.

## References

1. Svoboda, M., Marek, J.: Odhad úhlů pitch, yaw, roll pro kalibraci akcelerometru. Forum Statisticum Slovacum, IX (3), 117–122 (2013)
2. Widnall, S.: Lecture L3 - Vectors, Matrices and Coordinate Transformations, 16.07 Dynamics (Fall 2009), `http://ocw.mit.edu/courses/aeronautics-and-astronautics/16-07-dynamics-fall-2009/lecture-notes/MIT16_07F09_Lec03.pdf`
3. Stengel, R.: Flight Dynamics. Princeton University Press (2004)
4. Kubáček, L.: On a linearization of regression models. Applications of Mathematics 40(1), 61–78 (1995)
5. Kubáček, L., Kubáčková, L., Volaufová, J.: Statistical Models with Linear Structures. Bratislava, Veda (1995)
6. Kubáček, L., Marek, J.: Partial optimum estimator in two stage regression model with constraints and a problem of equivalence. Math. Slovaca 55, 477–494 (2005)

# Multi-Channel Multi-Objective Routing Metric for Vehicular Ad-hoc Networks

Peppino Fazio[1], Cesare Sottile[1], Mauro Tropea[1], Floriano De Rango[1], and Miroslav Voznak[2]

[1] D.I.M.E.S. Department
University of Calabria
87036, Rende, Italy
[2] Department of Telecommunications
VSB-Technical University of Ostrava
708 33 Ostrava, Czech Republic

**Abstract.** Nowadays, distributed mobile wireless computing is becoming a very important communications paradigm, due to its flexibility to adapt to different mobile applications. Routing operations assume a crucial importance in system optimization, especially when considering dense urban areas, where interference effects cannot be neglected. The implementation of new routing protocols becomes challenging in Vehicular Ad-Hoc NETworks (VANETs) so, at this aim, we propose a vehicular routing scheme in which the available channels are managed for optimizing a considered composite metric for multi-channel transmissions, which takes into account different parameters (multi-objective). Network Simulator 2 (NS2) has been employed to validate the Multi-Channel Multi-Objective Distance Vector (MCMO-DV), showing how it outperforms classical approaches in terms of throughput, packet delivery ratio, and overhead.

## 1 Introduction

In the last years, many efforts have been made in the mobile computing research field; in particular, the IEEE 802.11 standard completely dominates the market. In wireless networks, nodes are free to move randomly and organize themselves arbitrarily; thus, network topology may change rapidly and unpredictably. VANETs provide wireless communication among vehicles and vehicle-to-Road-Side-Units (RSU) equipments. RSU construct the infrastructure of the VANETs using wired and wireless communications among each other. Communication performance strongly depends on how the routing takes place in the network: the existing routing protocols for VANETs are not so efficient to meet the needing of every traffic scenario, since the high degree of mobility and propagation phenomena have a high impact on system performance. In this paper, the multichannel characteristic of VANET devices is considered, in order to improve system performance in terms of routing optimization. In fact, in a distributed multi-hop architecture, a mobile node may potentially find multiple routes for a given destination and, when it evaluates the network topology through its routing table, the

I. Zelinka et al. (eds.), *Nostradamus 2014: Prediction, Modeling and Analysis of Complex Systems*, Advances in Intelligent Systems and Computing 289,
DOI: 10.1007/978-3-319-07401-6_40, 

availability of different channels may enhance the quality of communication if properly exploited. The main aim of this work is to introduce this feature when considering classical routing metrics. In detail, a new routing protocol for interference reduction and link-duration enhancement is proposed for VANET environments, taking the advantage of a dynamic allocation of the Dedicated Short Range Communications (DSRC) spectrum, in order to reduce interference level among mobile nodes and to increase the overall link stability in the considered network. The proposed scheme can be integrated with different already-implemented routing protocols and its metric take into account the best values of Inter-Channel Interference (*ICI*), Link Duration Probability ($LD_{PROB}$) and Hop Count ($H_{CNT}$). So, MCMO-DV aims at choosing different channels along the path from a source to a destination, obtaining a global metric minimization for the considered connection. This paper is organized as follows: Section 2 presents an in-depth overview on state-of-the-art routing in VANET; Section 3 offers a deep description of the proposed idea, then Section 4 shows the obtained results. Finally, Section 5 concludes the paper.

## 2 Related Work

In literature for VANETs many authors have proposed some routing schemes, but most of them lack the employment of the multichannel availability of mobile devices. In [1], DIR protocol constructs a series of diagonal intersections between the source and destination vehicle. The DIR protocol is based upon the geographic routing protocol in which source vehicle geographically forwards the data packets towards the first diagonal intersection, second diagonal intersection and so on until the last diagonal intersection and finally geographically reaches to designation vehicle. In [2] the authors proposed the ROMSGP algorithm. It is an integration of the receive on most stable path (ROMSP), with the grouping of nodes according to their velocity vectors, as previously demonstrated, with certain modifications to suit it to the VANET scenario. For example, the non-disjoint nature of ROMSP is not considered due to the strict mobility pattern of VANET networks The effects of mobility are also considered in [3], in which a new metric is introduced in order to proactively adapt to a constantly changing topology. The scheme proposed by Sofra et al. considers the life-time of a link and the forwarding operation is carried-out on the basis of how much a link can be considered stable during routing operations. Link duration is evaluated by a precise mobility model, able to capture the trend of link degradation fluctuations. The authors have shown how introducing some fragmentation approaches, network performance can be improved, also in terms of delivery ratio. In our previous works [4], [5], [6], [7],[8] an enhancement of On-demand Distance Vector protocol has been proposed, in terms of metric optimization. In particular it has been modified in order to take consideration of the availability of different transmission channels with an integrated metric, which takes into account the interference level over the different channels. In particular, it allows the management of the multichannel capability of the WAVE standard at the routing layer through a higher-level channel selection, which is based on a interference-aware algorithm.

## 3 Problem Statement and Proposed Protocol Scheme

This paper focuses its attention on the enhancement of routing operations in VANETs, taking into consideration both neighbors' interference level and link duration, in addition to classical hop-count term. The proposed idea, called Multi-Channel Multi-Objective Distance Vector (MCMO-DV), is general and does not depend on the considered routing protocol. It can be integrated with the majority part of existing routing protocols and it is based on analysis of interference dynamics for choosing an appropriate transmission channel in order to minimize the interference; periodical refresh, in order to evaluate the updated interference value available on each channel; definition of Link Duration Probability (LDP), in order to choose more stable paths; transmission of synchronization packets in order to advise the receiving node of a new channel selection.

We consider a vehicular scenario in which each node participates to routing operations as in classical ad-hoc networks. Let us consider the VANET topology illustrated in Fig. 1. Let $V=\{v_1,\ldots,v_n\}$ be the set of vehicles (vehicular nodes or vertex) in the network, with $||V||=n$ and $E=\{e_1,\ldots e_m\}$ be the set of direct peer-to-peer links, with $||E||=m$ and $e_k=(v_h,v_k)$, $v_h,v_k \in V$. Let $CHAN=\{c_1,\ldots c_p\}$ be the set of available channels in DSRC spectrum, with $||CHAN||=p$. A path discovery phase is initiated each time a source node $v_i \in V$ needs to transmit to a destination node $v_j \in V$.

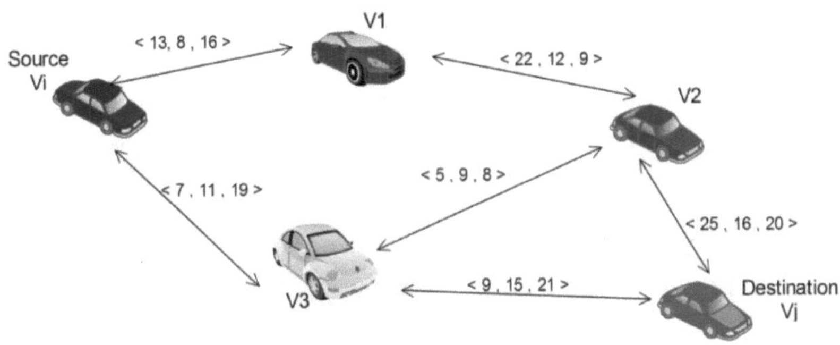


**Fig. 1.** An example of a typical VANET scenario

Our proposed protocol scheme aims at finding the best path in terms of interference, hop-count and link-duration (for example $v_i$-$v_2$-$v_4$-$v_6$-$v_j$), basing the choice of next-hops on a new defined metric. We hypothesize that:

- When a new node enters into the network it discovers its neighbors by broadcasting HELLO messages;
- The source node initiates the path-discovery phase by broadcasting a Route REQuest (RREQ) packet to its neighbors; if a neighbor has a route to the desired destination, then it sends a Route REPly (RREP) back to the source; otherwise the RREQ is forwarded again;

- All the nodes transmit at the same power level $P_t$, independently on the chosen channel and each node knows the power level of the received signal on each available channel.

When dealing with wireless communications (especially in vehicular environments), classical metrics become inadequate, since they do not consider all the negative effects that are present when paths from sources to destinations are built. If only the hop-count is considered, the obtained paths may suffer huge interference levels and/or short duration and, on the contrary, minimizing the interference may bring the considered protocol to obtain longer paths with scarce duration. In [3] and [4] the importance of link-duration and link-interference has been remarked, so we propose a new multi-objective metric that takes into consideration interference, link-duration and hop count parameters.

### 3.1 The Main Terms of MCMO-DV

This work focuses the attention on the proposal of a new multi-objective metric which combines inter-channel interference, link-duration probability and hop-count parameters. We are not interested now on how protocol signaling packets should be changed in order to take into account the new concepts; the main attention, instead, is focused on the definition of the key elements of a new metric. Each node $v_i \in V$ has to evaluate the metric related to each possible next-hop, so the terms are now defined.

#### 3.1.1 The Hop CouNT ($H_{CNT}$) Term

This is the classical term, used in the majority part of routing protocols. It simply counts the number of hops that belong to a particular path. If $Path_V(v_i,v_j)=\{v_i, v_2, v_3, \ldots, v_{p-2}, v_{p-1}, v_j\}$ with $//Path_V(v_i,v_j)//=p_{i,j}$ is a path (expressed in terms of vertex) from $v_i$ to $v_j$, then each node $v_k \in Path_V(v_i,v_j)$ knows the hop-count toward $v_j$ on $Path_V(v_i,v_j)$; in particular, node $v_i$ will know the $H_{CNT}$ on $Path_V(v_i,v_j)$, which can be expressed as:

$$H_{CNT}[Path_V(v_i,v_j)] = \| Path_V(v_i,v_j) \| = p_{i,j} \quad (1)$$

#### 3.1.2 The Link Duration Probability (LDP) Term

This is the term that takes into account the reliability of a link in terms of duration. If $Path_E(v_i,v_j)=\{(v_i,v_2),(v_2,v_3), \ldots, (v_{q-2},v_{q-1}),(v_{q-1},v_j)\}=\{e_1,\ldots,e_q\}$ with $//Path_E(v_i,v_j)//=q_{i,j}$ is a path (expressed in terms of edges) from $v_i$ to $v_j$, with $e_k \in E$ then, for each edge $e_k$, involving the couple of nodes on the link $(v_h, v_k)$, it can be written as in [9].

At this point, the sensitivity of the receiver can be defined though a threshold of the attenuation level $\beta_{th}$, that is to say a link among a couple of VANET nodes on the edge $e_k$ is still valid if $\beta^{e_k} \leq \beta_{th}$ and the probability of this event can be written as [9]:

$$P(\beta^{e_k} \leq \beta_{th}) = \frac{1}{2} + \frac{1}{2} erf\left( \frac{\beta_{th} - \alpha 10 \log\left( \frac{avg_s \cdot \tau}{L} \right)}{\sqrt{2}\sigma} \right) = LD_{PROB}(e_k) = LD_{PROB}[(v_h, v_k)] \quad (5)$$

Given $Path_E(v_i,v_j)$, then the link-duration probability on a path, $LD_{PROB}[Path_E(v_i,v_j)]$, can be evaluated as:

$$LD_{PROB}[Path_E(v_i,v_j)] = \prod_{k=1}^{q_{i,j}} LD_{PROB}(e_k), \ \forall e_k \in Path_E(v_i,v_j) \tag{6}$$

### 3.1.3 The Inter-Channel Interference (ICI) Term

The interference contribution is derived from the expression of the received power, for all the available channels. It strictly depends on the transmission power and radio propagation phenomena. Using the theory of [10], [11] for DSRC channels, it can be written that:

$$P_{loss}(v_i,v_j) = 40 \cdot \log(d_{ij}) - [10 \cdot \log(G_t) + 10 \cdot \log(G_r) + 20 \cdot \log(h_t) + 20 \cdot \log(h_r)] \tag{7}$$

which indicates the loss in signal strength (in dB) among the couple of nodes $v_i,v_j \in V$. The terms $G_t$ and $G_r$ are the *TX* and *RX* antenna gains respectively, while $h_t$ and $h_r$ are the *TX* and *RX* antenna heights. From eq. 7, the expression of the received signal strength (in dB) by node $v_i$, for the signal transmitted by node $v_j$, on channel $c_l \in CHAN$ can be easily written as follows:

$$P_r(v_i,v_j,c_l) = P_t - P_{loss}(v_i,v_j) \tag{8}$$

where $P_t$ is the transmission power (the same for each node on each channel). In real environments, the value of $P_r$ can be easily evaluated via *HW*, but for our simulation purposes, the expression of eq. 8 is very suitable. It can be used for accounting path-loss effects, that are dominant in VANET environments, because channel coding and frequency interleaving make the bit error performance of an OFDM link in a frequency-selective channel depend more on the average received power than on the power of the weakest subcarrier. From the value of eq. 8, the expression of the ICI term for node $v_i$ on channel $c_l$ is obtained as follows:

$$ICI(v_i,c_l) = \sum_{j}^{||adjacents(v_i)||} P_r(v_i,v_j,c_l) \tag{9}$$

where $adjacent(v_i)$ is the set of nodes adjacent to $v_i$. At this point, each node $v_i$ can evaluate the best value of ICI associated to a particular channel:

$$ICI_{MIN}(v_i) = \min_{c_l \in CHAN}\{ICI(v_i,c_l)\} \quad c_{l_{MIN}}(v_i) = index\left\{\min_{c_l}(ICI(v_i,c_l))\right\} \tag{10}$$

### 3.1.4 The Proposed Scheme for Routing in VANETs

As in the previous sub-sections, let us hypothesize that node $v_i$ has to transmit data to node $v_j$ ($v_i$, $v_j \in V$) and the considered routing protocol allows the utilization of *RREQ-RREP* mechanism to discover all the available paths from $v_i$ to $v_j$. When a *RREQ* packet is forwarded by an intermediary node $v_l \in V$ which participates to routing operations, the information about $ICI_{MIN}(v_l)$ evaluated in eq. 10 is inserted into the message. When a *RREP* is created by an intermediary node $v_l \in V$, which has knowledge of a path towards destination, the value of the $H_{CNT}$ is appended into the packet (our attention is not focused now on how a packet is modified in order to give each node the knowledge of the considered parameters), giving to the previous hop, indicated with $v_{I-1}$, the knowledge of $ICI_{MIN}$ and $H_{CNT}$. Thus, through the exchange of *RREQ* and *RREP* messages, a set of paths *P* from $v_i$ to $v_j$ is discovered and, for each of them, the source node can evaluate the hop-count $H_{CNT}=p_{i,j}$ (from eq. 1), the average $ICI_{MIN}$, denoted with $\underline{ICI}_{MIN}$, and the average probability of path duration, denoted with $\underline{LD}_{PROB}[Path_E(v_i,v_j)]$:

$$\underline{ICI}_{MIN}\left[P_V(v_i,v_j)\right]=\frac{\sum_{v_I\in P_V(v_i,v_j)} ICI_{MIN}(v_I)}{p_{i,j}} \qquad \underline{LD}_{PROB}\left[P_E(v_i,v_j)\right]=\min_{e_k\in P_E(v_i,v_j)} P\left(\beta^{e_k}\le\beta_{th}\right) \tag{11}$$

where $ICI_{MIN}(v_I)$ is the *ICI* value obtained through eq. 10 for the intermediate node $v_I$. Eq. 12 shows that $\underline{LD}_{PROB}$ is evaluated on $P_E(v_i,v_j)$ by considering the minimum probability, which represents the bottleneck of $P_E(v_i,v_j)$ in terms of link duration. After the node $v_i$ received all the RREPs from its neighbors, it knows the set of paths *P* toward $v_j$ and all the related values of $H_{CNT}$, *ICI* and $LD_{PROB}$, through the expressions of eq. 1, 11. At this point the metric for the MCMO-DV has to be defined adequately, as a multi-objective function:

$$\underset{p\in P}{m_{MCMO-DV}}=\gamma_1\cdot m_{H_{CNT}}(p)+\gamma_2\cdot m_{ICI}(p)+\gamma_3\cdot m_{LD_{PROB}}(p) \tag{12}$$

where $m_{H_{CNT}}$, $m_{ICI}$ and $m_{LDPROB}$ are three normalized terms related to the expression of the parameters, defined by using the following definitions:

$$ICI_{MAX}=\max_{Path_V(v_i,v_j)\in P}\{ICI[Path_V(v_i,v_j)]\} \quad H_{CNT_{MAX}}=\max_{Path_V(v_i,v_j)\in P}\left\{H_{CNT}\left[Path_V(v_i,v_j)\right]\right\} \tag{13}$$

$$LD_{PROB_{MAX}}=\max_{Path_V(v_i,v_j)\in P}\left\{LD_{PROB}\left[Path_E(v_i,v_j)\right]\right\} \tag{14}$$

So, the terms $m_{H_{CNT}}$, $m_{ICI}$ and $m_{LDPROB}$ in eq. 12 can be rewritten as follows:

$$\underset{p\in P}{m_{MCMO-DV}}=\gamma_1\frac{\underline{ICI}_{MIN}(p)}{ICI_{MAX}}+\gamma_2\frac{H_{CNT}(p)}{H_{CNT_{MAX}}}+\gamma_3\left(1-\frac{\underline{LD}_{PROB}(p)}{LD_{PROB_{MAX}}}\right) \tag{15}$$

where the terms in the metric have been normalized, in order to be comparable. At this point, when node $v_i$ has to choose among different paths to destination in the set *P*, it chooses the one for which:

$$\underset{p^*\in P}{m_{MCMO-DV}}=\min_{p\in P}\left(\underset{p\in P}{m_{MCMO-DV}}\right) \tag{16}$$

The proposed protocol, now, has three degrees of freedom ($\gamma_1$ called *ICI* weight, $\gamma_2$ called *HC* weight and $\gamma_3$ called *LDP* weight), which have to be set adequately. Next section shows some considerations about them. Before observing performance evaluation, it must be said that, due to the presence of mobility and wireless phenomena, some degradations are dynamically introduced into the system so, once the optimal channels have been chosen for data transmission, they have to be checked and refreshed each $\Delta$ amount of time, verifying if some better conditions (in terms of channels and paths) exist. If a change is needed, a node, which is aware of new channels conditions, sends a Change-REQuest packet (CREQ) to its neighbor, then waits for the acknowledgement Change-REPly (CREP).

## 4 Performance Evaluation

NS-2 has been used to integrate the proposed idea with different existing protocols. An example on how the signaling packets are changed in order to take into account the additional fields can be found in [4]. The C4R mobility generator, which represents a powerful extension of [12] with a user-friendly GUI, has been used to create mobility log-files. Differently from [13], we are not considering only a Manhattan scenario, but different urban scenarios have been taken into account, in order to deal with more effective maps: without loss of generality, we illustrate the results obtained for the centre of Rome. We considered a transmission rate of 3Mbps. The optimal values of some simulation parameters have been determined through different campaigns of simulation: the number of concurrent connections has been fixed to 15, $\Delta$ has been fixed to 60ms, while the values of $\gamma_i$ have been chosen by considering the following figures. In particular a first addicted campaign of simulations has been carried out, with a number of network nodes equals to 50, in order to evaluate protocol performance (in terms of Packet Delivery Ratio PDR and Throughput) by considering different values for $\gamma_i$ belonging to the set of values {0.2, 0.4, 0.6, 0.8}.It is demonstrated that, fixing the values of $\gamma_2$, there are some values of $\gamma_1$ and $\gamma_3$ that lead to maximum values on the surfaces. So, it can be concluded that if a higher weight is given to the Interference term and to the link duration probability ($\gamma_1$=0.2 $\gamma_2$=0.4 $\gamma_3$=0.8), then the system will observe a higher percentage of correctly delivered packets. Likewise the average system throughput, an optimal value can be obtained for the configuration $\gamma_1=0.2$, $\gamma_2=0.4$, $\gamma_3=0.8$, for which the maximum performance is obtained. The figures are not shown due to space limitations. At this point, the $m_{MCMO\text{-}DV}$ metric is completely defined and it can be used to evaluate the performance of the proposed protocol scheme. We compared the MCMO-DV scheme to AODV single (only one transmission channel is used as in the traditional definition), AODV multi (multiple transmission channels are chosen randomly), GPSR and OLSR (both with a single transmission channel). Simulation parameters are the same of the previous campaign, but in this case the number of mobile vehicles varies from 40 to 100. From Fig. 2 and Fig. 3, it can be noticed how the MCMO-DV outperforms the other protocols in terms of PDR and Aggregated Throughput (the sum of the throughputs of all connections): introducing a composite metric, interference level and link duration are taken into account, so more stable paths are chosen, reducing the probability of packet loss and retransmissions. So, this is evident when considering the percentage of correctly delivered packets and system throughput.

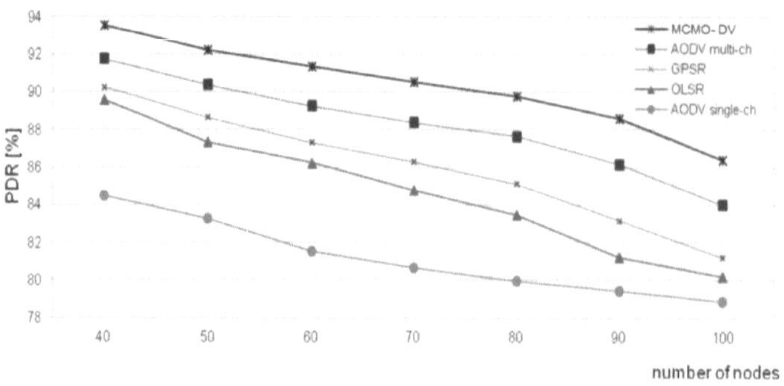


**Fig. 2.** Average PDR vs the number of mobile nodes

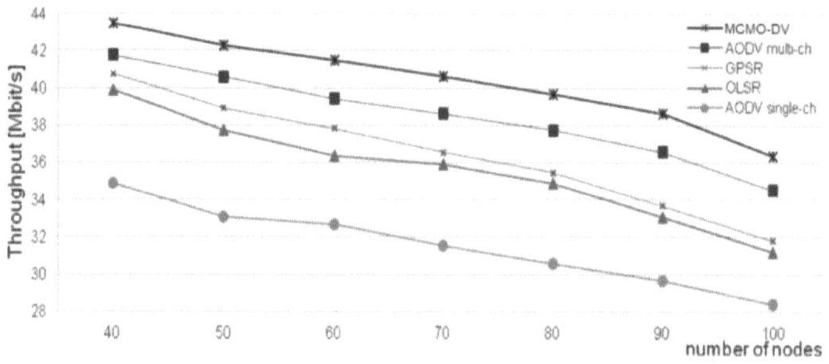


**Fig. 3.** Average Aggregated Throughput vs the number of mobile nodes

Referring to the overhead performance, as illustrated in Fig.4, the MCMO-DV protocol performs slightly worse than the other ones, because of the new signaling packets that are introduced into the network traffic for the construction of alternative paths; the difference on the number of transmission channels for AODV schemes is not evident because protocol messages remain unmodified: no new messages are introduced in the AODV Multi-channel case, but only a random selection of a transmission channel.

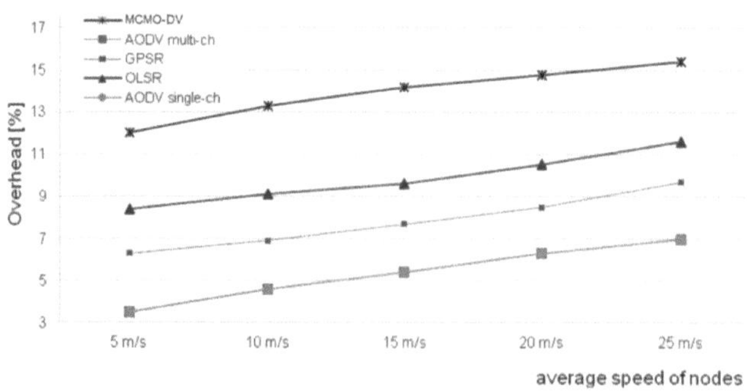


**Fig. 4.** Overhead vs the number of mobile nodes

# 5 Conclusions

In this work a new scheme for routing in VANETs, dedicated to the optimization of path-length, interference level and link duration is proposed. It is based on a dynamic allocation mechanism of the DSRC spectrum, aimed at the reduction of the inter-channel interference and maximization of link duration, two key issues in vehicular environments. A new composite multi-objective metric, based on the evaluation of interference levels, path lengths and link duration along the different links from sources towards destinations has been proposed. Through an NS2 implementation of the IEEE802.11p standard, with the simulation of vehicles mobility in a urban environment, it has been shown that the proposed idea enhances classical protocols performance in terms of throughput and packet delivery ratio, despite of a slight increase in protocol overhead and delay.

**Acknowledgement.** The research leading to these results has been supported by I-CONTACT project and partially by the project No. CZ.1.07/2.3.00/20.0217 "The Development of Excellence of the Telecommunication Research Team in Relation to International Cooperation" within the frame of the operation programme "Education for competitiveness" financed by the Structural Funds and from the state budget of the Czech Republic.

# References

1. Chen, Y.S., Lin, Y.W., Pan, C.Y.: A diagonal-intersection-based routing protocol for urban vehicular ad hoc networks. Telecommunication Systems 46, 299–316 (2010)
2. Taleb, T.: A Stable Routing Protocol to Support ITS Services in VANET Networks. IEEE Transactions on Vehicular Technology 56 (2007)
3. Sofra, N., Gkelias, A., Leung, K.K.: Route Construction for Long Lifetime in VANETs. IEEE Transactions on Vehicular Technology 60(7), 3450–3461 (2011)
4. Fazio, P., De Rango, F., Sottile, C.: An On-Demand Interference Aware Routing Protocol for VANETs. Journal of Networks (JNW) 7(11) (November 2012)
5. Fazio, P., De Rango, F.: A New Interference Aware On Demand Routing Protocol for Vehicular Networks. In: SPECTS 2011, The Nederland (2011)
6. Fazio, P., De Rango, F.: A New Channel Assignment Scheme for Interference-Aware Routing in Vehicular Networks. In: 2011 IEEE 73rd Vehicular Technology Conference (VTC Spring), pp. 1–5 (2011)
7. Fazio, P., De Rango, F., Sottile C., Manzoni P., Calafate C.: A distance vector routing protocol for VANET environment with Dynamic Frequency assignment. In: 2011 IEEE Wireless Communications and Networking Conference (WCNC), pp. 1016–1020 (2011)
8. Fazio, P., Tropea, M., Veltri, F., Marano, S.: A New Routing Protocol for Interference and Path-Length Minimization in Vehicular Networks. In: VTC 2012, Yokohama, Japan (May 2012)
9. Khayam, S.A., Radha, H.: Analyzing the Spread of Active Worms over VANET. In: Department of Electrical & Computer Engineering / 2120 Engineering Building, Michigan State University East Lansing, MI 48824, USA (2004)

10. Zang, Y., Stibor, L., Orfanos, G., Guo, S., Reumerman, H.J.: An Error Model for Inter-Vehicle Communications in Highway Scenarios at 5.9GHz. In: PE-WASUN 2005, Montreal, Quebec, Canada, October 10-13 (2005)
11. Yan, G., Olariu, S.: A Probabilistic Analysis of Link Duration in Vehicular Ad Hoc Networks. Intelligent Transportation Systems 12(4), 1227–1236 (2011)
12. Martinez, F.J., Cano, J.-C., Calafate, C.T., Manzoni, P.: CityMob: A Mobility Model Pattern Generator for VANETs. In: Communications Workshops, ICC Workshops 2008, Bejing, May 19-23 (2008)
13. De Rango, F., Veltri, F., Fazio, P., Marano, S.: Two-level trajectory-based routing protocol for vehicular ad hoc networks in freeway and Manhattan environments. Journal of Networks 4(9), 866–880 (2009)

# Artificial Intelligence in ISES Measureserver® for Remote Experiment Control

Michal Gerža[1], František Schauer[1,2], and Ivan Zelinka[1,3]

[1] Tomas Bata University in Zlín, Faculty of Applied Informatics, CZ-760 05 Zlín, Czech Republic
michal.gerza@email.cz
[2] University of Trnava, SK-918 43 Trnava, Slovak Republic
[3] VŠB - Technical University of Ostrava, CZ-708 33 Ostrava, Czech Republic

**Abstract.**The paper deals with the area of Internet School Experimental System (ISES) remote experiments in general and its core module called ISES Measureserver®. In particular ISES Measureserver® is, in fact, a finite state machine, serving for the measured data accumulation, processing and providing communication in the server-client system. Recently, we replenished ISES Measureserver® by a new functionality, namely diagnostics of the connected to the RE physical hardware, using the artificial intelligence solutions.

In the introduction, the state of the art of ISES remote experiments is described. In the next chapter a consideration for the applying of proper artificial intelligence method to improve the Measureserver® reliability is made. We focused on the cognitive Fault Diagnosis System (FDS) intended for distributed sensor networks. FDS makes advantage of spatial and temporal relationship among sensors connected to RE physical hardware to give the information for reduction of the influence of failures, ill effecting the Measureserver® functioning. The lower layer uses Change Detection Test (CDT) based on Hidden Markov models (HMM) configured to detect variations in the relationships among couples of sensors. Changes in the HMM are detected by inspecting the corresponding likelihood. The output information provided by the CDT lower layer is then passed to the cognitive higher layer collecting information to discriminate among faults, changes in the environment and false positive.

The intended improvement is the increase of the reliability, monitoring of the state and the fast remedy of the functioning of remote experiments in case of malfunction.Proposed diagnostics solution will contribute to improvement to remote experiments reliability and to a wider acceptance of this new ICT technology.

**Keywords:** ISES, Measureserver®, Fault Diagnosis System, Change Detection Test, Hidden Markov model, remote experiment, sensor network.

## 1 Introduction

The conventional approach of teaching methods, oriented at students at secondary schools and universities, are quite outmoded and not so broadly popular to understand taught scientific themes. The contemporary students demand higher level teaching

I. Zelinka et al. (eds.), *Nostradamus 2014: Prediction, Modeling and Analysis of Complex Systems*, Advances in Intelligent Systems and Computing 289,
DOI: 10.1007/978-3-319-07401-6_41, © Springer International Publishing Switzerland 2014

methods, which help them to perceive phenomena in better way in the field of physics, biology, chemistry and others. Educational materials accessibility is important as well, especially for distant students who frequently prefer studying scientific themes via Internet. These obvious advantages are provided by remote experiments (RE) built on ISES, which have been designed and developed for educational purposes by Charles University in Prague. After some time, ISES RE have been improved on a higher level educational tool by Tomas Bata University in Zlín called EASY REMOTE-ISES (ER-ISES) [1].

ISES RE consist of six constitutive hardware and software modules presented in Fig 1. More details about ISES RE are available in [1].

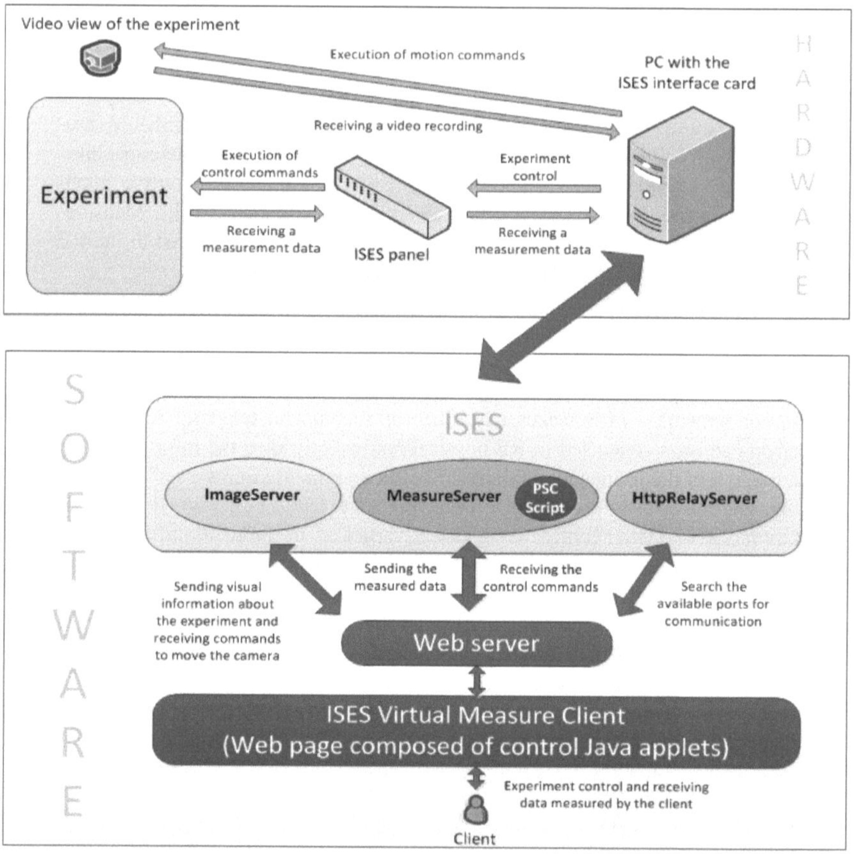


**Fig. 1.** Schematical arrangement of ISES RE [1]

## 2 Artificial Intelligence Implementation

Let us find a theoretical solution for the operative mechanism of ISES Measureserver® with respect to its diagnostics to reduce faults coming from sensors of the RE physical hardware, using the artificial intelligence (AI) approach.

Sensors monitoring a real environment are prone to faults or aging, ill affecting ISES RE functioning, or even fallout of the whole apparatus. In turn, the permanent or transient faults can influence sensors functioning already used, causing functional errors in the RE processing chain. Such erroneous information may exert strong side effects on the subsequent control chain leading to wrong decisions and inappropriate control actions.

Fault Diagnosis System (FDS) should play an important role of supervising the process operations for the purpose of detecting, isolating and identifying a potential fault and design possible accommodation actions. The main components of FDS are derived from the comparison of the running and model data of functioning. Model data, often unavailable, is often substituted by the data, generated by the RE physical hardware.

When a change with respect to the model is detected by applied FDS, the following situations might arise [2]:

- Model change: The used model is no longer representing the current data due to the model approximation deficiencies,
- Change in the environment: The environment is a time variable quantity and the trained model is no more able to explain the data,
- Fault: The sensor or its electronic unit is affected by a fault inducing an error.

Published solutions for FDSs of sensor networks do not generally allow distinguishing between faults and environment changes [5]. Moreover, in the original model bias is considered negligible, which is hardly acceptable hypothesis in many applications. However, a cognitive FDS may influence sensor data streams. This type of FDS already recognizes the model bias existence during measurements and proposes a method for discriminating between faults and changes.

Let us further attempt to propose a solution for FDS design, based on the AI processes, introducing functional dependency graphs and information related to spatial and temporal relationships among sensor data streams.

### 2.1 Functional Concept

Hidden a-priori information concerning spatial and temporal relationships among proposed sensor data streams is exploited, leading to a functional dependency graph where nodes are the used sensors and arcs are associated with the sensor-to-sensor functional relationship. In particular, for each used sensor couple, Hidden Markov model (HMM) is designed which gives the parameters of linear time invariant (LTI) model approximating the relationship. As such, spatial redundancy is modeled with HMM operating in the parameter space of linear time invariant dynamic models, embedding the time dependency. When the likelihood between the HMM-based learning machine and the new incoming data stream falls below a preset threshold (which can be inserted by the teaching process), a change is detected by the HMM-based Change Detection Test (CDT) at the detection layer. The cognitive layer of FDS, activated in response to a change, starts alarm raised by the CDT, discriminates time variant and bias faults using the dependency graph of the network. At the same time, it allows us for isolating the fault for a possible subsequent accommodation phase [2].

It is important to mention, HMM are the most popular means of temporal classification. They are effectively applicable in many areas like cryptanalysis, handwriting, gene prediction, and speech and gesture recognition. In other words, HMM is a variant of Finite State Machine (FSM), but in contrary to FSM they are not deterministic. A normal FSM generates deterministic symbol for every state, which forces deterministic transition to another state. HMM do not function deterministically, but both transition to the next state and emission is governed by the probabilistic model [3]. The basic theory of HMMs is also very elegant and easy to understand, making analyses and system development easier [4].

In following we describe the theoretical solution of a cognitive FDS in order to diagnose all sensors installed in RE physical hardware to reduce possible faults coming from the measurement subsystems. The proposed FDS is intended to be built in the next generation of the Measureserver® module for the AI on line diagnosis of the RE physical hardware.

## 2.2 Modeling Functional Relationships in Sensor Networks

Let us consider a sensor network composed of *N* fixed sensing units, which are deployed within the environment *P* depicted in Fig. 5. Each unit can host up to *M* sensors giving information on various physical properties of *P* space (for example, temperature, humidity, vibrations, rain intensity and change in slope). Each *j*-th sensor of the *i*-th unit acquires a scalar data stream $X_{i,j}$.

The FDS runs as a part of the Measureserver® module situated in the control room where the RE physical hardware is installed [2].

### A. Modeling the Network: The Dependency Graph

The cognitive framework for the fault diagnosis relies on the ability to model functional relationships among the acquired information on the space *P*. In detail, each functional relationship captures spatial and temporal dependencies from data provided by a generic couple of involved sensors. Fig. 2 shows an example of the sensor network with defined dependencies.

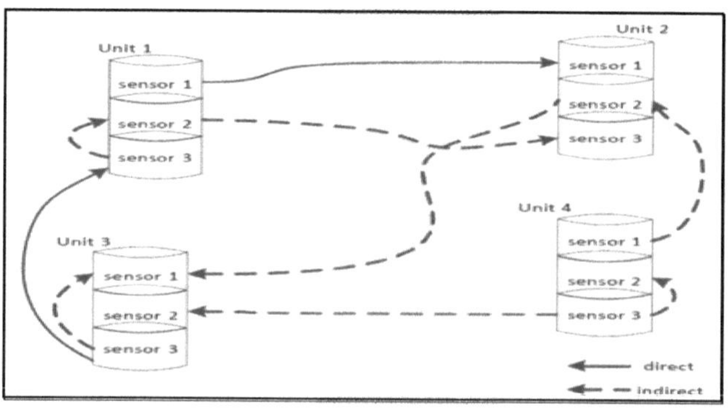


**Fig. 2.** Direct and indirect relationships in the network [2]

A direct relationship exists among couple of sensors of the same type like is usually temperature vs. temperature. If data streams $X_{i,j}$ and $X_{v,j}$, $i \neq v$ are correlated, then an arc linking the $j$-th sensor of unit $i$ with its counterpart of unit $v$ is introduced. For example, two clinometers insisting on the same connected structure are related; those deployed far apart probably are not. An indirect relationship can be introduced between two generic sensors by means of a third entity. Indirect relations are mitigated by the presence of compensation mechanisms. Information useful for the analysis must be extracted before compensation takes place.

In reality, direct and indirect relationships introduce a functional constraint among couples of sensors. Denote by $f\{(i,j),(u,v)\}$ the functional relationship between the generic $j$-th sensor of unit $i$ and the $v$-th sensor of unit $u$. The nodes of $G$ are the network sensors where the arcs represent the functional relationships among couples of those sensors. Given a network, not all the $(N{\times}M)(N{\times}M-1)$ relationships in $G$ are really relevant.

The *reduced dependency graph* is then derived from $G$ and defined as graph $GR=\{V,E\}$ where $V$ is the set of nodes of the graph representing the $N{\times}M$ sensors and $E$ a set collecting all arcs associated with functional relationships whose correlation is above a threshold. The level of dependency associated with relationship $f\{(i,j),(u,v)\}$ is here chosen to be the linear correlation index between two data streams $X_{i,j}$ and $X_{v,u}$. We remove from $G_R$ all isolated nodes. Fig. 3 shows the graph-based representation of the sensor network proposed in Fig. 2. We have 4 units; each unit is a sub-graph representing the sensors with bindings [2].

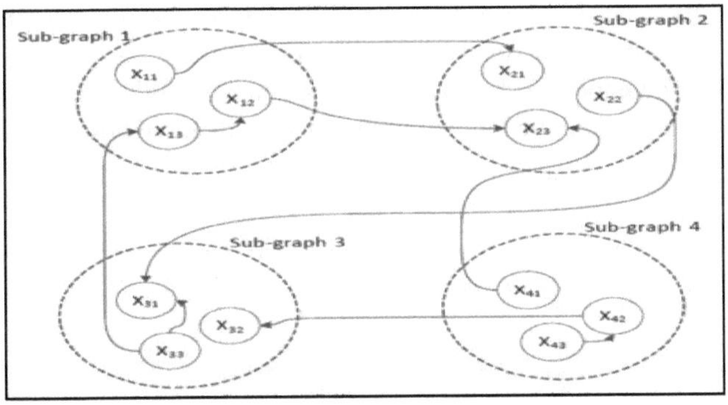


**Fig. 3.** The dependency graph of sensor network of Fig. 2 [2]

## B. Modeling the Relationship between Two Sensors by Hidden Markov Model

We assume that the relationship among couple of sensors $f\{(i,j)(u,v)\}$ can be modeled either as a time invariant (TI) dynamic system or as a finite sequence of TI dynamic systems satisfying the HMM hypotheses.

Let us imagine to model a $f\{(i,j)(u,v)\}$ with the Single-Input Single-Output (SISO) linear model. A given SISO locally approximates the output.

A training dataset is composed of $N_T$ *{input, output}* couples and a loss function whose minimization provides us an estimation of the optimal parameter.

Under the assumption that each *f{(i,j),(u,v)}* function satisfies the exponential stability for closed loop from [6].

It comes out that under the above assumption and a sufficiently large *N* the distribution underlying the parameter vectors is a multivariate Gaussian, with a mean and covariance matrix *P*.

HMM with parameters ruled by a mixture of Gaussians becomes a natural solution to approximate *f{(i,j),(u,v)}*. The HMM nodes of the represent in reality a probabilistic ensemble of used LTI models minimizing the model bias if a training set is sufficiently informative. By modeling parameters with HMM, we mitigate the effect of model bias and time variance provided that the training set is sufficiently informative as well and explores time variance and nonlinearity [2].

### 2.3 Cognitive Fault Diagnosis System

The proposed FDS is organized as the two-layer architecture of Fig. 4. The lower level is composed by a set of Change Detection Tests (CDTs) each of which monitoring the stationarity of a functional relationship associated with a couple of sensors in $G_R$. Each HMM-CDT works in the parameter space to detect variations in the relationship between two involved sensors. The CDT is not able to distinguish among changes induced by a fault in a sensor, an environmental change in *P* or a false positive generated by a model bias since such classes are indistinguishable. To address this issue the upper level of the FDS has been designed to be able to discriminate among faults, changes in *P* and false positives by exploiting information associated with the network graph $G_R$. The upper level of the FDS relies on a cognitive algorithm aggregating decisions and log-likelihood information provided by the HMM-CDT in the lower level [2].

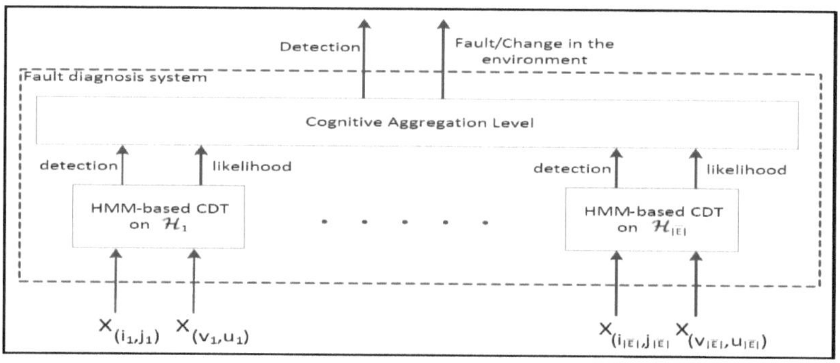


**Fig. 4.** Configuration of the proposed FDS [2]

### A. The HMM-Based Change Detection Test

The proposed HMM-CDT aims at evaluating, by means of HMM, the evolution over time of the estimated parameters approximating the relationship *f{(i,j),(u,v)}*; $X_{(i,j)}$ is the output and $X_{(u,v)}$ the input of the LTI. Estimated parameters are estimated on overlapping windows of $N_T$ data.

The HMM-CDT requires training of HMM devoted to model the relationship between sensors *(i,j)* and *(u,v)* is trained by the Baum-Welch algorithm [7].
During the operational life, the parameter is estimated on the *s*-th window of data and the log-likelihood that is computed with the Viterbi algorithm [8].

When the log-likelihood decreases below a threshold $T_h$, a change in the relationship is detected (the sequence of inputs is no more recognized by the learning machine). The threshold $T_h$ can be defined by the operator who exploits a-priori available information.

### B. The Cognitive Aggregation Level

The cognitive level aggregates the information coming from all sensor units to distinguish among faults, changes in *P* and false positives induced by model bias in the HMM-CDT. Differently from the HMM-CDTs executed sequentially, the cognitive aggregation level is activated only in response to a detection alarm raised by at least HMM-CDT. Detections and log-likelihoods of others CDTs are used to assess and, possibly, identify the change.

The motivating idea is that a change in *P* for a given type of sensors must be also perceived by a set of other CDTs, at least as a decrement in the log-likelihood values, which are not necessarily below the threshold. Differently, in the case of faults, only the CDTs associated with relationships that have either as input or output the faulty sensor are affected by the change. Finally, if a false positive occurs, other CDTs should not be affected.

To evaluate the reliability of the information coming from HMM-CDTs we introduce a reliability index for the HMM.

Weights are computed on the training set; the *weighted reduced graph* is the reduced graph that is augmented with the weight information.

Definitions are constructed as follows:

- Let $E^+$ be the set of functional relationships such that either the source or the target node of the arc is $X_{(i,j)}$.
- Let $E^-$ be the set of functional relationships such that either the source or the target node of the arc is $X_{(v,u)}$.
- Let $E^P$ be the set of functional relationships whose source or target node is neither $X(i_q,j_q)$ nor $X(v_q,u_q)$.

After a change detected in *f{(i,j)(v,u)}* the remaining functional relationships of the weighted reduced dependency graph are partitioned into sets $E^+$, $E^-$ and $E^P$. The reason for the partitioning is described as follows:

- A fault in sensor *X(i,j)* affects the relationships in $E^+$ but not in $E^-$ and $E^P$;
- A fault in sensor *X(v,u)* affects the relationships in $E^-$ but not in $E^+$ and $E^P$;

- A change in $P$ affects the relationships in $E^-$, $E^+$ and $E^P$;
- A model bias, affecting HMM, would mostly affect the relationship between *(i,j)* and *(u,v)* but not the relationships in $E^-$, $E^+$ and $E^P$ provided that approximating relationships are characterized by different bias contributions.

An example of partitioning is shown in Fig. 5a; in Fig. 5b a change is detected in relationship *f{(3,3)(1,3)}*. The definitions are

$$E^+ = \{f_{\{(3,3)(3,1)\}}\};$$
$$E^- = \{f_{\{(1,3)(1,2)\}}\};$$
$$E^P = \{f_{\{(1,1)(2,1)\}}, f_{\{(1,2)(2,3)\}}, f_{\{(3,1)(2,2)\}}, f_{\{(4,1)(2,3)\}}, f_{\{(3,2)(4,2)\}}, f_{\{(4,3)(4,2)\}}\}.$$

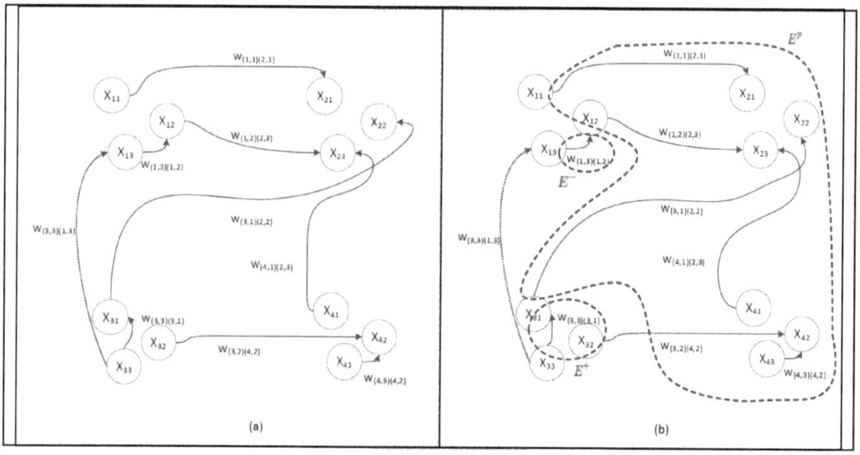


**Fig. 5.** The proposed cognitive aggregation level: a) the reduced weighted dependency graph; b) an example of arcs partitioning into $E^+$, $E^-$ and $E^P$ given a change detected in the functional relationship *f{(3,3),(1,3)}* [2]

Defined index of the data window where the HMM-CDT detected a change, the proposed aggregation level computes the normalized sum of the log-likelihoods, suitably weighted, of the arcs in $E^+$, $E^-$ and $E^P$.

The core of the cognitive aggregation level is thus the ability to compute $S^+$, $S^-$ and $S^P$ by exploiting information coming from all the relationships of the weighted reduced dependency graph. $S^+$, $S^-$ and $S^P$ measure how the change detected in the relationship *f{(i,j)(v,u)}* is perceived in other relationships. If a fault affects sensor *(i,j)*, $S^+$ should decrease, while $S^-$ and $S^P$ should not. Similarly, if a fault affects sensor *(u,v)*, $S^-$ should decrease, $S^+$ and $S^P$ not. If a change in $P$ occurs, $S^P$ should decrease as well as $S^+$ and $S^-$.

To detect decreases in $S^+$, $S^-$ and $S^P$ we rely on a simple thresholding mechanism that calculates thresholds $T^+$, $T^-$ and $T^P$ which can be scaled by a coefficient factor $c_2$ to increase the robustness with regard to false positives. We suggest selecting $c_2 > c_1$ since we want to detect decreases in the likelihood that did not yet raised an alarm. In reality, if we consider $c_2 \leq c_1$ we would require that the weighted average of the likelihoods decreases below the weighted average of the thresholds for change detection $T_h s$ but this is nonsense since relationships in $E^+$, $E^-$ and $E^P$ did not detect a change yet.

To sum up, the cognitive aggregation level acts as follows:

- If $S^P$ decreases below threshold $T^P$, a change in $P$ is identified;
- If $S^P>T^P$ and $S^+<T^+$ (or $S^-<T^-$), a fault in sensor $X_{(i,j)}$ (or in $X_{(v,u)}$) is detected;
- If $S^P>T^P$ and $S^+>T^+$ and $S^->T^-$, a false positive which is induced by a model bias is detected as the third alternative.

If both $S^+$ and $S^-$ are above their respective thresholds, we can raise the alarm fault in either $X_{(i,j)}$ or $X_{(v,u)}$ but we cannot isolate the affected sensor since not enough information is available.

## 3 Conclusions

This paper has suggested improvement of the Measureserver® module. Specifically, the goal was to minimize the influence of possible instabilities and faults in Measureserver® module of ISES RE, ill affecting remote RE based on ISES. We have analyzed several AI solutions to avoid such unwanted occasional events.

Our conclusions may be formulated as follows.

1. The experimentation based on ISES RE is a new approach of teaching and learning in comparison with traditional forms.
2. The cognitive FDS is a suitable AI approach integrated as a new component into the Measureserver® to avoid occasional faults incoming from sensors in the RE physical hardware, which negatively affect the ISES RE functioning.
3. The CDT is an important component of the cognitive FDS for detecting variations in functional relationships among couples of sensors. HMM may be effectively used as the basis for CDT for the purpose of working in the parameter space of linear time invariant dynamic systems approximating the relationship between two involved sensors over time.
4. We envisage building such the diagnostic system between the Measureserver® module and the client of the RE. This approach should secure a reliable connection in case of faults occurrence.

**Acknowledgment.** The paper was published thanks to the Grant of the Internal Agency of TBU No IGA/FAI/2014/044, Grant of the Kega Agency No. 10/TU/13 and Grant of the agency APVV project No. APVV 0096-11. The following grant projects are acknowledged for the financial support provided for this research: - the Grant Agency of the Czech Republic, GACR P103/13/08195S, - partially supported by the Grant of SGS No. SP2014/159, VSB, Technical University of Ostrava, Czech Republic, " Development of human resources in research and development of latest soft computing methods and their application in practice", - project, reg. No. CZ.1.07/2.3.00/ 20.0072 funded by Operational Programme Education for Competitiveness, cofinanced by ESF and the Czech Republic State budget.

## References


[1] Zeman, P.: Software environment for control of remote experiments. Ostrava: VŠB-Technical University of Ostrava (2011)

[2] Alippi, C., Ntalampiras, Roveri Manuel, S.: A Cognitive Fault Diagnosis System for Distributed Sensor Networks. IEEE Transactions on Neural Networks and Learning Systems 24(8), 1213–1226 (2013), http://ieeexplore.ieee.org/stamp/stamp.jsp?arnumber=06502725 (read on April 8, 2014), ISSN 2162-237X. doi: 10.1109 / TNNLS.2013.2253491

[3] Kadous, M.: Hidden Markov models. Australia: School of The University of New South Wales, Computer Science and Engineering (2002), http://www.cse.unsw.edu.au/~waleed/phd/html/node34.html (read on April 8, 2014)

[4] Kadous, M.: Advantages of HMMs. Australia: School of The Uni-versity of New South Wales, Computer Science and Engineering (2002), http://www.cse.unsw.edu.au/~waleed/phd/html/node35.html (read on April 8, 2014)

[5] Venkatasubramanian, Rengaswamy, Yin, Kavuri: A review of process fault detection and diagnosis: Part i: Quantitative model-based methods, Computers Chemical Engineering (2003), http://www.sciencedirect.com/science/article/pii/S0098135402001606 (read on April 8, 2014)

[6] Ljung, Caines: Asymptotic normality of prediction error estimators for approximate system models. In: 1978 IEEE Conference on Decision and Control including the 17th Symposium on Adaptive Processes, vol. 17, pp. 927–932 (January 1978)

[7] Rabiner, Juang: An introduction to Hidden Markov models. IEEE ASSP Magazine, 4–15 (January 1986)

[8] Durbin, Eddy, Krogh, Mitchison: Biological Sequence Analysis: Probabilistic Models of Proteins and Nucleic Acids. Cambridge University Press (1998), http://f3.tiera.ru/2/Cs_Computer%20science/CsBi_Bioinformatics/Durbin%20R.,%20Eddy%20S.R.,%20Krogh%20A.,%20Mitchison%20G.%20Biological%20sequence%20analysis%20%28CUP,%201998%29%28ISBN%200521629713%29%28O%29%28371s%29_CsBi_pdf (read on April 8, 2014)

# Artificial Intelligence Elements in Data Mining from Remote Experiments

Lukas Pálka[1], Franz Schauer[1,2], and Ivan Zelinka[1,3]

[1] Tomas Bata University in Zlín, Faculty of Applied Informatics, CZ-760 05 Zlín, Czech Republic
lukas.palka@prerov.eu
[2] University of Trnava, SK-918 43 Trnava, Slovak Republic
[3] VŠB-Technical University of Ostrava, 17. listopadu 15/2172, 708 33 Ostrava-Poruba, Czech Republic
ivan.zelinka@vsb.cz

**Abstract.** In spite of the fact that remote laboratories have been existing for at least three decades, virtually no attention has been devoted to the accumulated data analysis of this new means of education. The paper deals with the data analysis, gathered in the Datacentre (DTC) implemented with the Laboratory Management System (RLMS), connected in turn to remote laboratories and remote experiments. In particular, we concentrate and describe a new model of experiment data analysis, based on the principles of artificial intelligence, based on the criterion function in need. The leading idea of the model functioning is during the procedure of rig(s) recognition i.e Data weighting: Data recognition: Data preparation: Phenomenon modelling: Model and measurement data comparison: Result deployment, where the artificial intellingence is integrated with steps of Data weighting by association and regression using neuron network. Benefit of the suggested method is its speed and efficiency and thus using it for the optimization of individual remote experiments and ther efficiency. Paper may serve as an inspiring source for the development in the field of remote laboratories, but it may influence in the similar areas of data mining.

**Keywords:** ISES, analysis data, Measureserver, remote experiment, data mining

## 1 Remote Laboratories and Laboratory Management Systems – State of the Art

At the present stage of the development of ICT there are already plenty simulations and remote experiments for science and education purposes. Remote experiments and informatics resources are tools that are closely related and definitely need to process and store substantial amounts of data. A great deal of attention worldwide at universities and teaching institutions has been devoted to e-laboratories offering access to various real world remote experiments. Data, used with remote laboratories (RL), may have the form of simple queries, data analysis, comparative analysis, data mining for

I. Zelinka et al. (eds.), *Nostradamus 2014: Prediction, Modeling and Analysis of Complex Systems*, Advances in Intelligent Systems and Computing 289,
DOI: 10.1007/978-3-319-07401-6_42, 

associative analysis, extrapolation or predictive trend analysis. Surprisingly, in spite of the fact the RL have been existing for at least three decades, virtually no attention has been devoted to the the vital subject of security. In this connection we will use the term data warehouse (DW), referring to a complex system that allows to collect, organize, store and share consolidated data from all available operating systems, optimized for reporting, analysis, and data archiving. Users exploit the data warehouse for reporting, in this respect a synonymous for business intelligence technology, based on the use of the data and its accumulation, preservation and presentation. The working principle of the DW is that the data we need to process are first stored into the database in the raw state, then follows the data classifying and only subsequently comes their analysis for further use, e.g. experiments evaluation or research.

The present paper deals with the Data analysis using artificial intelligence in remote laboratories, especially that with Remote Laboratory Management System (RRLMS) for Internet School Experimental System (ISES).

The layout of the paper is following. First, the basic scheme of a net of remote experiments, built as the finite-state machine (FSM), based on the ISES physical hardware is described and the state of the art of its control program compiling is shortly described to introduce the type of the data generated and transferred by remote rigs [1].

The next chapter is devoted to the process of data mining and knowledge discovery models. The following chapters are then focused on the Methodology for Data Mining of the datawarehous (DW) of a university datacentre (DTC) with RRLMS for RL.

The following chapters are devoted to the Algorithms by Task of artificial intelligence in general (Ch.5) and next chapter to the particular appropriateness using these algorithms for remote laboratories in particular (Ch.6). The final chapter (Ch.7) is oriented on prospective future running analysis of data remote experiments, followed by conclusions.

## 2 Laboratory Management System – Tools Used

RRLMS is a system that manages remote experiments (RE) worldwide irrespective of their interfaces, but ISES RE manages with a great advantage, including diagnostics of its diagnostics. ISES RE generates a lot of data to be processed, analyzed and properly evaluated. For this purpose, it is appropriate to use the database processing, together with analytical functions needed. The data analysis of remote experimentation is the key issue of gaining and preserving knowledge from them and especially artificial intelligence (AI) solutions can process data in this respect, gaining valuable information.

As the RRLMS will process extreme bulks of data, a serious problem arisis stemming from the analysis of research bulks data [1]. In this article, we will focus primarily on data analysis from RE using AI solutions.

As all data are generated in RE by the ISES in general and Measureserver module in particular, let us give the short outline of the ISES system and RE built by ISES.

## 2.1 Internet School Experimental System – ISES

ISES is a powerful tool for RE control, data acquisition, and their collecting and processing in real time. Let us mention the basic features of the ISES system, more detailed description may be found elsewhere [2,4]. The basis of the system is ISES board, which is available in several versions, differing depending on the number of inputs/outputs and also on type of communication with the control PC (by PCI card, USB connector, Wi-Fi). To the board are may be plugged in various ISES sensors like A-meter, V-meter, thermometer, position sensor, ohmmeter, load cell, anemometer, microphones, sonar, light gate, pH meter, conductivity meter, heart rate monitor, etc. [2,4].

Due to its maximum signal transfer frequency (100 kHz) the system allows the study of sounds or other dynamic signals. The system allows simultaneous measurement, processing and displaying data via maximum eight input channels, as well as process control via two analogue and two binary output channels. But the uniqueness of this system is its possibility of using the same equipment both for experiments in the laboratory (so-called hands on experiments) and also for their RE versions [2,4]. The layout arrangement of the RE is in Figure 1. The most important component is the Measureserver module, functioning as finite-state machine (FSM) controlled by the controlling program of the PSC script file.The main feature of the Measureserver, is to communicate with the physical hardware and to check the setup of the ISES panel and its sensors/meters and to take care about their data collection and processing. Other parts of the system are ImageServer for life view of the remote experiment, Web server for the communication between RE and the client. Also, aprt of the RE is the communication web page as the interface communicating with the RE over the Internet by the client.

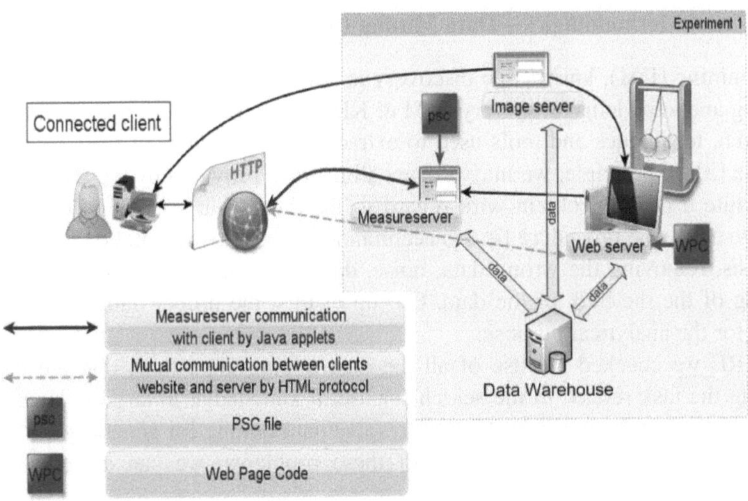


**Fig. 1.** Schematical arrangement of ISES remote experiment

The invevitable part of the RE system is the data warehouse for the storage of data for all above systems. It is a centralized repository service to Measureserver, web server, image server and other components of the solution.

In this article we will discuss this last part of the system with respect of data analysis using artificial intelligence but not only from the perspective a single RE, but of the whole RLMS.

Let us next discuss several methods, especially from AI branch, used for the RE data storing and processing. As noted earlier, to find the proper tool for the analysis of data from RE we need several AI database solutions.

### 2.2 Data Mining in RLMS Using SQL Server

Data Warehouse is a central unificated software (SW) system of database services, simplified for the use in remote laboratories, used for storage and analysis of general data, data storage for the web server, the Measureserver and the reservation system. This system includes a number of sophisticated instruments which together will apply the analysis of data obtained from remote experiments. The following part of the system is therefore necessary to describe and show how the data analysis works.

A database is an organized collection of the data for every part of RLMS. The data are typically organized to perceive the relevant aspects of reality in a way that supports processes requiring this information.

Service that provides databases is called SQL Server. SQL Server is a database platform for large-scale online transaction processing (OLTP), data warehousing, and e-commerce applications; it is also a business intelligence platform for data integration, analysis, and reporting solutions.

### 2.3 Data Mining and Knowledge Accumulation Process Models and Methodologies – Data Mining Concepts

Data mining (DM), knowledge discovery in databases (KDD) and the combined data mining and knowledge discovery (DM & KD) are terms used to refer to the results of research, techniques and tools used to extract useful information from large volumes of data [3]. In principle, we may process all the data provided by the RE, but this may constitute a heavy problem with regard to the data volume. This is the reason, we need to apply the procedure first to accommodate the experimental data to subsequent analysis, removing the wrong data, noise, data misprints, etc., with simultaneous lowering of the the bulk of the data. On top of this, the proper data conversion takes place for the analytical purposes.

In RE we checked the use of all the data mining algorithms that are capable of solving the task related to the search for useful knowledge in large volumes of data. Data mining can be in principle used to solve data mining for generally all laboratory data outputs. Based on the nature of these problems, we can arrange them into following data mining tasks.Let us develop the models of the data mining and knowledge discovery process and present and methodologies that provide the condensed knowledge from the data of stored in RLMS.

From the viewpoint of data mining methodologies and process models, the year 2000 marked the most important milestone, when the Cross-Industry Standard Process for Data Mining (CRISP-DM) was first published. CRISP-DM is the most widely used methodology for developing data mining projects. Since then, it is considered to be the standard in the field [3].This process describes the activities that must be done to develop a data mining project. Every activity is composed of individual tasks. For every task, generated outputs needed and inputs are detailed [3]. For the remote labs we chose an addopeted CRISP-DM methodology as it applies to the evaluation of the process model and to the model we want to explain furtheer.This model is ideal for the data analysis that posses predetermined structure (in XML format) as is the case in remote laboratories. Such data structure can be fundamentally changed without information loss and thus may be easily integrated into arbitrary process of knowledge acquiring.

### 2.4 Data Mining Methods by Neural Networks – Possibilities

After finishing the process of knowledge accumulation, we may approach to the DM process. For these ends, let us turn to the methods of AI using NNM, which analyze the data with respect to the knowledge contained. In the paper we are not in position to describe all the methods that we considered for use in ISES remote laboratories data analysis. Let us limit ourselves only to one significant method using AI - Prediction. In following we describe the AI method Prediction shortly and in the next paragraph want to deliver a simple model example.

## 3 Neural Networks in ISES RL Data Mining

In the previous chapter we choose the AI model Prediction for data mining from ISES RL, prediction. Once trained the Microsoft Neural Network model data Prediction looks like that in [5].

During prediction, the input nodes are populated with values derived from the input data. The values are linearly combined with the edges leading from input nodes to the middle (hidden) layer of nodes and the input vector is translated to a new vector, which has the same dimension as the hidden layer. The translated vector is then "activated" using the tanh() function for each component. The resulting vector goes through a similar transformation, this time from the hidden layer to the output layer. Therefore, it is linearly transformed to the output layer dimensionality (using the weight of the edges linking hidden nodes to output nodes). The result of this transformation is activated using the sigmoid function and the final result is the set of output probabilities. These probabilities are normalized before being returned as a result (in a call like PredictHistogram).

### 3.1 Practical Application Neural Network Using the Tools DMX Queries in MS Data Mining of OLAP in Remote Laboratories

Let us show the process of the data from individual RE analyzing within RLMS. In practice, the data for the analysis and the theoretical data are available as well as we know the way of entering the measured data into the system for the analysis. Using the DMX language used as a normal query in T-SQL, simplifies the evaluation of our data from remote laboraoties. Of course, it requires quite a good knowledge of the T-SQL topic, but for the present purpose, let us present only a simple example. Let us suppose we know the structure of the data arranged in tables from RE measurements and we have the data described by the metadata arrangement with the results of trials. These data will be analysed by AI, with the goal to get the predicted errors of measurements of the RE. This errors may be caused by the factors that we are not able to observe, but they can only be derived on the basis of input variables. E.g new users of RLMS will undertake unexpected trials resulting in unexpected errors. For the minimization of such errors the new users of RE will undergo testing and and experimental data will be specially chacked.

The easiest way to build a prediction query on a neural network model is to use the Prediction Query Builder, available on the Mining Prediction tab of Data Mining Designer in both SQL Server Management Studio and SQL Server Data Tools (SSDT). You can browse the model in the Microsoft Neural Network Viewer to filter attributes of interest and view trends, and then switch to the Mining Prediction tab to create a query and predict new values for those trends.

**Prediction Example of the Development of Faults on RELMS**

By expecting the data, collected during longer periods of time, you will find certain days of the week to have a high score of errors from RE and this logically correlate with the increased number of clients connected. Using the prediction query on the model to test a "what if" hypothesis and ask if by increasing the number of new clients on a low-volume day would result in the increased error number of the system. To do this, we create a query

```
SELECT Predict([use_rigs Default NN].[Number of errors]) AS [Pre-
dicted errors],
  PredictProbability([use_rigs Default NN].[Number of errors]) AS
[Probability]
  FROM [use_rigs Default NN]
  NATURAL PREDICTION JOIN
  (SELECT 'Tue.' AS [Day of Week],AS t
```

From the graphical output shown in Figure 2 showing the measured data errors intensity (measured during 5 days), we can predict the evolution of probability of errors in subsequent three days ( the red colour). The whole procedureserves for to the possibility of predicting the average number of errors in days to come. Data quality is an important factor for working with them and it is an integral part of the RLMS system work. The model example of errors prediction in Fig. 2 may serve e.g for the finding of the sources of the hidden causes of the falts in the system from the measured data and their subsequent elimination. This, in turn, may vastly improve the efficiency of the whole RLMS in long time perspective.

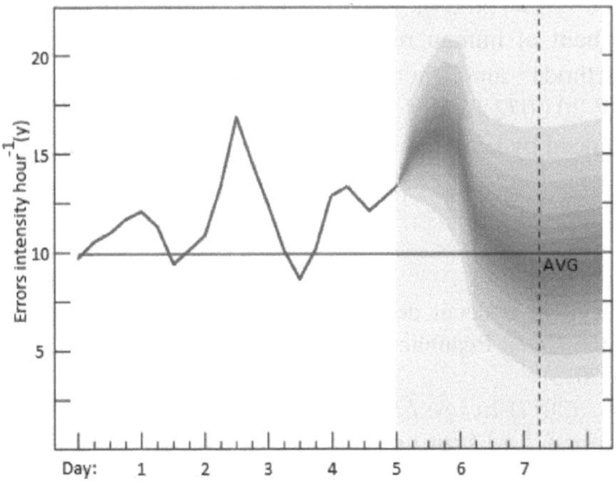


**Fig. 2.** Time representation of the measured errors occurrence intenisity ( in 5 cosecutive days) and predicted development probability of average number of errors in 3 days to come

# 4 Conclusions

The role of RLMS systems, introduced at present in USA in project e-Lab [6] and in Australia in project Sahara[7] and our built RLMS REMLABNET [8] that should serve similar goals in management and monitoring of remote experiments in frame of EU, has two main goals, i.e improve reliability and efficiency of the remote labs network. We want to contribute by this work to both aims by data mining of the data, accumulated during the functioning of the RRLMS. For this purpose we suggest to use the element sof artificial intellingence both for the refining the accumulated data using the suitable criterion function. As an example we present the prediction of errors from the past measured data, serving for the elimination of the hidden sources of the most execcive faults in the system.

The conclusions may be summarized as follows.

1. The whole process runs in three subsequent steps: accumulated data filtering, data analysis and prediction (last two steps may be denoted by data mining),
2. For the purpose of data mining the elements of artificial inteligence are very suitable and we suggest its use for both collected data analysis and prediction according the criterion function corresponding the goal of the process,
3. It turned out that the rough data filtering is absolutely necessary before the real data mining may start.
4. The whole suggested process may be used in many directions for optimizing the RLMS function.

**Acknowledgment.** The paper was published thanks to the Grant of the Internal Agency of TBU No IGA/FAI/2014/042, Grant of the Kega Agency No. 10/TU/13 and Grant of the agency APVV project No. APVV 0096-11. The following grant projects are acknowledged for the financial support provided for this research: - the Grant Agency of the Czech Republic, GACR P103/13/08195S, - partially supported by the Grant of SGS No. SP2014/159, VSB, Technical University of Ostrava, Czech Republic, " Development of human resources in research and development of latest soft computing methods and their application in practice",- project, reg. No. CZ.1.07/2.3.00/ 20.0072 funded by Operational Programme Education for Competitiveness, cofinanced by ESF and the Czech Republic State budget.

## References

1. The whole system is detail described in the project proposal Submitted Project Grant Agency of the Czech Republic: Informatics Means for Grid of e-Laboratories – Project Remlabnet (2013)
2. Krbeček, M.: (2012) http://Possible_utilization_of_the_artificial_intelligence_ elements_in_the_creation_of_remote_experiments (cited June 26, 2013)
3. A survey of data mining and knowledge discovery process models and methodologies. The Knowledge Engineering Review (2010), http://www.researchgate.net/../d912f50a8e22278bf8.pdf, (cited August 15, 2013), doi: 10.1017/S0269888910000032
4. Internet School Experimental System iSES. Internet School Experimental System iSES (2012), http://www.ises.info/index.php/en (cited August 11, 2013)
5. Microsoft Neural Network — Step-by-step Predictions, http://www.bogdancrivat.net/dm/archives/36 (cited December 31, 2013)
6. Harward, V.J., et al.: The iLab Shared Architecture: A Web Services Infrastructure to Build Communities of Internet Accessible Laboratories. Proceedings of the IEEE 96, 931–950 (2008)
7. Lowe, D., Murray, S., Lindsay, E., Dikai, L.: Evolving Remote Laboratory Architectures to Leverage Emerging Internet Technologies. IEEE Trans. Learning Technol. 2, 289–294 (2009)
8. Schauer, F., Krbecek, M., Beno, P., Gerza, M., Palka, L., Spilaková, P.: REMOTELABNET - open remote laboratory management system for e-experiments. In: Proceedings of the REV Conference, Sydney (2014)

# Fractal Models of Atoms and Molecules

Pavel Ošmera senior[1] and Pavel Ošmera junior[2]

[1] University of Technology VUT in Brno, Czech Republic
[2] St. Anne's University Hospital Brno, Pekarska 53, Brno, Czech Republic

**Abstract.** This paper is an attempt to attain a new and profound model of the nature's structure using a vortex-ring-fractal theory. Scientists have been trying to explain some phenomena in nature that have not been explained so far. The aim of this paper is the vortex-ring-fractal modeling of elements in the Mendeleev's periodic table, which is not in contradiction to the known laws of nature. We would like to find some acceptable structure model of the hydrogen as a vortex-fractal-coil structure of the proton and a vortex-fractal-ring structure of the electron. It is known that planetary model of the hydrogen atom is not right. The classical quantum model is too abstract. Our imagination is that the hydrogen is a levitation system of the proton and the electron. Structures of hydrogen and carbon atoms and a hydrogen molecule are presented too. Our goal is combination of the basic principle of grammatical evolution with vortex-ring-fractal structures of atoms to create new molecule nano-structures. This approach combines knowledge of evolutionary optimization with physical chemistry.

**Keywords:** fractal models, vortex, ring, fractal, graphene, vortex-fractal-ring theory.

## 1 Introduction

A scientific theory is a well-substantiated explanation of some aspect of the natural world that is acquired through the scientific method, and repeatedly confirmed through observation and experimentation. Typically, before a scientific theory can be created, a hypothesis must be developed which is a supposition or proposed explanation that is formed on the basis of limited evidence as a starting point for further investigation. In physics, the term theory is generally used for a mathematical framework, which is capable of producing experimental predictions for a given category of physical systems.

Fractals seem to be very powerful in describing natural objects on all scales. Fractal dimension and fractal measure are crucial parameters for such description. Many natural objects have self-similarity or partial-self-similarity of the whole object and its part [1], [2].

Most of our knowledge of the electronic structure of atoms has been obtained by the study of the light given out by atoms when they are exited. The light that is emitted by atoms of given substance can be refracted or diffracted into a distinctive pattern of lines of certain frequencies and create the line spectrum of the atom [3], [4].

I. Zelinka et al. (eds.), *Nostradamus 2014: Prediction, Modeling and Analysis of Complex Systems*, Advances in Intelligent Systems and Computing 289,
DOI: 10.1007/978-3-319-07401-6_43, 

The careful study of line spectra began about 1880. The regularity is evident in the spectrum of the hydrogen atom. The interpretation of the spectrum of hydrogen was not achieved until 1913. In that year the Danish physicist Niels Bohr successfully applied the quantum theory to this problem and created a model of hydrogen. Bohr also discovered a method of calculation of the energy of the stationary states of the hydrogen atom, with use of Planck's constant h. Later in 1923 it was recognized that Bohr's formulation of the theory of the electronic structure of atoms to be improved and extended. The Bohr's theory did not give correct values for the energy levels of helium atom or the hydrogen molecule - ion $H_{2+}$ , or of any other atom with more than one electron or any molecule [6].

During the two-year period 1924 to 1926 the Bohr's description of electron orbits in atoms was replaced by the greatly improved description of wave mechanics, which is still in use and seems to be satisfactory. The discovery by de Broglie in 1924 that an electron moving with velocity v has a wavelength $\lambda=h/m_{ev}$ [4], [6]. The theory of quantum mechanics was developed in 1925 with the German physicist Werner Heisenberg. Early in 1926 an equivalent theory, called wave mechanics, was independently developed by Austrian physicist Ervin Schroedinger. Important contributions to the theory were also made by the English physicist Paul Adrien Maurice Dirac. The most probable distance of the electron from the nucleus is thus just the Bohr's radius $r_B$; the electron is, however, not restricted to this distance. The electron is not to be thought of as going around the nucleus, but rather as going in and out, in varying directions, so as to make the electron distribution spherically symmetrical [2].

Matter is composed of tiny atoms. All the atoms of any elements are identical: they have the same mass and the same chemical properties. They differ from the atoms of all other elements. Twenties-century X-ray work has shown that the diameters of atoms are of the order 0.2 nm ($2x10^{-10}$ m). The mass and the positive charge are concentrated in a tiny fraction of the atom, called nucleus. The nucleus consists of protons *(p)* and neutrons *(n)*. Protons and neutrons are made up of smaller subatomic particles, such as quarks. Both protons and neutrons have a mass approximately 1840 times greater than an electron *(e)*. The more energy an electron has, the further it can escape the pull of the positively charged nucleus. Given sufficient energy, an electron can jump from one shell to higher one. When it falls back to a lower shell, it emits radiation in the form of photons [3]. [4], [6].

Main ideas and differences between a classical and the new vortex-ring-fractal model are presented on Fig.1 [5]. [10].

Allotropy is the property of some chemical elements to exist in two or more different forms, known as allotropes of these elements. Allotropes are different structural modifications of an element; the atoms of the element are bonded together in a different manner. For example, the allotropes of carbon include diamond (where the carbon atoms are bonded together in a tetrahedral lattice arrangement), graphite (where the carbon atoms are bonded together in sheets of a hexagonal lattice), graphene (single sheets of graphite), and fullerenes (where the carbon atoms are bonded together in spherical, tubular, or ellipsoidal formations) [6].

Previous papers, power point presentations and video demonstrations can be found on: http://www.pavelosmera.cz

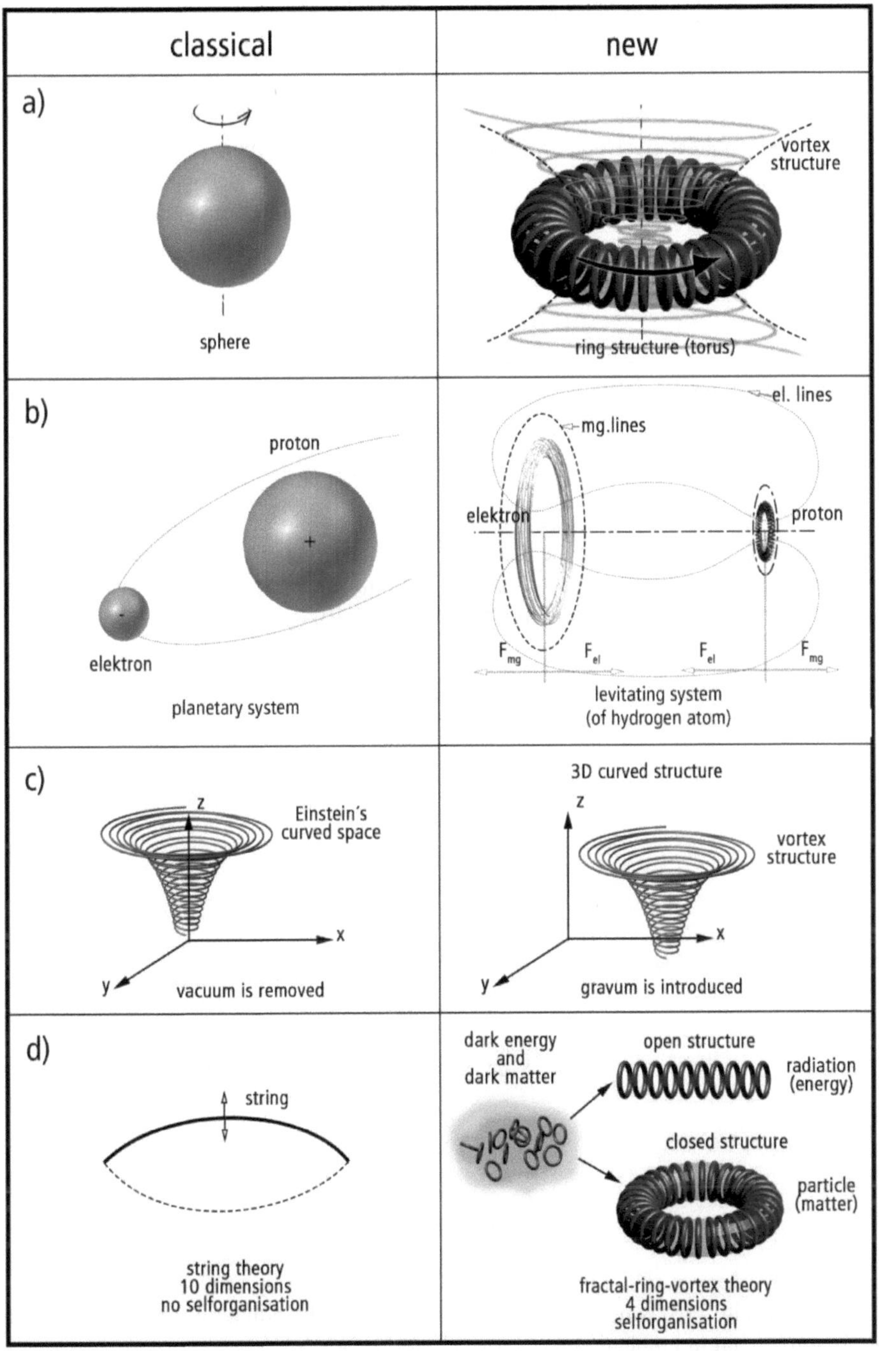


**Fig. 1.** Main ideas and four differences between a classical and the new VFR model (VFR is Vortx-Fractal-Ring theory).

## 2 The Spin of the Electron

Spin is a fundamental property of nature like electrical charge or mass. Spin comes in multiples of 1/2 and can be + or −. Protons, electrons, and neutrons possess spin. Individual unpaired electrons, protons, and neutrons each their structure possesses a spin of 1/2.

It was discovered in 1925 that the electron has properties corresponding to its spin S. The spin of the electron is defined as angular momentum $\vec{S}$:

$$\vec{S} = m_e(\vec{r}_e \times \vec{v}_e) \tag{1}$$

For the spin on axis z:

$$S_z = \pm N \frac{m_e}{N} r_e v_e = \pm m_e r_e v_e \tag{2}$$

where me is the mass of the electron, $r_e$ is the radius of the electron, $N$ is number of substructures (sub-electrons) inside the structure of the electron and $v_e$ is velocity of sub-electrons. For quantum number $n=1$ [2], [3], [6] :

$$r_{e1} = \frac{e^2}{8\pi\varepsilon_0 m_e} \cdot \frac{1}{v_{e-\max}^2} \tag{3}$$

$$v_{e-\max} = \frac{e^2}{2\varepsilon_0 h} \tag{4}$$

$$S_z = \pm m_e \frac{e^2}{8\pi\varepsilon_0 m_e} \cdot \frac{1}{v_{e-\max}^2} v_{e-\max} = \pm m_e \frac{e^2}{8\pi\varepsilon_0 m_e} \cdot \frac{1}{v_{e-\max}} =$$

$$= \pm m_e \cdot \frac{e^2}{8\pi\varepsilon_0 m_e} \cdot \frac{2\varepsilon_0 h}{e^2} = \pm \frac{1}{2} \cdot \frac{h}{2\pi} = \pm \frac{1}{2}\hbar = m_s \hbar, \tag{5}$$

where:

$$m_s = \pm \frac{1}{2} \tag{6}$$

The result in (5) is in coincidence with the generally equation for the spin, where $m_s$ is spin quantum number.

## 3 Quantum Model of the Electron

On the circumference of the double loop with the radius $r_e$ (see Fig.2) have to be n of de Broglie's wavelengths λ (n is quantum number) which are created by sub-electrons with mass $m_e/N$ [8]:

$$2\cdot 2\pi r_e = 4\pi \frac{e^2}{8\pi\varepsilon_o \frac{m_e}{N}} \frac{1}{v_e^2} = n\lambda = n\frac{h}{\frac{m_e}{N} v_e} \tag{7}$$

$$\frac{e^2}{2\pi\varepsilon_o} \frac{1}{v_e} = nh \tag{8}$$

where $v_e$ is velocity of the sub-electron with mass $m_e/N$ on quantum level $n$:

$$v_{en} = \frac{1}{n} \frac{e^2}{2\varepsilon_o h} \tag{9}$$

## 4 Model of the Hydrogen Atom and Hydrogen Molecule

In a new model of the hydrogen atom with a levitating electron [8], [9], [10] there is an attractive (electric) force $F_+$ and a (magnetic) repellent force $F_-$ :

$$F = F_+ - F_- = \frac{e^2}{4\pi\varepsilon_o}\left(\frac{1}{d^2} - \frac{d_o^2}{d^4}\right) \tag{10}$$

where $d$ is a distance between the electron and the proton. The hydrogen atom can have the electron on the left side or on the right side of the proton. The hydrogen molecule has two electrons between two protons (see Fig.3). The attractive force $F_+$ is Coulomb's force. The repellent force $F_-$ is caused with magnetic fields of the proton and the electron. The Bohr's radius $r_B$ has the same size as the distance $d_0 = 5.29\text{x}10^{-11}$ m in our vortex-fractal-ring model [6].

Energy $E_{rn}$ of rotation of the electron on quantum level $n$:

$$E_{rn} = \frac{1}{2} \frac{m_e}{N^2} N^2 \cdot v_{en}^2 = \frac{1}{n^2} \frac{m_e e^4}{8\varepsilon_o^2 h^2} \approx \frac{1}{n^2} 13.6eV \tag{11}$$

For quantum number $n=1$ ionization energy $E_{io}$ :

$$E_{io} = \frac{m_e e^4}{8\varepsilon_o^2 h^2} \approx 13.6eV \tag{12}$$

Energy $E_n$ of levitation from [4] for levitation distance $d_{on}$ on level $n$:

$$d = d_{on} = n^2 d_o \tag{13}$$

where

$$d_o = \frac{\varepsilon_o h^2}{\pi m_e e^2} = r_B \tag{14}$$

**Fig. 2.** Structure of the electron with spin ½

# 5 Graphene

The most interesting form to date is undoubtedly the graphene sheet. Graphene promises to be a super-material. For its successful isolation, scientists Andre Geim and Konstantin Novoselov, were awarded the 2010 Nobel prize for physics. Each sheet—one atom thick—consists of carbon atoms arrayed as hexagons, every hexagon completely surrounded by other hexagons, resembling the six-sided cells of a honeycomb—a kind of atomic chicken wire (see Fig.5 and Fig.8).

The carbon atom consists from two parts (every part has 3 protons and 3 neutrons, see Fig.6). A plane arrangement of electrons in the graphene (see Fig.7) has all electrons above the structure with parallel axes. These electrons have no gyroscopic problem with movement in the plane that is parallel with the plane of the graphene. It explains high mobility of these electrons and high conductivity of the graphene.

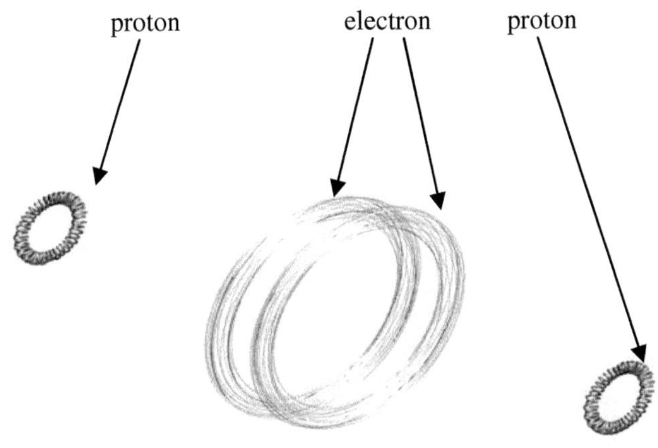


**Fig. 3.** Topological structure of the hydrogen molecule

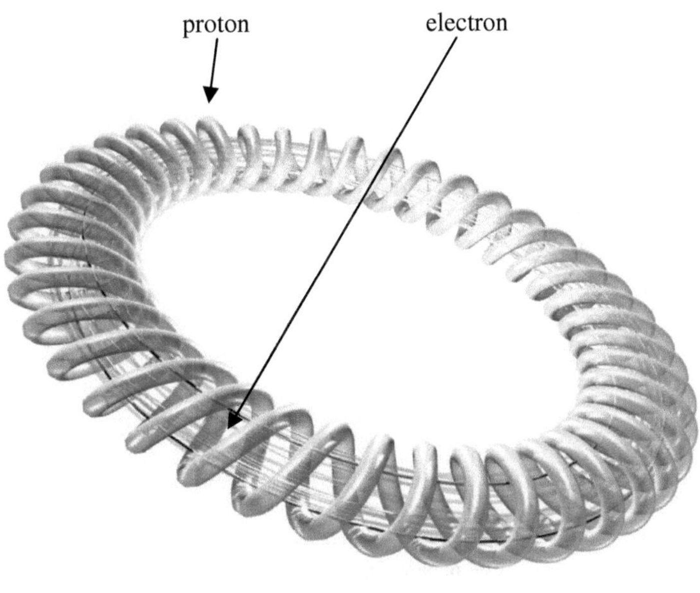


**Fig. 4.** Topological structure of the neutron

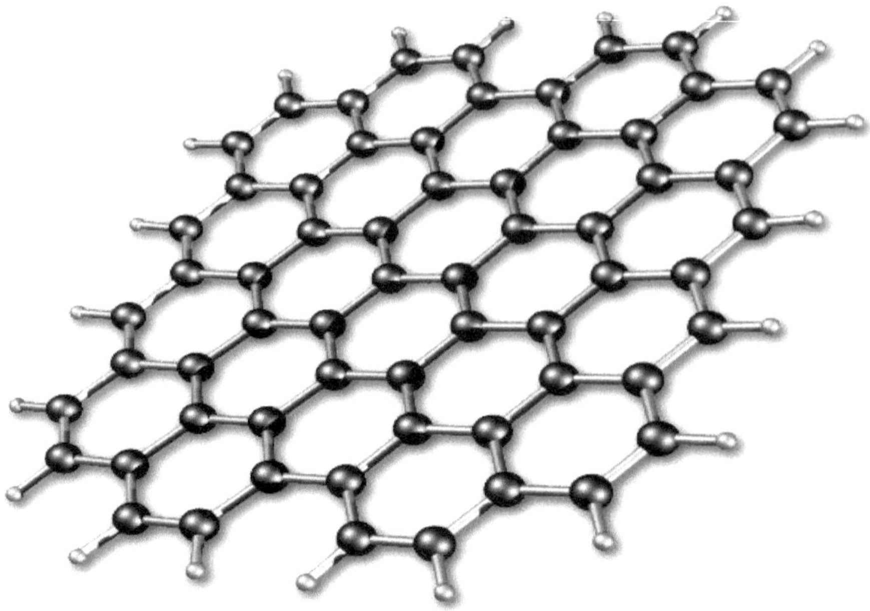

**Fig. 5.** Classical model of the grapheme

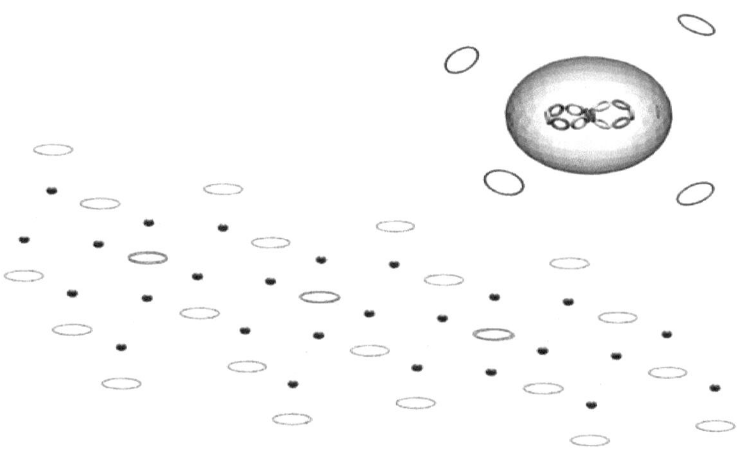

**Fig. 6.** New ring model of the graphene

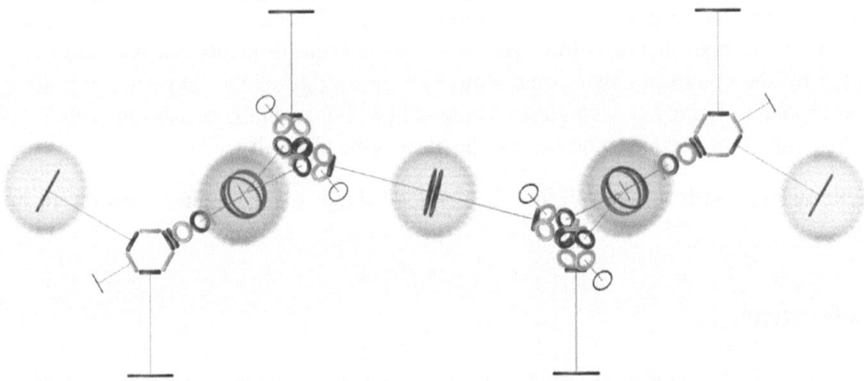

**Fig. 7.** Graphene as one atom layer with outer electron on parallel axes

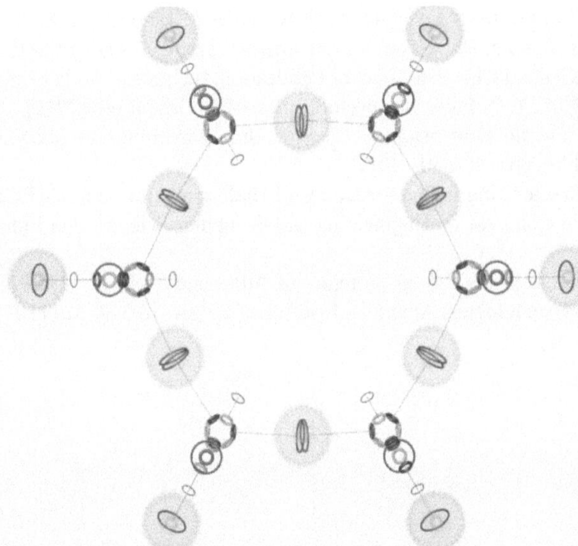

**Fig. 8.** In the graphene are carbon atoms bonded together in the hexagonal structure

## 6 Conclusion

To understand the structure of atom and molecules requires a high degree of imagination. The degree of imagination that is required is much more extreme than that required for some of the ancient ideas. The modern ideas are much harder to imagine. We use mathematical equations and rules, and make a lot of pictures. We can't allow ourselves to seriously imagine things, which are obviously in contradiction to the known laws of nature. And so our kind of imagination is quite a difficult game (or a puzzle). One has to have the imagination to think of something that has never seen before, never been heard before. At the same time the thoughts are restricted or limited by the conditions that come from our knowledge of the way nature really is. The problem of creating something which is new, but which is consistent with everything, which has been seen before, is one of extreme difficulty.

**Acknowledgment.** Pavel Werner is acknowledged for computer version of my figures.

## References

1. Zmeskal, O., Nezadal, M., Buchnicek, M.: Fractal-Cantorial geometry, Hausdorf dimension and fundamental laws of physics. Chaos, Solitons and Fractals 17, 113–119 (2003)
2. Zmeskal, O., Nezadal, M., Buchnicek, M.: Coupling constants in fractal and cantorian physics. Solitons and Fractals (2005)
3. Feynman, R.P., Leighton, R.B., Sands, M.: The Feynman Lectures on Physics. vol. I, II, III. Addison-Wesley Publishing Company (1977)
4. Duncan, T.: Physics for today and tomorrow. Butler & Tanner Ltd., London (1978)
5. Huggett, S.A., Jordan, D.: A Topological Aperitif. Springer-Verlag (2001)
6. Pauling, L.: General Chemistry, Dover Publication, Inc., New York (1988)
7. Smits, A.J., Lim, T.T.: Flow visualizatio. Imperial College Press (2012)
8. Osmera, P.: Fractal Dimension of Electron. In: Proceedings of MENDEL 2012, Brno, Czech Republic, pp. 186–191 (2012)
9. Osmera, P.: Vortex-ring fractal Structures of Hydrogen Atom. In: WCECS 2009, Proceedings of World Congress on Engineering and Computer Science, San Francisco, pp. 89–94 (2009)
10. Osmera, P.: Vortex-ring-fractal Structure of Atom and Molecule. IAENG Transaction on Engineering Technologies, American Institute of Physics 4, 313–327 (2010)

# Classification Methods Accuracy for Speech Emotion Recognition System

Pavol Partila, Jaromir Tovarek, Miroslav Voznak, and Jakub Safarik

Department of Telecommunications, Faculty of Electrical Engineering and Computer Science, VSB–Technical University of Ostrava, 17. listopadu 15, 708 33 Ostrava, Czech Republic
{pavol.partila, jaromir.tovarek, miroslav.voznak, jakub.safarik}@vsb.cz

**Abstract.** Emotional state classification of human speech and recognition accuracy of the classifiers is disclosed in this paper. Recent developments in speech recognition places more emphasis on the extraction of information about the speech source. This means obtain information about who and how it was said. This article describes research which seeks to recognize the information from speaking, emotional state in particular. Emotional state is recognized by using different classifiers and features of speech by nowadays known systems. Berlin database of emotional recordings was used to train and test the system. Mel-frequency spectral coefficients and dynamic coefficients were extracted from the audio signal of the database. For classification were used Gaussian Mixture Model, k-Nearest Neighbours and Artificial Neural Networks methods. The main effort of this research is to examine the accuracy and usability of classifying methods for detection of human stress status from his speech.

## 1 Introduction

The development of applications and services is trying to deploy natural interaction between man and computer. Specifying commands by voice and movements are very popular nowadays. Majority information is extracted from human speech with rather good accuracy. Human speech also includes secondary information, which holds properties of the speaker. Age, gender, emotional state, speech error and other features are contained in human speech. Mentioned source characteristics are highly valuable, because speech features can be simulated only by person with good acting skills.

As the title suggests, this article describes a system for classifying emotional state of human speech. Emotion is one of the characteristics of human which describes his mental condition affecting physiological changes in the human body. These changes are also reflected in the human speech. Information about the emotional state is requested in many fields. Statistical evaluation of customer satisfaction, his interest in the products is evaluated by affected emotional state. This information is a direct response to any stimulus. Call center agents can be evaluated with regard to their work and access to the customer. There is a chance to train new agents and teach them to correct the procedure of communication with the customer. Human body influenced by stronger emotions are getting stress. Positive stress is caused by emotions like happiness e.g.

I. Zelinka et al. (eds.), *Nostradamus 2014: Prediction, Modeling and Analysis of Complex Systems*, Advances in Intelligent Systems and Computing 289,
DOI: 10.1007/978-3-319-07401-6_44, 

from high prize win. Negative stress can be caused by psychological pressure, for example, fear of life. Sectors such as police, firemen and especially military generate the greatest emotional pressure on employees. Operational capability of a soldier, a police officer or firemen, is dependent on the coordinated orders from dispatching. Using the system for recognizing emotional state of human speech unable change the dispatching orders for a man in action. System for emotion recognition gives a possibility to manage orders and next steps for man in action. Applications required security often use speech such authorization key. Speech is affected by physiological changes caused by changing emotions. An authorized user can be denied because Authorisation unit recognizes the stress speech as a wrong key. Therefore, it is useful to detect secondary information from the speech that could avoid false denial of access. These are just the first examples of utilizations for speech emotion recognition systems. It is obvious that the system will have great application in human-machine interaction. Therefore it is appropriate to identify a classification ability of different classifiers for different emotional states. [1] [2] [3]

## 2 Speech Emotion Recognition System

System design consists of several blocks, which it distributed to major functions. Input values for the training and testing to create database of audio signals. Block diagram of the system is shown in Figure 1

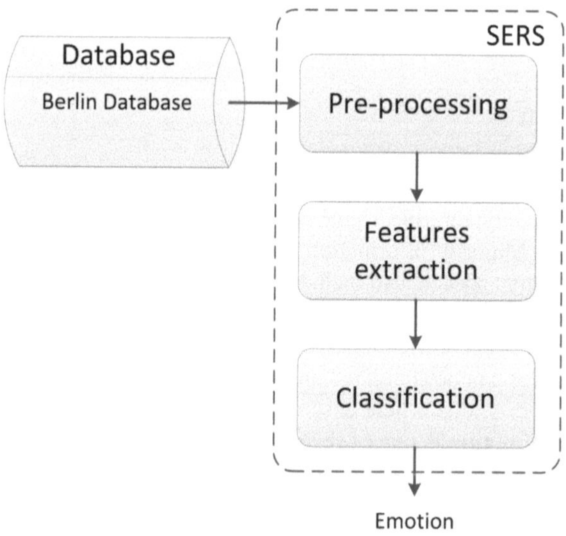


**Fig. 1.** Block diagram of Speech Emotion Recognition System (SERS). The system consists of a database that is used for training and testing and other blocks that describe the function of the algorithm.

Train respectively testing is the key step of the system. The quality of the input data, the audio signal in this case, has a direct impact on the classification accuracy. For this reason, it is used the Berlin database containing over 500 recordings of actors consists men and women. The database contains 10 sentences in the seven emotional states. This corpus of recordings is considered as a high-quality database, because it was created by professional actors in the sound studio. As mentioned, the speech signal has to be modified by routine operations such as removing the DC component, pre-emphasis and segmentation stochastic signal into quasi-periodic frames.

Speech recognition system is context-independent, that meaning take into account only signal parameters, not content informations. These parameters are the training and testing vectors for classifiers. The calculation parameters are represented by the Features Extraction block that extracts the Mel-Frequency Cepstral Coefficients (MFCC) and dynamic parameters (first and second derivative of MFCC). [4] [9] [10]

## 3 Classifiers

Individual research shows that cannot be said which classifier for emotion recognition is the best. Each classifier or their combination achieved some results accuracy, which depends on several factors. The success of classifier is directly dependent on the data. This is derived from the fact that the accuracy varies with the data character such as the quantity, density distribution of each class (emotions) and the language also. One classifier has different results with acted database, where the density of each emotion are equitable and different with real data from call centre where normal (calm) emotion state occupies 85 to 95 percent of all data. Appropriate choice of parameters has a considerable effect on the accuracy of these classifiers. The following subsections describe the used classification methods.

### 3.1 Gaussian Mixture Model

A Gaussian Mixture Model is a parametric probability density function represented as a weighted sum of Gaussian component densities. GMMs are commonly used as a parametric model of the probability distribution of continues measurements or features in biometric system, such as vocal tract, in speaker recognition systems as well. Probability distribution of the parameter vectors derived from human speech can be described using GMM.

$$p(o|\lambda^s) = \sum_{i=1}^{M^s} w_i^s \, p_i^s(o) \tag{1}$$

Where $M$ is number of components for $s$ class, $w_i$, $i=1, \ldots ,M$ are weights of components complying condition that sum of all weights is 1, $p$ means the probability density of the components represented by the mean value and covariance matrix $C_i$. Speaker model can be described mentioned mixture characterized by the equation below.

$$\lambda^s = \{w_i^s, \mu_i^s, C_i^s\}, \quad i, \ldots, M^s \tag{2}$$

The criterion of maximum likelihood is found lambda parameters with maximum $p$ probability density based on sequence parameters $O = o_1, o_2, \ldots o_n$ obtained from speech, seen below. [7] [8]

$$\lambda^s = argmax\, p(o|\lambda^s) \tag{3}$$

### 3.2 K-Nearest Neighbour

In pattern recognition, the k-Nearest Neighbour algorithm (KNN) is a method for classifying objects based on the closest training examples in the feature space. kNN is a type of instance-based learning, or lazy learning where the function is only approximated locally and all computation is deferred until classification. The kNN algorithm is amongst the simplest of all machine learning algorithms: an object is classified by a majority vote of its neighbours, with the object being assigned to the class most common amongst its k nearest neighbours ($k$ is a positive integer, typically small). If $k = 1$, then the object is simply assigned to the class of its nearest neighbour. The various distances between the vector $x_s$ and $x_t$.

$$d_{st}^2 = (x_s - x_t)(x_s - x_t)' \tag{4}$$

The neighbourhood distance is calculated through Euclidean metric. Given an m-by-n data matrix $X$, which is treated as $m$ (1-by-n) row vectors $x_1, x_2, \ldots, x_m$.

### 3.3 Artificial Neural Network

Our emotional state classification problem with high number of parameters can be considered as a pattern-recognition problem. In this case, it can be used two-layer feed-forward network. A two-layer feed-forward network, with sigmoid hidden and output neurons, can classify vectors arbitrarily well, given enough neurons in its hidden layer. The network is trained with scaled conjugate gradient (SCG) backpropagation.

Let's say that, MFCC and dynamic coefficients in this case, are the input vectors $x_i$ where $i = 1, \ldots, d$. The first layer of network forms $M$ linear combinations of these inputs to give a set of intermediate activation variables $a_j^{(1)}$

$$a_j^{(1)} = \sum_{i=1}^{d} w_{ij}^{(1)} x_i + b_j^{(1)} \quad j = 1, \cdots, M, \tag{5}$$

with one variable $a_j^{(1)}$ associated with each hidden unit. Here $w_{ij}^{(1)}$ represents the elements of first-layer weight matrix and $b_j^{(1)}$ are the *bias* parameters associated with the hidden units. Demonstration of such a network with 39 input parameters, 5 hidden layers and two output classes is shown in Figure 2.

SCG training implement mean squared error $E(w)$ associated with gradient $\nabla E$ and avoids the line-search per learning iteration by using Levenberg-Marquardt approach in order to scale the step size. A weights in te network will be expressed in vector notation.

$$w = \left(\ldots, w_{ij}^{(1)}, w_{i+1j}^{(1)}, \ldots, w_{N_1 j}^{(1)}, \theta_j^{(l+1)}, w_{ij+1}^{(1)}, w_{i+1j+1}^{(1)}, \ldots\right) \tag{6}$$

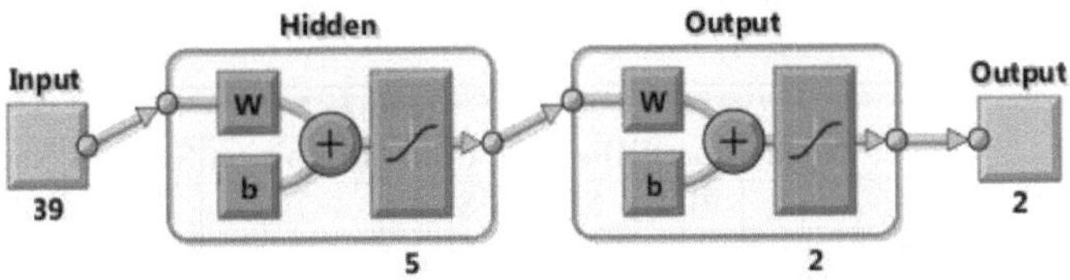


**Fig. 2.** Artificial neural network architecture with 5 hidden layers and 2 output classes.

The vector -deltaE points in the directionin which E(w) will decrease at the fastest possible rate. Weight update equation is shown bellow, where $c$ is suitable constant.

$$w(k+1) = w(k) - c \bigtriangledown E \tag{7}$$

The gradient descent method for optimization is very simple and general. Only local information, for estimation a gradient, is needed for finding the minimum of the error function. [5] [6]

## 4 Results and Discussion

The aim of the experiment was to clarify the significance and dynamic MFCC coefficients, as well as classification ability of selected classification methods for Speech Emotion Recognition System. Samples of examination were formed from recordings of human speech with various emotional character. The following settings and features were used in the experiment:

- Input samples - Berlin database of emotional utterances.
  - 10 different sentences recorded by 10 different actors (both genders).
  - Over 530 samples consisting with 7 emotions: *anger, bored, disgust, fear, happiness, sadness, neutral.*
- Feature Extraction - computing of input vectors (speech parameters).
  - 13 Mel-frequency Cepstral Coefficients $c_m$=[$c_m(0)$, ..., $c_m(12)$].
  - Dynamic coefficients of MFCC - $\Delta c_m$ and $\Delta^2 c_m$ (acceleration coefficients).
- Emotion Classification.
  - GMM - 64 mixture components.
  - K - Nearest Neighbours (set up 5 neighbours.)
  - Artificial Neural Network - Feed Forward Backpropagation.

Given the complexity of the solutions were compared two classes. In the first case, the classification accuracy was intended to recognize two emotional states with each other. Table 1-3 describes the accuracy percentage of three classifiers. Systems were trained by pairs of emotions and individual results in tables describe tested accuracy for a one of emotional state (left header of the table).

**Table 1.** Gaussian Mixture Model classification accuracy for different combinations of emotions.

| | | Train1 | | | | | | |
|---|---|---|---|---|---|---|---|---|
| | | Anger | Bored | Disgust | Fear | Happiness | Sadness | Neutral |
| Train2 / Test | Anger | - | 91.7 | 85.7 | 83.4 | 70 | 96.4 | 90.8 |
| | Bored | 76.3 | - | 64.9 | 66.2 | 71.9 | 65.3 | 59 |
| | Disgust | 59.1 | 64.3 | - | 62.5 | 56 | 78.5 | 64.7 |
| | Fear | 52.4 | 72.6 | 59.7 | - | 47.2 | 81.8 | 70.4 |
| | Happiness | 42.7 | 83.9 | 73.7 | 73.9 | - | 90.5 | 82.4 |
| | Sadness | 87.1 | 67 | 75.5 | 73 | 85.5 | - | 75 |
| | Neutral | 78.6 | 53.3 | 69.1 | 63.9 | 74 | 65.7 | - |

**Table 2.** K-Nearest Neighbours classification accuracy for different combinations of emotions.

| | | Train1 | | | | | | |
|---|---|---|---|---|---|---|---|---|
| | | Anger | Bored | Disgust | Fear | Happiness | Sadness | Neutral |
| Train2 / Test | Anger | - | 91.2 | 90.3 | 89.8 | 81.1 | 94.7 | 92.4 |
| | Bored | 92 | - | 70.4 | 67.7 | 75.9 | 61.6 | 64.6 |
| | Disgust | 49 | 55 | - | 59.7 | 51.8 | 71.2 | 56.6 |
| | Fear | 46.9 | 60.3 | 56.5 | - | 47.8 | 75.1 | 68.2 |
| | Happiness | 31.9 | 78 | 72.7 | 73.1 | - | 86.1 | 78.7 |
| | Sadness | 89.2 | 70.5 | 84.8 | 79.7 | 89.6 | - | 81.1 |
| | Neutral | 79.8 | 40.5 | 70.6 | 66 | 77.4 | 65.7 | - |

**Table 3.** Feed-Forward Backpropagation Neural Network classification accuracy for different combinations of emotions.

| | | Train1 | | | | | | |
|---|---|---|---|---|---|---|---|---|
| | | Anger | Bored | Disgust | Fear | Happiness | Sadness | Neutral |
| Train2 / Test | Anger | - | 95.8 | 93.7 | 92.4 | 87.9 | 98.1 | 96.7 |
| | Bored | 92 | - | 87.6 | 83.4 | 91.2 | 75.1 | 77.1 |
| | Disgust | 79.8 | 79.1 | - | 68.2 | 77.7 | 86.4 | 77.6 |
| | Fear | 69.6 | 70.6 | 73.7 | - | 68.2 | 82.2 | 76.2 |
| | Happiness | 32.3 | 88.3 | 79.8 | 83.9 | - | 95 | 88.9 |
| | Sadness | 97.9 | 81.5 | 92.6 | 95.3 | 97.5 | - | 85.2 |
| | Neutral | 93 | 49.9 | 86 | 85.1 | 88.2 | 52.4 | - |

## 5 Conclusion

Tabular results describe classification accuracy for a particular type of classifier that has been trained by a combination of two emotional states. All three classifiers showed the best recognition ability for the emotional state of rage. Emotional state of sadness was recognized with the evaluation very well. On the other hand, the worst-recognition ability of the system was the emotional state of fear (GMM and ANN) and disgust (ANN).

Much more interesting view on this topic, is recognizing the stressed out person, which means recognizing deviations from the neutral state. This state is not defined in the Berlin database. Therefore, it was necessary to assemble a set of data, so-called "stress cocktail" from defined emotional states. The stress of a person can be assembled from emotional states, other than neutral. Emotions Anger and Fear were used to compile the stress data set that because these emotional states are reflected most often when a person is exposed to stressful situations.

As before, the GMM, k-NN and the ANN were used to classify the stress versus neutral. Results for all three classifiers are shown in Fig. 3. The Receiver Operating Characteristic (ROC) is applied for better understanding the system. ROC curve is a tool for the evaluation and optimization of binary classification system (test), which shows the relationship between the sensitivity and specificity of the test or the detector for all possible threshold values.

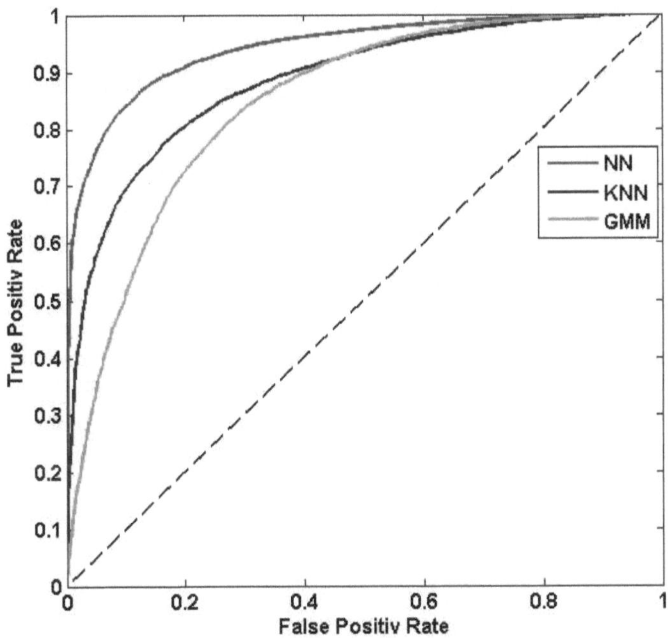


**Fig. 3.** Receiver Operating Characteristic of GMM, k-NN and ANN classifier for Neutral vs. Stress Recognizing.

The graph shows that the classification accuracy for the detection of stress is comparable to the classification of different emotional states. ANN achieved the best results in this experiment. Confusion Matrix for all classifiers is shown in Table 5. Explanation of cells for this matrix is described in Table 4. The results of Table 5 are presented for clearer representation of ROC curves.

**Table 4.** Confusion Matrix - description of cells

| | Classifier | | |
|---|---|---|---|
| Output classes | True Positive | False Positive | Positive Predictive Value |
| | False Negativ | True Negativ | Negative Predictive Value |
| | Senzitivity | Specificity | Precision |
| | Target classes | | |

**Table 5.** Confusion Matrix for ANN, k-NN and GMM classifiers and results for neutral vs. stress recognition

| | ANN | | | k-NN | | | GMM | | |
|---|---|---|---|---|---|---|---|---|---|
| Neutral | 7445<br>22.1% | 1901<br>5.7% | 79.7% | 4919<br>14.6% | 2209<br>6.6% | 69.0% | 6305<br>18.8% | 3507<br>10.4% | 63.3% |
| Stress | 1684<br>5.0% | 22587<br>67.2% | 93.1% | 4210<br>12.5% | 22279<br>66.3% | 884.1% | 2824<br>8.4% | 20981<br>62.4% | 88.1% |
| | 81,6% | 92.2% | 89.3% | 53.9% | 91.0% | 80.9% | 69.1% | 85.7% | 81.2% |
| | Neutral | Stress | | Neutral | Stress | | Neutral | Stress | |

This experiment shows that these classification methods can be used on the recognition of emotional state. At the same time, the question arises, what emotional states will characterize stress. The answer will probably depend on which system would be applied.

**Acknowledgement.** This work was supported by the Development of human resources in research and development of latest soft computing methods and their application in practice project, reg. no. CZ.1.07/2.3.00/20.0072 funded by Operational Programme Education for Competitiveness, co-financed by ESF and state budget of the Czech Republic and was partially supported by the grant of SGS No. SP2014/72 VB-Technical University of Ostrava, Czech Republic.

## References

1. Zarkowski, M.: Identification-driven emotion recognition system for a social robot. In: 2013 18th International Conference on Methods and Models in Automation and Robotics (MMAR), August 26-29, vol. 143, pp. 138–143 (2013)
2. Takeuchi, A., Nagao, K.: Communicative facial displays as a new conversational modality. In: Proceedings of the SIGCHI Conference on Human Factors in Computing Systems - CHI 1993, pp. 187–193. ACM Press, USA (1993), doi:10.1145/169059.169156.
3. Albrecht, I., Schröder, M., Haber, J., Seidel, H.: Mixed feelings: expression of non-basic emotions in a muscle-based talking head. In: Virtual Reality, vol. 8(4), pp. 201–212. ACM Press, New York (2005), doi:10.1007/s10055-005-0153-5
4. Burkhardt, F., Paeschke, A., Rolfes, M., Sendlmeier, W., Weiss, B.: A Database of German Emotional Speech. In: Interspeech 2005 Eurospeech, 9th European Conference on Speech Communication and Technology, Proceedings, September 4-8, pp. 1517–1520 (2005)

5. Knox, M., Mirghafori, N.: Automatic laughter detection using neural networks. In: Interspeech 2007, pp. 2973–2976 (2007)
6. Krajewsky, J., Kroger, B.: J., Using prosodic and spectral characteristics for sleepiness detection. In: Interspeech 2007, pp. 1841–1844 (2007)
7. Vlasenko, B., et al.: Combining frame and turn-level information for robust recognition of emotions within speech. In: Interspeech 2007, pp. 2249–2252 (2007)
8. Hu, H., Xu, M.-X., Wu, W.: GMM Supervector Based SVM with Spectral Features for Speech Emotion Recognition. In: IEEE International Conference Speech and Signal Processing, ICASSP 2007, vol. 4, pp. IV-413–IV-416 (2007), doi:10.1109/ICASSP.2007.366937
9. Voznak, M., Rezac, F., Rozhon, J.: Speech Quality Monitoring in Czech National Research Network. Advances in Electrical and Electronic Engineering 8(5), 114–117 (2010) ISSN 1804-3119
10. Partila, P., Voznak, M., Mikulec, M., Zdralek, J.: Fundamental Frequency Extraction Method using Central Clipping and its Importance for the Classification of Emotional State. Advances in Electrical and Electronic Engineering 10(4), 270–275 (2012) ISSN 1804-3119

# Complex Analysis of EEG Signal for Biometrical Classification Purposes

Jaromir Svejda, Roman Zak,
Roman Senkerik, and Roman Jasek

Tomas Bata University in Zlin, Faculty of Applied Informatics,
Nam T.G. Masaryka 5555, 760 01 Zlin, Czech Republic
{svejda,rzak,senkerik,jasek}@fai.utb.cz

**Abstract.** Aim of this article is to clarify the potential utilization of complex EEG signal in modern information age. Brain Computer Interface (BCI) represents the connection of brain waves with output device through some interface.

It was investigated whether the correlation analysis of the EEG signal may be used for finding appropriate classification parameters. EEG signal was measured in the idle state of mind of 3 subjects. Complex correlation analysis was performed for 16 samples of each obtained signal history. Moreover, the position of maximal correlation was also recorded.

## 1 Introduction

Many scientific disciplines deal with the human brain; for example numerical neuroscience, neuro-informatics, informatics or medicine. All of them bring theories, which could explain different brain activities. Numerical neuroscience provides mathematical and biophysical models, which are able to model basic processes in neurons and neural networks. The main goal of neuro-informatics is systematic development of database intended to collect information such as brain morphology, brain parts anatomy and their functional connection, brain electrophysiology, brain states obtained with magnetic resonance and their integration. Further, it seeks to develop tools for modelling, where the aim is the most accurate emulation of brain activity. In Informatics, complex networks are highly suitable to model a complex system among which the brain includes. The contribution of medicine is undisputable especially in brain anatomy research.

The human brain is a complex system, which is an object of our research. It is regarded as the most complex system in the universe. The modern science is currently attempting to understand the complex interconnection among individual parts of the brain (Sporns et.al. 2005). There are many publications, which deal with description of the brain (Adeli 2010; Damasio 1995; Sporns et al. 2005).

Currently there are many known applications of BCI technology, but not enough at each particular field of study. Signal that is sensed from the brain is the key element

I. Zelinka et al. (eds.), *Nostradamus 2014: Prediction, Modeling and Analysis of Complex Systems*, Advances in Intelligent Systems and Computing 289,
DOI: 10.1007/978-3-319-07401-6_45, © Springer International Publishing Switzerland 2014

in the BCI model; therefore the design of an appropriate algorithm for processing of the signal is the most discussed part of BCI model structure (Schalk et al. 2004).

Invasive methods of sensing the brain activity could provide very accurate data, but it is not both technically and user friendly; thus, it would not be further mentioned in this article. On the other hand, more accessible non - invasive methods can obtain relatively weak signal with amplitude ranging from units to hundreds of microvolts. Moreover, the signal is also prone to noise elements. Another disadvantage of this method is a summation of neuron signals; thus, obtained data are referenced to a specific group of neurons.

The brain itself is composed of several parts, without which his activity could not be possible. One of its basic structural parts is a neuron. The neuronal cells are characterized by the fact that electrical activity is carried out in them. These cells communicate with each other by electrical signals. According to the last estimate, there are approximately $10^{11}$ neurons in the brain. Every one of them is connected with thousands of other neurons. The main source of EEG signal is an electric activity of synapse - dendrites membrane located in the surface layer of the cortex. Each active synapse dispatches electromagnetic pulse to the environment during excitation. Due to the high number of these pulses, it is difficult to locate their source by means of multichannel sensor on the skin. This issue could be compared to full amphitheatre, in which there are chanting people and the task is to recognize from outside, which specific group of fans shouts. A different perspective on this issue may be such that the aim is to identify uniqueness of the signal for each individual subject. In the example shown above, it is as we would like to recognize the type of the stadium by the mass of chanting people. For example, there is noticeable difference between hockey and tennis fans. The biometric signatures are different for each creature on the planet Earth.

The aim of this article is to provide potential classification parameters through EEG signal analysis for neural network which could be then used for biometrical purposes.

## 2 Methods

There are several approaches for sensing brain activity. The most widely used is EEG technology, which belongs among the non – invasive methods. Devices based on EEG technology provide signal with very low voltage amplitude, because the signal has to pass through the relatively low conductive skull. The amplitude ranges from tens to hundreds microvolts.

### 2.1 Sensing the Brain Activity

Recently, we use Emotiv EPOC neuroheadset to obtain EEG signal from the human brain. Sensing of EEG by Emotiv EPOC neuroheadset has a number of advantages, because it already involves solved elementary issues in the processing of the measured signal. Due to this fact, it is not necessary to operate with raw data. It depends on the further usage of the data. Although the spectrum of this data could be used in many applications, it is not simple to understand the entire significance of the whole signal even if the proportion of the noise is minimal. This technology has the greatest expansion and certainly also the priority significance in diagnosis of various diseases in medicine (Adeli 2010).

Emotiv Corporation developed personal brain - computer interface for human – computer interaction using neuro-technology, which is based on processing of electromagnetic waves occurring in human brain. The interface has wide range of possible applications; for example in interactive games, intelligent adaptive environment, audio visual art and design, medicine, robotics and automotive industry. Moreover, it can be deployed in large amount of scientific research.

Emotiv EPOC neuroheadset (Figure 1) measures a signal wirelessly transferred to common personal computer. It is a device, which has a set of sensors intended for sensing the activity produced by human brain. Traditional EEG devices requires the use of conductive pasta to improve the conductivity between electrodes and hairs. On the other hand, the neuroheadset do not need any additional tools. It has 14 high resolution sensors, which are placed on optimal positions on the human head (Figure 2).

**Fig. 1.** Emotiv EPOC neuroheadset (Emotiv 2012)

Moreover, it also includes gyroscope for determinate the position in the area. Each channel has its own label based on its position on the head: AF3, F7, F3, FC5, T7, P7, O1, O2, P8, T8, FC6, F4, F8, and AF4. Internal sampling frequency of the neuroheadset is 2048 Hz. More information about neuroheadset can be found in (Emotiv 2012).

Example of complex nature of the measured signal is depicted in Fig. 3, which lends weight to the argument, that it is still hard task to analyse the output signals and to find a hidden complex dynamics behind. To find a promising approach, as to how

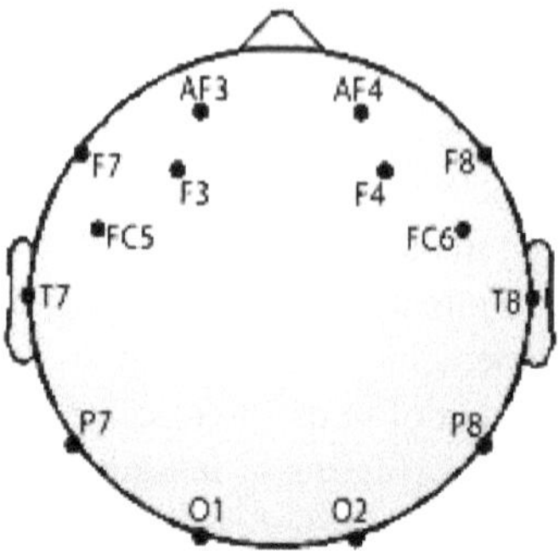


**Fig. 2.** Placement of electrodes of Emotiv EPOC neuroheadset

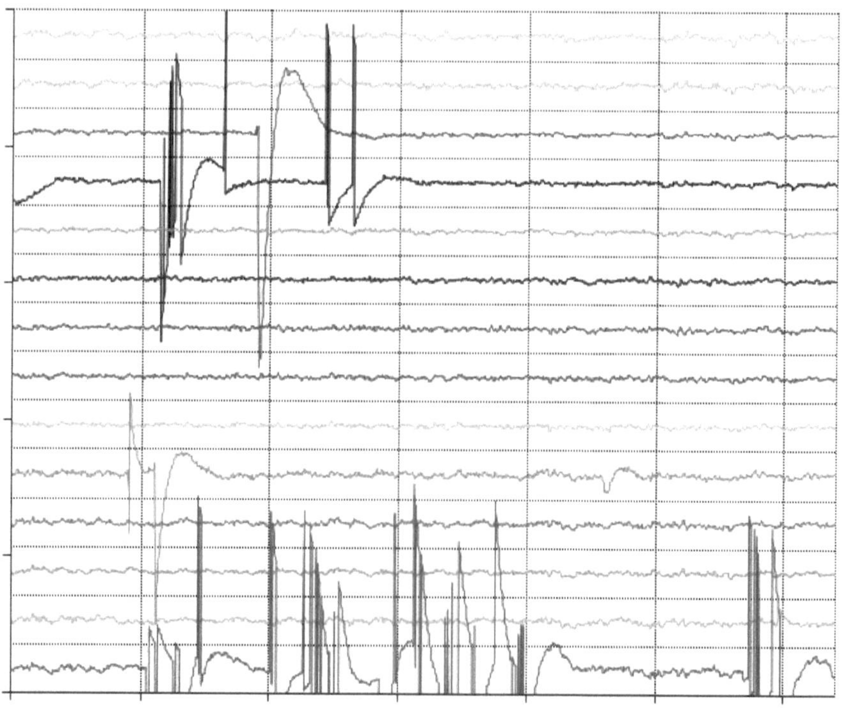

**Fig. 3.** Example of complex EEG output

reveal the hidden complex behaviour is challenging task, and many hybridization of modern techniques are often used for such an issue, e.g. connection of chaotic dynamics, neural classifiers, fuzzy logic, etc.

Emotiv provide basic software set containing many tools, which can be used for recording various signals such as electric potential from all 14 sensors, power spectrum of individual EEG channels in real time and rotational acceleration of the head in horizontal and vertical axis using data from gyroscope. All of these outputs are shown in graphs. Data are also available in raw form, which can be used for further analysis. If it is required special functionality, which is not provided by native software, it is desirable to develop own application using Emotiv SDK (Software Development Kit).

Native software consists of three classification suites. Each of them enables the usage of algorithm developed by Emotiv. First of them is Expressive suite, which contains identification system for recognition of facial expression such as smile, eyewink, etc. The muscle signals are used for this purpose. The sources for these signals are obtained by sensors, which are located around the face.

The second suite can be used to measure and identification of emotional state; for example nervousness, alertness, concentration, etc. Therefore, it is called Affective suite. Muscle signals and ocular signals are filtered by specially designed filters; thus, identification algorithm uses clear brain signal.

The last suite is called Cognitive. This classification mode uses whole measured signal, which contains both clear brain signal and muscle signal. Classification algorithm is based on artificial intelligence methods. Type and structure of applied neural network is patented by Emotiv Corporation; therefore, the specific information about the algorithm is protected.

If it is required other processing of the signal than the native software allows, it may be processed by another software application.

Measured raw data can be subjected to offline analysis to research.

## 2.2 Processing the Brain Activity

If person could be recognized by custom EEG traces, it would mean that the person could be uniquely identified by EEG signal and it could bring new ways of authorization routines. Critical phase lies in signal classification. Even if the meaning of both the waveform and the signal content is not very important for classification purposes, there is another issue which have to be considered. EEG device provides large amount of data which has to be effectively and quickly processed in order to perform correct classification of the subject from the signal in real time. Classification tasks could be realized by using neural networks. However, it is difficult to predict, which neural network could use its cognitive potential for classification task mentioned above (Hazrati and Erfanian 2010). Investigation of the most appropriate classifier requires testing of many subjects. Furthermore, it is necessary to find algorithm with the shortest response time with respect to the credibility of obtained output. Another issue is to determinate which output is suitable. There could be considered the theory of large numbers; thus, maximum possible number of subjects needs to be tested. Therefore, it is more important to select key parameters of the signals that are different for each person. Even though the parameters are different for different people, the question remains whether the parameters remain constant in different time frames for the same person.

# 3 Results

Idle state of mind was chosen as the optimal for the measurement. The relative idle state is such that the EEG signal does not contain any artefacts. Some artefacts are already filtered out before the signal processing. That mostly includes elements in signal that related to some physical responses such as eye blinking, motion and muscle activity, heartbeat etc. Furthermore, external artefacts interfering the signal are primarily eliminated in analog-to-digital conversion.

In order to perform classification, it has to be set unique characteristics of the signal (hereinafter called as classification parameters). That task is the first step of our research. The most important is to find appropriate classification parameters. However, the device returns fourteen channels with various amplitudes. Therefore, it is necessary to normalize the signal by applicable algorithm. This procedure count with sampling frequency of the neuroheadset, which is 128 Hz. Prepared set of data is ready for another mathematical or statistical analysis.

Firstly, it was performed a correlation analysis between channels of the first subject's EEG signal. All combinations of signals were tested in order to find which pair of channels influence each other. Further aim of the analysis was to find out whether another subjects have different relations between the channels. Correlation was calculated for each compared pair of channels.

Data was obtained for three different subjects. Minimal number of samples was around 27 000 (corresponds to approximately 3.5 minutes of EEG record). Tables (Tab. 1, Tab. 2, Tab. 3) contain values of maximal correlations and position of electrodes which related to the correlation. Both parameters were counted for different parts of EEG signal. A length of individual parts were set to 10000 samples (78, 125 s). Time shift between parts was set to 1000 samples (7,8125s).

**Table 1.** Correlation values and positions of electrodes for various time shifts for Subject 1

| ID | First sample | Last sample | Max. correlation | Position of electrodes |
|---|---|---|---|---|
| 1 | 1 | 10000 | 0.952552 | AF3 F3 |
| 2 | 1001 | 11000 | 0.948659 | AF3 F3 |
| 3 | 2001 | 12000 | 0.950475 | AF3 F3 |
| 4 | 3001 | 13000 | 0.94621 | AF3 F3 |
| 5 | 4001 | 14000 | 0.946543 | AF3 F3 |
| 6 | 5001 | 15000 | 0.914729 | P7 O1 |
| 7 | 6001 | 16000 | 0.909592 | P7 O1 |
| 8 | 7001 | 17000 | 0.901304 | P7 O1 |
| 9 | 8001 | 18000 | 0.900297 | P7 O1 |
| 10 | 9001 | 19000 | 0.892232 | AF3 T7 |
| 11 | 10001 | 20000 | 0.923422 | AF3 F4 |
| 12 | 11001 | 21000 | 0.916445 | AF3 F4 |
| 13 | 12001 | 22000 | 0.869922 | AF3 F4 |
| 14 | 13001 | 23000 | 0.84772 | T7 P7 |
| 15 | 14001 | 24000 | 0.839024 | T7 P7 |
| 16 | 15001 | 25000 | 0.901458 | T7 F8 |

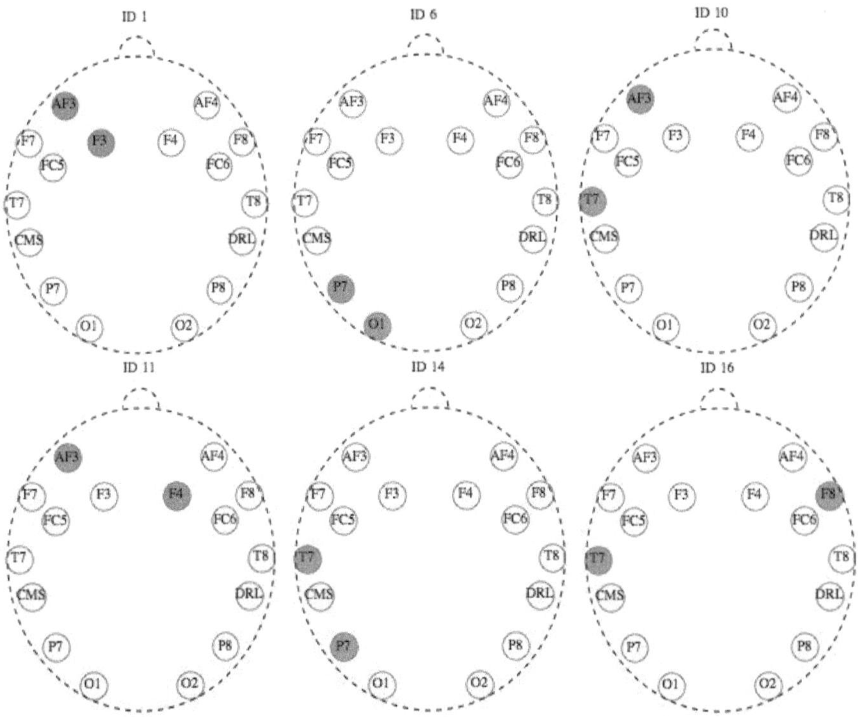


**Fig. 4.** The chosen positions of electrodes with the highest correlation for Subject 1

There are three figures (Fig. 4, Fig. 5, Fig. 6) which are connected to the tables mentioned above. Each of them illustrates a distribution of channel pairs with the highest correlation. These channels are highlighted. It can be observed that some pair combinations occur multiple times; therefore, the figures always contain only one illustration for each pair combination. Each ID number is related to the first occurrence of corresponding pair. Electrode markers are taken from Emotiv EPOC neuroheadset documentation.

**Table 2.** Correlation values and positions of electrodes for various time shifts for Subject 2

| ID | First sample | Last sample | Max. correlation | Position of electrodes |
|---|---|---|---|---|
| 1 | 1 | 10000 | 0.846923 | O2 P8 |
| 2 | 1001 | 11000 | 0.817771 | O2 P8 |
| 3 | 2001 | 12000 | 0.859306 | O2 P8 |
| 4 | 3001 | 13000 | 0.969112 | O2 P8 |
| 5 | 4001 | 14000 | 0.961076 | O2 P8 |
| 6 | 5001 | 15000 | 0.959598 | O2 P8 |
| 7 | 6001 | 16000 | 0.95466 | O2 P8 |
| 8 | 7001 | 17000 | 0.95149 | O2 P8 |
| 9 | 8001 | 18000 | 0.947908 | O2 P8 |
| 10 | 9001 | 19000 | 0.939777 | O2 P8 |
| 11 | 10001 | 20000 | 0.942595 | O2 P8 |
| 12 | 11001 | 21000 | 0.939619 | O2 P8 |
| 13 | 12001 | 22000 | 0.933342 | O2 P8 |
| 14 | 13001 | 23000 | 0.686855 | O2 P8 |
| 15 | 14001 | 24000 | 0.69902 | O2 P8 |
| 16 | 15001 | 25000 | 0.652335 | O2 P8 |

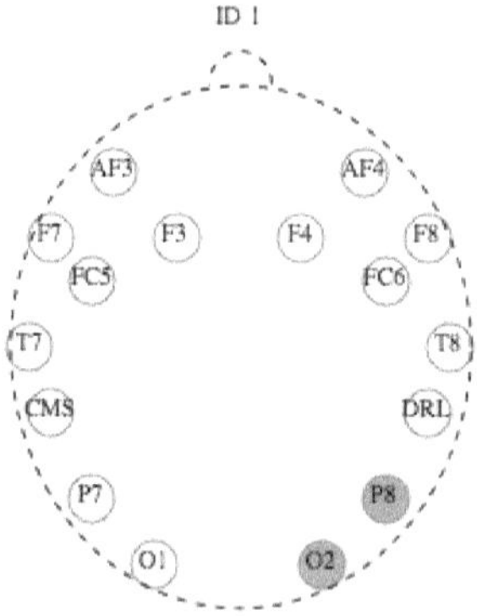


**Fig. 5.** The chosen positions of electrodes with the highest correlation for Subject 2

**Table 3.** Correlation values and positions of electrodes for various time shifts for Subject 3

| ID | First sample | Last sample | Max. correlation | Position of electrodes |
|---|---|---|---|---|
| 1 | 1 | 10000 | 0.88404 | P7 O1 |
| 2 | 1001 | 11000 | 0.840489 | P7 O1 |
| 3 | 2001 | 12000 | 0.811294 | P7 O1 |
| 4 | 3001 | 13000 | 0.736946 | P7 O1 |
| 5 | 4001 | 14000 | 0.732744 | P7 O1 |
| 6 | 5001 | 15000 | 0.697164 | P7 O1 |
| 7 | 6001 | 16000 | 0.632396 | P7 O1 |
| 8 | 7001 | 17000 | 0.54687 | F3 T7 |
| 9 | 8001 | 18000 | 0.548756 | F3 T7 |
| 10 | 9001 | 19000 | 0.601577 | FC6 F8 |
| 11 | 10001 | 20000 | 0.578018 | FC6 F8 |
| 12 | 11001 | 21000 | 0.589425 | FC6 F8 |
| 13 | 12001 | 22000 | 0.599788 | P7 O1 |
| 14 | 13001 | 23000 | 0.673097 | FC6 F8 |
| 15 | 14001 | 24000 | 0.685173 | FC6 F8 |
| 16 | 15001 | 25000 | 0.696007 | FC6 F8 |

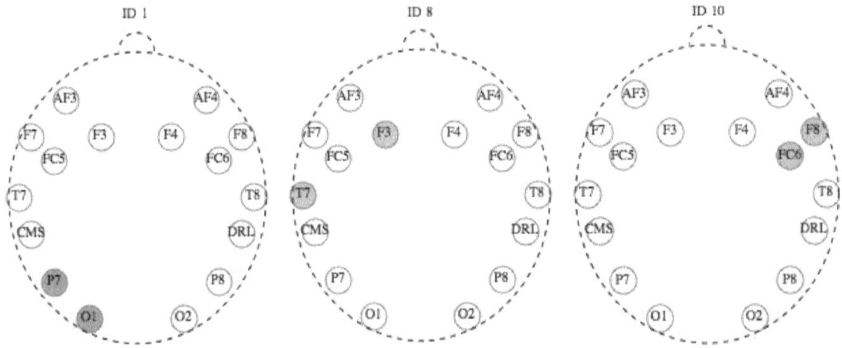


**Fig. 6.** The chosen positions of electrodes with the highest correlation for Subject 3

# 4 Discussion

Human brain is the most complex known system in the universe. Study of its activity is extremely important mainly due to the more precise diagnosis of brain diseases and their treatment. Furthermore, acquired knowledge could be used in modern technologies with BCI systems, where an interaction between brain and computers appears.

We are currently performing the measurement of the EEG signal in our research. One of our aim is to discover interesting regularities in the EEG signal waveform, which could contribute to the improvement of current approaches of brain activity simulation. Moreover, these regularities could be used to recognize some specific states of the brain, which can be then used to control the software or equipment connected to the computer.

There are many approaches to analysis of data. Moreover, EEG signal belongs to group of biometric signals which are usually very complex. The question remains whether it is possible to involve significance of all characteristics and signal history of EEG signal to classification process.

Biometrical data are typically represented as an image or a quantification of measured physiological or behavioural characteristics. As these data should refer to very complex human behaviour or describe very precisely physiological characteristic (typically iris scan, fingerprint, palm vein image, hand scan, voice, walk pattern etc.) these data can easily became very large and hard to process. For this reason a modern ways of data processing and classification are applied for biometrical data. The leading method is the usage of neural networks (Tangkraingkij 2009).

Correlation analysis demonstrates that there may exists relations between individual EEG channels. Further, it shows that the highest value of correlation was mostly found between neighbouring electrodes. Subject 1 has the highest correlation between electrode AF3 and F3 which are both located in frontal region of the brain. On the other hand, the subject 2 has the highest correlation between electrodes O2 and P8, which are located in rear regions of the brain. Both subject were measured in the idle state of mind. Subject 3 has the highest correlation between electrodes P7 and O1, which are located in rear regions of the brain. That behaviour of the subject's brains should be proven on more measurements of the same subjects. If the behaviour remained the same, it would mean, that it could be set as another classification parameter.

Interesting conclusion can be found for subject 2. Unlike other subjects, the position of electrodes with maximal correlation did not change during the shifting from the first data set to the last data set for subject 2. The question remains whether it is possible to determine the person from the position of electrode pairs with maximal correlation. Another question is whether it is possible to use binary representation to record the electrode position in the form, which can be then used as classifier i.e. input of the neural network. The specific trend of maximal correlation cannot be observed from obtained data, because its value seems unstable. It could be an advantage in order to use the average of maximal correlations as one of the classification parameters.

From obtained results we concluded that our future research could possibly answer the question which statistical characteristics are the most suitable for usage in classification algorithm based on neural network. For example, difference between individual subjects is the feature, which could be used as another classification parameter. Results described in this article are the first part of future extensive research.

**Acknowledgments.** This work was supported by Internal Grant Agency of Tomas Bata University under the project No. IGA/FAI/2014/31. Further it was supported by European Regional Development Fund under the project CEBIA-Tech No. CZ.1.05/2.1.00/03.0089.

## References

1. Adeli, H.: Wavelet-Chaos-Neural Network Models for EEG-Based Diagnosis of Neurological Disorders. In: Kim, T.-H., Lee, Y.-H., Kang, B.-H., Ślęzak, D. (eds.) FGIT 2010. LNCS, vol. 6485, pp. 1–11. Springer, Heidelberg (2010)
2. Damasio, H.: Human brain anatomy in computerized images. Oxford University Press (1995)
3. Emotiv I EEG System I Electroencephalography (2012), http://www.emotiv.com/index.php
4. Hazrati, M.K., Erfanian, A.: An online EEG-based brain–computer interface for controlling hand grasp using an adaptive probabilistic neural network (2010), doi:10.1016/j.medengphy.2010.04.016.
5. Schalk, G., Mcfarland, D.J., Hinterberger, T., Birbaumer, N., Wolpaw, J.R.: BCI2000: A General-Purpose Brain-Computer Interface (BCI) System (2004), doi:10.1109/TBME.2004.827072.
6. Sporns, O., Tononi, G., Kötter, R.: The human connectome: a structural description of the human brain. PLoS Computational Biology 1(4), e42 (2005)
7. Tangkraingkij, P., Lursinsap, C., Sanguansintukul, S., Desudchit, T.: Selecting Relevant EEG Signal Locations for Personal Identification Problem Using ICA and Neural Network. In: Eighth IEEE/ACIS International Conference on Computer and Information Science, ICIS 2009, pp. 616–621 (2009)

# Author Index

Andziulis, Arunas 21, 31

Barot, Tomáš 185
Bednář, Hynek 11
Bielik, Marian 347
Bielinska, Ewa 241
Bobál, Vladimír 281
Bououden, S. 205
Briš, Radim 327

Chadli, M. 205
Chalupa, David 251
Clementis, Ladislav 291

Daniel, Honc 385
Dao, Tran Trong 1, 123
Davendra, Donald 79, 89, 99, 141, 161
De Rango, Floriano 401
Dewi, Reisa Rahmatu 131
Dobrovsky, Ladislav 219
Dolezel, Petr 229, 301
Dostál, Petr 195, 377
Drábik, Peter 153
Drungilas, Darius 31
Dušek, František 175, 385
Dvorský, Jiří 65
Dzemydiene, Dale 21, 31
Dzik, Petr 55

Fazio, Peppino 401

Gerža, Michal 411
Gricius, Gediminas 21, 31
Grunt, Ondřej 327

Heckenbergerová, Jana 393
Hoang, Duy Vo 1, 123
Honc, Daniel 175
Hotař, Vlastimil 43

Jakovlev, Sergej 21
Janošek, Michal 307
Janoška, Zbyněk 65
Jasek, Roman 449
Jehlička, Vladimír 317

Kim, Tae-Hyong 131
Kochana, Oskar 241
Kominkova Oplatkova, Zuzana 99
Kotyrba, Martin 261, 271
Kříž, Radko 365
Ksiazek, Patrycja 241
Kubalčík, Marek 185
Kurmis, Mindaugas 21

Lampart, Marek 355
Lukosius, Zydrunas 21

Marek, Jaroslav 393
Mariška, Martin 229, 301
Matousek, Radomil 219
Matúšek, Ondřej 43
Metlická, Magdalena 141
Mikšovský, Jiří 11
Minar, Petr 219
Mouralova, Katerina 219

Nespurek, Stanislav 55

Olszak, Michal 241
Ošmera junior, Pavel 429
Ošmera senior, Pavel 429

Pálka, Lukas 421
Partila, Pavol 439
Pluhacek, Michal 79, 89, 99, 161
Pospíchal, Jiří 251

Raidl, Aleš 11
Řehoř, Adam 111

Safarik, Jakub 439
Salač, Petr 43
Šaloun, Petr 123, 153
Schauer, František 411
Schauer, Franz 421
Senkerik, Roman 79, 89, 99, 123, 161, 449
Sharma, Rahul 175, 385
Skanderova, Lenka 1, 111, 123
Škapa, Stanislav 377
Sottile, Cesare 401
Svejda, Jaromir 449
Svoboda, Martin 393

Talaš, Stanislav 281
Tovarek, Jaromir 439
Tropea, Mauro 401

Vantuch, Tomáš 337
Vesely, Michal 55
Vojtesek, Jiri 195
Volna, Eva 261, 271
Voznak, Miroslav 21, 31, 401, 439
Vraná, Marie 153
Vrobel, Jan 307

Zak, Roman 449
Zapomĕl, Jaroslav 355
Zelinka, Ivan 1, 79, 89, 99, 123, 153, 161, 205, 411, 421
Zmeskal, Oldrich 55

MIX
Papier aus verantwortungsvollen Quellen
Paper from responsible sources
FSC® C105338

If you have any concerns about our products,
you can contact us on
**ProductSafety@springernature.com**

In case Publisher is established outside the EU,
the EU authorized representative is:
**Springer Nature Customer Service Center GmbH**
**Europaplatz 3, 69115 Heidelberg, Germany**

Printed by Libri Plureos GmbH
in Hamburg, Germany